Polluted Earth

Polluted Earth

The Science of the Earth's Environment

ALEXANDER GATES
Department of Earth and Environmental Science
Rutgers University
Newark, New Jersey, USA

Library of Congress Cataloging-in-Publication Data applied for
Paperback ISBN: 9781119862529

Cover Design: Wiley
Cover Image: © overcrew/Shutterstock

Set in 9.5/12.5pt SourceSansPro by Straive, Pondicherry, India

SKY10043817_030223

Contents

Preface

Most environmental textbooks are written in a scientific style and approach the subjects in a traditional manner. However, recent changes in society and especially in the outlook of young people have emphasized concern over environmental justice. Damage to the environment commonly includes injustice. Most environmental textbooks, however, deal only peripherally with this aspect and therefore do not engage students as well as they could. In contrast, this textbook introduces environmental justice, including environmental racism, in Chapter 2, thereby giving instructors the opportunity to set up the course to be viewed through that lens. Once students understand how environmental damage can be focused in areas and situations where socially and economically underprivileged people and groups are taken advantage of, they will be able to recognize it in all of the other chapters and examples in the book. This puts the importance of the course material in a much broader and more pertinent context. This allows the course to satisfy personal, department, and institutional priorities on social issues. Otherwise, instructors may skip Chapter 2 and present the course in a more traditional manner.

In addition, introductory environmental courses are primarily attended by non-science majors. These students are commonly not accustomed to the organizational and writing style of science textbooks and consequently have a difficult time with them. Much of the problem is that chapters must be read sequentially in one sitting or students lose their train of thought and are forced to start over. Chapters typically involve increasing complexity of concepts. Wading through scientific information at college level is foreign to most of these students.

This book contains the chapters and organization needed for any environmental course but most of the content is delivered through a plethora of exciting case studies. These are short, self-contained, and well-illustrated stories of specific pollution disasters and their outcomes that are highly engaging for both science and non-science majors. The stories incorporate the science into the description so students appreciate and remember it as part of the story. By relating the event to the impact on society and human lives, this method also places the science into a context that is important to the student. The case studies are self-contained, so chapters can be read in pieces without losing continuity. However, the chapters are organized to deliver the material in the logical manner of the topical development. Therefore, the book can be used as a stand-alone resource and be effective for exam preparation.

In addition, a list of readily available videos is compiled for all but a few of the case studies. They either provide interviews with victims of the pollution to give readers a look into the emotional aspect of being subject to pollution or they illustrate complex situations to give a better understanding. Reinforcement of reading the case studies with viewing a video leads to a better understanding and appreciation of the situation. It is recommended to watch the videos after reading the case study. The videos are primarily divided into short presentations, typically 3–5 minutes but less than 12 minutes in length. Long videos are typically 30 minutes or longer and are documentaries. These videos also enhance the impact of the course.

The book is organized to introduce the scientific basis of the polluted media either early in the chapter or in a separate leading chapter. These are interspersed with case studies or followed by a group of case studies to give the reader a flavor of the character of the processes and damage. This organization coupled with the exciting and highly illustrated stories is intended to enhance interest and understanding.

The other important feature is the visual support of the case studies. A number of artistic, highly detailed illustrations of regional pollution sources serve as lead-ins to the case studies. The pollution types include point source versus non-point source, groundwater pollution, surface water pollution, and ocean pollution. These illustrations show the spatial and temporal relations of the various sources. Each case study then describes a specific example of the pollution source and how it became a pollution disaster. The case study can then be related back to the illustration to place it in context. Case studies on interesting pollution disasters involving each of these is presented in two-page stories illustrated with several photographs of the event and results.

There is also a compilation of pollutants, their health and/or environmental impact, and the regulated limits on exposure. In this way the students have a single location to view these pollutants and their attributes rather than having them repeated multiple times in the book.

Acknowledgments

The production of this book benefitted from testing ideas with Environmental Geology and Planet Earth classes at Rutgers University, Newark. Suggestions by Dr. David Valentino made excellent suggestions that helped guide the book. Dr. Robert Blauvelt identified some of the better case studies that are used in this book. Drs. Ismael Calderon, Cy Stein and Francisco Artigas reviewed a few chapters. Hydrogeologist Richard Fox performed a critical review of the entire book and provided helpful suggestions that improved it. Support and encouragement from family members including Dr. Colin Gates, Dr. Jill Stein and Thomas Gates are also appreciated.

About the Companion Website

This book is accompanied by a companion website:

www.wiley.com/go/gates/pollutedearth

This website includes:

- Powerpoints of all figures and tables from the book for downloading
- Web links from the book for downloading

Humans and the Environment

CHAPTER OUTLINE

Polluted Earth: The Science of the Earth's Environment, First Edition. Alexander Gates.
© 2023 John Wiley & Sons, Inc. Published 2023 by John Wiley & Sons, Inc.
Companion website: www.wiley.com/go/gates/pollutedearth

Words you should know:

Cryptocurrency – A digital or virtual money or asset that is built into a blockchain verification system. Mining of cryptocurrency is environmentally costly.

Ecology – The branch of biology covering the interrelations of organisms and their relations to their surroundings.

Environment – The interrelations of the chemical, biological, and physical natural world on a global or local level.

Point source pollution – A single source of pollution that can be identified.

Non-point source pollution – Pollution for which the source is so diffuse or distributed that it cannot be absolutely identified.

Rachel Carson – The pioneer and most impactful person on the American environmental movement.

1.1 Human Impact on the Environment

Aside from a possible major war or very deadly pandemic, the greatest challenge humans are facing and will face for the foreseeable future center around pollution of the natural environment. Climate change as a function of human impact has taken center stage over the past few decades. US Vice President Al Gore and, more recently, Swedish student activist Greta Thunberg have become media sensations by bringing climate change to public attention. Society has even made an attempt to address it with hybrid and electric cars, solar panels, and wind farms. However, climate change is only one small aspect of the horrific problems plaguing the natural environment. Dead zones in the oceans from overuse of fertilizers, loss of pollinators from overuse of pesticides and habitat destruction, invasive species destroying native species and resulting massive extinctions are some of the less prominent but just as deadly issues. The new environmental spokespeople are helping to bring attention to some of these issues but all pale in comparison to the original and true environmental champion in America and indeed the world, Rachel Carson. She has been called the "mother of the American environmental movement" but this title does not nearly reflect the magnitude of her impact. During a time when few gave more than a fleeting thought to the natural environment, Rachel Carson brought her serious concerns to the highest level of the government and to the attention of the general public both in the United States and internationally. Her work started the entire study of the environment. She is a true hero.

Humans are notoriously damaging to the natural environment. Even in primitive cultures, groups of humans would cut down trees, make fires, move and break rocks, leave excessive waste, and strip areas of game and vegetation. It took several decades of natural processes for the areas to recover. Once humans established permanent settlements, the environmental damage became very localized but much more intense. The permanent structures, agriculture, and concentrated waste completely disrupted the natural environment and would take centuries to recover once abandoned. The first serious chemical pollution coincided with the Bronze Age and Iron Age which added mining, smelting, and forging of metals. These caused poisoning of air, soil, and water, and death of plants and animals in localized areas.

As civilization advanced, the building of stone structures on ever grander scales led to even greater environmental disruption. Settlements grew ever larger into cities, requiring larger agricultural efforts to support them. Forests were removed to support the production of food. Cities and structures built thousands of years ago still scar the landscape even if they are abandoned. Several inventions occurred over the centuries that accelerated the environmental damage. Gunpowder and arms that utilize it was a major change in environmental damage in addition to the humans killed by them. Paints, dyes, glass, burning of coal, and other chemical developments furthered this damage.

The Industrial Revolution brought a drastic change to the level of environmental impact. The first major pollutant was from mining and use of coal. It produces soot from burning but also from mining and storage. Coal powered the Industrial Revolution and it was spread widely by trains. In addition to dust, the impurities such as sulfur, mercury and other heavy metals, and coal caused widespread soil pollution by emissions and dumping of waste. Dumping dangerous and unsightly waste and emissions was the norm for the Industrial Revolution. As more dangerous contaminants were developed, the pollution intensified and increased. It was not until the environmental movement of the 1950s through 1970s that these practices were even slightly controlled.

Concurrent with the advancement of technology and largely as a result, the human population ballooned. The world population increased slowly from less than one half billion in 1000 AD to 1 billion in 1800. The accelerated growth began soon after and the population reached 2 billion by 1930 (Figure 1.1). In 1950, the acceleration of world population became exceptional, growing from about 2.5 billion to 7.8 billion over the next 70 years. This is far more than the Earth can handle and most environmental systems, both chemical and biological, are being overwhelmed. So many people are contributing to pollution problems that they will be difficult to control, if they are controllable at all. It will take a cooperative effort to keep the run-away overuse of resources and disruption of the environment not to destroy the human race and the planet.

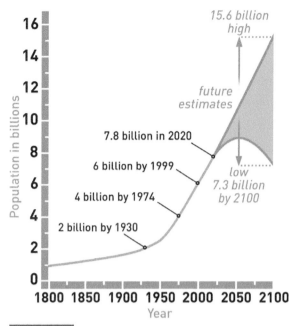

FIGURE 1.1 Human population growth curve from 1800 to 2020 with projected possible growth to 2100 *Source:* Data from United Nations.

Pollution has now caused irreparable damage to the Earth for the foreseeable future. The result of human impact is that the Earth is experiencing what is widely believed to be the sixth major mass extinction event in the past 500 million years. It is estimated that more than 25% of all the known species will go extinct by 2050. It is not only the environment that is impacted by pollution but groups of humans are chosen to be more impacted than others based on economic status, whether they are in a developed country or not, and on race and ethnicity. This inequality has evolved into the field of environmental justice.

CASE STUDY 1.1 Rachel Carson (1907–1964)

Mother of the Environmental Movement

Rachel Carson is regarded as the mother of the American environmental movement (Figure 1.2). It was through her efforts that the first banning of pesticides and especially DDT occurred. Her work also led to the formation of the US Environmental Protection Agency. It is impressive that such a private, unassuming woman could influence such a major movement. Carson captured the interest of the American public and the top levels of government despite the efforts of major chemical companies to suppress her work and to professionally and personally discredit her. She testified before the US Congress about the dangers of pesticides. President Kennedy read her book *Silent Spring*, and as a result, he ordered an investigation of the pesticides that Carson identified as dangerous. They would be banned less than a decade later.

Rachel Carson was born on 27 May 1907, in Springdale, Pennsylvania. She grew up on a small farm, where she developed her love of nature. In 1925, she enrolled in the Pennsylvania College for Women, which later became Chatham College. She began her studies as an English major but soon changed to biology and graduated magna cum laude in 1929 with a bachelor's degree. Carson received a scholarship to continue her education at Johns Hopkins University where she earned a master's degree in zoology in 1932. Her thesis was

FIGURE 1.2 Portrait of Rachel Carson. *Source:* Courtesy Everett Collection / Everett Collection / Adobe Stock.

"The Development of the Pronephros during the Embryonic and Early Larval Life of the Catfish." She then taught zoology at the University of Maryland and conducted research at the Woods Hole Oceanographic Institution, Massachusetts during the summers.

Rachel Carson began a part-time position at the US Bureau of Fisheries in 1935, writing science scripts for the radio show *Romance Under the Waters*. She also wrote articles on natural history for the *Baltimore Sun*. Her part-time position transitioned in 1936 into a full-time position as junior aquatic biologist. Carson was the first woman in the United States to take and pass the civil service exam. She rose through the ranks of the US Fish and Wildlife Service, the succeeding division, over the next 15 years, ultimately becoming the Chief Editor of its publications. Rachel Carson also wrote several books that were successful enough to allow her to retire in 1952 at age 45. She built a cottage on the Sheepscot River near West Southport, Maine, and also kept a residence in Silver Spring, Maryland. In retirement, she completely devoted herself to writing.

Rachel Carson's life included extensive personal tragedy. In 1931, the Carson family had to give up the farm because of pressures of the Great Depression but also because large chemical plants were built on both sides of the farm. They reduced the value of the property as well as its productivity. In 1935, Rachel's father died suddenly which left her to care for her mother. Her sister died the next year leaving Carson and her mother to raise her two orphaned children. Her niece then became ill in 1959 and died and Carson adopted her orphaned son. Her mother died that year and Carson was diagnosed with breast cancer. She battled the illness but succumbed to it on 14 April 1964 in Silver Spring, Maryland, at 56 years of age.

Rachel Carson faced many hurdles to publishing her work. Her first national article was published in *Atlantic Monthly* magazine in 1937 and entitled "Undersea." Her first book was published in 1941 and entitled *Under the Sea-Wind*. It received positive reviews but went largely unnoticed because America was focused on the start of World War II. Carson's writings became more geared toward environmental activism as her interest grew. Her writing on the origin and properties of the oceans was rejected by 15 magazines, including the *Saturday Evening Post* and *National Geographic Magazine*. It was finally published in the *New Yorker* magazine as a collection under the title *A Profile of the Sea*. Parts of it were also published by *Nature*, *The Yale Review*, and *Science Digest*. In 1951, the full collection was published as the book *The Sea Around Us*. It sold more than 200 000 copies in hard cover in its first year and was on the *New York Times* bestseller list within 2 weeks. It remained on the list for a record 86 weeks, 39 of which it was the top seller. As a result, it won the John Burroughs Medal, the 1952 National Book Award, and it was the Outstanding Book of the Year in *The New York Times* Christmas Poll. Carson was awarded honorary doctoral degrees from Oberlin College and Pennsylvania College for Women for these achievements. As a result of this popularity, *Under the Sea-Wind* was rereleased and made *the New York Times* bestseller list. The next book, *The Edge of the Sea*, was released in 1955 and was also on the bestseller list. The National Council of Women of the United States named it the Outstanding Book of the Year, and Carson received the Achievement Award from the American Association of University Women as a result. Carson also published additional articles including "Help your Child to Wonder," in *Woman's Home Companion* in 1956. The book *The Sense of Wonder* was released posthumously in 1965. She wrote her final and most famous book *Silent Spring* as a result of her environmental activism from 1957 to 1961. The writing was first published in installments in June of 1962 by the *New Yorker* before it was published as a book later that year. It was at the top of the bestseller list within two weeks and remained on the list for many years.

It was the book *Silent Spring* that elevated Carson to the position of leader of the environmental movement. In 1992, *Silent Spring* was declared the most influential book of the past 50 years by a US Congressional panel. It was ranked among the 25 greatest science books of all-time by *Discover Magazine*. It has been compared to Harriet Beecher Stowe's *Uncle Tom's Cabin* and Charles Darwin's *On the Origin of the Species* in terms of overall impact. *Silent Spring* declared that chlorinated hydrocarbons and organophosphates were the most dangerous pesticide pollutants and it identified specific compounds, in particular DDT. It also proposed the term "ecosystem", which is now an important concept in environmental research.

Carson reported the death of hundreds of songbirds in Massachusetts as the result of spraying for mosquitoes, damage from spraying DDT in Long Island, New York, poisoning of workers and farmers in the south from pesticide use on fire ants, and the banning of cranberries for Thanksgiving in 1959 because of overuse of pesticides. These events caused a public furor, setting the stage for *Silent Spring*. Several attempts to discredit Rachel Carson by chemical companies served to confirm her as the leader of the American environmental movement. Her writings were added to the US Congressional Record and she received an award from the Secretary of the Interior in 1962. Carson was introduced as the originator of the environmental movement when she testified to the US Congress in 1963. As a result of her work, in 1962, over 40 bills were introduced to regulate pesticide use in several states. On 3 April 1963, Carson appeared on national television to explain pesticide dangers. Her death from breast cancer in 1964 is suspected to have resulted from exposure to the chemicals she was attempting to protect the public from. Most of the pesticides identified by Carson were eventually banned. *Time* magazine named Carson as one of the most influential people of the twentieth century. In 1980, President Jimmy Carter posthumously awarded Carson the Presidential Medal of Freedom. Rachel Carson is a true champion of the environment.

1.2   Earth's Environments

The Earth has a multitude of environments (Figure 1.3) which can be divided by a number of factors both geographical and ecological. Only some of the major environments are described here. The main division is between marine and non-marine or terrestrial. Marine environments are those not on land, so in oceans, seas, and gulfs. There are coastal environments at the marine-continental boundary, primarily beach environments and including barrier islands with the lagoons landward of them. They also include deltas where rivers build land into the ocean and estuaries where ocean water floods back up the river. Areas that flood during high tide but are otherwise land are tidal flats. There are also ocean islands and atolls far from the shore.

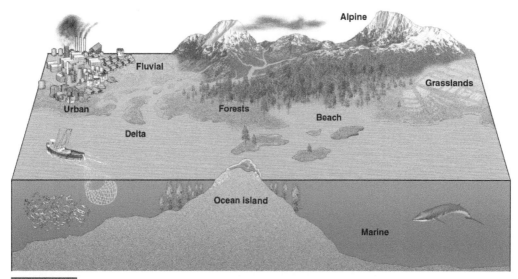

FIGURE 1.3 Diagram illustrating several environment types.

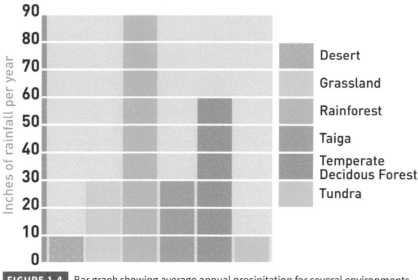

FIGURE 1.4 Bar graph showing average annual precipitation for several environments.

Within the ocean itself, shelf areas are near-shore and shallow. These are areas of continental crust covered by ocean. Sunlight can penetrate the entire water column in these areas. This is the photic zone and highly productive, including marine organisms inhabiting the seafloor. In contrast, the deep ocean only has the photic zone in the top part of the water column, with minimal life on the seafloor. The mid-ocean ridge is a volcanic chain in the middle of the ocean that supports a variety of life below the photic zone, primarily on the seafloor.

Terrestrial environments are numerous and both climatic and geographic. Alpine environments are in the mountains and are typically colder, can have glaciers and only support limited life above a certain elevation. Deserts have less than 10 in. (25 cm) of precipitation per year (Figure 1.4). They can be tropical and hot or tundra/polar and cold. The hot deserts are the classic deserts with sand, caliche, and distinct biota. Alluvial environments are also deserts but typically occur in rugged terrains with a variety of geomorphic features and biota. Tropical rainforests typically have rainfall amounts between 69 in (175 cm) and 79 in (200 cm) but sometimes in excess of 390 in (1000 cm). Rainforests are home to 40–75% of all biotic species on Earth. They can be tropical or temperate and each contains a unique biota.

Temperate climates typically contain grasslands, savannas, and forests, though these biomes can exist in the tropics as well. As the name implies, grasslands contain virtually all grasses and are typically dry, with precipitation of 20–35 in. (50.8–88.9 cm) per year with broad temperature ranges. They can support herds of large grazing animals. Savannas are slightly wetter with 30–40 in. (76.2–101.6 cm) of precipitation per year and warmer temperatures than most grasslands. Savannas contain both trees and grasses but the trees are less dense than forests. Deciduous forests receive 30–60 in. (75–150 cm) of precipitation per year, with average temperatures ranging between −22 °F (−30 °C) and 86 °F (30 °C). They are typically composed of densely packed, primarily broad-leafed trees and support a variety of animals. Aquatic environments in temperate climates are lacustrine, wetlands, and fluvial-riparian.

Lakes and ponds are lacustrine and they have broad expanses of still, open and variably deep water with free swimming fish, vertebrates, and other biota. Wetlands include swamps, bogs, and fens and also contain still water. Most only have minor and small fish, if any, and are dominated by plants, microorganisms, birds, and insects. Rivers and streams are fluvial and are marked by flowing open water capable of supporting fish. The areas along rivers are called riparian.

Very cold regions have several environments as well depending upon temperature and precipitation. Glacial ice forms in polar regions as well as alpine. Polar regions are the coldest regions and are also very dry having between 4 and 40 in. (10–100 cm) of precipitation per year. However, the higher precipitation amounts are at the edges of the glaciers near open water and, by far, precipitation is less than 10 in. (25 cm) per year making it a desert. The tundra is slightly warmer than polar regions and not covered by ice. Tundras are also deserts with less than 10 in. (25 cm) of precipitation per year but contain some permanent vegetation like grasses, moss, and lichen. The taiga is a bit warmer and wetter, with 12–20 in. (30–50 cm) of precipitation per year, 6 months of the year near or below freezing, and a cover of coniferous trees.

1.3   Point Source and Non-Point Source Pollution

Pollutants emitted to the natural environment, to air, water, or soil are either from point sources or non-point sources (Figure 1.5). The difference between them is that point source pollution is from a single identifiable source whereas non-point source pollution is from diffuse sources that might not be easily identified. Point source pollution is usually at much higher concentrations than non-point source pollution and it can result in disasters, including the loss of human life. On the other hand, they can be addressed quickly and directly. Non-point source pollution is usually a long-term effect that can slowly modify the environment to the point of large-scale changes and problems. They are much more difficult to address.

1.3.1   Point Source Pollution

Point source pollution can be disastrous to the environment and public health but it is generally not aerially extensive. The concentration of the pollutant may be high but the volume of pollutant release is small compared to non-point sources. Point source pollution is common in groundwater and can be from a number of sources such as landfills, leaking underground storage tanks (USTs) in gas stations, broken pipelines, overflowing septic systems, industrial spills, military facilities, waste pits and lagoons, mines and others where the pollution comes from a single identifiable place or group of related places. The chemistry of the pollution is unique for several of these sources.

Petroleum is a common point source pollutant in groundwater primarily home heating oil, diesel, or gasoline. It typically leaks from USTs at gas stations and homes with oil heat. Other point source pollution in groundwater can be industrial chemicals. BTEX is shorthand for benzene, toluene, ethylbenzene, and xylene, which are volatile organic compounds (VOCs) that commonly occur together. Other common groundwater pollutants include PCE (perchloroethylene) or perc, which is mainly from dry cleaners, and TCE (trichloroethylene), which is a solvent used in degreasing among many others.

Petroleum is also a common point source pollutant in surface water. Fuel spills from oil tanker accidents, oil transfer spills, and even some oil well leaks or blowouts, are more common in marine waters but raw sewage is more common in lakes and rivers. In the United States, raw sewage in lakes and rivers is less common today but in many less developed countries it is still a major problem. Animal waste from stockyards is generally a problem only during storms and floods in developed countries. Acid mine drainage (AMD) is common from active and abandoned mines and is tainted by sulfuric acid that lowers the pH of the surface water. AMD dissolves ions out of mine waste and can contain heavy metal contamination as well. Mines and mineral processing facilities can pollute surface waters with suspended particles of heavy metals and other pollutants. Gold processing can contain cyanide and mercury as point source pollutants.

Soil pollution from point sources commonly coincides with point source groundwater pollution because the liquid pollutant filters through soil to reach the groundwater and some adheres to the soil. Soil pollution is extreme around certain mines and is the source of pollutants in surface water runoff. Animal stockyards form soil pollution point sources. Soil may also contain point source pollution that does not impact water. In the past, most landfills were buried directly into the soil without any protection. Many industrial plants simply dumped waste materials behind or around the facilities. Metals like chromium, mercury, lead, and nickel and radioactive elements like radium were dumped into holes or on the surface.

Point source pollution into the air includes smokestack emissions from incinerators, coal-fired powerplants, oil refineries, steel and metal refineries, and other sources. These sources produce particulate, polycyclic aromatic hydrocarbons (PAHs), benzene, CO, CO_2, nitrogen oxides, sulfur oxides, and vaporized pollutants like mercury. There are many other point sources of air pollution like chemical and other manufacturing plants that use organic chemicals. Evaporation of these VOCs yield air pollutants that are abundant in the atmosphere. Particulate blown from mine spoils, waste piles at manufacturing facilities, trains and dump trucks in the

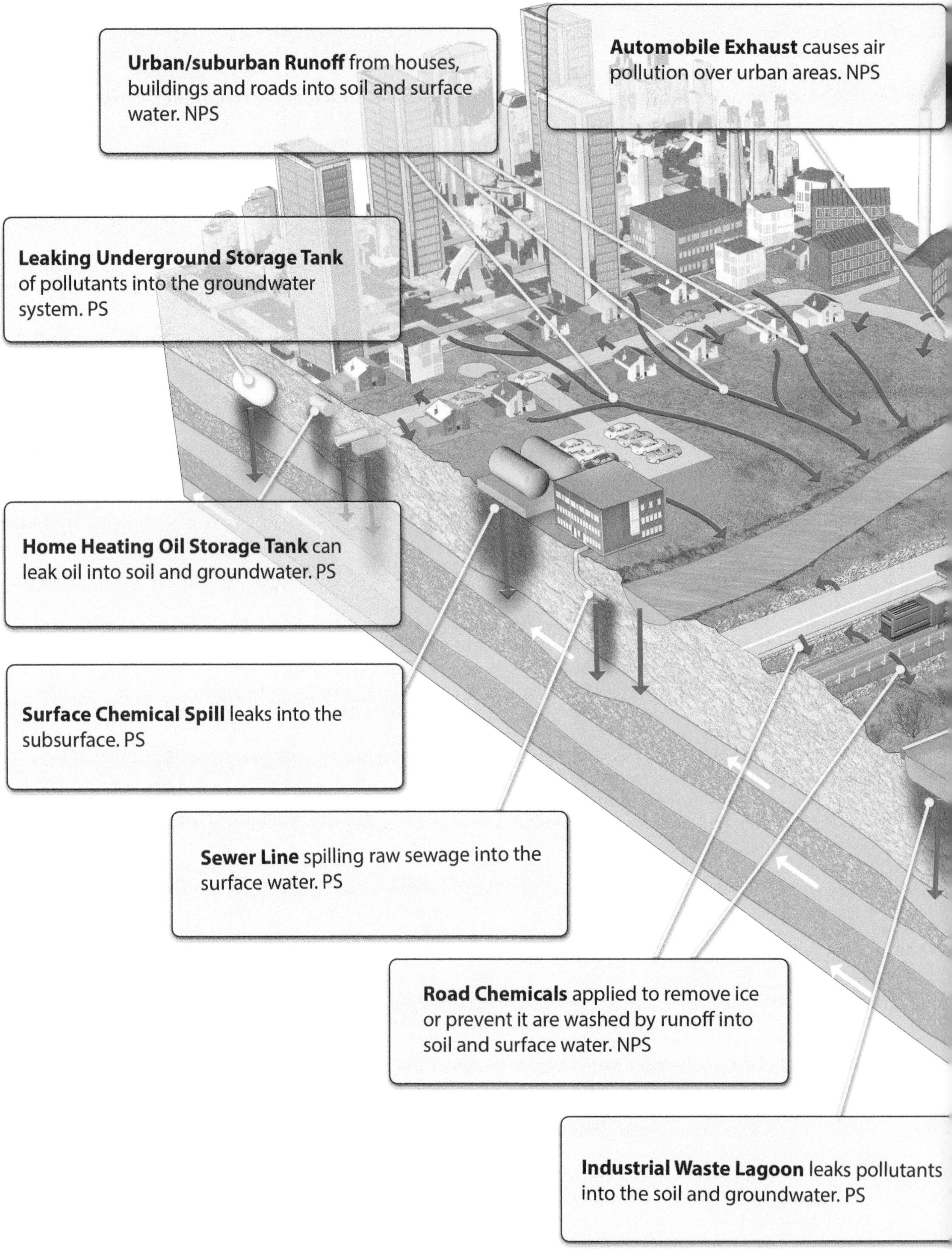

Urban/suburban Runoff from houses, buildings and roads into soil and surface water. NPS

Automobile Exhaust causes air pollution over urban areas. NPS

Leaking Underground Storage Tank of pollutants into the groundwater system. PS

Home Heating Oil Storage Tank can leak oil into soil and groundwater. PS

Surface Chemical Spill leaks into the subsurface. PS

Sewer Line spilling raw sewage into the surface water. PS

Road Chemicals applied to remove ice or prevent it are washed by runoff into soil and surface water. NPS

Industrial Waste Lagoon leaks pollutants into the soil and groundwater. PS

FIGURE 1.5 Diagram showing several point (PS) and non-point sources (NPS) of pollution.

Smokestack Pollution is released from industrial processes and municipial incineration polluting the air. PS

Agricultural Runoff of fertilizers and pesticides into the surface water and soil. NPS

Air Pollution Fallout deposition of soot and other pollutants on the surface. NPS

Waste Disposal Injection Well puts toxic waste deep underground but can leak into groundwater. PS

Septic Tank leaks raw home waste into soil and groundwater. PS or NPS

Landfill that is unlined and leaking leachate into the soil and groundwater. PS

Leaking Gas Station Tank spills gasoline into soil and groundwater. PS

form of dust also produces air pollution. They can contain heavy metals, asbestos, lime, phosphates and other fertilizers, sulfur, and sulfides among others. In addition, there are natural point sources of pollution like volcanoes and forest fires.

Point source pollution can also come from disasters such as industrial and transportation accidents, war and other attacks, or natural disasters like earthquakes, hurricanes, and volcanoes. If the human toll of these events is great, whether death, injury, loss of property, or just threat, the environmental impact is commonly overlooked. If the toll is not great, then environmental damage will be considered. The exception to this generality is if the environmental impact involves human health and is extensive. An example of this is the 9/11 World Trade Center disaster in which the first responders to Ground Zero have subsequently suffered extensive long-term health effects.

CASE STUDY 1.2 9/11 World Trade Center Disaster New York

Collateral Air Pollution

The idea to build the world's tallest buildings on the 16-acre (6.5-ha) lot in lower Manhattan was planned in the early 1960s to revitalize the area (Figure 1.6a). Construction began in 1966 by the Port Authority of New York and New Jersey. Five blocks were closed and 160 buildings were demolished to make room for the twin 110-story high buildings. The towers were open in 1972 and quickly became the symbols of American trade and commerce. By 2001, 50 000 people worked in the World Trade Center and tens of thousands of visitors passed through it daily.

On 11 September 2001, terrorists hijacked two commercial passenger airplanes and crashed them into both the North and South Towers of the World Trade Center (Figure 1.6b). More than 2800 people were killed in the crash, fire, and collapse, and victims were still being identified using DNA through 2007. The attacks led the United States to take military action in Afghanistan and Iraq that lasted for decades. The death, destruction, and retaliation dominated the American attention. However, the explosion, fire, and collapse of the towers produced toxic smoke, particulate, and gas that engulfed the disaster area and damaged the health of tens of thousands of residents, emergency response personnel, and construction workers. Approximately 250 000–400 000 people were exposed to the toxic emissions.

The buildings that were directly destroyed included the 110-story North and South Towers of the World Trade Center, the observation deck, and Seven World Trade Center, a 47-story office building. The number three World Trade Center, Marriott Hotel was crushed by the tower collapses. Two nine-story office buildings at four and five World Trade Center and the seven-story building at six World Trade Center were badly damaged and demolished later. The disaster caused the deaths of 343 firefighters, 66 police officers, and 148 passengers and crew on the airplanes.

The outpouring of responders to assist in the rescue and relief of the disaster was immense. They included federal, state, and local government agency personnel, and volunteers from private organizations, as well as private citizens. Included among the responding federal agencies were the Environmental Protection Agency, Federal Emergency Management Agency (FEMA), Centers for Disease Control and Prevention (CDC), Agency for Toxic Substances and Disease Registry (ATSDR), Occupational Safety and Health Administration (OSHA), National Institute for Occupational Safety and Health (NIOSH), Federal Bureau of Investigation (FBI), US Marshals Service, Department of Energy, National Institute of Environmental Health Sciences (NIEHS), Public Health Service Commissioned Corps, Substance Abuse and Mental Health Services Administration (SAMHSA), US Coast Guard, National Park Service, New York State agencies including the Department of Environmental Conservation, Emergency Management Office, National Guard, Office of Mental Health, Department of Health, and New York City offices including the Fire Department (FDNY) and emergency medical services (EMS), Department of Health and Mental Hygiene, Police Department (NYPD), Department of Design and Construction, Department of Environmental Protection, Department of Sanitation, Office of Emergency Management, as well as the American Red Cross, and Salvation Army. These responders were exposed to the air pollutants of the disaster.

A thick cloud of smoke and dust flooded the streets of lower Manhattan to an altitude of more than 1000 ft (305 m) high from the collapse of the North and South Towers (Figure 1.6c). The fleeing survivors and responding emergency personnel were exposed to a mixture of partially burned hydrocarbons and pulverized plastics, metals, cement, and sheetrock in addition to numerous other toxic materials. The pile of debris that resulted from the 9/11 disaster was six stories high and covered a number of city blocks. The pile was composed of 1 000 000 tons (907 185 mt) of concrete, 200 000 tons (181 437 mt) of steel, and 600 000 square feet (55 742 m^2) of glass and paper, computers, office furniture, and airplane wreckage. The particulate generated by the pulverizing of these materials was suspended, settled, and resuspended several times until 16 September when a rainstorm finally stabilized the dust. The fires burned in the rubble until 20 December which finally ended the smoke and gases.

The destruction dust and smoke from fires contained hundreds to thousands of chemical compounds, some of which were dangerous. The main health concern was exposure to asbestos. It was used extensively in fireproofing of the World Trade Center buildings, mixed into cement, and sprayed onto steel structural supports and beams. Over 12 000 samples of dust were analyzed for asbestos by the Environmental Protection Agency (EPA) which indicated high concentrations of asbestos at Ground Zero and in settled dust in nearby apartments. Health effects of asbestos inhalation include asbestosis or scarring of the lungs, and mesothelioma, a cancer of the lungs, in addition to abdominal cavity, pericardium, and gastrointestinal, kidney, and liver cancer. The EPA banned most uses of asbestos in 1989 as a result.

Particulate also includes aerosol liquids and solids as smoke and dust. Particulate larger than 10 μm (PM_{10}) is filtered by nasal hair and mucous membranes of the throat. The larger particles caused what was known as World Trade Center (WTC) cough among construction workers and response personnel. The large dust particles were primarily pulverized fragments of concrete, sheetrock, and fiberglass, which irritated throats, sinuses, and nasal passages.

Particles between 10 and 2.5 μm were largely trapped in the upper airways of the lungs and were not a health risk to most people depending on composition. Particles with very high or low pH caused scarring of bronchial tubes and possible chronic bronchitis. The pulverized sheetrock was paper-coated gypsum, which is a sulfate mineral. Once in the lungs, it forms sulfurous and sulfuric acid that irritates or scars them. Some particles were from partially burned jet fuel and plastics such as polycyclic aromatic hydrocarbons (PAHs), which are known human carcinogens. Exposure to this size particulate can produce severe to fatal reactions in at-risk people such as the very young and old and ailing people.

Inhalation of particles smaller than 2.5 μm (PM$_{2.5}$) penetrated deeply into the lungs, bronchi, and alveoli of the workers. They were absorbed directly into the bloodstream through the lungs. They were commonly aerosols and included harmful and carcinogenic organic compounds like PAHs. These particles are linked to increased asthma, lung cancer, other respiratory diseases, and premature death.

Workers at Ground Zero within four to eight hours of the collapse were exposed to very high levels of PM. By mid-October, PM levels in lower Manhattan returned to normal. PM$_{2.5}$ levels were much lower with distance from Ground Zero. Within 3–10 blocks from Ground Zero, they were normal. The EPA identified PM as the source for increased risk of chronic health problems for the workers who were most exposed. PM$_{2.5}$ concentrations were not a significant risk for people who remained farther away.

(a)

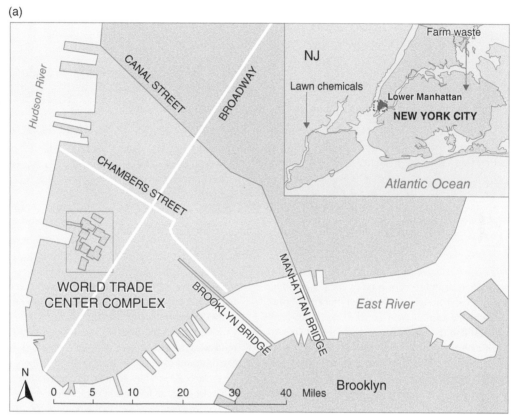

(b)

FIGURE 1.6 (a) Map of the World Trade Center complex at Ground Zero and the restricted access area in lower Manhattan. Inset shows the location of the area in New York City and surroundings. *Source:* World Trade Center Health Registry. (b) Photo showing the collision of the second aircraft with a World Trade Center tower while the other tower burns from the first collision. *Source:* Robert J. Fisch / Flickr / CC BY-SA 2.0. (c) Aerial photo of the dust cloud from the collapse of the second tower. *Source:* 9/11 photos / Flickr / CC BY 2.0.

(c)

FIGURE 1.6 (Continued)

EPA analysis of dust samples from window ledges and building roofs indicated that individuals within a few blocks of Ground Zero likely inhaled lead and chromium in excess of regulatory levels. Between 23 September and the end of November, Ground Zero fires increased airborne dioxin levels to between 10 and 150 picograms per cubic meter, which is over 1000 times higher than normal. Dioxins are classified as a known human carcinogen, and exposure to dioxin also results in disfiguring chloracne. However, the EPA concluded that health risks from exposure to dioxin at Ground Zero were not significant because exposure through inhalation is very small. Low levels of PCBs (polychlorinated biphenyls) were also found in smoke samples. Health effects from exposure to PCBs are skin rashes and chloracne. Both the EPA and IARC (International Agency for Research on Cancer) classify PCBs as probable human carcinogens. The highest concentration of PCB at Ground Zero was 153 nanograms per cubic meter compared with typical urban air which contains 1–8 nanograms per cubic meter and the maximum acceptable level for American workers is 1 mg per cubic meter. By the middle of October, the PCB increase had subsided. The EPA concluded that there was no increased cancer risk associated with this PCB exposure. Potential synergistic impact of exposure to PAH, PCBs, metals, and other airborne pollutants have not been determined. By January 2002, ambient air quality around Ground Zero returned to normal.

Within 48 hours of the disaster, about 90% of the 10 116 firefighter and EMS responders were suffering from acute cough and gastrointestinal issues. Most recovered from them but others did not and some died from them. Nine months after the disaster, over 46% of responders were still experiencing at least one pulmonary symptom and 52% were experiencing ear, nose, and throat symptoms. By March 2004, 380 of the responding firefighters were suffering from debilitating respiratory illnesses. The incidence of COPD (chronic obstructive pulmonary disease), irritant-induced asthma, and WTC cough was a function of the time of arrival at the site, duration of exposure, and location. Residents, however, also experienced negative health effects from even

more remote exposure. Between 25 October and 2 November, it was found that 66% of nearby residents reported nose or throat irritation and 47% reported a persistent cough.

Right after the disaster, the director of the EPA, Christine Whitman released a statement that there was no direct health threat from the dust and gases. When it was clear that there was a threat, a cover-up ensued. Apparently, the EPA findings had been downplayed by the White House staff to prevent panic. They wrote Whitman's statement. The controversy escalated to the point of congressional investigations and a White House admission of poor handling of the environmental and public health impacts.

Senators and Representatives from the New York area convinced the government to allocate of $12 million in December 2001 establishing the World Trade Center Worker and Volunteer Medical Screening Program run by Mount Sinai Hospital. They later secured an additional $90 million to expand the number of eligible workers and volunteers. They then lobbied Congress and obtained $125 million in additional funding for long-term medical and mental health treatment. This effort was expanded over the years and involved celebrity spokespeople to finally obtain more than $7 billion to be awarded to 97% of the impacted survivors and families. In addition, in February 2022, President Joe Biden dedicated a $3.5 billion fund from frozen assets of the government of Afghanistan to the victims of the 9/11 attacks.

More than 60 types of cancer and about 24 other health conditions have been linked to Ground Zero exposures. At least 4627 first responders who enrolled in the World Trade Center (WTC) Health Program have died though not all from pollutant exposure. By 2022, 257 firefighters who were first responders had died prematurely compared with the 343 who perished in the disaster. From the data to date, 9/11 first responders have a 25% increase in prostate cancer, a doubling of thyroid cancer and a 41% increase in leukemia compared to the general public. This illustrates how devastating environmental impacts of a disaster can be on humans alone.

1.3.2   Non-Point Source Pollution

Non-point source pollution is not as damaging to the environment as point source pollution in the short term but it can cause major and lasting environmental and public health damage in the long term (Figure 1.5). The releases are generally in low concentrations that are unlikely to be a public health hazard individually or even collectively in most cases. The problem is that the sheer volume of pollutants released by the overwhelming number of non-point sources is immense. Collectively, these pollutants are capable of causing major and possibly irreparable damage to the environment in a region if not the entire planet. This non-point source pollution that everyone contributes to is responsible for the most pressing environmental problems. Individuals may consider their part to be small, but collectively the huge human population makes even small individual contributions devastating.

Non-point source pollution is from infinite small sources to which most or all humans contribute. Non-point source air pollution is mainly from automobile exhaust. Automobiles emit NOx (nitrogen oxides), CO_2, CO, SOx (sulfur oxides), benzene, particulate, and possibly some partially burned hydrocarbons. Other very common non-point combustion sources include home heating furnaces, fireplaces, and grills. The main air pollutants from these are CO_2, CO, PAHs, and particulate. The CO_2 from non-point combustion is the main greenhouse gas in human-generated climate change. The NOx and SOx from non-point sources are the primary producers of acid precipitation. The other main pollutants are vapors from VOCs. If gasoline, diesel, or other fuel is exposed to the air, some of it evaporates into vapor. Drying oil paints and finishes, solvents, glues, and other adhesives, fabric waterproofing, some cleaners, and many other everyday items also emit VOCs, meaning that they are present in most air near populated areas. VOCs react with the NOx from automobile exhaust in the atmosphere if exposed to strong sunlight to produce ozone, the primary component of urban smog. Ozone is a public health hazard to people with respiratory issues and is very damaging to plants as well.

Surface water is also polluted by non-point sources. In urban and suburban areas, chemicals on the surface are carried into surface water bodies by runoff during rain and thaws. Fallout from air pollution coats exposed surfaces and is easily lifted. Lawns and gardens are sprayed with pesticides, herbicides, and chemical fertilizers that are flushed by rain. Septic systems and some sewer lines can leak or overflow of raw sewage into surface water, especially during spring thaw and wet periods. Virtually all roads contain numerous pollutants including road salt and de-icing compounds, and leaking oil, gasoline, and antifreeze from automobiles. Chemicals on houses, driveways, sidewalks, parking lots, businesses, and litter are washed in the untreated runoff into sewers and surface water bodies. Organic input produces eutrophication of lakes and wetlands and even some rivers. The fallout and washout of NOx and SOx from air pollution produces acid precipitation that can be a public health threat, degrade infrastructure, and destroy the ecology of lakes and ponds.

Non-point agricultural pollution sources produce large-scale damage to the environment. Agricultural fields are covered with chemical fertilizer before seeds are planted and then sprayed with pesticides. They are used in excess to ensure success and productivity of the plants. Rain dissolves the fertilizer and/or it is carried as particles by runoff to lakes and rivers. Overfertilization causes eutrophication of the lakes and ponds and rivers carry the excess nutrients to the oceans. If input is to a restricted body like a bay, gulf, sound, or some seas, eutrophication can take place and produce "dead zones." The dead zones are oxygen-poor to the point of not being able to support life beyond jellyfish. The fish and mobile invertebrates escape from the area and the rest die.

Even though pesticides have saved millions of human lives by reducing disease-carrying and crop-destroying insects, they also produce damage to animals and beneficial insects. In the 1960s and 1970s, there was so much pesticide in human breast milk that it was dangerous to babies. As a result, many of the more persistent pesticides were banned and more selective, potent, and shorter-acting pesticides were developed. This solved one problem but created others. There is a crisis in honeybee populations called colony collapse disorder (CCD), in which whole hives perish and pesticides are strongly suspected. Nearly one third of the greatly diminished honeybee population disappeared in 2007 alone. The problem is that most pollination is done by honeybees making our supply of fruit, nuts, and most vegetables in jeopardy without them.

1.4   Pollution Affects Everything

The problem with pollution is the number of people contributing to it. Even practices with small per capita impact are causing a devastating impact on the country and global scale. This means that even though it feels for most people like they have no power over the situation, they actually do. Many take the view that the only way to overcome the vast environmental problems we face is for the government to fix them. However, governmental action is typically slow and with the varying opinions, may not address the issues in a unified approach, if at all. It will take the collective will of the people to make any impact on the pollution problems.

Primarily, action includes understanding each type and aspect of pollution and its impacts. If the cause and effects of each type of pollution are appreciated, then appropriate action can be taken. Popular solutions like installing solar panels, purchasing hybrid and energy-efficient vehicles, recycling waste, and switching to energy-efficient appliances make a huge difference on even an individual family basis. However, there are many practices that are very damaging to the environment that may not be obvious and require some research to understand why they should be avoided. Many domestic products have been changed several times as a result of environmental problems. Paint has changed from lead-bearing, primarily VOC-bearing, mercury-bearing, etc. as the new formulations were found to be unhealthy to humans or the environment. Imported items can be manufactured under less stringent regulations than those in developed countries. Pet food from China was manufactured with melamine which caused sickness and death among many pets in the United States. Currently, mining cryptocurrency is making many people rich but abuses the energy grid and is causing extreme environmental impacts.

CASE STUDY 1.3 Environmental Impact of Cryptocurrency

Broad Reach of Environmental Damage

Cryptocurrency is virtual money that exists purely in electronic form. There is no central regulating authority for cryptocurrency meaning that there are no transaction fees, they are simple and everyone can use it. Transactions are recorded but anonymous meaning that tax evaders, criminals, and terrorists can misuse cryptocurrencies. Perhaps surprisingly, cryptocurrency strongly degrades the environment. The situation with cryptocurrency, however, is in a rapid state of flux and there is much misinformation. This case study contains the best current information but the situation is changing and currently accepted information could be in error. It may take additional research to fully understand the situation at the time you read this.

At first, cryptocurrencies could not ensure the security of transactions or that tokens could not be used more than once. Bitcoin was developed in 2008 including a ledger system with transactions recorded in blockchain. A blockchain is a database available across a network with transactions recorded in blocks that are linked together. The system allows 21 million total bitcoins to be created. An estimated 18.8 million bitcoins are currently in circulation and the rest will be released by 2140. Bitcoin is the largest cryptocurrency in the world, with more than 50% of the total value. It is used to buy cars, furniture, vacations, and others. The total value of bitcoins is currently $903 billion.

New bitcoins are created through mining, which is accomplished by validating and recording new transactions in the blockchain. Miners verify the validity of bitcoin transactions bundled into a block. This involves verifying 20–30 separate variables including address, name, timestamp, and adequate account value among others. Miners compete to have their validation accepted first. They solve problems to find a nonce or "number used once", a unique, identifiable 64-digit string of letters and numbers. The new block is linked to the previous block and all blocks are uniquely chained together, making the network essentially tamper-proof. The successful miner acquires a new bitcoin.

More miners compete for higher priced bitcoins and the problems are accordingly more difficult. Bitcoin is designed for blocks of transactions to be mined every 10 minutes. If there are more miners competing using more computing power, the probability of determining nonce in less than 10 minutes increases. In response, the system makes it more difficult to find nonce. With less computing power operating, the system makes the problems easier. The bitcoin system adjusts the difficulty of mining regularly to keep blocks to 10 minutes. The bitcoin value for miners is reduced by half after every 210 000 blocks. Mining a single block earned 50 bitcoins in 2009 when bitcoin mining began. In 2011, each bitcoin was worth $1 (US) and reached a maximum value of $65 000. In March of 2022, a block earned 6.25 bitcoins. The price at this time was about $39 000 per bitcoin, earning each miner $243 750 for completing one block. With such a rich payout, there are about one million bitcoin miners competing for blocks.

The main environmental impact of cryptocurrency is the electricity used in the mining. It is becoming progressively more difficult to mine new bitcoin blocks, requiring progressively more computational power.

It is now estimated that bitcoin mining uses more electricity per year than many countries, including the Netherlands and Pakistan. Even in the early days of the practice, mining of a single bitcoin block was using about 2292.5 kWh of electricity, which an average household in the United States uses in 78 days. Other estimates are that mining uses 121.36 terawatt hours per year. This adds carbon dioxide to the atmosphere from fossil fuel burning and worsens climate change. The United States hosts about 35% of the bitcoin mining and about 60% of the electricity is generated from burning of fossil fuels. This power usage is increasing at phenomenal rates. Between 2015 and 2021, energy consumption for bitcoin mining increased about 62-fold. Many mining operations are moved to locations where electricity is cheaper to reduce mining costs.

In 2020, more than 65% of the global bitcoin processing power was controlled by China. Miners were taking advantage of the cheap electricity from dirty coal-fired power plants and hydropower. However, China cracked down on cryptocurrency mining because of the financial risks and enormous energy consumption. As a result, most of the Chinese bitcoin miners moved their operations to countries with cheaper electricity and less constraints, like Kazakhstan. Many of these countries rely on fossil fuels for electricity making the climate change impact even worse. Before this change, it was estimated that by 2024 bitcoin mining in China alone would generate 143 million tons (130 million mt) of CO_2.

The power plants needed for mining of cryptocurrency can even have a direct impact on the surrounding environment. Millions of gallons of water are required to cool the plants while they are heating up because of producing such high levels of electricity for mining. Some plants discharge some of the hot water into local lakes, increasing the water temperature by as much as 30–50 °F (17–28 °C) above normal locally in one case. This heat has a disastrous effect on the aquatic life and local wildlife.

Electronic waste is also damaging the environment as well as causing massive shortages from computers to automobiles and anything else that uses computer chips. Computers, graphics cards, mining rigs, networks, and other computer hardware are used for mining. Having increased computing power gives miners an advantage in obtaining bitcoin blocks. As a result, miners upgrade and throw away even slightly outdated computer equipment at an ever-increasing rate. It is estimated that they produce between 30 000 tons and 115 kt (27 216–104 326 mt) of electronic waste per year.

Non-fungible tokens (NFTs) are a new phenomenon in cryptocurrency and the main controller is Ethereum, the second-largest cryptocurrency dealer. NFTs are digital files of artwork including photos, music, videos, and other forms that are stamped with unique strings of code. Any user can view or even copy NFTs, but the one unique NFT for each item is the property of a buyer. It is obtained using the same energy-intensive mining process as bitcoin and stored on the blockchain. NFTs typically sell for hundreds of thousands of dollars but one single NFT sold for more than $69 million.

The average NFT generates about 440 pounds (200 kg) of carbon to the atmosphere. A digital artist estimated that the carbon footprint of average NFTs is more than a European's month electricity consumption.

1.5 What Can I Do About It?

Many people feel that the environmental problems are so immense that there is nothing they can do to help address it. More than 100 million people vote for President of the United States, so individual votes mean almost nothing, yet most people vote and many even fight over their choices. The same attitude can be applied to the environment. Further, many people think that the government will not take action even if they petition or pressure them. However, President Richard Nixon did not support environmental protection but was forced to sign a bill creating the US Environmental Protection Agency because the pressure from the American public was so great. People have the power to force the government and private corporations to adjust their practices to be more environmentally friendly if they regard it as a priority.

In addition to pressuring elected officials, government agencies, and private corporations by using their votes and pocketbooks, citizens can make changes on a personal level. Collectively, these individual changes make global change. These individual changes involve three levels; understanding, education, and taking action. Fully understanding all aspects and collateral impacts of an action allows a person to decide whether they approve or disapprove of it. This understanding allows them to decide if they want to pursue further action on it. If they feel strongly enough, they may wish to inform their family and friends about it and possibly even colleagues at work. They might post it on social media for an even broader impact. Finally, if they feel strongly enough, they might choose to take action. This could be making personal changes, family changes, or even changes to work practices and routines.

There are typically extensive environmental impacts in any accident, attack, natural disaster, or development project that may be otherwise overlooked. The public health impact of the 9/11 disaster brought these effects to the attention of the public. However, even the building of a housing complex involves clearing the site, change of runoff and drainage, increased chemicals from the building, landscaping and de-icing, increased automobile exhaust in the air, extra volume in the sewer system leading to possible overflow, extra stress on electrical generation and water supply, etc. This cumulative environmental stress could impact local ecosystems and even public health. In response, a person might attend town meetings, inform their neighbors of potential problems or collect signatures and submit a petition against the project. These actions would have to be weighed against the benefits of the project.

If a person was unhappy to learn about the cost of cryptocurrency to society, they could research it and learn more. They could certainly not participate in bitcoin, NFTs, or any other system. They could caution their friends and family about the cost of participation regardless of benefits. They could use social media to dissuade people from participating. They could talk to people at their work about the company disenfranchising from cryptocurrency. Finally, they could support political candidates who propose regulation of cryptocurrency at whatever level they deem appropriate.

Environmental Justice

CHAPTER OUTLINE

Words you should know:

Climate justice – The view of climate change as an ethical, legal, and political issue, rather than purely environmental or scientific.

Ecocide – Essentially, genocide of the natural environment or some portion thereof.

Environmental justice – The fair treatment of people regardless of race, ethnicity, or income, in the development, implementation, and enforcement of environmental laws, regulations, and policies.

Environmental racism – Targeting of a community for unjust treatment in terms of exposure, disposal, and/or legislation of hazardous materials based on their predominant race, ethnicity, or gender. In some definitions, economic status is included.

Sacrifice zone – An area that is so profoundly polluted through needed activity that it is considered unrestorable. Also known as a National Sacrifice Zone or Area.

2.1   Environmental Justice

By definition, environmental justice strives for the fair treatment of all people regardless of race, color, national origin, or income level, in the development, implementation, and enforcement of environmental laws, regulations, and policies. Although this is a simple and reasonable definition, the field of environmental justice is actually quite complex and constantly evolving, especially with the current predominance of divisive political views. The origins of the environmental justice movement lie within the American civil rights movement of the 1960s but was relegated to a secondary position in favor of the more pressing issues of the time. Therefore, it really had origins in environmental racism in the United States. More recently, it has become a global issue intertwined with climate justice and even with issues that had their roots in colonialism. It would take an entire textbook to explain all of the relationships and intricacies and the course should be taught in a Political Science program. In reality, many of the case studies in this book could fall under the topic of environmental justice and might be revisited in that context. To best set the stage for such an understanding, this chapter will cover several of the salient topics of environmental justice and illustrate them through examples.

The actual beginning of the environmental justice movement is attributed to the North Carolina polychlorinated biphenyls (PCB) pollutant protests in 1982. Cancer-causing PCB-contaminated soil was dumped in a predominately African American community and it resulted in massive protests that led to more than 500 arrests. As a result, the inordinate placement of hazardous waste in African American communities was investigated. This led to further research and resulting lawsuits on hazardous waste being deposited in poor, predominantly African American communities in the United States. This narrow definition quickly expanded. As restrictive environmental legislation increased in the 1970s and 1980s, so did the cost of hazardous waste disposal in the United States and other developed countries. In response, new companies emerged that would dispose of this waste for a price. Many of them chose to dispose of this waste in the area termed the "Global South" because there was less environmental restriction and the costs were far less (Figure 2.1). These practices escalated in the 1980s and 1990s, effectively reducing the hazardous waste disposal and pollution burden in the developed areas like the United States, Canada, and Europe where it was generated and transferring it to the countries of the Global South.

This division of the Global North versus the Global South is another way of grouping countries by socioeconomic and political traits very similar to the industrialized versus third-world designations. The Global South refers to lower-income countries rather than an exclusively geographical term. The Global South consists of countries in Africa, Latin America, the Caribbean, Pacific Islands, and the less advanced countries in Asia and the Middle East. The majority of these countries are in or near the tropics. In contrast, the Global North includes the United States, Australia, Canada, Europe, Russia, Israel, Japan, New Zealand, Singapore, South Korea, and Taiwan. These developed countries generated the hazardous waste that was disposed of in the Global South.

This practice and inequity was quickly recognized and battled on a local level in a very disorganized manner by environmental action groups and investigative reporters, all acting independently and without much clout. Many government officials were happy to accept the additional income and even bribes that came with receiving the hazardous waste for disposal. The groups opposing the practice commonly had very little power and could do little more than report the more outrageous incidents, educate residents on identifying dangers, and collect petitions to present to officials and regulatory agencies in countries where they were not residents. The only place the efforts against environmental injustice could be effective at this time was in organizing marginalized and impacted groups such as the poor and certain minority groups in developed countries to compel their legislators to take action. Even this was unevenly effective.

The first real global recognition of environmental justice was at the 1991 First National People of Color Environmental Leadership Summit in Washington, DC. More than 650 delegates from the United States, Mexico, Chile, and other countries adopted 17 principles of environmental justice. These were presented at the 1992 Earth Summit in Brazil. This effort expanded the definition of environmental justice to include public health, worker safety, land use, transportation, among other issues. It was later further

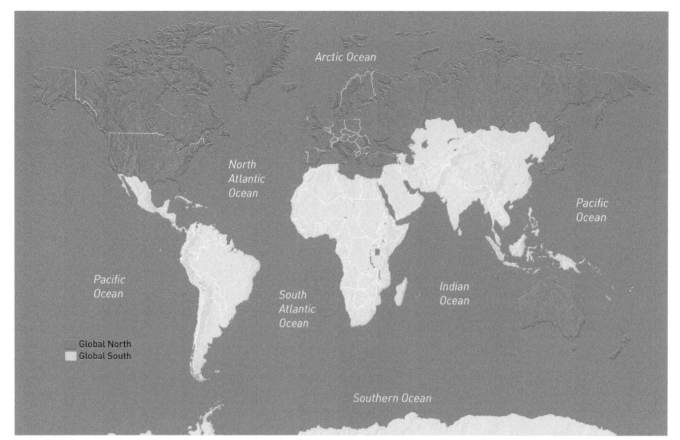

FIGURE 2.1 Map of the world showing the Global North and Global South countries.

expanded to include gender issues, international injustices, and inequalities within disadvantaged groups. More recently, more advanced concepts have been included such as climate justice, corporate accountability, ecocide, ecological debt, environmental racism, sacrifice zones, and others.

Climate justice addresses the social and ethical impacts of climate change. This has taken on a life of its own separate from mainstream environmental justice and heavily overshadows it. With cultural revolutions sparked by people like Vice President Al Gore and Swedish youth activist Greta Thunberg directly opposed by others including President Donald Trump, climate justice has become a complex issue that includes a great deal of public opinion. As a result, it approaches climate change as an ethical, legal, and political issue, rather than its purely environmental or scientific nature. It relates the causes and effects of climate change to environmental justice and social justice including equality, human rights, and historical responsibilities. Climate justice usually includes the legal actions on climate change as well.

The term "ecological debt" refers to the inferred or calculated debt of wealthier, industrialized countries over some period of time for having plundered poorer countries by seizing or utilizing their resources with little or no compensation. It also applies to degrading their natural habitats, or unethically discharging hazardous waste on their lands and waters. These actions have their roots in colonialism but continue today. Basically, whether directly or indirectly, some countries take advantage of political and economic situations in other less developed countries to obtain natural or human resources at a much lower cost than they would otherwise pay or degrade their environment in ways that they would be unable to under the system they are bound by. This includes flexing of military or economic muscle to achieve these ends.

Ecocide is the term for the extreme degradation of the environment at a criminal or elimination level. Destructive environmental practices that directly impact human health and safety are crimes in all countries and are enforced. However, activities that substantially damage or destroy ecosystems or harm the health and well-being of a particular species, especially over the long term, are less clear. Ecocide in the environment is viewed like the term "genocide" for the deliberate elimination of certain human ethnic or racial groups. Humans are committing ecocide through industrial effects on the global environment. Each type of life that is systematically being driven toward extinction is experiencing ecocide, such as birds, insects, and certain marine life. It can also be applied at the species level for all plants and animals that have been driven toward extinction. The ultimate source of ecocide is the ever-growing demand from consumers combined with a near total disregard for the long-term ecological damage.

2.2 | Environmental Justice of Industrial Impacts

The basic tenet of environmental justice is the equitable distribution of environmental risks and benefits. That means there should be the same degree of protection from environmental and health hazards regardless of the group. No group should bear disproportionate environmental consequences of industrial or government practices. However, certainly, this is not the case. Industrial companies work to maximize profits. Therefore, they tend to site plants where, in addition to having convenient appropriate transportation and a trained workforce, the cost is the least possible. This means that land, construction, supplies, and labor costs should be minimized. Depending upon the product, this can be in a less expensive area in an industrialized country such as those occupied by the poor and possibly minority groups. If possible, companies can site industrial plants in Global South countries where costs are less.

Companies are under ever-increasing pressure to maximize profits. This commonly translates to cutting corners and underpaying employees. In some industries, this just amounts to shoddy products and eventual reduced sales. With others, such as chemical companies or nuclear powerplants, this can pull the operation to the precipice of disaster. This occurs even when companies are profitable and operating properly. Many walk the fine line between safe and unsafe. However, if companies begin to falter by suffering profit declines or losses or if specific product lines or initiatives become unprofitable, safety practices may become less rigid. This can lead to significant industrial accidents. Although rare, they can cause profound environmental damage and loss of human life. When these plants are located in places where they might hurt people that are impoverished or of a certain racial or ethnic group or where the company might skirt jurisdiction for their actions, it falls in the purview of environmental justice. There are many examples of such accidents, but perhaps the worst is the Bhopal, India disaster.

CASE STUDY 2.1 Bhopal Chemical Disaster

Negligent Poisoning

The accidental methyl isocyanate (MIC) release from a chemical plant in Bhopal, India on 3 December 1984, is possibly the world's worst industrial accident. Its history began in the 1960s when India was having trouble feeding its people and maintaining political stability. In 1969, the Indian government announced that Union Carbide Corporation from the United States would construct a new agricultural chemical plant in Bhopal in central rural Madhya Pradesh (Figure 2.2a). Union Carbide worked with the Indian government, who had a one third ownership, to build a plant to produce pesticides and fertilizers for India's farmers, who were beginning to utilize Western agricultural methods. The plant opened just outside this small, 120-square-miles (308-km^2) densely populated urban area in the economically depressed Jai Prakash Nagar. It employed about 600 people in three shifts. India provided tax and other economic incentives to convince Union Carbide to come to Bhopal. They were originally hesitant to build the plant because of concerns over the inexperience of the local work force but bowed to the pressure from India.

The manufacture, transport, and processing of MIC require stringent handling and precautions. It is one of the main ingredients in the manufacture of carbamate pesticides. It is highly volatile, evaporating quickly at temperatures below 102.4 °F (39.1 °C), reacts violently with water, and quickly heats to boiling producing isocyanic gas. MIC is highly flammable and explosive. If released, it moves downhill following topography because it is heavier than air. Exposure to MIC causes serious negative health effects. If inhaled at concentrations as low as 0.4 parts per million (ppm), it induces coughing, shortness of breath, and chest pain, as well as eye, nose, and throat irritation and skin damage. At exposure of 20 ppm, health impacts include pulmonary or lung edema, vascular hemorrhage, bronchial pneumonia, and death. At concentrations of 2 to 4 ppm, MIC causes tearing, but at 21 ppm, it burns corneas and causes permanent blindness.

The plant was producing pesticides and fertilizers from ingredients made in the United States by the mid-1970s. By the late 1970s,

through work force training and quality control, the plant became more independent, manufacturing the chemicals directly from raw materials. In 1980, a very effective carbamate pesticide called *carbaryl* was developed and produced. This pesticide was widely adopted in the United States. The manufacture of carbaryl required large quantities of MIC. Union Carbide was reluctant to allow the Bhopal plant to handle and store this hazardous chemical. Their MIC was manufactured and stored at a plant in Charleston, West Virginia, under tightly controlled procedures. However, the Indian government wanted the additional jobs and expanded plant operations and pressured them. Reluctantly, Union Carbide agreed to process MIC at Bhopal.

A MIC production unit was installed at Bhopal as well as three partially buried stainless steel storage tanks, with a capacity of 15 000 gal (68 000 l) each. By 1984, however, the Bhopal plant was losing money. India's agriculture had improved and was less dependent on Western methods, reducing the need for expensive fertilizers and pesticides.

On 2 December 1984 at 3:00 p.m., the Bhopal plant reopened after a week-long shutdown, and about 100 workers began an eight-hour shift. Because of reduced demand, the plant was struggling financially. As a result, important maintenance had been delayed, including repair of faulty valves in the MIC tank system. Additionally, several experienced managers and staff were replaced with less well-trained staff. At 9:30 p.m., a worker was ordered to clean a 23-ft (7-m) section of pipe that filtered MIC before it filled the storage tanks. The worker disconnected a section of pipe and connected a water hose to it. For the following three hours, water seeped past a faulty valve and reacted with MIC in a processing tank.

By 11:45 p.m., the pressure and temperature in the tank was more than five times the normal levels and the worker's eyes were tearing. As the heat and pressure continued to build, the supervisor and operation crew took a tea break. By 12:40 a.m., the MIC fumes were overpowering and the half-blind and nauseous workers evacuated the plant, as a relief valve burst and the vaporized MIC shot 130 ft (40 m) into the air from a vent pipe.

(a)

(b)

FIGURE 2.2 (a) Map of India area showing the location of Bhopal. Inset shows position of India in southern Asia. (b) Photo of toxic cloud overtaking Bhopal. *Source:* Twitter / @Hardeep S. Puri. (c) Photo of the mon_ment to the victims of the Bhopal chemical disaster. *Source:* Bhopal Medical Appeal / Wikimedia commons / CC BY-SA 2.0.

(c)

FIGURE 2.2 (Continued)

Alarms sounded as MIC streamed from the stack. The plant fire department attempted to spray a water curtain around the vent and wash the MIC out of the air, but their equipment could only spray water to 100 ft (31 m) height, and the MIC flow was 130 ft (40 m) high. A gas scrubber inside the stack should have pumped soda into the vent pipe which would have neutralized the MIC. However, it failed to function and no soda was released. The emergency tanks intended to contain MIC in case of plant system failure could not be used because they were full of MIC. Plant emergency procedures required them to be empty. A flare system that burned escaping gas could not be activated because a replacement part was on order from the United States. The plant sounded sirens to alert nearby residents of the release, but they were turned off by plant operators to avoid panic. No calls were made to local officials to warn them about the deadly gas flowing toward Bhopal. More than 40 tons (36 mt) of MIC was vaporized and released to the air before the reaction terminated.

Once out of the plant, MIC was trapped under a layer of cool air and pushed along by a gentle breeze. It flowed downhill and enveloped the sleeping residents of Bhopal (Figure 2.2b). Many died in their sleep. Others awoke and unable breathe, struggled to escape outside. Tens of thousands of people quickly filled the streets, struggling to escape the toxic cloud. Many were blinded by the gas and subsequently killed by speeding cars and buses filled with people desperately trying to escape. By the morning, local hospitals were filled to capacity and officials gave up trying to count the dead. Thousands of bodies filled the streets. By 3:00 p.m. the following afternoon, the gas cloud had essentially dissipated, and recovery began.

It is estimated that about 3787 people were directly killed by the gas, and nearly 600 000 were injured, many severely (Figure 2.2c). Several studies estimated that 10–15 people continued to die every month

for the next decade or more from damaged lungs and gastrointestinal systems, bringing the death toll to at least 8000. Some claim that it was closer to 16 000 deaths. More than 7000 animals were killed, and the local economy suffered losses up to $65 million.

The Bhopal Union Carbide plant operations were terminated and it was shut down permanently by the Indian government. In 1989, on behalf of more than 500 000 accident victims, the Indian government settled with Union Carbide for $470 million in damages. Union Carbide employees were also exempted from criminal prosecution. However, in 1991, the Indian judiciary rejected the immunity granted to Union Carbide employees. Warrants were issued for the arrest of the Union Carbide CEO when he and other executives failed to appear in court to face the criminal charges.

Victims and their families began to be compensated in 1993 after having been delayed as a result of legal battling between Union Carbide and the Indian government. By then, the settlement had been increased to $700 million. Relatives of the deceased received $3200 per death and more for multiple deaths within the same family. Seriously injured survivors received $3000 each and those with minor injuries received a few hundred dollars. In August 1999, Union Carbide was acquired by Dow Chemical, making it the second-largest chemical company in the world.

The Bhopal disaster caused chemical companies to re-evaluate their processing and manufacturing practices and instituted new stringent safety systems. They also developed new site selection criteria, especially for plants that managed dangerous materials. Government regulatory agencies around the world enacted stringent laws requiring companies to disclose the types and quantities of materials they managed and to document plant safety procedures were adequate and followed.

Bhopal is a case of a large United States corporation building and operating a plant in a Global South country. However, direct involvement is not the only way large companies from industrialized countries can impact environmental disasters in Global South countries. These countries have reduced construction costs, cheaper labor, cheaper resource costs, and fewer environmental, health and safety, and labor restrictions. As a result, production is far easier and less expensive than in industrialized countries. Yet the demand for cheap products in industrialized countries is high.

Poverty is so rampant in many third-world countries that the residents are willing to do almost anything for whatever payment they can get. Their very survival depends on the meager wages they earn in the factories despite the poor conditions. Companies in the Global South countries take advantage of this situation to reduce costs and maximize profits. The working conditions constitute one type of injustice but there is damage to the environment from these companies that the underpaid residents must endure, and that falls in the realm of environmental justice. This situation appears to be a problem internal to a poor country, and in many cases it is. However, in many other cases it is not so simple. Many large companies outsource production of their products to these companies to reduce costs. They provide the instruction, some of the equipment, and any required resources and put their logos on the products to be sold in Global North countries at large mark-ups. In this way, the large company can skirt the environmental regulations of their home country and indirectly contribute to pollution in the other country. As an alternative, they can operate the company as an external owner. Both of these are very common practices by companies in industrialized countries. The secondary aspect of this practice is reducing the ability of small companies in the home country to compete with the cheap imports. They are restricted by the much more stringent environmental, labor, and safety laws.

CASE STUDY 2.2 Citarum River, Indonesia

Public Poisoning from Surface Water

The Citarum River has been called the "world's most polluted river." However, it plays an integral role in life of the residents of its catchment area, which has a population of 40 million. Many of these residents live in abject poverty and suffer greatly as the result of government inaction and influence of industry to keep their costs and regulations to a minimum. This is an extreme case of environmental injustice.

The Citarum River basin is located in Bandung, Indonesia and covers an area of about 5020 square miles (13 000 km²) (Figure 2.3a). It is the longest and largest river in West Java. The river is used for irrigation and drinking water, and has three dammed reservoirs. These three dams, Saguling, Cirata, and Jatiluhur, were built in 1962 and include hydroelectric plants that generate 5 billion kWh/year of electricity, which is equivalent to 16 million tons (14.5 million mt) of fuel per year. It provides as much as 80% of the water needs of Jakarta, the capital and major city in Indonesia, and irrigates farms that supply 5% of Indonesia's rice from 1.04 million acres (420 000 ha) of land as well as fishery and livestock resources. In addition to Jakarta, it is the source of drinking water for the cities of Bandung, Bekasi, Cimahi, Cianjur, Karawang, and Purwakarta. In all, the Citarum delivers 20% of the gross domestic product of Indonesia.

The 185 miles (297 km) river flows across 12 districts in West Java that are inhabited by more than 18 million people, 11.3 million of whom live in its riparian and related areas. By 2018, there were 2700 industrial sites in the Citarum basin with about 1000 situated along the river and utilizing river water for processing. Many of these plants are for textiles and garments which employ a large number of the residents. In the upstream area, there is widespread deforestation for agriculture.

The problem is that the factories and residents are causing inordinate pollution of the Citarum River. Until recently, only 47% of the factories had wastewater treatment facilities and many of them were substandard (Figure 2.3b). The rest dumped their unprocessed wastewater into the river. In addition to the massive industry waste, household, livestock, fishery, and agricultural unprocessed waste were also dumped into the river in massive amounts. Sediment in runoff from erosion of agricultural and deforested lands caused the water to be brown and muddy and threaten the downstream areas with siltation.

In many stretches of the river, the water surface is entirely covered by waste making it invisible (Figure 2.3c). This garbage includes plastic bottles, Styrofoam, plastic bags, wood, and other trash thick enough to walk across. The source of the trash is many of the small villages along the Citarum River with no public garbage collection or landfills. Residents either burn their trash or throw it into the river. Further, there are no centralized sewer systems available for all but 5% of the residents also reflecting the poverty of the area. Instead, sanitation facilities are mainly septic tanks or cubluks at each house or, more commonly, raw sewage is dumped directly into the river.

The amount of waste dumped into the river is a matter of speculation. One estimate is that 440 tons (400 mt) of livestock waste is dumped into the river daily. Further, as much as 32 700 cubic yards (25 000 m³) of household waste and 309 tons (280 mt) of industrial waste are dumped into the river each day. It is estimated that more than 20 000 tons (18 144 mt) of waste and 340 000 tons (308 443 mt) of wastewater are disposed into the river from the textile factories on a daily basis. Other estimates are that more than 2800 tons (2540 mt) per day and that almost 653 975 cubic yards (500 000 m³) of trash are dumped into the Citarum River annually.

As a result of all of this discharge, the quality of the Citarum River water was deemed unhealthy to the point of dangerous. The river water is black and slick with chemicals and dyes. Lead was found at levels more than 1000 times the USEPA limit for drinking water. Some estimates are that lead is 25 000 times acceptable levels. Testing also found aluminum at 97 parts per billion (ppb) compared to the world average of 32 ppb, manganese at 195 ppb versus a 34 ppb average, and iron concentrations of 194 ppb versus 66 ppb. Testing also determined the dangerous pollutants cadmium, chromium, pesticides, mercury, arsenic, sulphites, nonylphenol, phthalates, PCBs, paranitrophenol, and tributylphosphate were at excessive levels.

The raw sewage increases the organic content of the river. Water containing 5000 times allowable fecal coliform levels were found in some locations. E-coli bacteria levels were also elevated, reaching

(a)

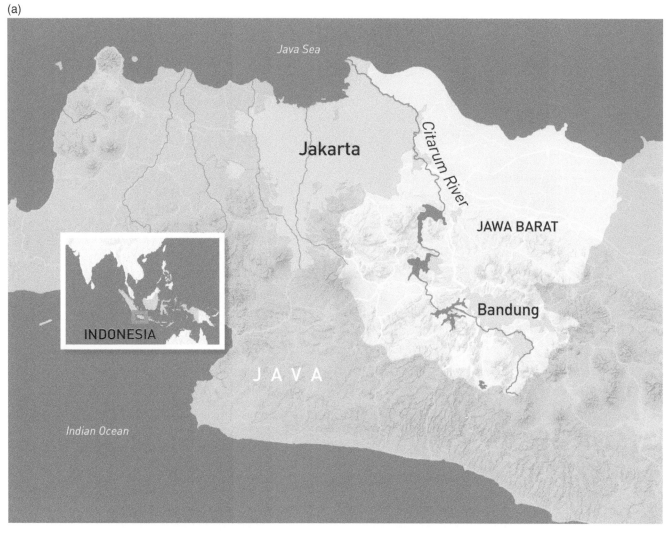

(b)

FIGURE 2.3 (a) Map of the Citarum River and basin and proximity to Jakarta. Inset shows position in Indonesia. (b) Photo of untreated waste being released into the Citarum River. *Source:* Greenpeace Indonesia / CC BY 2.0. (c) Photo of the Citarum River through a city *Source:* MNN galleries / Wikimedia Commons / CC BY-SA 4.0.

(c)

FIGURE 2.3   (Continued)

1600/100 ml, with biological oxygen demand (BOD) of 10 mg/l or 9 times the maximum allowable. It was further found that 64% of BOD in the river was produced by domestic and municipal activities, and 36% was from industrial or agricultural activities combined. All of this pollution has killed off an estimated 60% of the fish population in the river since 2008.

The problem is that a large proportion of the population in the lower Citarum live in abject poverty and have no choice but to depend on the river for their livelihood regardless of the danger. They scavenge the floating and deposited trash for items they can use or sell. They eat the remaining fish from the river even though they are loaded with toxins and even drink the water. As a result, about 60% of residents living around the river suffer from persistent skin infections as well as dermatitis from contact with the polluted water. Many suffer from intestinal problems from consuming the water or fish from it. These residents also suffer from chronic bronchitis, as well as more serious conditions like renal failure, dengue hemorrhagic fever, and a significant incidence of tumors. Children suffer significant delays in development. It is estimated that 120 000 people become seriously ill and 50 000 people die per year from river water pollution in Indonesia including 20 000

children. The main contributor to these grisly statistics is the Citarum River, the most polluted in Indonesia and, indeed, in the world.

There has been some effort to address this situation. On 5 December 2008, the Asian Development Bank granted a $500 million loan to Indonesia to clean up the river. In November 2011, a river revitalization project began, with an estimated cost of $4 billion over 15 years. The project covers a distance of 112 miles (180 km) along the most polluted part of the river but it barely made a dent in the pollution. In February 2018, the President of Indonesia began a seven-year project to clean the entire river vowing to achieve clean drinking-water status. He ordered 7000 soldiers to clean up several sections of the river. However, the lack of money, lack of responsibility coordination, bribes from factory owners to avoid change, and upstream soil erosion from deforestation made this effort ineffective as well. With internet publicity and government enforcement, foreign consultants are joining the effort by helping with upstream changes and launching local awareness and anti-plastics campaigns. New efforts to remove and recycle plastics appear to be having some success, but the Citarum River is still regarded as the most polluted in the world and has a long way to go to be even marginally acceptable.

2.3   Environmental Racism

The environmental justice movement evolved from the concept of environmental racism but it expanded to embrace broader issues on an international level. However, the environmental justice movement has retained the original roots of racism in its definition. This makes discerning the two more complicated. Born in the civil rights movement, the concept of environmental justice evolved throughout the 1970s and 1980s, primarily in the United States. The term "environmental racism" was first proposed in 1982 by Benjamin Chavis, former director of the United Church of Christ (UCC) Commission for Racial Justice during the North Carolina PCB dumping and protest incident. As used, it described racial discrimination in environmental policies, enforcement of laws and regulations, targeting of African American communities for toxic waste disposal, sanctioning of poisons and pollutants in African American communities, and the exclusion of African Americans from leadership roles in the environmental and ecology movements.

The main factors leading to environmental racism are poverty and the lack of affordable land, political power, and mobility for the residents. These factors are largely related to economics making it difficult to absolutely determine whether actions are truly racial or because of economic limitations. Environmental racism is facilitated by practices that might not seem racist outwardly like

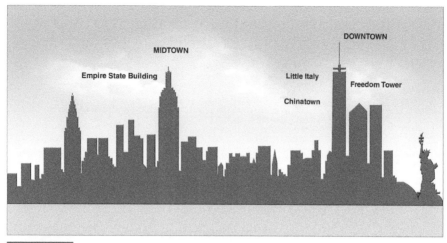

FIGURE 2.4   A silhouette of the New York City skyline showing relative height of buildings.

suburbanization, gentrification, and decentralization. In suburbanization, however, wealthy and primarily non-minority residents flee industrially impacted areas for safer, cleaner, and less expensive areas in suburban locations. Gentrification is the opposite process, where wealthy people purchase urban properties and renovate them, gradually converting the area to affluence and thereby forcing out the original residents by driving up property taxes and other costs.

Documenting and prosecuting environmental racism can be difficult because discerning whether a situation is environmental racism or economic injustice is usually confusing if even possible. Consider the New York skyline (Figure 2.4). There are tall, magnificent buildings in lower Manhattan around the Freedom Tower and opulent skyscrapers in midtown like the Empire State Building and Chrysler Building. Between these areas, the buildings are smaller and cheaper. It was the impression that the placement of the tall expensive buildings is because bedrock is exposed in these areas, providing better anchoring for the tall buildings. However, tall buildings have long been able to be built in areas of deeper bedrock. The real reason is that the intervening area was originally a swamp, loaded with mosquitoes and other biting insects, unpleasant odors, frequent flooding, and a predominance of mud. Wealthy people did not want to live in this area so it was occupied by poor ethnic, racial, and certain other groups who could not afford to live anywhere else and, as a result, contains Little Italy, Chinatown among others. Wealthy people can afford not to live in environmentally less desirable areas making the causal relationships difficult to prove. It can be argued that racial minority groups were prevented from accumulating enough wealth to avoid the areas or were not sold to because of race which is certainly valid but it further complicates the area of study. Having the political power to prevent companies from disposing waste in certain neighborhoods and to prevent legislation from being passed that adversely affect certain areas or groups has more of the chance to be proven as racially motivated. The actions of elected and other government officials are easier to evaluate and prosecute. However, these ambiguities have allowed the practice of environmental racism to persist.

CASE STUDY 2.3 Flint, Michigan Water Crisis

Economic and Racial Environmental Injustice

A recent, high-profile case of environmental injustice occurred in Flint, Michigan (Figure 2.5a). Drinking water in the city was tainted with lead, exposing many residents to this health hazard and causing a health crisis. A number of high-level officials were prosecuted as a result. There are many claims that it is a clear case of environmental racism because of the high percentage of African American residents impacted. Others claim that it is purely economic and the demographics have no bearing on the problem. Regardless, Flint has now gone from relative obscurity to a prime example of injustice, whether racial or economic or both.

Incorporated in 1855, Flint became famous in 1908 for being the home of a new company, General Motors. As a result, the city boomed with high employment rates and wealth. It was about this time that many public water lines were installed, primarily between 1901 and

1920. The water mains were made of cast iron but the lines from the mains to homes and within homes were lead, which was the common practice at the time. By the mid-twentieth century, Flint was home to nearly 200 000, as people flocked to the area to work in the booming automotive industry.

During this boom, the Flint River, flowing through the center of the city, became horribly polluted. Companies along the shore dumped treated and untreated waste into the river with impunity. These companies included carriage and car factories, meatpacking plants, and lumber and paper mills among others. It also had raw sewage from the city's waste treatment plant piped into it and had agricultural and urban runoff, including leachate from landfills adding to the pollution. The Flint River twice caught fire and had to be extinguished.

This boom ended in the 1970s–1980s when high oil prices and an increasing imported automobile market resulted in auto plant closures

(a)

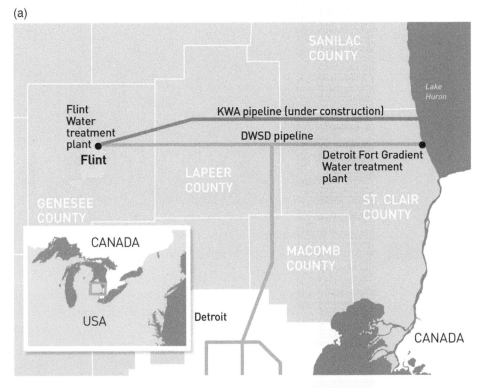

(b)

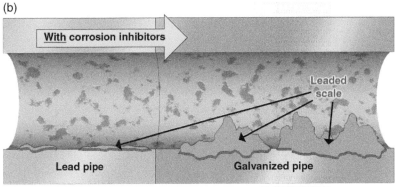

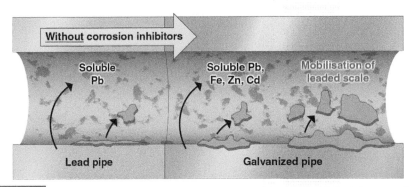

FIGURE 2.5   (a) Map of the Flint, Michigan area showing the proposed and available water lines to Lake Huron and the Flint River. Inset shows the location of Flint in Michigan and surroundings and second inset shows position of Michigan in the United States. (b) Diagram showing the difference between corrosiveness of drinking water with (top) and without (bottom) treatment with corrosion inhibitors. (c) Photo of bottles of tap water taken at different times from Flint during the water crisis showing the levels of lead as a bar scale next to each. The red color in the water is the result of corroded iron from pipes. *Source:* Pieper et al., 2017. (d) Map of Flint showing the distribution of lead in drinking water tested during the crisis. The EPA designates below 5 PPB of lead as safe and recommends that at above 150 PPB water filters may not work. *Source:* Mark Brush, Michigan Radio.

(c)

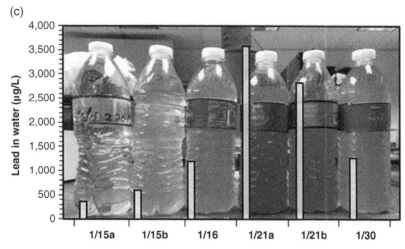

(d)

FIGURE 2.5   (Continued)

and massive lay-offs. As a result, the population dropped by half, including the wealthy citizens, and the city became a poverty and crime-ridden slum. One in six houses were abandoned. This greatly reduced municipal coffers and, as a result, most public projects were abandoned. This included replacing any of the 43 000 water lines in the city including 3500 lead lines, 9000 galvanized steel lines, and 9000 lines of unknown type. Beginning in 1967, Flint received its water from the Detroit Water and Sewerage Department (DWSD), out of Lake Huron with the Flint River as a water source in emergency situations. This arrangement ended in 2013 because of a catastrophic fiscal crisis. By switching water suppliers, the city would save about $5 million over 2 years.

The succeeding arrangement began on 25 April 2014. The city had arranged with another company to construct a new water line from Lake Huron but that would take 30 months to complete (Figure 2.5a). Further, DWSD legally contested the plan further slowing the project. As a temporary solution, Flint decided to obtain all of its drinking water from the Flint River. Residents complained about the color and smell but that was minor in comparison to the public health crisis it would cause.

The city treated the Flint River water with high amounts of chlorine in an effort to reduce bacteria. However, this caused an increase in trihalomethanes which are carcinogenic. The city learned of this problem by later in the summer but did not make it public until 21 January 2015. The chlorine also did not adequately remove bacteria. An outbreak of Legionnaire's disease apparently began in June and the use of river water is suspected. The outbreak lasted through November 2014, infected 87 people and caused the death of 12. However, this too was not announced until 2016. The city also detected bacteria in the water and issued a boil water advisory from 14 to 20 August. Another warning was issued in September.

The high levels of chlorine in the water began corroding engine parts at Flint's General Motors Truck Assembly plant, causing them to discontinue using it in October 2014. This still did not convince the city to correct the water issues. The turning point came on 26 February when the EPA detected lead levels seven times the maximum limit in one home. The reason for the increase was that city officials did not add needed corrosion inhibitors to the treated Flint River water (Figure 2.5b). The corrosive tap water was leaching lead out of the old lead pipes as well as iron which discolored the tap water (Figure 2.5c). Alarmed, Flint council members voted to reconnect to DWSD on 23 March but the emergency manager vetoed the decision. This began a well-publicized battle between some city officials against scientists and the public over the safety of Flint's water.

On 24 June the EPA reported extremely high lead levels in four Flint homes and in response on 9 July the Flint mayor drank tap water on television to prove its safety. Further, on 13 July a Michigan Department of Environmental Quality (MDEQ) official reported that the Flint water quality was perfectly fine on a radio interview. However, on 8 September a more extensive research study was released that found 40% of Flint homes having elevated levels of lead (Figure 2.5d). Further, on 24 September a Hurley Medical Center pediatrician released a study showing that Flint children had increased levels of lead in their blood since the supply was switched.

All of this evidence finally turned the tide of public opinion in Flint. On 15 October the Michigan Governor signed a bill to reconnect Flint to DWSD for $9.35 million and it was enacted the next day. This did not end the crisis, causing the Flint mayor to declare a state of emergency on 15 December, and the Michigan National Guard was called in to distribute water on 12 January 2016. Two days later, the governor requested that President Obama declare Flint a disaster area, which he did on 16 January, which included $5 million in aid.

In all, it is estimated that leached lead from aging pipes exposed all of the approximately 100 000 Flint residents to dangerous levels. This exposure included between 6000 and 12 000 children. The clear mishandling of the crisis caused four city, state, and federal officials to resign or be fired. In January 2021, the former Michigan Governor and eight officials were charged with 34 felonies and seven misdemeanors, including two being charged with involuntary manslaughter. On 20 August 2020, a settlement of $600 million was awarded in the case, with 80% for the families of impacted children.

Flint's bad image was not helped by the crisis. The population in 2010 was 102 434 but by 2020 it had dropped to 81 252. This crisis has been classified as environmental racism because the proportion of African Americans of Flint constitutes 54.1% of the population. Therefore, African Americans were primarily impacted by the lead exposure. On the other hand, the poverty rate of Flint is 38.8%, meaning that the poor were also inordinately exposed to lead. In this case, the injustice could be on an economic basis as well.

In cases where companies release hazardous waste that can impact any group but choose to release it in an African American or other community based on race or ethnicity, then there is no doubt that it qualifies as a case of environmental racism. These are rare because, if it is documentable, it exposes the company to likely legal action. This is also dependent on the viewpoint of the officials toward prosecuting abusers and the willingness of the impacted residents to pursue the company in court. In most areas, well-documented infractions are pursued quickly, which is why many of the examples are not clear-cut. That is how cases of environmental racism can persist.

A geographic area that is permanently impaired by such repeated severe environmental impact is called a sacrifice zone or area or a national sacrifice zone or area. These are areas damaged by locally unwanted land use (LULU) including excessive chemical pollution, substance pollution (asbestos, particulate) or radiation. The idea of sacrifice zones was developed during the Cold War, where fallout from nuclear detonations, whether in testing or war, could render an area uninhabitable. The idea was also applied to the long-term effects of coal strip mines in the American West. Sacrifice zones have a much more likelihood of being in areas populated by communities of predominantly racial minorities. These dangerous areas can be developed while the communities are present through illegal disposal and releases or the communities can be forced into these areas through economic, societal, or legislative pressure. These are the most heinous examples of environmental racism.

CASE STUDY 2.4 Cancer Alley, Louisiana

Sacrifice Zones and Environmental Racism

The 85-miles-long industrial corridor along the Mississippi River between Baton Rouge and New Orleans, Louisiana has been come to be known as Cancer Alley because of the inordinate incidence of cancer (Figure 2.6a). It is the largest hotspot of carcinogenic air pollution in the United States and it is steadily worsening as the chemical industry in the area grows. Further, the area is primarily populated by African American and low-income communities. Despite the proximity and toxic burden borne by this population, few of the residents are actually employed by the many chemical companies. These many factors combined make this area among the most convincing cases of environmental racism. In fact, on 2 March 2021, the United Nations (UN) Human Rights Committee strongly condemned the area for environmental racism.

The land along the Mississippi River between Baton Rouge and New Orleans primarily constitutes St. James Parish and parts of other parishes. Petrochemical and related chemical industries began building the first plants in this area during the 1940s. Once begun, this area was quickly developed. During the 1950s, plants were expanded and new ones built because of the ready availability of petrochemicals, ease of transportation on the Mississippi River, lax environmental restrictions, low taxes, and low population density. The plants soon expanded into the nearby majority-African American communities that characterized the area (Figure 2.6b).

The population of the full area is about 45 000 residents of whom 48.8% are African American and 0.3% from other minority groups. It also includes 16.6% living in poverty. The residents live among more than 150 petrochemical plants, plastics plants, and oil refineries. This

(a)

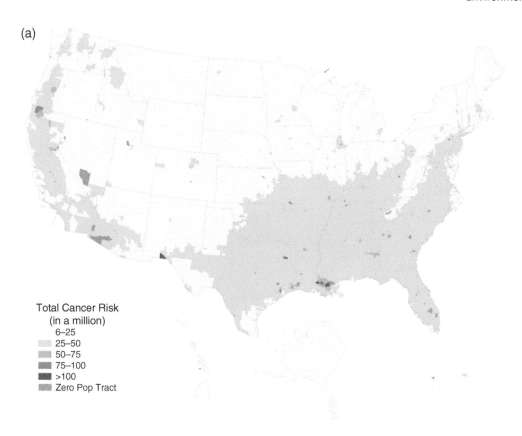

Total Cancer Risk
(in a million)
6–25
25–50
50–75
75–100
>100
Zero Pop Tract

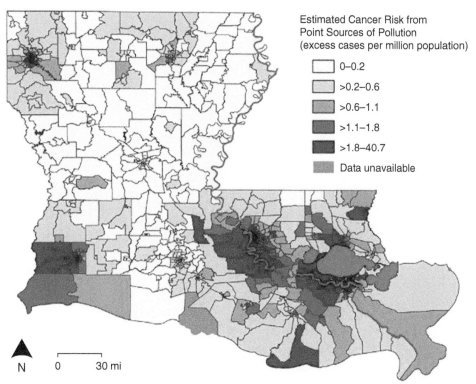

Estimated Cancer Risk from
Point Sources of Pollution
(excess cases per million population)

0–0.2

>0.2–0.6

>0.6–1.1

>1.1–1.8

>1.8–40.7

Data unavailable

N   0   30 mi

FIGURE 2.6   (a) Map of "Cancer Alley" in Louisiana as shown by the band of increased cancer incidence along the Mississippi River between Baton Rouge and New Orleans. Inset shows the location of Louisiana and "Cancer Alley" in the United States with national cancer risk shown. *Source:* National Air Toxics Assessment (NATA), 2014/ EPA / Public Domain. (b) Aerial photo of a section of "Cancer Alley" along the Mississippi River showing several large chemical plants. *Source:* Jackson Jost / Unsplash / Public Domain. (c) Map of "Cancer Alley" showing the location of ethylene oxide producing plants and the amount of onsite disposal of the carcinogenic chemical in each.

(b)

(c)

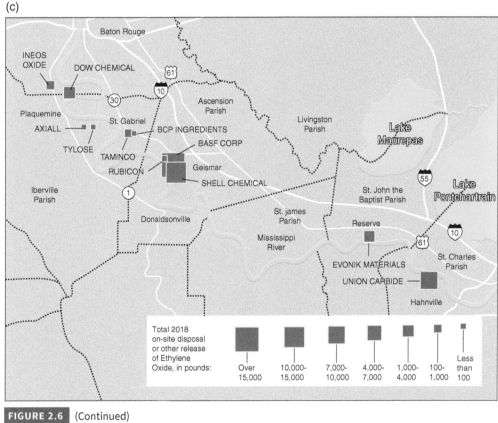

Total 2018 on-site disposal or other release of Ethylene Oxide, in pounds:	Over 15,000	10,000- 15,000	7,000- 10,000	4,000- 7,000	1,000- 4,000	100- 1,000	Less than 100

FIGURE 2.6 (Continued)

small corridor produces about 25% of the petrochemicals in the United States so is essential to the economy. These industries emit and spill some of the most dangerous pollutants exposing residents to risks considered unacceptable by the EPA.

The danger of the chemical exposure in the area was recognized quickly, with the term "Cancer Alley" being applied much later in 1987. Despite this known danger, expansion of the industries causing it

continued. In 1996, the chemical company Shintech announced three new polyvinyl chloride (PVC) plants in Convent, Louisiana. PVC is a known carcinogen. The residents within five miles (8 km) of the plants are 81% African American compared to the average 49%. This meant that these largely African American populated areas were chosen purposefully.

By 2002, the state of Louisiana had the second-highest death from cancer rate in the United States largely driven by the extreme rate in Cancer Alley. In Cancer Alley, 46 people per each million will develop cancer annually, compared with the national average of 30 per million. However, locally, it is 105 per million. At that time, seven of the 10 plants with the largest environmental pollutant releases in Louisiana were located in Cancer Alley.

In a time when the rest of the United States was trying to reduce public exposure to toxic pollutants, these shocking statistics still did not slow the growth of the chemical industry or the releases. The number of plants reporting toxic releases in Louisiana, increased from 255 to 320 in the past three decades, which is a 25% increase compared to the 16% decrease on a national level over the same period. By 2012, it was estimated that residents of predominantly African American areas had a 16% higher risk of developing cancer than resident from mostly white neighborhoods. Further, residents from low-income areas had a 12% higher risk than in higher-income areas.

Seven large new petrochemical facilities and expansions in Cancer Alley have been approved since 2015, and five others are awaiting approval. One of these was the proposed and permitted Formosa Plastics Sunshine Project. This $9.4 billion complex was estimated to become the single greatest environmental threat of all petrochemical and plastics plants. It was estimated that it would release 13.6 million tons (12.4 million mt) of greenhouse gas annually. It would cover 2500 acres (1012 hA) and, shockingly, be located one mile (1.6 km) from an elementary school. This was finally too much for local residents and environmental groups and on 15 January 2020, four groups joined together to sue the Trump administration for permitting the complex.

This pressure caused the US Army Corps of Engineers to suspend the Sunshine Project permit on 4 November 2020.

One of the released chemical pollutants of concern is ethylene oxide, which is manufactured by 13 plants in Louisiana, 11 of which are in Cancer Alley (Figure 2.6c). It is used for sterilization and as an ingredient in antifreeze, detergents, synthetic fibers, and plastic bottles. This proven carcinogen causes breast cancer and lymphoma. In one town near a plant, the risk of developing cancer from exposure is 1 in every 210 people, or 47 times the EPA acceptable risk level. In two other towns near plants, it has been estimated at 26 and 32 times acceptable levels. There are also other chemicals that are as dangers such as chloroprene. This carcinogen is used in the production of synthetic rubber but can cause liver cancer. Around plants that produce it, the cancer risk is as much as 32 times the EPA acceptable risk level. In the area around one Denka/Dupont neoprene plant in Cancer Alley, the likelihood of cancer from inhalation of pollutants is estimated by the EPA at more than 700 times the national average.

In the study of environmental justice, Cancer Alley is the type example of an environmental sacrifice zone. By definition, this is an area severely and dangerously contaminated by chemical pollution. This term developed from the term "National Sacrifice Zones" which described areas that were severely polluted from the mining and processing of uranium for the building of nuclear weapons for defense during the Cold War. The definition of the term was since expanded to include all areas where exposure to dangerous pollutants is disproportionate. It is especially applied to low-income and predominantly African American, Hispanic, and/or Native American communities.

Reference

Kelsey J. Pieper, Min Tang, and Marc A. Edwards (2017). Flint Water Crisis Caused by Interrupted Corrosion Control: Investigating "Ground Zero" Home. Environ. Sci. Technol., 51, 2007–2014.

CHAPTER 3

Architecture of the Earth and Atmosphere

CHAPTER OUTLINE

Words you should know

Asthenosphere – The soft, ductile upper mantle that circulates and drives the lithospheric plates.

Atmosphere – The gaseous envelope held to the Earth by gravity.

Core – The iron/nickel center of the Earth including a solid inner core and a liquid outer core.

Crust – The outermost chemical layer of the Earth including both relatively light, thick, and old continental crust and relatively dense, thin, and young ocean crust.

Hydrosphere – The layer of liquid and solid water covering the outer part of the Earth, including freshwater, salt water, ice, and in some definitions the water in the atmosphere including clouds and humidity as well.

Lithosphere – The outer mechanical layer of the solid earth that compose the tectonic plates. It contains a lower layer of rigid mantle attached to an upper layer of crust.

Mantle – The chemical layer of the Earth between the core and the crust and including the lower mantle, the asthenosphere, and the rigid mantle that makes up the lower part of the lithosphere.

Stratosphere – The layer of the atmosphere above the troposphere.

Troposphere – The lowest level of the atmosphere.

3.1   Solid Earth Spheres and Systems

The Earth system forms a series of spheres that are contained within each other. The Earth is primarily solid and lesser amounts of liquid. The solid earth is contained in a layer of largely liquid water called the hydrosphere and layers of gases called the atmosphere. The solid earth is composed of the core in the center overlain by the mantle and then surrounded by the crust (Figure 3.1). The core is about 2175 miles (3500 km) thick and composed of an iron-nickel mix whose density ranges from 9.9 to 13 g/cm³. The core has a solid center but a liquid outer layer. The mantle is 1802 miles (2900 km) thick of iron and magnesium silicate minerals with a density of 5.5 g/cm³. The crust is 3.1–50 miles (5–80 km) and composed of two types, oceanic and continental. Ocean crust is 3.1–6.2 miles (5–10 km) thick and composed of dense iron and magnesium silicate and oxide minerals giving it a density of about 3 g/cm³. Continental crust varies from about 12.4 miles (20 km) thick under the ocean to as much as 50 miles (80 km) under mountains. It is composed primarily of light silicate minerals giving it an average density of 2.65 g/cm³ but as high as 3.0 g/cm³ or more locally.

The Earth also has mechanical layers that coincide with the largely chemical layers described above (Figure 3.2). The separation of a liquid core surrounding a solid core makes up the innermost mechanical layers. The mantle surrounding the liquid core is solid so the boundary is both chemical and mechanical. The mantle takes on a gum-like consistency in the shallower part which defines the next layer called the asthenosphere. The boundary between the lower solid mantle and the asthenosphere is diffuse but may be as deep as 430 miles (700 km) below the surface. However, the top of the mantle is rigid again and attached to ocean or continental crust above it. The sandwiched crust and rigid mantle is the lithosphere. The boundary between the crust and the rigid

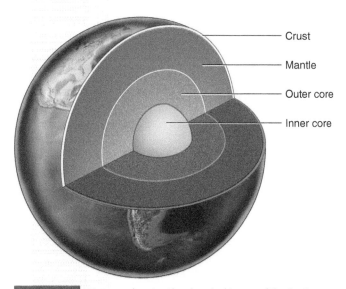

Crust

Mantle

Outer core

Inner core

FIGURE 3.1 Diagram showing the chemical layers of the Earth.

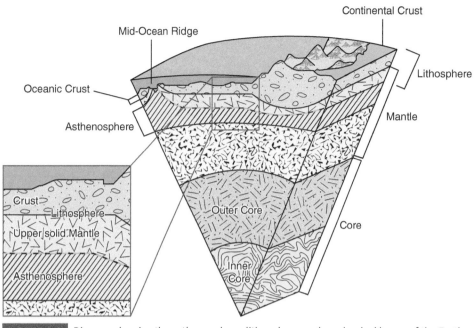

FIGURE 3.2 Diagram showing the asthenosphere, lithosphere, and mechanical layers of the Earth.

mantle is the Mohorovicic Discontinuity or "moho." The depth to the top of the asthenosphere varies by the thickness of the lithosphere. The lithosphere is broken into fragments or plates of various sizes and composition. These rigid lithospheric plates float on the softer asthenosphere beneath them.

The density and thickness of the plates controls the height that they sit on the Earth. The balance among density, thickness, and surface elevation is called isostasy and it controls surface elevation and plate thickness. A hardboiled egg with the shell cracked makes a good illustration of the relationship. The rigid shell fragments imitate the lithospheric plates and which can be pressed down into the rubbery egg white which represents the asthenosphere. However, the positions of the shell fragments on the hardboiled egg are fixed whereas the seven major and several smaller lithospheric plates move around the Earth. They move because they are carried on the mantle flowing underneath them. The flow of the mantle is driven by thermal currents. The flowing mantle forms large convection cells where hotter areas rise up because the hot material is less dense. When the hot material reaches the top of the asthenosphere, it spreads out away from the hot area and cools. Cooling causes it to become denser and it sinks down into the deeper mantle. There, it is reheated and goes through the convection cell again in a repetitive process.

The numerous mantle convection cells drive the lithospheric plates in many directions so they grind against each other at their margins (Figure 3.3). The source of the solid earth natural disasters are where plates grind against each other. There are three types of interactions among plates: convergent, divergent, and transform. Convergence results from mantle flow pushing plates into each other in a collision, divergence results from plates being driven away from each other in a rift and transform results from plates sliding past each other. There are three types of convergent margins depending upon the types of crust being driven together. They primarily consume ocean crust at subduction zones where they form volcanic and magmatic arcs. Where two plates of continental crust collide, they form the highest mountains on Earth. Divergent margins form basins and valleys on continental crust and eventually develop into ocean basins. Transform margins can produce hills or valleys but generally have minor topographic signatures.

The uppermost part of the crust is the only part of the solid earth impacted by human pollution. The crust forms the bedrock of the Earth and can be infiltrated by contaminated fluids from the surface or shallow burial. They can only penetrate tens to hundreds of feet at most and become diluted as they percolate down. Fractures and cave systems locally allow deeper penetration with minimal filtering. These areas can be much more polluted and present difficulty in finding large sources of potable water. Deep injection wells can add severe human pollution to thousands of feet depth but are very expensive and have other repercussions like causing earthquakes. They are therefore rare and highly localized. Deep injection of fracking produce or return fluids changed Oklahoma from seismically quiet to among the seismically most active states in the 2010s.

The solid bedrock is largely mantled by soils and sediments at the surface. Soils form from weathering of the surface rocks and sediments and are generally thin at less than 82 ft (25 m) in thickness. These soils are thicker where weathering is more intense, especially in moist tropical areas. The weathering here is so chemically intense that residual soils can be stripped of most chemicals leaving them enriched in aluminum which is more stable. These soils are called laterites. In dry areas, the lack of water limits the depth and type of weathering. These soils are calcium-rich, forming hard, cemented surfaces in some areas, called pedocals. In northern latitudes, advancing glaciers stripped off all of the old soils. The soils in these areas are consequently young.

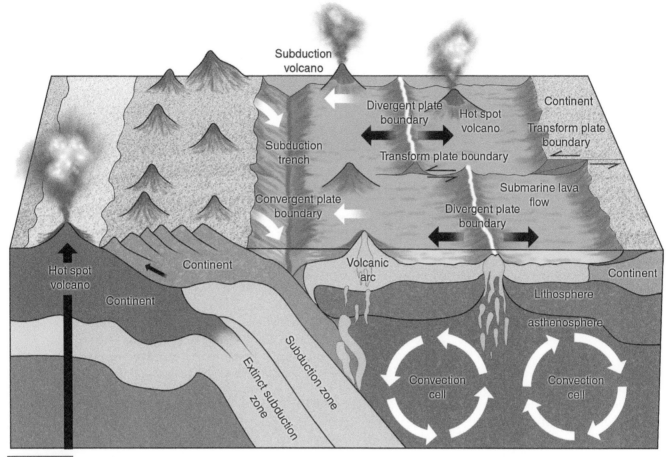

FIGURE 3.3   Diagram showing convection cells in the mantle driving lithospheric plates and forming plate margins.

In addition to soils as the result of weathering, there are sediments on the Earth's surface. They are especially thick around coastal areas and composed largely of clay, sand, and gravel. River valleys also have significant deposits of sand and gravel in the channels and clay in the floodplain. In mountains and higher latitudes, they were impacted by glaciers. They are covered by glacial sediments of multiple grainsizes from clay to boulders. Deserts and other arid areas can be covered with thick sediments, primarily sand, but also of a variety of grain sizes depending upon the area. Loess deposits of windblown dust can be thick downwind of deserts but coarse rock rubble forms at the base of mountain slopes.

Most buried and spilled waste impacts soil and surface sediments. Soil may be the most polluted of the Earth resources. Many industries dumped waste onsite with chemical, radioactive, and physical contaminants. The dumping of radium on soil in New Jersey produced indoor radon problems in homes throughout the area. Non-point source pollutants are applied to lawns throughout suburban areas worldwide including herbicides, chemical fertilizer, and pesticides. Mining operations dump waste onto soils, contaminating them with a variety of heavy metals among other chemicals or physical contaminants like asbestos. Vermiculite mining in Montana contaminated the area with asbestos, distributed throughout by the wind. Many industrial sites occur along rivers and dump waste into the water which spreads it throughout the channel and floodplain deposits. The Hudson River's sediments are contaminated with PCBs for 200 miles (320 km) of its length.

3.2   Liquid Spheres and Systems

The hydrosphere is primarily liquid, overlies the lithosphere, and is the smallest of the systems (Figure 3.4). It includes the oceans, surface freshwater, glaciers, groundwater, and in some definitions all of the moisture in the atmosphere as well. This constitutes about 332.5 million cubic miles (1386 million km^3) of water. The thickest and most continuous layer of hydrosphere is the oceans and composed of salt water. It covers 71% of the Earth and penetrates the sediments on the ocean floor and some of the solid rock beneath it. The oceans alone contain 96.5% of all water on the Earth. Freshwater constitutes the remaining 3.5% of the water on the planet. Freshwater in the liquid form in lakes, streams, wetlands and groundwater has a total volume of 2 551 000 cubic miles (10 633 450 km^3). Freshwater in glacial ice both in alpine glaciers and continental glaciers in polar regions has a total volume of

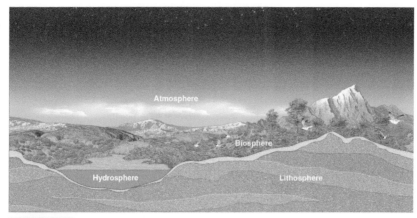

FIGURE 3.4 Diagram showing the lithosphere, the hydrosphere, the biosphere, and the atmosphere.

about 7 million cubic miles (29.2 million km³). This constitutes about 68% of the freshwater on the planet surface. Groundwater makes up about 30% of freshwater with a volume of about 2 million cubic miles (8.4 million km³). Water in the form of vapor in the atmosphere has a volume of about 3100 cubic miles (12 900 km³).

Although the hydrosphere may be small, it is likely the most important of the systems. This is because the biosphere is completely contained within the hydrosphere. The biosphere is where all life on Earth resides and it overlaps the lithosphere and lower atmosphere (troposphere) in addition to the hydrosphere. Most larger forms of life are in the hydrosphere or on the surface of the lithosphere. The lower atmosphere and in the subsurface below several feet only contain microorganisms. Humans have extended the biosphere outside of its normal range but this is on an occasional basis such as space travel.

The hydrosphere can be profoundly polluted. Surface fresh water is typically the most contaminated. It contains anything that humans spill on the surface or that settles to the earth from the air with little to no filtering. Only dilution with other fresh water reduces the concentration. Surface water that filters through mine tailings further dissolves contaminants and incorporates them. These pollutants are typically from point sources but non-point source pollution can be equally dangerous. Runoff of pesticides, herbicides, and fertilizers into surface waters cause severe degradation. Fertilizers have caused eutrophication of most surface water, causing hypoxia and reducing productivity. This overfertilization can be enhanced by sewage overflow and leakage. Acidity from air pollution fallout and runoff from coal and sulfides reduces the pH and can cause dead lakes like those in the Adirondacks of New York. Pesticide runoff can reduce the productivity of aquatic life and even insects that reproduce in water. In addition, runoff from roads and buildings can contain road salt, oil, and industrial and household wastes that also impact surface water.

Oceans are also impacted by human pollution from dumping, spills, river input, and air pollution fallout among others. The most devastating individual events are oil spills, which commonly involve tankers along shorelines. Crude oil is a complex mix of chemical compounds that can be directly refined into everything from jet fuel to tar or manufactured into most of what humans use from plastics to pharmaceuticals. The most toxic components of crude oil are volatile and evaporate quickly upon exposure to air to form volatile organic compounds (VOCs) in air pollution. These are at peak concentration when first spilled, making shoreline tanker accidents highly damaging to the environment.

CASE STUDY 3.1 Exxon Valdez Oil Spill, Prince William Sound, Alaska

Impaired Piloting

The Exxon Valdez oil tanker was the largest ship ever built on the West Coast of the United States at the time. It was 987 ft (300 m) long and 166 ft (50 m) wide with a capacity to carry 1.5 million barrels (238.5 million L) of crude oil from Valdez, Alaska to refineries in the continental United States (Figure 3.5a). The Exxon Valdez could travel at about 19 miles per hour (30.6 km/h) fully loaded. The tanker met all international and US safety standards and had a strong uni-welded hull,

10 cargo tanks, four ballast tanks, and two "slop" tanks designed to provide maximum stability and cargo management in the event of an accident. It also had a double bottom, a major safety feature required of all new tankers at the time. A double bottom ship has two watertight layers in the hull separated by several feet. The inner hull prevents the contents from spilling or seawater from entering the tanker in the event of an accident.

The Exxon Valdez was put into use on 6 December 1986. The tanker regularly transported oil from the Alaskan oil fields through the

Port of Valdez to refineries along the West Coast. Valdez was the site of the terminus of an 800-mile (1288-km) long Trans-Alaska Pipeline that transported crude oil from Prudhoe Bay, near the Arctic Ocean. It was loaded onto tankers there because it was considered an "ice-free" harbor as the water did not routinely freeze during the winter so oil loading and shipping could take place all year.

The Trans-Alaska Pipeline System was one of the largest, privately financed construction projects ever. ExxonMobil has a 20% interest in the pipeline. Environmental concerns slowed approval of the project, but the OPEC oil embargo of 1973 removed these obstacles and construction was completed between March 1975 and May 1977. Crude oil from Prudhoe Bay oilfields began arriving at Valdez on 28 July 1977, and the first oil tanker left the Valdez Harbor on 1 August 1977. At peak production, 70 tankers entered and departed Valdez every day.

The Exxon Valdez tanker arrived at Port of Valdez on 23 March 1989. The captain was Joseph Hazelwood, a very qualified pilot with almost 20 years of experience. The Exxon Valdez was the newest and most modern tanker in the fleet and it was loaded with 1.3 million barrels (206.7 million L) of crude oil. Captain Hazelwood ordered the ship underway at 9:00 p.m. where it ferried southward under the guidance

(a)

FIGURE 3.5 (a) Map showing the location of Valdez, Alaska. Inset shows the location of Alaska in North America. (b) Map of the Exxon Valdez oil spill area in the Gulf of Alaska. (c) Photo of ship attempting to unload oil from Exxon Valdez tanker. Black area is oil slick. *Source:* Courtesy of NOAA. (d) Photo of worker steam cleaning oil off of rocks. *Source:* Courtesy of NOAA. (e) Photo of team of workers cleaning the shoreline of oil. *Source:* Courtesy of NOAA.

(b)

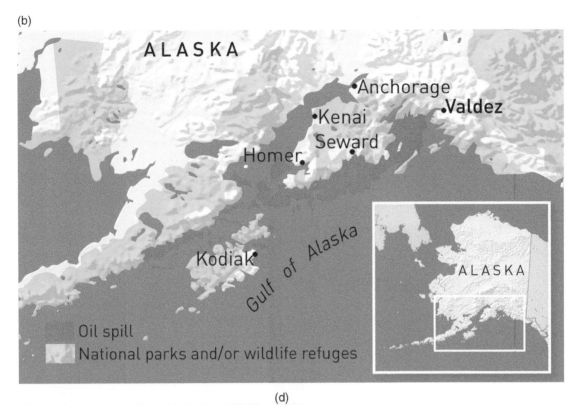

(c)

(d)

(e)

FIGURE 3.5 (Continued)

of the harbor pilot. It passed safely through the very tight, 1700-yard (1.6-km)-wide Valdez Narrow entrance to the harbor and out into the more open waters of Prince William Sound, Alaska.

Prince William Sound is an arm of the Gulf of Alaska that contains an open expanse of water about 100 miles (160.9 km) wide. There is approximately 3000 miles (4828 km) of shoreline around the sound and it contains small islands, submerged reefs, and fjords that form breakwaters and shelters. The Columbia Glacier lies along the northwestern side of the sound. The fjords, reefs, and islands form an important ecological habitat for numerous seabirds, marine life, and commercial fisheries. Most of the land surrounding Prince William Sound was designated as National Forest since 1907. Among these lands, the Chugach National Forest is the second-largest National Forest in the United States. The US Congress passed the Alaska National Interest Lands Conservation Act in 1974 in an attempt to preserve the wilderness quality of this area.

After the Exxon Valdez cleared the Narrows and the harbor pilot departed, it entered southward shipping lanes. These lanes are clearly marked and lie along the western side of Prince William Sound. The weather was clear with 4 miles (6.44 km) of visibility. Captain Hazelwood ordered the ship to about 20 knots (37 km/h) and at 11:30 p.m., he asked permission from port traffic control to divert eastward into the northbound lanes to avoid reported floating ice. Port traffic control gave the captain permission to change course and the tanker steered eastward. Hazelwood's plan was to sail through a one mile (1.6 km) wide gap in the icebergs, between a series of submerged rocks along the eastern edge of the northbound shipping lanes. Once through the ice hazard, the Exxon Valdez would turn westward and back into the southbound lanes. This maneuver required expert timing and navigation for such a large ship. However, Captain Hazelwood departed the bridge and turned over control to a relatively new third mate, who did not have the necessary rating to pilot the vessel.

This junior officer was instructed to turn back toward the southbound shipping lane when a navigational beacon on Busby Island was spotted. However, his attention was divided between watching for the leading edge of the ice flow on radar and worrying about when to turn. The third mate missed the Busby Island light. When a crew member reported quickly approaching the flashing buoys of Bligh Reef, the third mate realized his mistake. He ordered an immediate evasive maneuver but it was too late. The Exxon Valdez tanker plowed 600 ft (182.9 m) into the rocks of Bligh Reef at almost full speed just after midnight on Friday, 24 March 1989.

Between the collision, and subsequent efforts by Captain Hazelwood to drive the Exxon Valdez back from the reef, holes were torn in the four main cargo tanks. The double-hulled bottom was ripped open along the side of the ship and oil poured into the sea. Within 30-minutes of the accident, more than 110 000 barrels of oil (17.4 million L) spilled into the protected waters of Prince William Sound. Within eight-hours, more than 215 000 barrels (34.1 million L) had spilled and in the end, almost 260 000 barrels (41.6 million L) or 20% of the total cargo of oil was in the sea (Figure 3.5b).

The Exxon Valdez accident was the first real test of the oil spill response plans that had been developed for the area. The emergency response ship was in dry-dock at the time, damaged several months earlier and awaiting repairs. The emergency oil booms and skimming equipment could not be mobilized because they were in a warehouse, and only one equipment operator was available to operate the fork lift

and crane needed to load them onto the ship. The damaged response ship was used despite a crack in its bow. With all of the problems, it took 14 hours for the ship to reach the accident where the slick had covered nearly 20 square miles (51.8 km²). The Exxon Valdez spill was so close to shore that environmental officials were hesitant to use dispersants because of the potential for long-term damage to the local ecology. Offloading of the remaining oil to other ships was delayed because the tanker could capsize and sink if the cargo was removed too quickly (Figure 3.5c). In addition, the portable pumps to transfer the oil were lost. The calm weather kept the spill from spreading at first but strong winds and rough seas set in and pushed the oil toward the shore. In the end, more than 1200 miles (1931.2 km) of rocky coastline were oiled.

Exxon employed at least 10 000 workers to conduct the cleanup during the summer of 1989. Spill response costs were estimated at approximately $2 billion. Several methods were attempted to remove the oil on the shoreline, including limited burns and dispersants (Figure 3.5d). However, most of the removal activities were accomplished by workers with mops, buckets, and steam cleaners (Figure 3.5e). They attempted to physically collect the oil or to dissolve it so it would wash back into the sea to be captured by skimmer ships. This approach was not quick and experts claim that pressure washing damaged the sensitive local ecology.

Of the 11 million gallons (41.6 million L) spilled, 6% or 660 000 gal (2.5 million L) were recovered by skimming, 18% or 1.9 million gallons (7.2 million L) evaporated or were photodegraded by sunlight, 28% or 3.1 million gallons (11.7 million L) were emulsified and dispersed into the seawater, and 48% or 5.3 million gallons (20.1 million L) washed up on the shoreline of Prince William Sound. More than 250 000 birds were killed, as were almost 3000 sea otters and 200 harbor seals. Fishing for herring and salmon was banned for the next year, and these species never recovered to predisaster levels. Few traces of the spill are visible today but a layer of oil is present a few inches below the rocky cover along several shorelines.

The main cause of the accident was that the third mate did not make the westward course correction as ordered. This failure was caused by fatigue because he only had a few hours of sleep after supervising oil loading. The other main issue was the lack of command on the bridge. Captain Hazlewood was drinking prior to departure and did not ensure correct navigation of the ship before leaving the bridge. Captain Hazlewood should have been relieved of command for previous incidents of substance abuse. Exxon had previous knowledge of this problem but took no action. The last issue was that the radar installations of the Coast Guard and Port of Valdez traffic control were antiquated and inadequately staffed.

As a result, the Coast Guard improved its radar system and changed their rules, strictly forbidding ships from changing lanes except in an emergency, and a 10 knot speed limit is strictly enforced. Tanker captains are now given a mandatory sobriety test one hour before departing. The US Congress passed the Oil Pollution Control Act (OPA) of 1990, which forbids any ship that ever spilled more than one million gallons (3.8 million L) of oil to enter US waters. Captain Hazlewood had his Masters License suspended and was found guilty of negligent release of oil. He was fined $50 000 and sentenced to 1000 hours of community service, which he served over five years by picking up trash along Alaskan highways and working in a soup kitchen.

Oil is not the only dangerous pollutant humans have dumped into the ocean. Garbage scows have transported waste of all types from ports along most coastlines in developed areas around the world on a daily basis. They dumped this waste as close to the shore as possible without impacting coastal communities with complete disregard for the local marine ecology. The military has dumped all kinds of unwanted ordinance from basic ammunition and supplies to nerve gas and nuclear waste. Fallout from air pollution onto the ocean surface adds more pollution. Most ocean-going vessels from small motorboats and jet skis to luxury ocean liners at least leak fuel and lubricant into the water but more commonly dump refuse into the ocean. Transfer of fuel and chemicals into ships also includes accidental spills into the ocean. Ocean productivity has drastically declined over the past century, largely because of the complete disregard of humans toward this precious resource.

Even glacial ice contains pollutants though in much lower concentrations than surface waters. The pollutants are deposited on ice by air pollution fallout. Sulfur-rich layers in the ice record major volcanic eruptions through this process. Human-produced chemical contaminants are also contained in ice layers but primarily in the Northern Hemisphere because there is less industry and resulting pollution in the Southern Hemisphere. Polar glacial ice is less plentiful than in the past because melting is taking place so fast especially in the north. This fast melting is, in part, the result of human activity. Not only is climate change accelerating melting, but dark-colored particulate, especially from China, is being swept onto the northern ice sheet. The dark color absorbs the sun better than reflecting white ice, which heats it up and melts it.

3.3 The Atmosphere

The atmosphere is the envelope of gases held to the Earth by the force of gravity. These gases are denser closest to the planet surface and less dense farther upward (Figure 3.6). This gradation is somewhat stepwise, resulting in the subdivision of the atmosphere into spheres, the troposphere, the stratosphere, the mesosphere, the thermosphere, the exosphere, and outer space. The most important of these layers is the troposphere because all surface weather takes place there. The stratosphere is the second most important because aircraft can fly in it. The rest are less impactful.

The troposphere extends from the surface to about 7.5 miles (12 km) altitude. It contains all of the air used for photosynthesis and respiration in the biosphere. It also contains 99% of the water vapor and aerosols. Temperatures decrease with elevation in the troposphere. It is the densest layer of the atmosphere because it is compressed by the weight of the atmosphere above it. Virtually all weather happens there with the exception of some supercell thunderstorms, whose tops rise into the lowest parts of the stratosphere. Most aviation takes place in this layer.

The stratosphere lies from approximately 7.5–31 miles (12–50 km) high. It contains the ozone layer, which protects the surface from harmful ultraviolet radiation. Because of UV radiation, the temperature increases with altitude in the stratosphere. Polar stratospheric clouds are sometimes present in its lowest, coldest altitudes, but this is uncommon. It is the highest part of the atmosphere used for aviation.

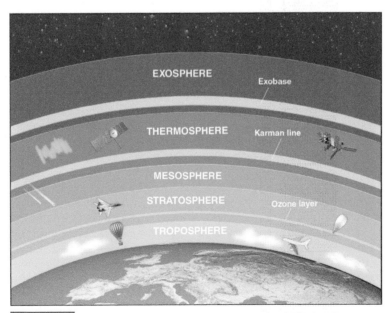

FIGURE 3.6 Diagram showing the layers of the atmosphere.

The mesosphere is about 31–50 miles (50–80 km) high and gets progressively colder with altitude. The top of the mesosphere is the coldest place in the atmosphere, with an average temperature of about minus 120 °F (−85 °C). The minor water vapor forms noctilucent clouds at the top of the mesosphere. Most meteors burn up in the mesosphere. The thermosphere is about 50–440 miles (80–700 km) high and contains the ionosphere in its lowest part. Temperatures increase with altitude. The aurora borealis can occur in this layer and the International Space Station orbits in it. The exosphere is about 440–6200 miles (700–10 000 km) high. The exosphere is the highest layer in the atmosphere and merges with the solar wind at its top. Most satellites orbit in the exosphere.

CASE STUDY 3.2 Hole in the Stratospheric Ozone Layer

Potential Remediation of a Major Environmental Problem

Ozone is oxygen that is changed from normal O_2 to the more reactive form of O_3. It is formed naturally in the lower part of the stratosphere from ultraviolet radiation from the Sun (Figure 3.6). It causes some O_2 molecules to break apart and rejoin to form O_3. This is stratospheric ozone. Ozone can also form through electrical current discharges in air, such as during a lightning storm or around electrical generation plants and devices. The discharge causes oxygen in the air to split and part of it reforms as ozone.

Ozone that forms in the atmosphere is not very stable. The ultraviolet rays that break apart the O_2 molecules also split the O_3 molecules into single oxygen atoms and double oxygen molecules. Ozone also is destroyed by chemical reactions with nitric oxide (NO), which is naturally present in the atmosphere. NO forms when the very common nitrous oxide (N_2O), enters the stratosphere from the troposphere. Once in the stratosphere, N_2O reacts with oxygen atoms that are excited by ultraviolet rays to form NO, which then combines with ozone in the reaction:

$$NO + O_3 <---> NO_2 + O_2$$

The reactions that create and destroy ozone are naturally in equilibrium. If the ultraviolet radiation from the Sun increases, the concentration of ozone in the stratosphere increases. If ultraviolet radiation decreases, ozone concentrations decrease. These ozone formation and destruction processes take place in the 13 miles (21 km) thick ozone layer in the lower stratosphere. However, ozone is sparse within this layer. There are only three ozone molecules in every 10 million molecules of air.

Ozone shields the Earth from the Sun's harmful ultraviolet radiation. Without enough ozone in the stratospheric ozone layer, humans might not survive. Normal O_2 filters most ultraviolet radiation with wavelengths shorter than 240 nm. It is this radiation that breaks or change O_2 molecules into single oxygen atoms (O). However, it cannot absorb ultraviolet radiation with wavelengths between 240 nm and 290 nm. This longer wavelength ultraviolet radiation is absorbed by the stratospheric ozone layer, causing the O_2 molecules and O atoms to bond into O_3.

Without or with a thin ozone layer, there is increased exposure to ultraviolet radiation (UVR) from the Sun. Increased exposure to UVR increases the likelihood of skin cancer. With exposure, UVR damages the DNA in skin cells eventually beyond the body's ability to repair it. These damaged skin cells grow and multiply at a quick rate and skin cancer develops. Skin cancer generally first appears in the outermost layer of skin, the epidermis, which allows early detection of most skin cancers. The disease can be divided into two basic types, non-melanoma skin cancer, which includes basal-cell cancer and squamous-cell cancer, and melanoma. If left untreated, they can be fatal. Melanoma accounts for about 5% of the skin-cancer cases in the United States but causes 75% of skin-cancer deaths. About 54 000 cases of melanoma are diagnosed each year, and it proves fatal for about 7800 people annually.

The ozone layer protected life on Earth for millions of years, contending only with output from some stronger volcanoes. However, the need for refrigeration drove the invention of Freon in 1928. Freon was the first of a new group of chemicals called chlorofluorocarbons or CFCs that could be used for cooling. These organic compounds are colorless, odorless, nonflammable, nontoxic, noncorrosive, nonreactive, and very stable, so are safe for humans and other life. Their boiling point is low (77 °F, or 25 °C), which makes them excellent coolants for refrigerators and air conditioners. In refrigeration, as liquids evaporate, they absorb heat through latent heat transfer creating cooling or freezing conditions. CFCs have also been used in fire extinguishers, as aerosol propellants, solvents, and foaming agents.

The health and safety aspects of Freon in addition to its effectiveness as a cooling agent quickly made it extremely popular very quickly. House temperatures could be controlled in hot climates and perishable food stored effectively. Cars could be used in comfort on the hottest days. The demand was so high that CFCs became one of the highest volume manufactured chemicals in the United States by the 1960s.

Unfortunately, there is a negative aspect of CFCs. The problem is that it degrades ozone and was reaching the ozone layer in the stratosphere. The culprit in this unfortunate situation is the chlorine. Dissociated chlorine (Cl) can breakdown ozone (O_3) by the reaction:

$$Cl + O_3 ---> ClO + O_2$$

The chlorine can then be dissociated if ClO interacts with free oxygen atoms:

$$ClO + O ---> Cl + O_2$$

This cycling of Cl reactions can continue as long as two years. During this time, a single chlorine atom can destroy 100 000 precious ozone molecules in the stratosphere. Finally, the chlorine forms either hydrogen chloride or chlorine nitrate, sinks into the troposphere, and is washed out by rain or other forms of precipitation.

Most forms of chlorine do not reach the stratosphere. Chlorine from most common sources combines with water vapor and is washed to the ground. Only large fires and some marine life are able to produce a form of chlorine that could reach the stratosphere. However, CFCs make up about 85% of the chlorine in the stratosphere and they can penetrate 15 to 25 miles (24–40 km) above the Earth's surface without breaking down. At that point, ultraviolet radiation dissociates them, producing several Cl ions from each CFC.

By 1974, scientists had realized the potential danger of CFCs to the ozone layer and predicted catastrophic effects. Then, in 1985, scientists discovered an "ozone hole," or depletion in the stratosphere over the Antarctic of as much as 70% (Figure 3.7a). With further investigation, it was found that the highest rates of depletion were over Antarctica, but also to a lesser extent over the North Pole as well. The hole was quite large and growing. It alarmed the general public and even though the ozone at mid-latitudes was not depleted nearly as much, people began staying out of the sun and becoming concerned about sunscreen protection and skin cancer. The holes in the ozone layer were the bases of several broad publicity and educational campaigns by scientists and environmentalists that were aimed at policy makers and the public. The efforts were effective and led to bans on the use and release of CFCs and other types of ozone-depleting chemicals.

(a)
Southern Hemisphere ozone levels (October 7, 2021)

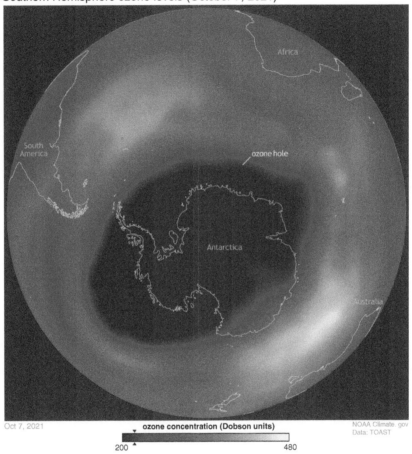

(b)

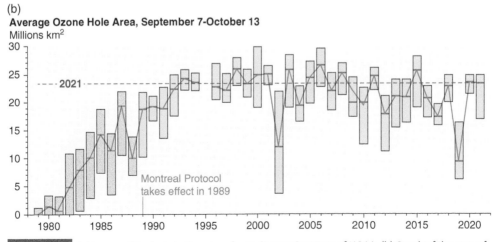

FIGURE 3.7   (a) Image of the hole in the ozone layer. *Source:* Courtesy of NOAA. (b) Graph of the area of the ozone hole from 1979 to 2020. *Source:* Courtesy of NASA.

In 1987, the United Nations sponsored an international treaty called the Montreal Protocol that banned the use of CFCs and all of the most dangerous chemicals to stratospheric ozone by 1996 in most developed countries. It was signed by 198 countries and phased out 99% of the ozone-depleting chemicals. However, it will take a long time for the Montreal Protocol to have an effect. Even this many years later, ozone-depleting gases are still infiltrating the stratosphere, and it will take more than 100 years before the ozone layer will reach pre-CFC levels. Hydrochlorofluorocarbons (HCFCs) or hydrofluorocarbons (HFCs) were chosen as the immediate solution to replace CFCs. They have less chlorine and as such are less destructive to stratospheric ozone. However, the Montreal Protocol requires that they must be phased out by 2030 as well. A more ozone-friendly replacement for all these compounds will be required in the near future.

The shrinking of the hole in the ozone layer is not a continuous process (Figure 3.7b). The 2019 ozone hole over the Antarctic reached its peak extent of 6.3 million square miles (16.4 million km^2), which was the smallest since 1982. During normal weather conditions, the ozone hole normally has a maximum area of about 8 million square miles (20.7 million km^2). However, the 2021 ozone hole reached a maximum area of 9.6 million square miles (24.8 million km^2) ranking it the 13th largest since 1979.

CHAPTER 4

Climate Change: Crisis of our Time

CHAPTER OUTLINE

Polluted Earth: The Science of the Earth's Environment, First Edition. Alexander Gates.
© 2023 John Wiley & Sons, Inc. Published 2023 by John Wiley & Sons, Inc.
Companion website: www.wiley.com/go/gates/pollutedearth

Words you should know:

Climate change – A long-term change in global or regional weather or climate patterns. Current usage of global warming.

Drought – The absence of water in an area including (i) meteorological droughts, (ii) hydrological droughts, (iii) agricultural droughts, and (iv) socioeconomic droughts. Meteorological droughts are 75% or less of normal precipitation relative to a 30-year average.

El Niño – Reversal of equatorial Pacific Ocean circulation that causes worldwide changes in weather.

Global warming – Climate change.

Greenhouse gas – A gas in the atmosphere that permits the input of sunlight, but which traps heat in the troposphere.

Industrial effect – The replacement of radiogenic carbon by dead carbon from the burning of fossil fuels.

Sea-level rise – The increase in the height of the sea level because of melting ice caps.

Wildfire – An unplanned, unwanted, uncontrolled, and destructive fire in vegetation and especially forests.

4.1   The Climate Change Controversy

Climate change is considered by many to be the most serious environmental issue of our time and, as a consequence, it has taken the spotlight. It is certainly the most controversial of the current environmental issues. Acceptance of climate change as a problem has been split down political and ideational lines. The people who do not believe in climate change argue that the climate is constantly changing. This is true. Until 12 000 years ago, continental glaciers extended southward from the North Pole to the northern United States. New York City was covered by a glacier at least 2000 ft (600 m) thick and a mere 10 miles (16 km) north of the city, it was one mile (1.6 km) thick (Figure 4.1). Tundra conditions extended through Virginia. In addition, there was a cold period from about 1400–1850 known as the Little Ice Age.

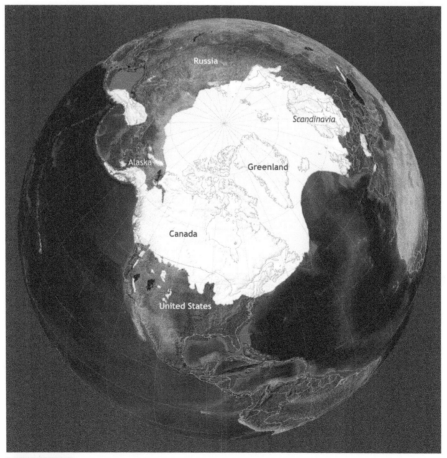

FIGURE 4.1 The Northern Hemisphere of the Earth showing the extent of continental glaciers during the last ice age. *Source:* Courtesy of NOAA.

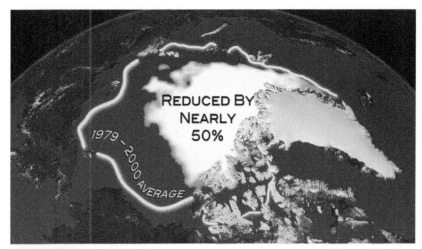

FIGURE 4.2 Map of the Arctic showing the average extent of ocean ice between 1979 and 2000 (yellow line) versus the extent in 2020. *Source:* Courtesy of NOAA.

This allows opponents to claim that climate changes are normal. They also argue that it is better to be too warm than too cold. The difference in this period of climate change is that it is much faster than natural changes and there is irrefutable evidence that it is being caused by human activity. It can be dangerous enough when climate changes are caused by natural processes but when humans cause climate change, the final outcome is unknown and potentially devastating to life on the planet.

The other danger of climate change is the paired result of global sea-level rise. The height or depth of the oceans too have varied throughout geologic history. Between 500 and 250 million years ago, oceans covered the land masses to varying degrees for most of the period. However, there were no humans on the planet at the time. During the last ice age between 22 000 and 12 000 years ago, sea level was about 200 ft (60 m) lower than it is today. However, there was no modern civilization at the time and, being nomadic, humans could simply migrate to an area that best suited their needs.

The changing of the climate coupled with the excessive amount of particulate, or soot, produced by humans is melting all glacial ice on the planet at an extremely quick pace. Increasing temperature certainly melts ice but the black color of the soot also absorbs more sun by increasing the surface albedo. This heats the ice from the surface causing it to melt quicker. There are horrifying time sequences showing the disappearance Arctic Sea and land ice over the past 40 years that document the change (Figure 4.2). Alpine glaciers are similarly disappearing. Glacier National Park had about 80 glaciers in 1850 but by 2015 there were just 26 named glaciers. Some of these are now too small to be considered glaciers. Some of them decreased in size by 80% between 1966 and 2015.

All of this melting water is causing global sea level to rise catastrophically. It is already doing damage. As our coastal cities flood, it will cost us exceptional amounts of money. The estimates are rough and vary greatly but by 2050, it is estimated that climate change will cost the world's economy between $23 and $50 trillion and continue to cost about $1.7 to $1.9 trillion per year through 2100. It will displace a tremendous amount of the world's population and cause many cities like Venice, Italy and Miami, Florida to be abandoned permanently. Whole countries like the Netherlands and Bangladesh will barely have any remaining land. Global climate change will disrupt our entire society in these and many ways we have yet to fathom.

4.2 Basics of Climate Change

Carbon dioxide (CO_2) occurs naturally in the atmosphere in concentrations less than many trace gases. In 2021, the concentration of CO_2 was 416 ppm in comparison to argon which is about 9300 ppm. However, unlike argon and most other trace gases, CO_2 plays a vital role in the survival of life on Earth. It is an essential component in the respiration of the planet. Humans and other animals inhale air that is rich in oxygen and exhale air with much more CO_2. The carbon molecules are used to provide energy to consumer forms of life. In contrast, CO_2 is respirated by producers like plants, marine plankton, and certain bacteria that convert the carbon back into stored energy and oxygen which is emitted back into the atmosphere. This process is photosynthesis.

CO_2 is also instrumental in maintaining the climate balance of the planet as a greenhouse gas. The greenhouse effect results because solar radiation enters the atmosphere, heats it up, and the heat cannot escape forming a situation similar to a greenhouse (Figure 4.3). The reason that the heat gets trapped is because incoming solar radiation has a short wavelength and penetrates all gases in the atmosphere. It is absorbed by the Earth's surface and is converted into long wavelength radiation or heat. Gases like oxygen allow long-wave radiation to pass back out of the atmosphere. Other gases, like carbon dioxide, absorb long-wave (infrared)

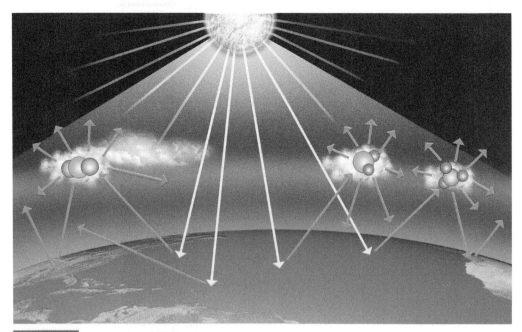

FIGURE 4.3 Illustration of sunlight entering the troposphere, being converted to heat (orange arrows) and being reflected back and absorbed by greenhouse gas molecules.

radiation instead of allowing it to pass back into space. This absorption of radiation heats up the atmosphere and causes global warming. This is why CO_2 is a greenhouse gas. It is the atmospheric gas of most concern because it is increasing at such a quick rate. However, there are stronger greenhouse gases. For example, methane, or natural gas, is also increasing in the atmosphere as the result of human activity and is 23 times more powerful a greenhouse gas than carbon dioxide. The amount of methane in the atmosphere is much less than carbon dioxide.

In the ideal, steady state system, if increased CO_2 in the atmosphere causes temperatures to rise, then plant and bacterial growth increase and the excess CO_2 is removed at increased rates by photosynthesis. Excess CO_2 will also be absorbed by the ocean further removing CO_2 from the atmosphere. As the CO_2 is removed, temperatures will decrease, slowing photosynthesis, and the diffusion of CO_2 from the oceans into the atmosphere will increase. CO_2 levels increase and global temperatures rise. This process is how a balance is maintained.

The problem is that human activity has radically changed the CO_2 balance on Earth. The change began long ago with extensive wood burning for heat and energy and deforestation in many developed areas but in recent history, it has accelerated radically (Figure 4.4). At the beginning of the Industrial Revolution, the amount of CO_2 in the atmosphere was about 280 ppm By 2021, the CO_2 levels had reached 416 ppm and were accelerating. Humans are currently adding 4.4 million tons (4 million mt) of CO_2 to the

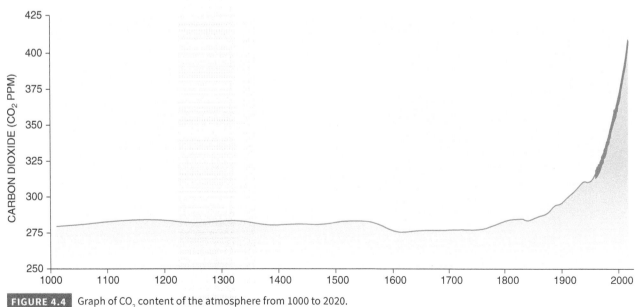

FIGURE 4.4 Graph of CO_2 content of the atmosphere from 1000 to 2020.

atmosphere per year. CO_2 from the burning of petroleum and natural gas, is estimated to represent 82% of total industrial emissions in the United States. Transportation produces 27% of the greenhouse gas emissions, electricity produces 25% and industry produces 24% of all emissions.

How we can tell which CO_2 is anthropogenic and which is from respiration or other natural sources? In the 1950s, Hans Suess found that natural CO_2 includes radiogenic carbon whereas fossil fuel produced CO_2 does not. Radiogenic carbon is produced in the atmosphere from radiation and is contained in all natural CO_2. If carbon is removed from the atmosphere, like in oil and natural gas, it decays to a non-radiogenic isotope that is "dead." This means that air sampled anywhere on the Earth away from combustion of fossil fuel shows the amount of CO_2 added by humans. Suess and Roger Revelle first coined the phrase "the industrial effect" to describe this but later it was changed to the greenhouse effect. Revelle has been recognized as the "grandfather of the greenhouse effect" by most climate change proponents.

CASE STUDY 4.1 Revelle, Roger R. D.

Grandfather of the Greenhouse Effect

Roger Revelle received the National Medal of Science from President George H. W. Bush in 1990, for being the "grandfather of the greenhouse effect" and no one disagreed (Figure 4.5). The New York Times named him "one of the world's most articulate spokesmen for science." In the 2006 Oscar-winning documentary "An Inconvenient Truth" by former Vice President Al Gore, Roger Revelle is credited as the discoverer of global warming and the greenhouse effect. Revelle is a widely revered scientist and the recognized pioneer of climate change.

Roger Revelle was born in Seattle, Washington in 1909 and raised in Pasadena, California. Revelle enrolled in Pomona College, California in 1925 at 16, intent on a career in journalism. However, he changed his major to geology after an inspiring class. He married Ellen Virginia Clark in 1931 who was the grandniece of Scripps College's founder, Ellen Browning Scripps. Revelle graduated in 1929 and entered graduate school at the University of California at Berkeley in geology. He specialized in oceanography and received a prestigious research assistantship, at the Scripps Institute of Oceanography in La Jolla, California where he also served as an instructor. Revelle earned a PhD in 1936 and a postdoctoral fellowship at the Geophysical Institute in Bergen, Norway. He returned in 1937 and obtained a research associate position at Scripps Institution.

FIGURE 4.5 Photo of Roger Revelle.
Source: Courtesy of World Meteorological Organization.

Roger Revelle enlisted as a reserve officer in the US Navy. He was called to service and assigned to the Bureau of Ships in 1942, where he conducted oceanographic research. He was later on the staff of the Commander of Amphibious Forces for the Pacific Fleet and helped plan the invasion of Japan. Revelle later participated in the first post-war atomic test on Bikini Atoll. He led the research program to study effects on marine life and remnant radiation in the impacted rocks and sediment. Revelle transferred to the Office of Naval Research in 1947 and served as the head of the Geophysical Branch. He established a program to support research at universities, which was later used at the National Science Foundation. In 1948, Revelle retired from the Navy at the rank of commander.

That year, Revelle returned to Scripps Institution and advanced to director by 1951. Scripps developed into among the top oceanographic institutions in the world under his leadership. He expanded the research vessel fleet, participated in research cruises, and was involved in many of the discoveries. In 1959–1960, Revelle established Scripps as part of the University of California at San Diego and served as dean. Revelle was on leave in 1962–1963 and served as science advisor to Secretary of the Interior Stuart Udall. He was instrumental in helping to gain recognition and acceptance of the work of Rachel Carson. Her book "Silent Spring" was released that year.

Before the 1950s, the scientific community believed that CO_2 produced by industry would be absorbed by the oceans and keep atmospheric levels relatively constant. In 1955, Hans Suess developed a new technique of measuring carbon isotopes which allowed him to distinguish natural carbon from that produced by industry. Revelle invited Suess to Scripps Institution and they collaborated on a groundbreaking article in 1957 using Suess's techniques. It clearly demonstrated that CO_2 was increasing in the atmosphere as the result of human activity. In 1956, Revelle brought Charles Keeling to Scripps, and the two systematically measured CO_2 in the atmosphere at Mauna Loa, Hawaii, and Antarctica. This research was the foundation for the study of greenhouse gases and global warming.

In 1965, Revelle served on the US President's Science Advisory Committee Panel on Environmental Pollution. Through his leadership, CO_2 from burning fossil fuels was recognized as a potential global problem. Revelle also served on many international panels, informing the world about global warming. He is best known for an article published in Scientific American in August of 1982. It explained the increase in atmospheric CO_2 and described the effects on the Earth including melting glaciers and ice caps, shifting climate belts, thermal expansion of the ocean and changes in circulation patterns.

Roger Revelle accepted a position as the Richard Saltonstall Professor of Population Policy at Harvard University in 1964, where he also served as the director of the Center for Population Studies. He studied food production in relation to population growth and became an expert in the field. In 1976, Revelle returned to the University of California at San Diego as a professor of science and public policy. Roger Revelle died on July 15, 1991, in La Jolla, California, at the age of 82.

Roger Revelle was one of the most decorated scientists. In addition to numerous awards, Revelle served on the National Academy of Science from 1958 and was awarded the Agassiz Medal in 1963. He received the 1986 Balzan Prize for Oceanography and Climatology by President Francesco Cossig of Italy. The American Geophysical Union awarded Revelle the Bowie Medal in 1968 and the Tyler Medal in 1986.

CO_2 in the atmosphere increased dramatically in the late twentieth century and into the twenty-first century. In the documentary "An Inconvenient Truth," Al Gore shows this dramatic increase and refers to it as the "hockey stick" in the graph because of the sharp change (Figure 4.4). The more CO_2 added to the atmosphere, the warmer the planet will become. This is the reason for calling the increase "global warming" or climate change. It will lead directly to major changes in the planetary temperatures, precipitation patterns, ocean levels, storm intensity, desertification, deglaciation, and numerous other impacts that we will only understand with time. Exacerbating this situation is the rapid global deforestation, especially in the rainforests of Central and South America, thereby reducing total photosynthesis which could remove some of the additional CO_2. In addition, ocean pollution through dumping of waste, runoff of chemical fertilizers and pesticides, air pollution fallout, and acidification through equilibration with the higher CO_2 levels could reduce the phytoplankton productivity and removal of CO_2 by marine organisms.

4.3 Climate Change Impacts

As the atmospheric temperature rises through climate change, significant changes to global climate will occur. Temperatures are clearly increasing in step with CO_2 increase (Figure 4.6) showing a 2.5°F (1.4°C) between 1910 and 2020 which may not sound like much but it is tremendous on a global basis and it is accelerating. Further, the majority of the rise is from 1975 to 2020 (1.8°F or 1°C) which is particularly concerning because it is so fast and bodes ill for the future. These changes will result in major disruptions to the worldwide real estate, transportation, trade, agriculture, public health, and sustainability of communities. These will result from a number of complex climate and related changes. In addition to global temperature increase, there will be sea-level rise, increase in the ocean surface temperature, reduction of land and sea ice, and shifts in ocean circulation and weather patterns. These will also cause changes in vegetation patterns, range and survival of biota, changes in diseases and their geographic ranges, and other unforeseen changes.

4.3.1 Sea-Level Rise

As a result of melting of glacial ice, the average sea level has risen about 9 in. (23 cm) in the past century (Figure 4.7). Although 9 in. (23 cm) may not seem like much, the pace of sea-level rise is increasing quickly meaning that it will be much faster in the future. Further, coastal urban areas are already suffering from this rise. This is expected in areas that already include waterways like Venice, Italy or southern Bangladesh or most of the Netherlands but it even impacts hilly coastal areas. For example, the New York metropolitan area has some of the most expensive real estate in the world. As a result of sea-level rise, many low-lying areas are now

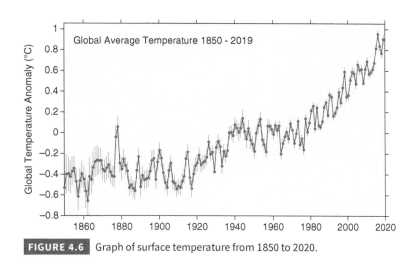

FIGURE 4.6 Graph of surface temperature from 1850 to 2020.

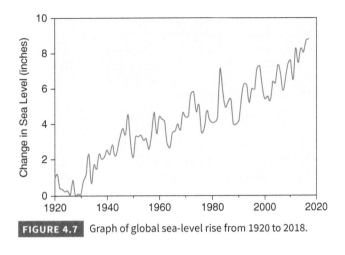

FIGURE 4.7 Graph of global sea-level rise from 1920 to 2018.

under threat of flooding (Figure 4.8). The cost of this threatened real estate is astronomical. Currently, it is estimated at less than $10 billion but will exceed $25 billion by 2040 and in the hundreds of billions by 2100.

Even worse is the threat to commerce. All three of the major airports in the New York/New Jersey metropolitan area are near sea level and will be disabled with continued sea level rise. This is one of the top three major air traffic hubs in the US. Elizabeth, New Jersey is the home of the regional port and has the second most commerce of all ports in the Western Hemisphere. It is naturally at sea level and will not survive sea-level rise. The amount of rail and truck freight traffic through this area is the highest in the United States and among the top in the world. The major train and truck routes including tunnels into New York City are all under threat of sea-level rise. Unless well anticipated and planned, sea-level rise will cause a major disruption to the entire American economy by virtue of the impact on New York alone and there are many other port cities that are under threat like Los Angeles, New Orleans, Miami, Seattle, and Houston in the United States alone. The other aspect is that poor and minority communities tend to live in the most vulnerable locations and yet are least financially capable of relocating or planning for these changes. Sea-level rise will add to the inequity of environmental justice.

Stronger Tropical Storms

Evidence that increasing atmospheric temperatures is leading to stronger tropical storms that strike the coastal areas is building. The theory of increased storms resulting from global warming is that the increased air temperatures will also lead to increased ocean surface temperatures. Surface ocean temperatures have been steadily increasing by an amount similar to air or 2.5 °F (1.4 °C) between 1910 and 2020 (Figure 4.9). The increase has also been much more pronounced since 1975. Hurricanes and other tropical storms increase their power if the ocean has surface temperature of 79 °F (25 °C) or greater. With increasing surface temperatures, hurricanes have greater areas to develop and strengthen leading to more and stronger storms. If warmer temperatures spread farther northward such as up the eastern seaboard of the United States, stronger hurricanes will be able to strike farther northward which, up until now, has been generally protected by cooler ocean temperatures. In 2012, Superstorm Sandy struck the New York metropolitan area and all of the areas threatened by sea-level rise (Figure 4.8) were impacted. This storm caused $68.7 billion dollars making it among the costliest in American history. What would happen if these became a regular occurrence?

Have tropical storms, hurricanes, typhoons, and cyclones been increasing? Between 1878 and 2020, there has been a general increase in Atlantic storms (Figure 4.10). There has been some variability but since 1980, the increase is consistent with the all-time

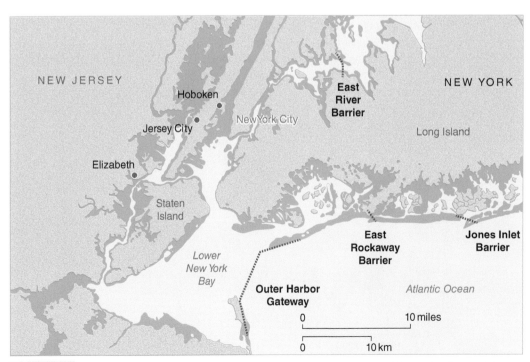

FIGURE 4.8 Map of the New York metropolitan area showing the areas threatened by sea-level rise in blue.

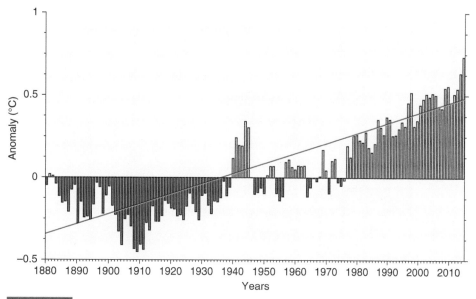

FIGURE 4.9   Graph of ocean surface temperature increase from 1880 to 2015.

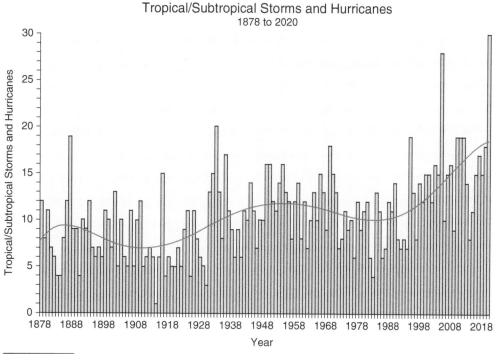

FIGURE 4.10   Bar graph showing Atlantic hurricanes and tropical/subtropical storms per year from 1878 to 2020.

record set and reset during this period. This is a global trend but considering changes in frequency and intensity of storms at the local level yields even more startling results. For example, the Philippines sits in the middle of the primary Pacific Ocean typhoon track. In addition, it has an inordinate amount of coastal area relative to the size of the country because it contains numerous islands. For this reason, it is prone to typhoons and badly impacted by them. The number of typhoons striking the Philippines has increased dramatically in recent history (Figure 4.11). It was struck by about 20 major storms per decade in the 1950s and 1960s but in the 2010s, it was struck by 91 typhoons and several were devastating supertyphoons. The 2013 Supertyphoon Haiyun or Yolanda is among the most powerful storms ever. Could there be more of these in the future?

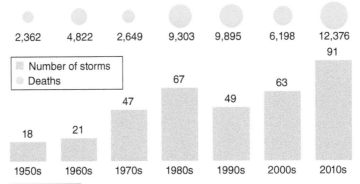

FIGURE 4.11 Bar graph of the number of tropical storms and typhoons to strike the Philippines per decade and the resulting number of deaths as circle size.

CASE STUDY 4.2 2013 Super Typhoon Haiyan

Stronger Pacific Typhoons

A large area of low pressure developed a few hundred miles east-southeast of Pohnpei, Micronesia on 2 November 2013 (Figure 4.12). It tracked westward and quickly strengthened. On 4 November it became a tropical storm named Haiyan. It then underwent rapid intensification, reaching typhoon intensity by 5 November The intensification continued and on 6 November it reached super typhoon status with the wind speed of a Category 5 storm on the Saffir–Simpson Hurricane Scale. The maximum 10-minute sustained winds of the storm were measured at an incredible 145 mph (230 km/h) on 7 November but they were estimated to have reached 180 mph (285 km/h). One-minute maximum sustained winds were measured at an astonishing 195 mph (315 km/h), which made Haiyan the strongest typhoon and tropical cyclone of any kind on record (Figure 4.12a). Haiyan made its first landfall early on the morning of 8 November at 5:40 a.m. in Guiuan, Eastern Samar, Philippines at full intensity. It made five other landfalls as it crossed the Philippines before entering the South China Sea on 9 November There it turned northwestward and weakened.

Super typhoon Haiyan is also known as Super typhoon Yolanda in the Philippines. It directly caused the death of 6300 people with 1061 missing and 28 689 injured. This level of casualties made it the second most devastating typhoon to strike the Philippines. The ferocious winds caused great damage, but the storm surge at first landfall was unique.

Super typhoon Haiyan made landfall and crossed the small islands southeast of Samar Island. It then moved westward across the Leyte Gulf to Tolosa on Leyte Island and traveled northward through the funnel-shaped Bay of San Pedro off Leyte Gulf. The shape of coastlines greatly affects the power of the waves coming ashore. Convex coastlines disperse wave energy but funnel-shaped coastlines focus the energy (Figure 4.12b). The storm surge of Haiyan was 10–14 ft (3–4 m) in the Leyte Gulf but at 5:30 a.m. at the north end of Bay of San Pedro, there was a sea level drop of 3.3 ft. (1 m). However, the storm surge then returned into the cities at tsunami speeds (Figure 4.12b). The storm surge was 16 ft. (5 m) high along the shore of the bay but 23–26 ft. (7–8 m) in the city of Tacloban, and up to 46 ft. (14 m) in the city of Guiuan.

The Haiyan wave was not a true tsunami because it was not caused by an earthquake or submarine landslide. Tsunami waves produced by storms are called meteotsunamis or meteorological tsunamis. The storm surge waves caused most of the deaths and damage in Haiyan. Damage was estimated at $2.2 billion making it the costliest typhoon disaster in Philippine history at the time (Figure 4.12c). A documentary on Super Typhoon Haiyan by Discovery Channel is entitled "*Megastorm: World's Biggest Typhoon.*" Large storms are now increasing in the Philippines, ostensibly as the result of climate change.

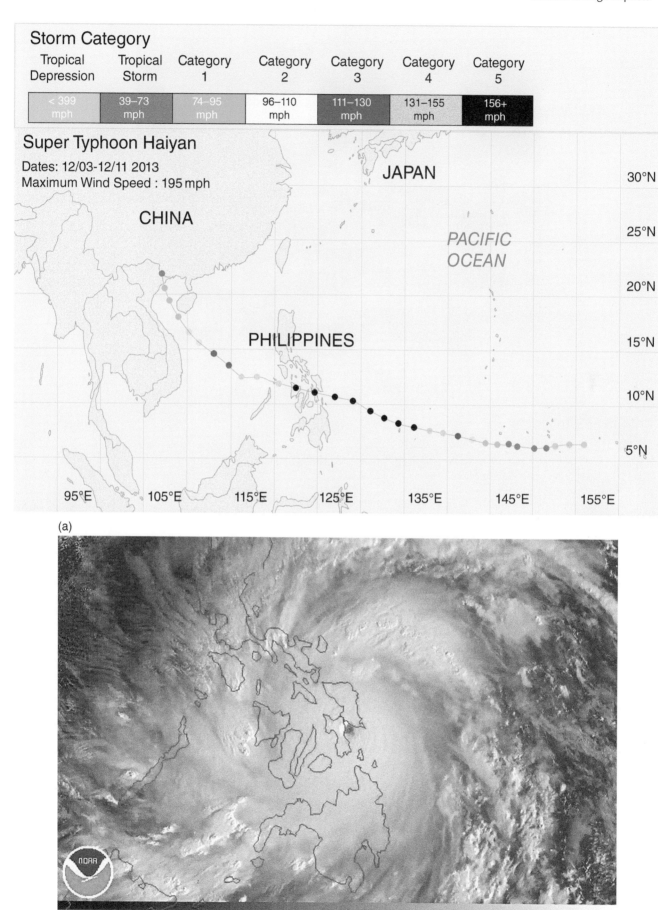

Storm Category						
Tropical Depression	Tropical Storm	Category 1	Category 2	Category 3	Category 4	Category 5
< 399 mph	39–73 mph	74–95 mph	96–110 mph	111–130 mph	131–155 mph	156+ mph

Super Typhoon Haiyan

Dates: 12/03-12/11 2013
Maximum Wind Speed : 195 mph

(a)

FIGURE 4.12 Map showing the track and intensity of Super Typhoon Haiyan. (a) Satellite image of Super Typhoon Haiyan in the Philippines. *Source:* Courtesy of NOAA. (b) Map diagram showing how Super Typhoon Haiyan formed a meteotsunami. (c) Photo of damage done to the Philippines by Super Typhoon Haiyan *Source:* DFID – UK Department for International Development / Flickr / CC BY 2.0.

(b)

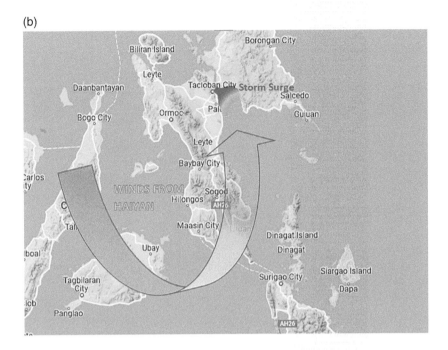

(c)

FIGURE 4.12 (Continued)

4.3.2 Changes in Ocean Currents

The movie "The Day After Tomorrow" credits a shift in the North Atlantic Ocean current for causing the Earth to enter a new ice age in just a few weeks. This may be science fiction but ocean currents can apparently be changed through global warming. Surface ocean currents can greatly impact weather. Hurricanes, typhoons, and cyclones require warm surface waters to spawn and grow. Therefore, warm surface currents can impact areas of generation and paths of the storms. There are also areas with upwelling of deep, cold ocean waters in addition to the laterally moving currents. The upwelling areas have high biologic productivity. There are also downwelling areas marked by descending waters. The currents can strengthen, weaken, and shift locations. They also significantly control weather and climate patterns. The El Niño and La Niña Pacific Ocean current phenomena cause marked changes in temperature, precipitation, and weather patterns, especially in the Americas. The cause of these changes is shifting of the Pacific Ocean circulation system and they affect weather and climate worldwide. Climate change is causing especially rapid and unpredictable changes in atmospheric and ocean circulation.

The El Niño and La Niña oscillations are regular swings in Pacific Ocean circulation (Figure 4.13). Over the past 70 years, during the rapid increase in air and ocean surface temperatures, there have also been changes in these oscillations. The La Niña conditions were much stronger and longer in duration during the 1950s through 1970s than they are now. In contrast, El Niño conditions are not more common but they are becoming much stronger. The past few events have caused significant worldwide disruption.

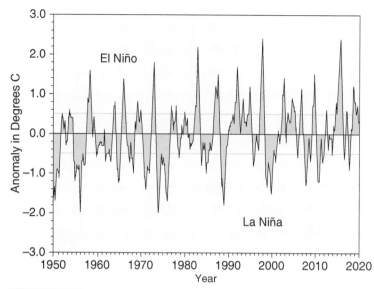

FIGURE 4.13 Graph showing El Niño (orange) versus La Niña (blue) oscillations from 1950 to 2020.

CASE STUDY 4.3 1982–1983 El Niño

Increasing Ocean Current Changes

An El Niño event is marked by unusually high ocean surface temperatures in the equatorial region of the Pacific Ocean. In contrast, La Niña events are marked by unusually cold surface ocean temperatures in the same area. El Niño is an ocean–atmosphere oscillation in the Pacific Ocean that impacts worldwide weather. The strongest El Niño on record lasted from 1876 to 1878 and caused devastating droughts that triggered massive famines in Asia, Africa, and South America. It killed more than 50 million people globally which was 3% of the world population at the time. In recent years, El Niños have been getting stronger and more frequent possibly as the result of global warming.

Normally, trade winds blow from South America toward Asia from east to west along the equator in the Pacific Ocean. The winds drive warm surface waters toward the Asian coast which raises sea level near Indonesia about 1.5 ft (0.5 m) higher in elevation than by Ecuador on the other side of the Pacific Ocean. The ocean surface temperature near Indonesia is 14 °F (8 °C) warmer than near Ecuador. This is partly because ocean currents push cold water from deeper levels toward the surface near South America. The cold water that is upwelled is nutrient-rich, and supports diverse marine life, including major fisheries. This makes the weather around the cold South American waters also cold and dry. In contrast, the rising air over the warm water near Indonesia produce clouds and rainfall.

During an El Niño, the Pacific Ocean trade winds decrease and reduce the piling up of warm water near Indonesia and reduce the upwelling of deep waters near South America (Figure 4.14). As a result, the sea surface temperature warms near South America and the supply of nutrient-rich water is cut off to the main productivity zone. As a result, there is a sharp decline in productivity. The weakened easterly trade winds shift weather patterns making the rain move eastward with warm water. This shift results in flooding rains in Peru but

substantial droughts in Indonesia and Australia. It also causes changes in the global atmospheric circulation which translates into weather changes around the world.

In 1982–1983, the strongest and most devastating El Niño of the century, to date, struck the Pacific Ocean. It was so intense that the trade winds reversed their direction and blew backwards. The ocean thermocline west of South America deepened to a 500 ft. (124 m) depth. At Paita, Peru, the sea surface temperature increased a phenomenal 7.2 °F (4 °C) in just 24 hours on 24 September 1982. The El Niño was so strong that the waves of warm water it produced were still detectable 12 years later in the Pacific Ocean. The oscillation produced weather disasters all around the world. In Africa, Australia, and Indonesia, there were severe droughts, dust storms, and wildfires. The heaviest rainfall in history fell in Peru. Several areas with average seasonal rainfalls of 6 in. (15 cm) received as much as 11 ft (3.4 m) of rain. This caused flooding of rivers as much as 1000 times their normal flow. There were 1300–2000 deaths and more than $13 billion (US) in damage that were directly attributable to the 1982–1983 El Niño.

This massive El Niño produced a slew of related problems. The warm, wet spring on the North American East Coast increased the mosquito populations resulting in encephalitis outbreaks. The cool, wet spring in New Mexico increased rodent populations and resulting fleas causing a rise in bubonic plague cases. Above-average temperatures in Northwestern Canada and Alaska reduced the salmon population. Hawaii was struck by one of the strongest typhoons on record, but the Atlantic hurricane season was relatively calm.

The 1982–1983 El Niño oscillation was considered an anomaly but just 14 years later, the 1997–1998 El Niño was more powerful. It increased the global air temperature by 2.5 °F (1.5 °C) killing some 16% of the reef systems in the world. It was the warmest year in history at that point. It caused similar widespread droughts, massive flooding,

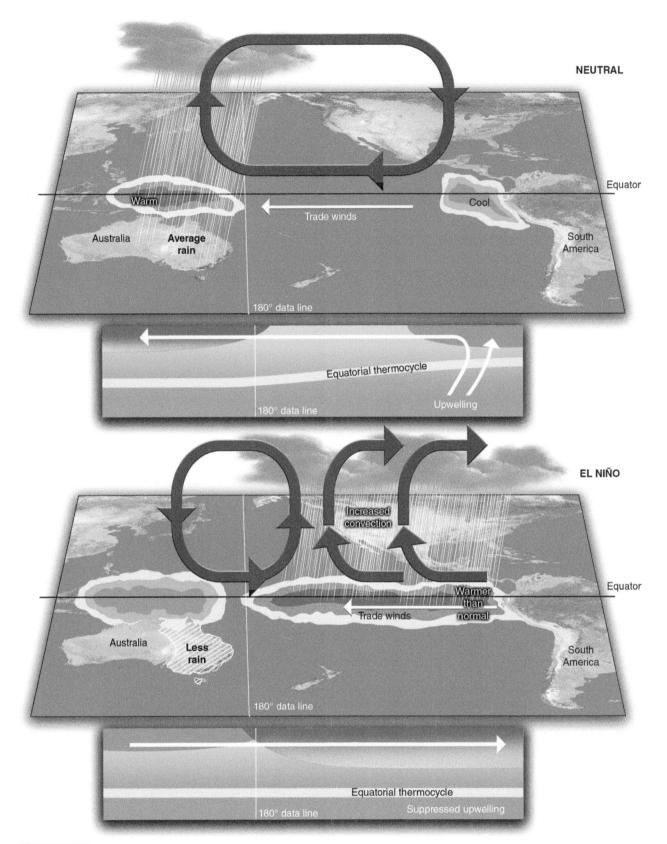

NEUTRAL

Equator

Warm
Cool

Australia
Average rain

Trade winds

South America

180° data line

Equatorial thermocycle
Upwelling

180° data line

EL NIÑO

Increased convection

Equator

Australia
Less rain

Warmer than normal

Trade winds

South America

180° data line

Equatorial thermocycle
Suppressed upwelling

180° data line

FIGURE 4.14 Diagram showing ocean surface, subsurface and atmospheric views for normal equatorial Pacific Ocean water circulation (top) and El Niño conditions (bottom).

and other natural disasters worldwide. Heavy rainfall in Somalia and Kenya resulted in a severe outbreak of Rift Valley fever. In contrast, Indonesia suffered one of its most devastating droughts ever. Surface temperature of sea water off Peru was 19.8 °F (11 °C) above average.

Another impact was that in 1997–1998, there were a record of 11 super typhoons in the western Pacific Ocean. Ten of these attained Category 5 intensity. The estimated cost of damage for the 1997–1998 El Niño ranged from $32 to $96 billion (US).

4.3.3   Droughts

Increased air temperature results in increased evaporation. This causes the land and covering vegetation to dry out much quicker than under cooler conditions. This means that even if weather patterns remain the same, areas will be more subject to droughts and their dangerous effects. Shifts in weather patterns can even reduce the amount of precipitation, further drying the land surface.

The lack of precipitation relative to drying causes a drought. There are several kinds of droughts including meteorological droughts, hydrological droughts, agricultural droughts, and socioeconomic droughts depending upon the area of water shortage. Meteorological droughts are of the most concern and defined as 75% or less of normal precipitation relative to the 30- year average for the area. They also need to last for a certain amount of time depending upon type. Drying of the soil results in the vegetation holding the soil together to slowly die off and be removed. As a result, wind directly abrades the exposed soil, removing the fine material. The danger to non-desert areas is removal of the top layer of the soil because it contains the nutrients and makes the soil productive. Its removal severely limits the ability of the soil to produce food. In windstorms, surface material can be removed as particulate and it can create a dust storm.

Increased frequency and intensity of droughts ultimately leads to desertification. This is defined as the conversion of productive area into desert conditions which are unproductive and nearly devoid of life. Desertification is preceded by intense droughts and commonly megadroughts. Megadroughts are droughts that are severe for the area and persist for at least 20 years. There are several megadroughts around the world at any given time. Multiple tree- ring studies, indicate numerous megadroughts in the North American Midwest and West over the past 1200 years (Figure 4.15). There were four major megadroughts during the 900 s AD, the 1100s–1200s, and the late 1500s. The megadrought from 940 to 985 AD, lasted 45 years, the 1100–1247 drought, lasted 147 years, the 1340–1400 drought, lasted 61 years, and the 1550–1600 drought lasted 50 years.

The American Midwest and far West are currently in a megadrought. Drought conditions between 2000 and 2018 were the second driest of all of the 19-year dry periods over the past 1200 years. The drought lapsed in 2019 with a wet year but returned in 2020. This current dry period led to 2020 having the worst wildfire season for Colorado and California ever and other nearby states were at near record levels as well. Smoke from these fires filtered the sun all across the country including the East Coast. The only reason that these fires did not develop into disasters is the advancement in wildfire fighting techniques.

4.3.4   Wildfires

One of the potential problems of climate change is forest fires. Increased temperature and higher evaporation rates exacerbate wildfire conditions. However, weather patterns could yield higher amounts of precipitation on land which could potentially decrease the chance of wildfires in some areas. This, however, is not the case in general.

Controlled fires were traditionally used as a method to clear land for agriculture. Many of these fires got out of control and burned unintended areas. Further, firefighting in remote areas was not well organized and methods were not that effective. For this reason, wildfires were common and destroyed large amounts of forests. This situation slowly changed as efforts were improved,

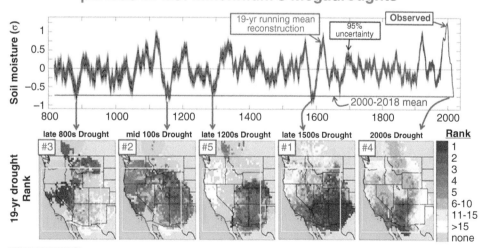

FIGURE 4.15   Top: Graph of the results of the tree-ring and interpreted soil -moisture study for the past 1200 years in the Western United States. Bottom: Maps showing the distribution of low soil moisture during the interpreted five worst droughts over the past 1200 years. Source: Park Williams / Columbia University.

(a)

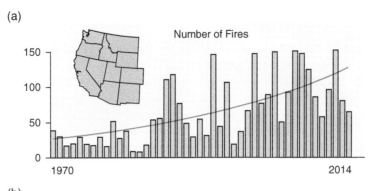

(b)

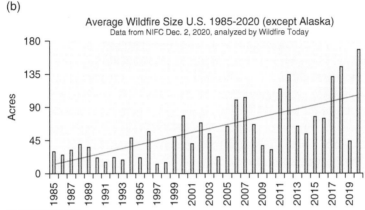

FIGURE 4.16 Bar graphs of wildfires in the United States: (a) number of wild fires in the western states per year from 1970 to 2014 and (b) area burned by wildfires per year for the continental states from 1985 to 2020.

especially with the addition of aircraft to deliver water and fire retardants. As a result, the number and extent of wildfires in the United States decreased dramatically into the 1950s and remained low into the 1970s. However, despite ever more effective techniques, coordination and fire extinguishers, in the early 1980s the number and extent of wildfires began to increase steadily and dramatically (Figure 4.16). Unlike the older wildfires that destroyed cities and claimed many human lives, the current wildfires have mainly been destroying woodlands and impacted communities have been successfully evacuated before humans are killed. However, if the increase in fires continues, they may begin to encroach on populated areas and, once again, claim human lives.

The United States is not the only place to have had an increase in wildfires. There have been some enormous wildfires worldwide over the past several decades. The 1987 Daxing'anling Wildfire or Great Black Dragon Fire struck northeast China and into Siberia between 6 May and 2 June 1987. It killed at least 200 people, injured more than 250, and is reported to have burned more than 32.1–37 million acres (13–15 million ha) of woodland. The 2003 Siberian Taiga Fires burned another 47 million acres (19 million ha) of taiga woodlands. From 7 February to 14 March 2009 more than 1.1 million acres (0.45 million ha) of Australian bush were burned, destroying 3500 structures and killing 173 people and injuring 414. In the 2014 Northwest Territories of Canada Fires, 8.4 million acres (3.4 million ha) were burned. The Amazon fires burn huge areas every year but in 2019, 40 000 fires burned 2 240 000 acres (906 000 ha), which was declared to be the worst in more than a decade. All of these examples, however, were dwarfed by the 2019–2020 Australian Bush fires which burned approximately 60 to 84 million acres (24.3–33.8 million ha) and killed 479 people.

CASE STUDY 4.4 1871 Peshtigo Wildfire

Wildfire Disaster

Droughts dry out vegetation and make it more flammable, increasing the threat of wildfires. The lack of moisture allows fires to spread much quicker making them much deadlier and less controllable. They were historically among the deadliest threats in droughts and a potentially reemerging threat in climate change. The 8–10 October 1871 wildfire outbreak was the worst in United States history. The Upper Midwestern states lost millions of woodland acres and thousands of human lives in the fire. The deadliest fire was the Peshtigo Fire, Wisconsin, but the most famous was the Great Chicago Fire (Figure 4.17). There have been larger wildfires but modern early warning systems and firefighting techniques have made them less deadly.

The Peshtigo and other fires in Michigan and Chicago were caused by several factors. The summer of 1871 was hot and dry with a prolonged drought including no precipitation in August or September.

There was also extensive logging and land clearing for agriculture, leaving dried-up and flammable vegetation and logging debris throughout the area. In addition, land clearing included the use of hundreds of small fires. On 8 October a deep low-pressure system developed over the central plains of the country and produced powerful southwesterly winds beginning at about 2 p.m. Among the gale-force winds, warm temperatures, and dry conditions the small fires were whipped together producing an uncontrolled huge blaze.

The Peshtigo Fire began on Sunday afternoon, 8 October in dense forests of Wisconsin, and spread quickly into Sugar Bush village which killed every resident. The gale-force winds sent 200-ft (61-m)-high flames toward Peshtigo. The flames built a firestorm with temperatures reaching 2000 °F (1093 °C), which caused trees to literally explode into flames. The fire entered Peshtigo in the evening without any warning.

Survivors reported that it moved through Peshtigo in a fire tornado, hurling houses and railcars into the air. The searing flames

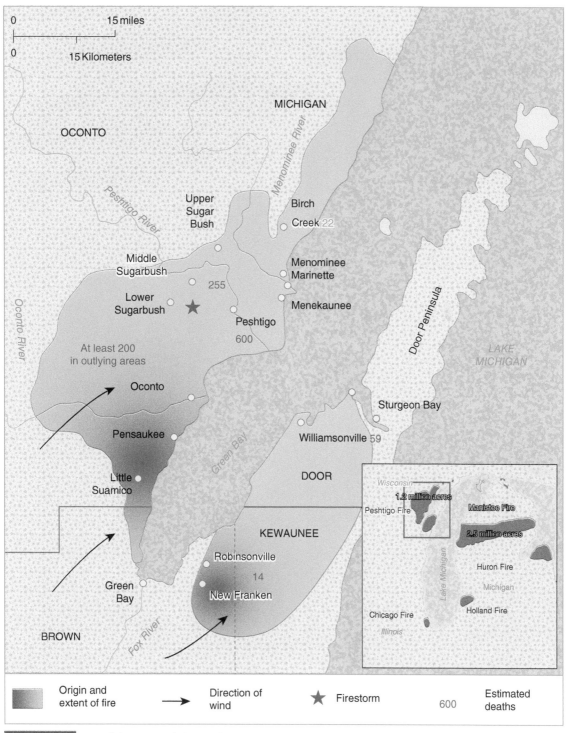

FIGURE 4.17 Map of the 1871 Peshtigo Fire in Wisconsin. Inset map showing location other fires in Michigan, Wisconsin, and Illinois on the same day. (a) Photo of the 1871 Great Chicago Fire. *Source:* Courtesy of US Library of Congress. (b) Photo of the part of Chicago burned in the 1871 fire. *Source:* Courtesy of US Library of Congress.

(a)

(b)

FIGURE 4.17 (Continued)

consumed all available oxygen which suffocated some fleeing people and caused other people to burst into flames. As many as 200 people were killed in a single tavern and another 75 people perished while taking shelter in a boarding house. Three people sheltered in a water tank but were boiled to death as flames heated the tank. Many people tried to escape to the Peshtigo River. However, flames crossed the river and burned both sides of town. In response, people attempted to escape the heat by submerging themselves in the river water but some drowned and others died of hypothermia. The firestorm passed through Peshtigo in less than one hour.

The death toll in Peshtigo was estimated at 800 people. A mass grave of at least 350 people was dug because the bodies were charred beyond recognition. Only two buildings in the entire town were not destroyed by the fire.

After Peshtigo, the firestorm burned a swath of forest 10 miles (16 km) wide and 40 miles (64 km) long destroying Brussels and damaging 16 other towns in just two hours. The fire burned some 1875 miles^2 (4860 km^2) or 1.2 million acres (490 000 ha) of forest. The death toll of the firestorm was 1152 confirmed and 350 believed dead. Some sources claim that it caused as many as 2400 deaths. At least 1500 people were injured and over 3000 people were made homeless. Property damage was estimated at $5 million (1871 US dollars) as well as losing 2 000 000 trees and scores of animals. Others claim damage was more than $160 million.

The Great Michigan Fire was at the same time as the Peshtigo Fire and it included a several individual wildfires in Michigan. These fires were fanned by the same gale-force winds that drove the Peshtigo Fire. Several cities and villages, such as Alpena, Holland, Manistee, and Port Huron, were damaged or destroyed. In addition to wildfires that originated in Michigan, the Peshtigo Fire

crossed the Menominee River and burned Menominee County in Michigan, causing the deaths of at least 128 people. Thousands of houses, barns, stores, and mills were destroyed but no trees/lumber were left to rebuild. This left hundreds of families homeless. The Port Huron Fire burned cities, towns, and villages including White Rock and Port Huron, and forested areas totaling 1.2 million acres (4850 km^2) in the "Thumb" region of Michigan. At least 50 people died in the Port Huron Fire. More than 2.5 million acres (1 million ha) were burned in Michigan. The death toll for the Great Michigan Fire was at least 500 people.

The most famous of the 8–10 October 1871 wildfires was the Great Chicago Fire. At 2 p.m., on 8 October the wind in Chicago turned southwest and intensified. Fire alarms were sounded when fire was spotted in the forested area south and southwest of the city. The wind increased into the evening until it was at hurricane force by midnight and the firestorm spread quickly into the city (Figure 4.17a). The fire brigades could not get the inferno under control before it destroyed at least 4 miles^2 (10.4 km^2) of the city, including at least 17 000 buildings. The damage for the Great Chicago Fire was estimated at $169 million and the death toll was at least 300 victims with 90 000 of 500 000 people left homeless (Figure 4.17b).

4.4   Is it Really that Bad?

Opinions on climate change vary wildly around the world. There are people who deny that it exists at all while others accept that climate change exists but maintain that it is naturally occurring. Still others accept that humans may be changing climate but climate changes all of the time anyway so it is nothing to worry about. Some people even think that it is a positive development. With temperatures increasing, moving back into an ice age is less likely. Our current climate is classified as interglacial because there were so many ice ages in recent geologic history. A climate that is too warm is preferable to one that is too cold for human survival. Further, as the Arctic Ocean sea ice melts, it will allow cargo ships to cross from the Pacific Ocean to the Atlantic Ocean much quicker than going through the Panama Canal (Figure 4.2). This will reduce costs and increase trade. Also, as climate change heats the planet, the climate belts will shift poleward and there is more land to be farmed farther north which could provide more food.

Besides all of the negative consequences described and many others that were not described, the main problem with climate change is that we cannot predict the ultimate outcome. Adding CO$_2$ back to the atmosphere after natural processes sequestered it could lead to many unanticipated consequences like flourishing diseases, microbes, plants, and other forms of life that are dangerous to human health. Further, altering earths chemical systems could set them cascading in directions that result in catastrophe. It has happened in the past, especially in the Permian period when almost 95% of all life went extinct. There have been proposals to add aerosols to the stratosphere to diffuse incoming sunlight so the Earth can cool, thereby offsetting climate change. How do we know that adding aerosols will not make to Earth too cold and set it into a worldwide ice age? Experimenting with the Earth climate system is a dangerous proposition.

CASE STUDY 4.5   Permian Extinction Event

Catastrophic Global Warming

The public concern about global climate change may not be an overreaction. The great Permian extinction event was the greatest mass extinction of life ever in Earth history and it was probably the result of climate change that cascaded into a global catastrophe (Figure 4.18). It resulted from collapse of Earth's biogeochemical systems and caused the extinction of 96% of marine life and nearly 75% of terrestrial life. Erupting volcanoes were the ultimate cause of the catastrophe but a specific set of global conditions facilitated the extinctions.

The Permian period was at the end of the Paleozoic Era. When the Permian ended about 245 million years ago, there was only a single supercontinent Pangea and a single superocean Panthalassa (Figure 4.18a). Circulation in the ocean was weak and slow. A major volcanic province developed first with the Emeishan basalt province of South China which covers about 96 525 miles^2 (250 000 km^2). The Siberian flood basalt province or "Siberian Traps" immediately followed this activity and it filled the West Siberian Basin covering about 1.5 million miles^2 (3.9 million km^2), or 15 times the area of the United Kingdom (Figure 4.18b). The province is made up of 2.2-miles (3.5-km)-thick basalt

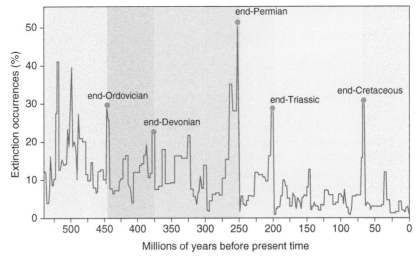

(a)

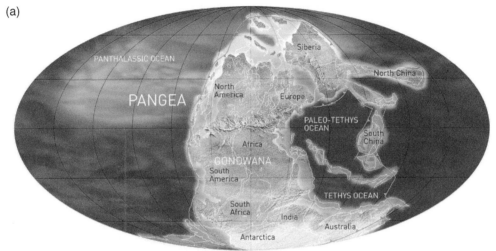

(b)

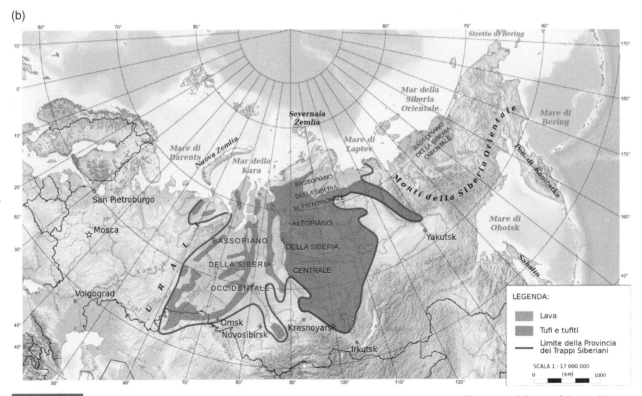

FIGURE 4.18 Graph showing the major mass extinctions and their intensity over the past 600 million years. (a) Map of the world at the time of the great Permian extinction showing Pangea. *Source:* NASA (2018), Recovering from the End-Permian Mass Extinction/ Public Domain. (b) Map of the area of Permian volcanism including the China and Siberia volcanic provinces. *Source:* Wikimedia Commons/ CC BY 3.0. Retrieved from https://commons.wikimedia.org/wiki/File:Extent_of_Siberian_traps-it.svg.

and pyroclastics intruded by gabbro. The volume of magma in this sequence was about 0.3–0.6 million miles³ (1.2–2.5 million km³).

Extrusion of the Siberian Traps released enough CO_2 alone to cause major global warming. One of the 96-miles³ (400-km³) basalt flows released more than 7 gigatons (6.4 gigaMT) of CO_2 in a decade. The entire Siberian Traps released about 11 000 gigatons (10 000 gigaMT) of CO_2. This is an equivalent of nearly 5000 ppm of CO_2 being added to the atmosphere compared to the current atmospheric concentration of 416 ppm The volcanism and CO_2 release occurred over 165 000 years, which is geologically quick. The Siberian volcanism alone appears to have doubled atmospheric CO_2 causing a global temperature increase of 2.7–8.1 °F (1.5–4.5 °C) in addition to the volcanism in China. Increased ocean temperatures caused the release of methane that was trapped on the ocean floor.

Volcanism also released toxic gases including chlorine, sulfur, and fluorine which produced acid precipitation. This would have stressed the Earth's biotic systems. The incoming solar radiation was also reduced as a result of aerosols and volcanic haze. This geochemistry reduced oxygen solubility in the seawater and released methane hydrates from the seafloor, reducing the global carbon sinks. The slow ocean circulation allowed unhealthy areas to build up without being dissipated. Early algal blooms were followed by oxygen-consuming bacteria caused large areas of ocean hypoxia to anoxia. The result was death of many marine organism species and later, all marine life was catastrophically impacted. Only organisms that were able to tolerate low oxygen conditions survived this. These catastrophic marine extinctions occurred in a short 10 000–30 000 years meaning that the Earth's biologic systems collapsed and the regulatory systems failed.

The cause of terrestrial extinctions is less certain but it shows a strong interdependence of the marine and terrestrial biologic systems. The reduced sunlight and acid precipitation stressed and reduced terrestrial plant life. Before this, vegetation was in its greatest boom in Earth history. The lack of food from the oceans caused a mass extinction of insects. Vegetation and insects were the primary diet of the larger animals at the time. Without food, many species went extinct, especially the larger herbivores which depended on the vegetation. The changing of the global ocean–atmosphere system chemical balance like what is happening today caused this catastrophe and provides a good lesson for humans.

4.5 What Can I Do About It?

The main way to reduce CO_2 emissions is to reduce the amount of fossil fuel burned. In transportation, the best solution is using mass transit as much as possible. A train has an equivalent efficiency to cars getting over 400 miles per gallon (644 km/L). Otherwise, people can buy more efficient, electric or hybrid vehicles. In addition, trips can be reduced by combining errands, eliminating pleasure driving, and carpooling. Working remotely from home also reduces automotive use. Otherwise, keep the tires inflated and the car tuned and drive conservatively.

Keeping the house cooler in the winter and wearing a sweater and keeping the house warmer in the summer reduces energy consumption. Pulling down shades keeps heat out. Turn off lights if not needed, unplug power bricks, and power down computers and other devices. Plant trees around your house to consume CO_2 and reduce paper and wood usage. Install solar panels on roofs, if possible. Solar panels shading areas that could have more vegetation should not be installed. Vegetation removes CO_2.

The best way to heat and cool your house to reduce CO_2 is using a cold-wet geothermal system. These systems circulate water through pipes underground where the temperature is a constant 55 °F (12.8 °C) or thereabout depending upon the area. The water is pumped into the home heating/cooling system as a radiator and provides a starting point for the temperature. This temperature is great for summer cooling and blowers are all that are needed to cool the house. In the winter, heating from 55 °F to the desired temperature is much better than heating from freezing outdoor air. The problem with geothermal systems is that they are expensive and can be difficult to retrofit onto an older house. However, they pay for themselves in time through utility bill savings.

The other ways to reduce CO_2 are probably less popular. Recreational use of gasoline, and especially in boating, consumes a lot of energy without much gain. Power boats are measured in gallons per mile rather than miles per gallon in comparison to sail or row boats which consume no gasoline. Unless used for work, jet skis, snow mobiles, quads, and dirt bikes consume a lot of energy with no real benefit and can be reduced. Driving off road vehicles in the woods instead of walking or riding a mountain bike is also less helpful. Tightly cropped lawns consume less CO_2 than larger plants. Broad grassy areas can be allowed to grow larger, be replaced with plantings, or be allowed to be overgrown by larger, native plants.

On a larger scale, local and government officials can be encouraged to invest in clean energy. Companies where you work can also be encouraged to invest in solar panels and energy-efficient equipment. Another potentially unpopular effort is encouraging more use of nuclear power. It produces no CO_2 and even small plants produce a lot of power. Accidents are rare and new efficiency efforts reuse spent fuel minimizing waste.

Natural Pollution

CHAPTER OUTLINE

Words you should know:

Dust storms – Densely concentrated windblown dust driven by a weather change that is at unhealthy levels.

Hot springs – Highly mineralized groundwater that is driven to the surface by underground heat.

Particulate – Solid material and liquid droplets suspended in the air. TPM is total particulate suspended matter between 100 and 0.1 µm; PM10 is particles between 10 and 2.5 µm; PM2.5 includes particles between 2.5 and 0.1 µm.

Protozoa – Microorganisms of a distinct type that can cause health issues in drinking water.

Radon – A naturally occurring, colorless, and odorless radioactive gas that is a health hazard if concentrated in indoor air.

Reactive minerals – Fresh, unweathered minerals that readily chemically react with surface and groundwater and precipitation.

Smoke – Clouds of carbon-rich particulate produced by burning.

Volcanoes – Vents on the surface that release damaging lava, particulates, and gases to the environment.

5.1   Pollution from Naturally Occurring Materials

Although most people think of the greatest pollution threats to be from humans, there are quite significant pollution threats to air, soil, and water that are natural. In many cases, human activity has exacerbated these threats. Many of the naturally occurring threats are included in the chapters that cover the area and situation where they occur. Also, some of the greatest threats to water, air, and soil quality are from natural disasters. These are described in that chapter though they are recognized here for their devastating impact.

The most important point is that most human-made pollutants are from naturally occurring substances. Heavy metals occur naturally in rocks and minerals that are exposed at the surface. However, the surface weathering removes and dilutes them enough to make them less dangerous. If these rocks and minerals are mined, they have fresh surfaces with potentially high concentrations of dangerous compounds that are readily released to the environment and can produce serious health hazards.

Crude oil is also a naturally occurring substance that can reach the surface naturally through seeps. This is the source of the famous La Brea tar pits in California. However, fresh oil pumped from a well contains its most toxic substances in maximum concentration. Most of the more toxic components are volatile organic compounds (VOCs) and evaporate away with time. This is why fresh oil spills are so deadly. Most of the dangerous organic pollutants we face are refined from crude oil.

Radiation and radioactive materials are considered to be among the most hazardous pollutants. However, the vast majority of rocks, soils, and sediment on the Earth produce radiation and contain radioactive materials. Besides indoor radon, unless concentrated by some process, almost all of them are completely safe and nothing to be bothered about. It is only the concentration and refinement of them by humans that makes them so dangerous. Even the sun and outer space produce radiation that can be dangerous with high exposures. Some people guard against these dangers by limiting exposure to the sun or wearing protection.

These are just three examples but additionally, wildfires, dust from deserts, salt spray from oceans, rising sea level, bacteria and molds, and some plants can produce dangerous pollutants that are unrelated to natural disasters or direct human activity. This chapter will focus on these situations.

5.2   Natural Water Pollution

Surface water can be contaminated to the point of causing health hazards by multiple sources. The most prevalent of these are diseases caused by the activity of humans and animals. Using surface water for waste from bodily functions can introduce fecal coliform, E. coli, typhoid, and cholera to the system and cause local problems to epidemics. Most of these microorganisms are bacterial and can be transmitted from one or group of animals or humans to surface water and thereby impact other groups. If animals die in surface water, the decomposing corpse can contribute other diseases that can infect humans and animals. These diseases can be even more dangerous. Even if the fecal matter and corpses are on land, heavy rain can wash these bacteria into surface water bodies, including drinking-water reservoirs. In most cases, the water decontamination plants are equipped to remove the contaminants. If they cannot, a caution to boil water can be issued. However, smaller water systems can be more readily overwhelmed. Private wells, cisterns, and well houses are generally not decontaminated and subject to more easy and common bacterial contamination.

FIGURE 5.1 Two photos of mineral pools at Yellowstone National Park, Wyoming. *Source:* Courtesy of US National Park Service.

More recently, private and public water supplies have been found to contain dangerous protozoa contamination. These microbes cause sickness and can even be fatal. Two common protozoa in water supplies that cause sickness are giardia and cryptosporidium. These two compromise water quality in most countries including the United States. Cryptosporidium is particularly difficult to eradicate. They form microscopic cysts that even can resist boiling for as long as two minutes. Wells that draw water from fracture systems that reach the surface can yield relatively unfiltered water.

Natural rock and soil can contain excessive levels of pollutant depending upon composition. Mining prospects are identified by high concentrations of target metals or other elements usually at or near the surface, many of which are health risks. Water that filters through this material can dissolve and carry a number of these health risks, meaning that groundwater in the area can also be enriched in them. If the elements are heavy metals, the groundwater in the area could be tainted. Fresh volcanic deposits are especially reactive and prone to leaching. Hot springs can contain exceptional amounts of dissolved dangerous compounds of all types. For example, some contain enough sulfur to yield naturally very highly acidic water. Hot springs in Pozzuoli, Italy have pH of 1.5.

A good example of hot springs are the thermal waters at Yellowstone National Park, Wyoming (Figure 5.1). These waters have pH from super acidic at less than 1 to very basic at 10. With surface temperatures up to 199 °F (93 °C), they can contain very high concentrations of chloride, sodium, silica, hydrogen sulfide, sulfate, alkalinity, and arsenic in solution. The acidic waters are termed "acid-sulfate" and it is meteoric water (from rain or snow) infused with high-temperature carbon dioxide and hydrogen sulfide gas. These waters contain high sulfate concentrations and low chloride concentrations. Acid-sulfate waters can contain high concentrations of ammonium and mercury.

The other situation that can naturally produce contaminated water is active tectonism. Rapid uplift or subsidence exposes fresh rock surfaces that can contain highly reactive minerals, some of which can contain toxic compounds. Examples of fresh surfaces can be from fault scarps, avalanche scars, alpine glacial scars, and even just rapid weathering and erosion from active uplift. A famous case of naturally occurring inorganic pollutants from erosion of rapid uplift was in the Himalayas, in which sediments were spread over a large area of Bangladesh. Arsenic-rich sediments occur in large areas of the country, which contaminate the groundwater system. Arsenic poisoning in the country is severe, with many serious health effects including blackfoot disease affecting large numbers of residents. It is estimated that 35–77 million Bangladeshis are at risk for arsenic poisoning.

Enhanced natural groundwater pollution occurred in New Jersey in a tectonically inactive area. Residents of central and southern New Jersey draw their drinking water from Atlantic Coastal Plain sediments. Through routine testing, 600 wells were found to have elevated levels of mercury. There were no obvious sources for pollution such as landfills or manufacturing plants. It was found that infiltrating acidic water was encountering a widespread mercury-rich clay in the area. Exchange reactions released naturally occurring mercury into acidic groundwater.

CASE STUDY 5.1 Bangladesh Arsenic in Soil and Groundwater

Natural Water Pollution

Bangladesh is in a low-lying area with 90% less than 30 ft (9.1 m) above sea level, The Ganges (Padma), Brahmaputra (Jamuna), and Meghna rivers converge into the Bay of Bengal in Bangladesh forming extensive marshes and jungles in the delta deposits (Figure 5.2a). The monsoon season, which lasts from June through October, provides extensive rainfall to the region. Bangladesh has a population of 167 million people, making it the eighth most populous and the ninth most densely populated country in the world.

(a)

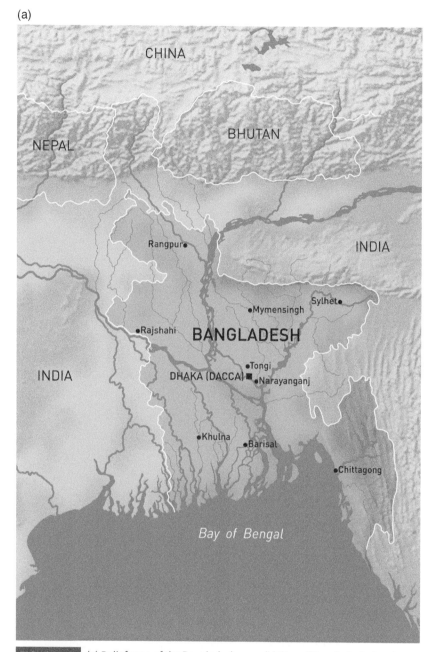

FIGURE 5.2 (a) Relief map of the Bangladesh area. (b) Map of Bangladesh showing the percentage of wells with arsenic concentrations above the limit. (c) Bangladeshi man suffering from arsenic poisoning through water. *Source:* Courtesy of the USGS. (d) Map of arsenic concentrations in soils in the continental United States *Source:* USGS / Public Domain.

The soil of Bangladesh is filled with organic nutrients and irrigated by monsoon rains, making it very fertile. As a result, the principal occupation is farming for about two thirds of the population. The country is very poor, with the per capita annual income at about $2500. The country is also regularly beset by natural disasters like droughts when the monsoons do not come and floods from some of the deadliest cyclones in history. This makes life in Bangladesh difficult. The United Nations has made several attempts to improve economic conditions in Bangladesh, which is one of the largest recipients of their aid.

Bangladesh has long struggled with providing clean drinking water to its residents. When it became a country in 1971, the rural population relied primarily on ponds, streams, and small creeks as its source of drinking water. This surface water, however, is prone to bacterial contamination from both human and animal waste. As a result of waterborne cholera and dysentery, 250 000 children were dying annually. To address this, the United Nations Children's Fund developed a program to provide safer, more bacterially free, groundwater sources as a replacement for the surface water sources. About 8 million tube wells were installed in the shallow aquifer systems of the central and southern parts of Bangladesh. These tube wells or driven wells, were made by pressing or hammering a perforated PVC drainage pipe into the ground until it reached the groundwater. The pipes are 4 to 6 in. (10.2–15.2 cm) in diameter and as much as 20 ft (6.1 m) long. They were installed without specialized drilling equipment, using just a metal tripod, a pulley,

(b)

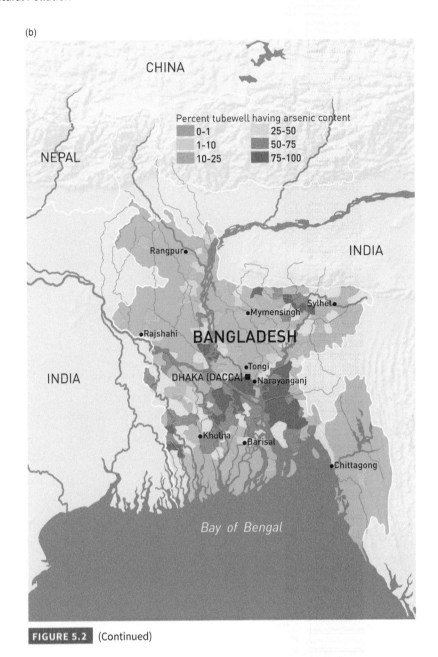

FIGURE 5.2 (Continued)

and some weights. The perforated pipe has a sharp, hardened tip that can be driven into the soil using drive weights as hammers and additional sections can be added to reach the desired depth.

When the tube reached the desired depth, a small pump was installed to bring water to the surface. These pumps operate on a treadle system so a person simply steps on the pedals and "walks in place" to bring water to the surface. More than 6 million wells were installed in Bangladesh in the 1970s and 1980s, and by the late 1990s more than 90% of the population was drinking well water. The wells were transformative for Bangladesh's agricultural and public health systems. Irrigation was more reliable and, as a result, crop yields increased dramatically. Wells were installed in villages and farms where needed, eliminating the need to transport it. Infant mortality from waterborne diseases plummeted.

Although the tube wells seemed to be the answer to Bangladesh's dilemma, they actually caused another problem. The Ganges (Padma), Brahmaputra (Jamuna), and Meghna rivers eroded rocks and sediments in the highlands of Bangladesh which are rich in arsenic-bearing minerals. The rivers carried arsenic-bearing sediments in addition to the fertile soil (Figure 5.2b). These sediments were deposited in deltas and floodplains and the arsenic was leached into the groundwater system. Even low levels of arsenic cause significant health effects, including skin lesions, called blackfoot disease, central nervous system damage, cancer, and death (Figure 5.2c). The World Health Organization (WHO) and US Environmental Protection Agency (EPA) established the limit of arsenic in drinking water at 10 parts per billion (ppb). The Department of Environment in Bangladesh established a much higher limit of 50 ppb for drinking water, which is deemed dangerous by most other countries. Even at this unsafe limit, more than 20% of the drinking water in Bangladesh contains arsenic concentrations that are deemed unsafe. There have been reports of arsenic in drinking water in excess of 2700 ppb, and it has been estimated that 20–60 million people regularly drink water with levels greater than 50 ppb.

The solution proposed for the chronic arsenic poisoning or arsenicosis in Bangladesh is to encourage a diet high in protein, preferably from meat and vitamins to flush the arsenic from the system. The problem with this is that the rural poor in Bangladesh has the most elevated arsenic levels in drinking water and their diets are low in protein, much less meat protein, because it is expensive and difficult to preserve. The United Nations estimates as many as 20 000 people per year will likely die from the effects of drinking arsenic-contaminated water in Bangladesh.

(c)

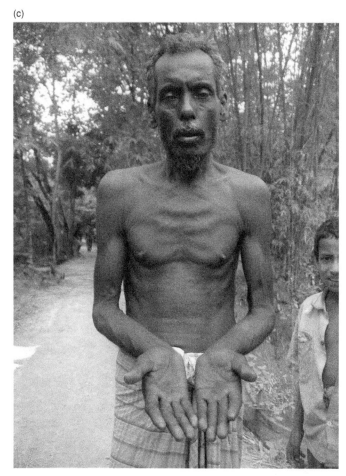

(d)

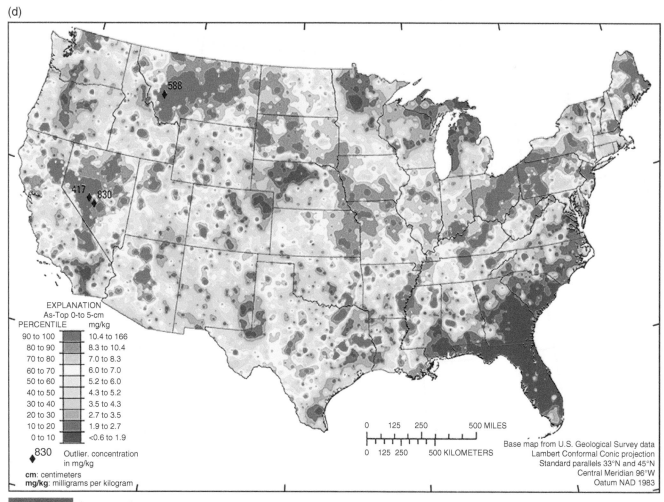

EXPLANATION
As-Top 0-to 5-cm

PERCENTILE	mg/kg
90 to 100	10.4 to 166
80 to 90	8.3 to 10.4
70 to 80	7.0 to 8.3
60 to 70	6.0 to 7.0
50 to 60	5.2 to 6.0
40 to 50	4.3 to 5.2
30 to 40	3.5 to 4.3
20 to 30	2.7 to 3.5
10 to 20	1.9 to 2.7
0 to 10	<0.6 to 1.9

◆ 830   Outlier. concentration
        in mg/kg

cm: centimeters
mg/kg: milligrams per kilogram

0   125   250   500 MILES

0   125 250   500 KILOMETERS

Base map from U.S. Geological Survey data
Lambert Conformal Conic projection
Standard parallels 33°N and 45°N
Central Meridian 96°W
Oatum NAD 1983

FIGURE 5.2 (Continued)

Widespread naturally occurring arsenic in groundwater also occurs in Taiwan, Chile, Vietnam, and Switzerland among others. There is band of elevated arsenic-contaminated soil and water across the United States from New York/New Jersey to Montana through Pennsylvania, Ohio, the Dakotas, and parts of many states in addition to other isolated elevated areas in Maine and Nevada among others (Figure 5.2d). In these areas, as many as 50% of the water wells can exceed the 10 ppb arsenic limit established by the EPA. This exposes many residents to potential health threats from arsenic exposure.

5.3 Natural Soil Pollution

The arsenic in ground problem in the United States applies to the soil in the same area. This coincidence of natural pollutants both in soils and groundwater is common in many areas. Volcanic soils can contain numerous natural pollutants depending upon the rock type. Many of the heavy metals are less soluble than the other components of dark-colored volcanic rocks such as basalt and can be concentrated in residua. Ash is also more reactive in percolating water because of its high porosity and permeability which can result in enriched unhealthy compounds. The same is true in fresh rock and sediments in rapidly uplifting areas or glacially eroded areas just like they are in groundwater.

Deeply weathered terrains also concentrate certain elements. Laterites are the end-product of deep weathering in tropical areas. Most elements are dissolved away but aluminum is left in residuum because it is more stable in those conditions. The soil is so aluminum rich that it can be compacted into bauxite and be mined. Aluminum is usually not dangerous but in certain situations it can be undesirable.

Finally, the top layer of soil is organic-rich in many areas. As such, it is typically highly reactive. As water percolates through it, any pollutants can react with it and form compounds. The water has much fewer pollutants after it passes through. This is how a drinking-water filter works except that it uses activated charcoal as the filtration material. With time, the organic-rich soil can become contaminated with these compounds and elements. There was a case in Maine where a filter was put in a home water supply system to remove radioactive elements. The technicians did not calculate the amount of radionuclides that would be removed with time, and in a short period the filter became so contaminated that it qualified as high-level nuclear waste. Fortunately, in most natural systems this does not occur.

5.4 Natural Air Pollution

There are numerous instances of natural air pollution both similar to natural pollution of water and soil and in separate processes. Certainly, volcanoes can cause unprecedented air pollution that can cause major health impacts as well as climate change. There are numerous examples of such catastrophic events. Evaporation of volatile components from oil seeps or tar pits but also from certain trees can also cause local air pollution. Certain microbes can be airborne in addition to waterborne and impact health.

Perhaps the worst of the natural air pollutants is particulate. Much of the worst particulate pollution is anthropogenic but natural particulate can be very dangerous. Particulate is designated as one of the six "criteria" air pollutants by the EPA, primarily because of human-generated material. Particulate is divided into types, and it has been subject to more stringent regulations. It is responsible for an estimated 65 000 deaths per year in the United States and as many as 200 000 deaths per year in Europe. This translates to a 17% mortality risk increase. New legislation has been enacted to address this but urban and suburban residents in the United States with respiratory or cardiac problems must be cautious about exposure to particulate pollution.

About 99% of inhaled particulate is exhaled or trapped in the upper respiratory tract and exhaled. Coarse particulate can lodge in lung tissue and slow oxygen and carbon dioxide exchange in the blood, causing shortness of breath. Fine particulate can be absorbed into the bloodstream. This is very dangerous to people with heart or respiratory diseases like emphysema, bronchitis, chronic obstructive pulmonary disease (COPD), and asthma. Long-term exposure to particulate risks lung diseases such as emphysema, pulmonary fibrosis, bronchiectasis, cystic lungs, and lung cancer.

Any suspended solids or small liquid particles in air constitute particulate. They can be composed of any of hundreds of elements and compounds. Some of these may be hazardous including heavy metals and asbestos. Particulate can be total particulate matter (TPM), which is all suspended matter between 100 and 0.1 μm. PM10 include particles between 2.5 and 10 μm, and PM 2.5 are less than 2.5 μm. The main reason is medical as PM10 are respirable and PM 2.5 are absorbed directly into the bloodstream through the lungs.

Natural particulate includes rock and soil dust, sea salt, pollen, mold, fungi, yeast, bacteria and viruses, smoke ash, volcanic ash, and droplets of aerosols from biologic activities and volcanoes. Particulate of 50–100 μm is heavy and settles close to the source or is removed by precipitation. Coarse and fine particulate can be transported great distances by wind before settling. Particulate of volcanic ash is primarily glass, which is highly abrasive. Inhalation can cause scarring of the lung tissue and upper respiratory system as well as damaging eyes. These, however, are not common.

Smoke is also composed of particulate. In contrast to volcanic ash, smoke from burning is far more common both locally and on the large scale with the increase in wildfires in recent years. These fires are commonly caused by humans though they can be from lightning. Climate change has caused some of the drying that is leading to more fires, which is also human-induced. It is estimated that particulate is 30–50% organic and elemental carbon as the result of burning (soot) but this includes industrial, incineration, heating, personal, and transportation sources as well.

The blowing of sand and dust from deserts and other arid areas causes pollution of soil, water, and air (Figure 5.3). Sand is eroded from an exposed surface onto productive land polluting the soil. In some cases, whole fields of sand dunes can migrate over productive areas and drown the soil in sand. Wind lifts sand and dust from deserts, damaging air quality with particulate. The particulate can cause severe health effects. Devastating dust storms are common near deserts and growing more common with human impact. These storms can cover large areas and even cross oceans. Windblown dust from the Sahara Desert regularly crosses the Atlantic Ocean from Africa to the Americas causing health effects there (Figure 5.4). On occasion, dust even crosses the Pacific Ocean from China to North America. Thick deposits of loess, or dust from deserts, show that this process has been operating for a long time. China contains thick loess deposits from the Gobi Desert.

FIGURE 5.3 Dust storm approaching a town. *Source:* Courtesy of NOAA.

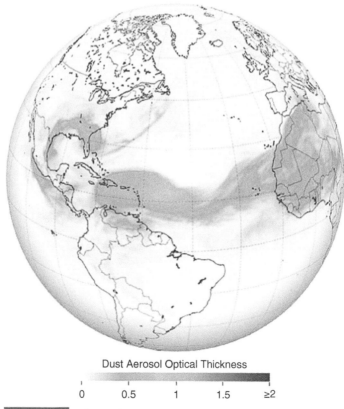

Dust Aerosol Optical Thickness

0   0.5   1   1.5   ≥2

FIGURE 5.4 Saharan dust crossing the Atlantic Ocean and impacting the Caribbean and southern United States. *Source:* NASA.

CASE STUDY 5.2 2021 Mongolia-China Dust Storm

Natural Source of Particulate

The most notable dust storm is the 1935 Black Sunday storm during the Dust Bowl of the Midwestern United States but it is described in Chapter 16. Although dust storms are naturally occurring pollution events, they have been exacerbated by human activity. This was certainly the case in the 2021 Mongolia-China Dust storm. Dust is particulate that has been swept into suspension by the wind. Particulate includes solid and liquid particles suspended in air. They can be composed of hundreds of elements and compounds, some of which may be hazardous.

As forecasted by the weather service, a major continental cyclone developed over Mongolia and produced violent winds from 14 to 15 March 2021 (Figure 5.5a). This major storm followed an extended warm and dry period that left loosened dry soils on the surface

(a)

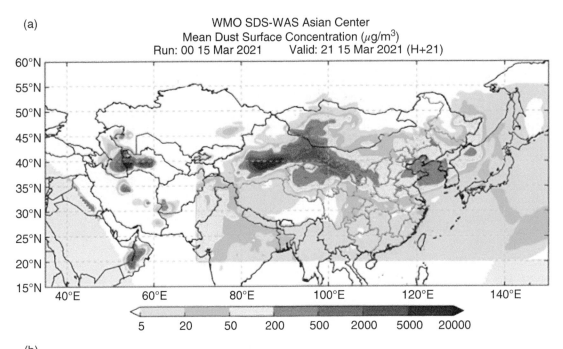

(b)

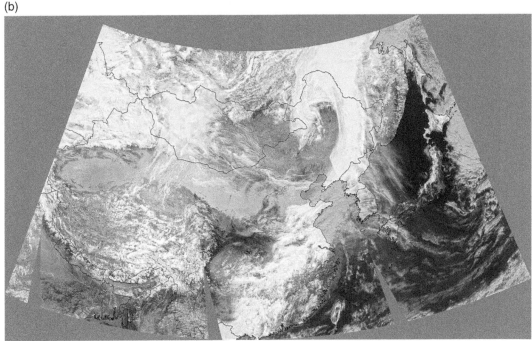

FIGURE 5.5 (a) Diagram showing the dust concentrations across Mongolia and China during the 2021 dust storm. *Source:* World Meteorological Organization. (b) Satellite image showing the band of airborne dust in tan across China and Mongolia (middle of map). *Source:* Courtesy of NASA. (c) Photograph of the 2021 dust storm entering a town in Mongolia *Source:* Greenpeace/Wikimedia Commons/CC BY 2.0.

(c)

FIGURE 5.5   (Continued)

in addition to the dust and sand of the huge Gobi Desert. The deep cyclone produced wind speeds of 40–76 miles per hour (18–34 m/s) in Uvurkhangai, Umnugovi, and Bulgan provinces, 49–90 miles per hour (22–40 m/s) in Dundgovi province and 36–63 miles per hour (16–28 m/s) in Arkhangai, Bayankhongor, Govi-Altai, Dornod, Tuv, Khentii, Sukhbaatar, and Dornogovi (Figure 5.5b). The Mongolian Government classifies wind speeds exceeding 40 miles per hour (18 m/s) as a disaster and speeds exceeding 54 miles per hour (24 m/s) as a catastrophic phenomenon.

After the winds passed through Mongolia, they continued southeastward into China, north of the Yangtze River, covering 600 miles (1000 km). They continued through the week before dissipating. The Chinese Meteorological Administration had issued a yellow alert on 19 March before the storm impacted the northern Chinese provinces of Shanxi, Liaoning, and Hebei. Citizens were advised to stay home if possible. When the winds reached Beijing on Sunday, air pollution levels were already at 500 ppm, the maximum allowable level in China. The sun appeared blue in Beijing as the dust storm arrived, and it left visibility near zero as it struck. As a result, schools canceled all outdoor events, and people with respiratory diseases were advised to stay indoors. More than 400 flights at Beijing's main airports were canceled.

Compared with the World Health Organization classifications of PM 10 levels of 0–54 micrograms per cubic meter ($\mu g/m^3$) for good, 55–154 for moderate and >155 for high, the levels in this storm were extreme (Figure 5.5c). Hourly PM10 concentrations exceeded 1500 $\mu g/m^3$ in the cities of Zhengzhou, Baoding, Tianjin, Luoyang, and Zhoukou. PM10 exceeded 2000 $\mu g/m^3$ in Handan. PM10 concentrations exceeded 3900 $\mu g/m^3$ in Hohhot, Lanzhou, Alashan, Changzhou, and others. On 15 March they exceeded 7000 $\mu g/m^3$ over the provinces of Gansu, Inner Mongolia and Ningxia, in the cities Ordos, Jinchang, Wuwei,

and Zhongwei. The peak PM10 in Beijing exceeded 9000 $\mu g/m^3$. Satellite images showed that the impacted areas exceeded 173 745 square miles (450 000 km^2). Data also showed that the dense layer of dust extended to an altitude of 5 miles (8 km).

The storm caused flight cancelations and delays in northern China, and in Inner Mongolia. At Chifeng Airport, more than 60% of flights were canceled and 50% were canceled at Baotou Airport one day. When the dust storm first struck, 341 people and 1.6 million livestock went missing. In the end, there were 10 human deaths but all of the livestock were lost, many of them were buried in the sand.

Since 1992, Mongolia has enjoyed high economic growth as the result of mineral and agricultural exports. As a result, overgrazing, gold, coal, and copper mining, and climate change have resulted in severe land degradation. Mean temperature in Mongolia increased by 4 °F (2.24 °C) between 1940 and 2015, and annual precipitation dropped by 7%. This resulted in higher aridity across the area with more than one quarter of lakes of 0.4 square miles (1.0 km^2) or more having dried out between 1987 and 2010. Another factor is overgrazing with the number of Mongolian livestock animals having tripled in the last 30 years. Mongolian goats produce more than 40% of the world's cashmere. The number of goats has increased from 5 million to 27 million and they eat twice as much grass as sheep. As a result, the Gobi Desert is growing, with the desert expanding into northern Mongolia at 75 miles (121 km) per year.

The March dust storm was the strongest in China in more than 10 years. It resulted from a combination of extreme weather, climate change, and environmental degradation. It was the first of a series of eight Mongolia-China dust storms through March, April, and May. These storms destroyed animal herds, worsened respiratory problems, and canceled numerous flights.

Environmental radon is an even more dangerous natural air pollutant. The US EPA regards radon as the most dangerous environmental hazard. On their 2007 Priority List of Hazardous Substances, radon ranks number 105 of the 275 most dangerous pollutants. Radon is a step in the decay series from uranium 238 to lead 206, as is radium 224 (Figure 5.6). The process of alpha recoil can eject the daughter radon 222 atom from radium atoms off the grain surface and into soil pore space. It can be a problem if there is a house with a basement over soil or sediment containing excessive radon in soil gas. Basements typically are not that tightly sealed, containing French drains, sumps, and cracks in the walls that allow soil gas into the house (Figure 5.7). In winter, the heated air in the house rises, which pulls air from the basement and draws soil gas through the walls and floor and into the house. Forced air heating and cooling systems can do this all year long.

Radon is radioactive and changes to another element in the decay series. The half-life of radon decaying to its daughter, polonium 220, (the time for one half of the mass of radon to change to polonium) is just 3.82 days. Polonium is a radioactive solid that reacts quickly. Polonium undergoes an alpha decay which emits a short-lived, heavy particle, producing 7.6 million eV in a few seconds. Alpha decays or particles are generally not dangerous. Alpha particles have four atomic mass units and are too large to penetrate glass or paper. They only travel a half inch (1 cm) before using up their energy as a result of their heavy mass. If they are adhered to lung tissue, however, the alpha decay can cause the lung cell to mutate. Lung tissue can dispose of random mutated cells but if there are too many mutated cells, the immune system can be overwhelmed, resulting in lung cancer.

The potential for radon being a risk to human health was first discovered in uranium mines in Sweden, where miners were found to have a greatly increased risk of lung cancer. The radon level in the Swedish mines was approximately 700 picocuries per liter (pCi/L), which was considered to be extreme compared to indoor levels. This has since been found to not be the case. Radon in indoor home air is measured using a charcoal canister or an alpha track. The most common is the 3-day charcoal canister test and it is always performed by building inspectors.

Radon can also occur in groundwater. It is released from the water if agitated and can contribute to indoor radon. Water radon poses a problem for homes relying on well water. Radon is released from running the tap, dishwashers, showers, and washing. Water can hold very high concentrations of radon. The record for water is 2.5 million pCi/L, but levels in the tens

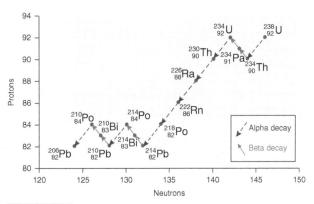

FIGURE 5.6 Uranium decay series showing the steps and daughter product elements from uranium to lead including alpha and beta decays. Radon is Rn-222.

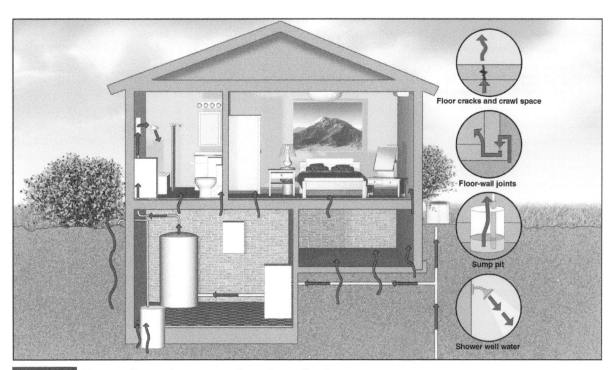

FIGURE 5.7 Diagram of how radon can enter a house from soil and water.

to hundreds of thousands are common. The EPA set an action level of 10 000 pCi/L for indoor water. One study found that people drinking water with moderate levels of radon were exhaling air with significant radon levels after a 20-minute metabolic period. The radon passed through their stomach and into their blood, circulated through their body, and passed out of their lungs. All organs and blood would be exposed to radioactive radon and its daughter products according to this.

CASE STUDY 5.3 Radon in Boyertown, Pennsylvania

Natural Indoor Air Health Hazard

Radon in indoor air was unheard of for decades. It was known to be common in uranium mines, where it was shown to increase lung cancer incidence, but there was no evidence that it could occur in indoor air. Radon is a naturally occurring colorless, odorless, and tasteless radioactive gas. It is one step in the 14 step decay series from uranium-238 to lead-206. Solid radium-226 undergoes an alpha decay to form radon-222 gas. If the radium is on the surface of a soil, sediment, or rock, the decay is powerful enough to expel the radon into the air or water that surrounds it. Radon is a noble gas so cannot react with the material it passes through or be filtered through standard techniques. It can pass through cracks in basement walls and accumulate in indoor air.

In indoor air, radon can circulate in any room. It can be inhaled but has no real health effects. However, radon decays relatively quickly (3.82-day half-life) to polonium-218, which is a solid and highly charged. If the radon is in a person's lungs when it decays, the polonium attaches directly to lung tissue. If the polonium attaches to dust and then is quickly inhaled, it can also attach to the lung tissue. The polonium decays within seconds in a powerful alpha decay to lead-214, which is also radioactive. In addition, the succeeding radon daughter products emit gamma radiation, which is damaging.

This health issue was historically only the worry of uranium miners because radon levels could be greater than 700 picocuries per liter (pCi/l). Most surface deposits have too little uranium to produce much radon and houses were too drafty to accumulate radon. This was the conventional wisdom. However, the oil crises of the 1970s compelled builders to seal houses as tight as possible to save energy without thinking about consequences.

On 19 December 1984 at about noon, the senior consulting health physicist at Philadelphia Electric Company's Limerick Nuclear power plant made a confused call to the Bureau of Radiation of the Pennsylvania Department of Environmental Resources. The plant was not yet operational and yet the radiation leak detector was being set off on a daily basis. They determined that every time one of the consulting engineers left the plant, he tripped the radiation monitor. The plant personnel swept him with a gamma spectrometer and found that he had enough gamma- and beta-producing radon daughters in his clothes that he was setting off the alarm.

No one had experience with residential radon, so no equipment was available to confirm the finding or measure the amounts. Philadelphia Electric Company engineers devised an instrument and brought it to the home of the consulting engineer in Boyertown, Pennsylvania (Figure 5.8). They measured indoor radon concentrations up to 20 working levels (100 pCi/l per working level of radon) or 2000 pCi/l in the home. These results astonished the Bureau of Radiation who took over the investigation. They urged an immediate evacuation of the house, investigated the house, and set up radon detectors on 26 December 1984. They returned on 2 January 1985 to find the family back in the house. The findings were 2535 pCi/l in the basement, 2441 pCi/l in the family room, and even 114 pCi/l in the garage. The current EPA maximum for indoor radon is 4 pCi/l. On 5 January a letter from Secretary of the Pennsylvania DER was sent to the family, strongly advising them to move out and they complied.

The engineers at the Philadelphia Electric Company took on the challenge of remediating the house. They excavated all around the foundation of the house. Once exposed, they sealed cracks, wrapped the foundation with an impermeable membrane, and sealed the sump. The next phase involved breaking up and removing the slab. A passive venting system of PVC pipes and stack was installed under the basement that vented the air outside. After the four phases of this remediation, the basement had 0.4 pCi/l and the family room had 0.9 pCi/l.

The house was located in the Reading Prong province of Pennsylvania, New Jersey, New York and Connecticut (Figure 5.8). This province is composed of crystalline rock that is locally rich in uranium. It became infamous for indoor radon, especially in Pennsylvania. About 40% of homes in the state have radon levels above 4 pCi/l but most impressive are the records for indoor levels. The 2600 pCi/l in Boyertown held the record in the United States until a house in Annandale, New Jersey measured 3500 pCi/l in 1988. This was the record until 2014, when a home in Lehigh County, PA had a concentration of 3715 pCi/L, more than 900 times the action level of 4 pCi/L. It was thought that this would be the record forever, but in October 2016 a vacated home in Center Valley, PA showed a level of 6176 pCi/l, more than 1500 times the limit. This is 10 times the level found in uranium mines.

Pennsylvania is not the only state with radon. The average indoor radon in the top 10 states is Alaska at 10.7 pCi/l, South Dakota at 9.6 pCi/l, Pennsylvania at 8.6 pCi/l, Ohio at 7.8 pCi/l, Washington at 7.5 pCi/l, Kentucky at 7.4 pCi/l, Montana at 7.4 pCi/l, Idaho at 7.3 pCi/l, Colorado at 6.8 pCi/l, and Iowa at 6.1 pCi/l. It is estimated that one out of every 15 homes in the United States has an elevated radon level. North American Prairie residential construction still displays a very high and worsening radon exposure. Recent construction is moving toward greater square footage, fewer storeys, greater ceiling height, and reduced window opening, all of which increase radon. Canada is no better with 6.9% of residents living in homes with radon levels above the national limit of 5.4 pCi/l. One study showed a 31.5% increase in radon levels in buildings constructed after 1992 compared to older buildings.

The EPA regards indoor radon as the greatest environmental health hazard. They estimate that 21 000 Americans die each year from radon-induced lung cancer. The US Surgeon General warned that radon is the second leading cause of lung cancer in the United States. In Canada, the estimate is 3000 deaths per year. For smokers, the risk

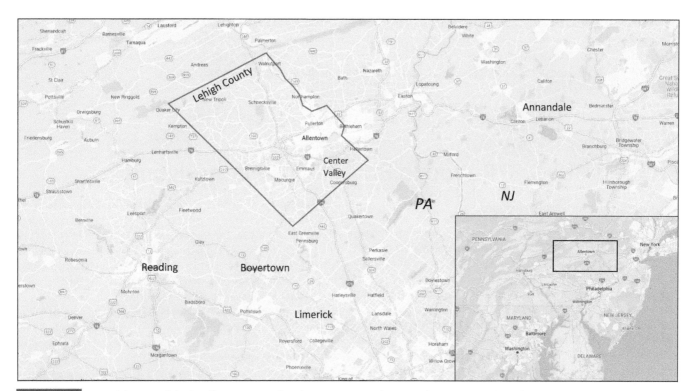

FIGURE 5.8 Map of the area of the Reading Prong province in eastern Pennsylvania and New Jersey where the record indoor radon concentrations were found. Inset map shows the location of the area in the Mid-Atlantic states.

is far greater because radon acts synergistically with smoking. As a result, smokers are at 25 times greater risk from radon impacts than non-smokers. It is not a problem restricted to North America. On a worldwide basis, radon is deemed the leading cause of lung cancer in non-smokers and the second leading cause in smokers and is responsible for an estimated 3–20% of lung cancer deaths. European studies estimate the risk of lung cancer increases by 16% per 2.7 pCi/L increase in radon concentration exposure.

CHAPTER 6

Environmental Perspective on Rocks and Minerals

Polluted Earth: The Science of the Earth's Environment, First Edition. Alexander Gates.
© 2023 John Wiley & Sons, Inc. Published 2023 by John Wiley & Sons, Inc.
Companion website: www.wiley.com/go/gates/pollutedearth

Words you should know

Asbestos – Needle-like mineral crystals that damage the lungs when inhaled.

Clastic rocks – Rocks formed by existing rocks eroded into particles that are then physically transported and deposited before being lithified.

Fracking – The opening and propping open of fractures in rock by pumping high-pressure fluids down wells to enhance production of oil or gas.

Igneous rocks – Rocks that are cooled or crystallized from melted rock, either magma if underground or lava if on the surface.

Metamorphic rocks – Preexisting rock that has mineral, compositional, and/or textural changes by changes in temperature, pressure, and/or fluids.

Minerals – Naturally occurring inorganic solid compounds or elements with a specific chemical composition and crystal structure.

Rocks – Naturally occurring solid masses or aggregates of one or more minerals, organic material, and/or mineraloid matter such as volcanic glass.

Sedimentary rocks – Rocks produced by the chemical and/or physical erosion of existing rocks, transport of the material, deposition of the material, and lithification.

6.1 Classification of Minerals

Minerals are naturally occurring inorganic solid compounds or elements with a specific chemical composition and crystal structure. Basically, they are solid forms of chemicals that can react the same way as those in a chemistry lab. The difference is that they do not include any organic molecules or any other form besides solids. However, the minerals can react with fluids, gases, or each other both at the surface or underground. These reactions produce different minerals.

The elements in minerals include those that are both safe and hazardous to human health. They include heavy metals such as lead, arsenic, chromium, mercury and others bonded to other elements or themselves to form minerals. Heavy metals are hazardous to human health. They can also contain radioactive elements such as uranium and radium among others and elements that can form dangerous compounds, such as sulfur which can form sulfuric acid. Even elements that are relatively benign in low concentrations can be concentrated in minerals to the point of being dangerous.

Some minerals are safe to humans and animals but unsafe for other forms of life. The best example is halite or table salt that humans eat daily with no adverse effects. On the other hand, salt is deadly to slugs, many microbes, and many plants.

Some minerals are hazardous not because of their chemistry but because of their form. Asbestos is actually a crystal form that is possible in several mineral types. Chrysotile is the most commonly used asbestos but crocidolite is the most dangerous. Asbestos can cause the lung cancer mesothelioma. Even common minerals like quartz can be broken into sharp fragments that can damage eyes and lungs if driven by the wind.

6.1.1 Chemistry

As bonded atoms of elements, minerals follow the rules of chemistry. Atoms are the smallest sample of an element and consist of a nucleus surrounded by an electron cloud (Figure 6.1). The nucleus contains protons and neutrons, usually in equal number. The number of protons is the atomic number and the mass of the neutrons plus protons is the atomic mass. If the neutrons and protons in the nucleus are not in equal number, it is an isotope of the element and has a different atomic mass. The protons have a +1 charge and must be balanced by electrons which have a –1 charge to keep a neutral charge on the atom.

Electrons fill orbitals around the nucleus like planets are in orbits around the Sun. However, the electrons must fill the orbitals in a specific geometry. The orbital closest to the nucleus can have two electrons and the next one surrounding it can have 8 electrons to fill it. Each orbital has a specified number of electrons to fill them. The number of electrons in the outer orbital is the basis for constructing the periodic table of the elements.

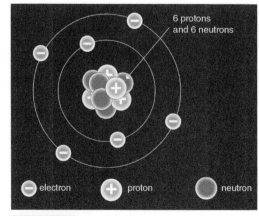

6 protons and 6 neutrons

electron   proton   neutron

FIGURE 6.1 Diagram of an atom with particles.

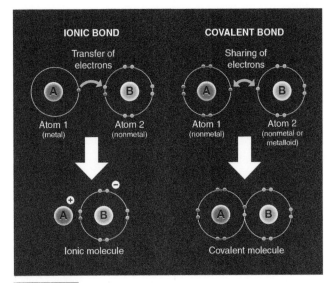

FIGURE 6.2   Diagrams showing ionic and covalent bonding.

If the outer orbital is almost full but missing an electron or two, it will act as an anion and take on a negative charge in reactions. If the atom has an electron or two in the outer shell, it will act as a cation and take on a positive charge in reactions. These reactions bond atoms together in compounds which are minerals. There are three types of bonds, ionic, covalent and metallic, and a force that can hold compounds together called Van der Waals forces (Figure 6.2). In ionic bonds, an anion is bonded to a cation by excess outer shell electrons filling those in the deficient atom to reach full electron orbitals. Halite or table salt is a good example, where cation Na+ is bonded to Cl-. In covalent bonds, outer shell electrons are shared to reach a stable configuration. Metallic bonding is where electrons form a cloud around bonded atoms. If an electron is added to one end of the cloud, another will eject out of the other end of the cloud. This is why metals can conduct electricity. These can involve inner shell electrons. In Van der Waals forces, excess charges on one side of bonded atoms in a compound attract excess charges on another side of bonded atoms.

The end result of bonding is to form a stable compound. Noble gases to the far right of the periodic table have all of their orbitals completely filled so they do not react or form compounds (Figure 6.3). The left side column next to noble gases is halogens, which are strong anions and missing one electron. They tend to bond ionically to the elements in the column to the far right, which are Alkali metals and containing one extra electron, or sometimes the next column to the left, which is the Alkaline-earth metals, which contain two extra electrons. The columns to the left of the halogens, topped by elements boron (B), carbon (C), nitrogen (N), and oxygen (O), tend to form anion groups and be involved with covalent bonding to Alkali, Alkaline-earth, and transition metals to form compounds. The transition metals can also bond to atoms of the same element in metallic bonding. There are plenty of exceptions to these generalities.

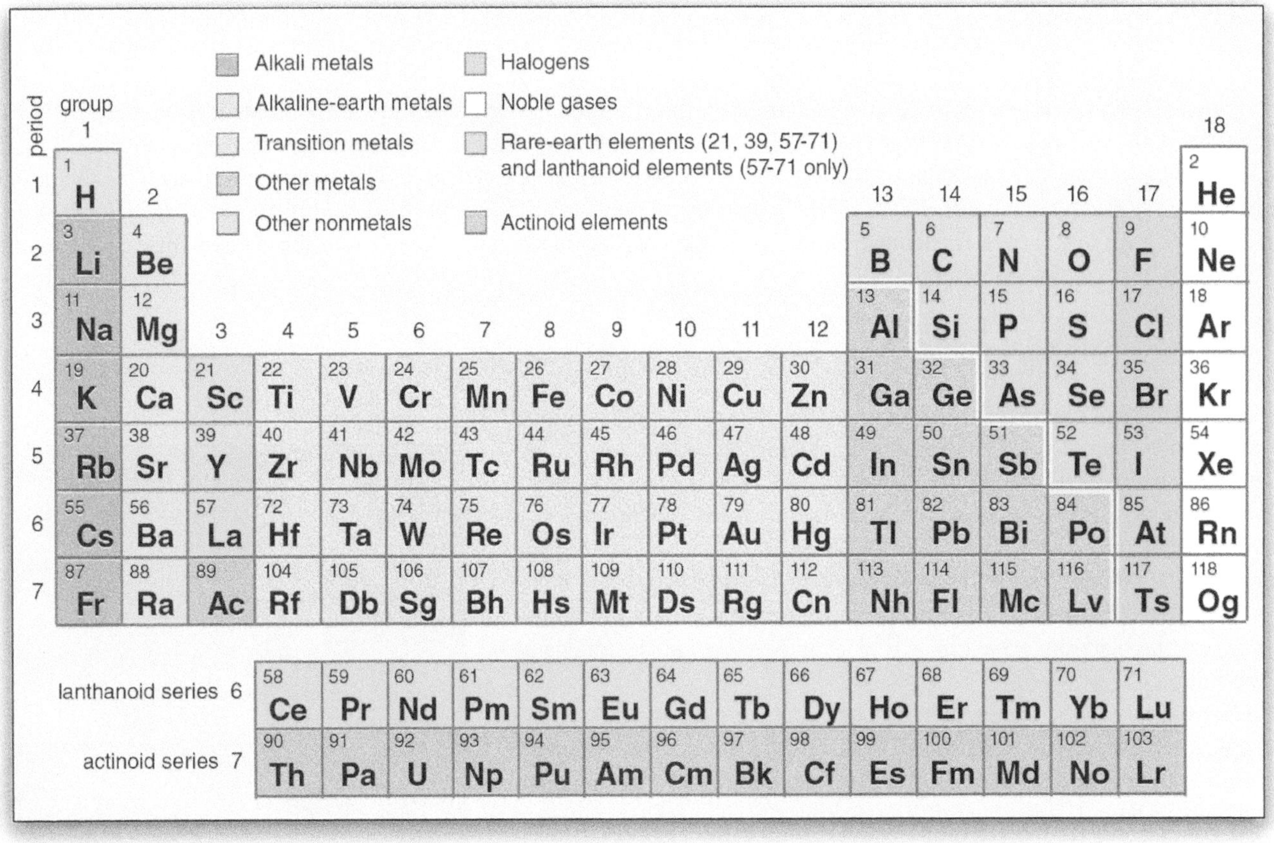

FIGURE 6.3   Periodic table of the elements including the classes of elements based on the filling of electron shells.

All minerals are classified by the anions or anion groups that they contain. There are seven major groups of minerals:

Elements – Minerals composed of single elements bonded together such as diamond, native copper, etc.

Oxides – Minerals made of cations bonded to oxygen like magnetite, hematite, etc.

Sulfides – Minerals made of cations bonded to sulfur like galena, pyrite, etc.

Halogens – Minerals made of cations bonded to elements from the halogen group such as halite, apatite, etc.

Sulfates – Minerals having cations bonded to sulfur-oxygen (sulfate) anions like gypsum, etc.

Carbonates – Minerals having cations bonded to carbon-oxygen (carbonate) anions like calcite and dolomite.

Silicates – Minerals having cations bonded to silicon-oxygen (silicate) anions like quartz and feldspar.

By far the most common of these groups is silicates, which make up 90% of all minerals. There are so many that they too are subdivided for classification. Silicon is bonded in a geometry called a tetrahedron. The number and geometry of silica tetrahedra bonded together defines the various groups (Figure 6.4):

Single – Only one tetrahedron is bonded in the basic compound; olivine is an example.

Double (pairs) – Two tetrahedra are bonded in the basic compound.

Chain – Chains of tetrahedra are bonded in the basic compound. There can be single chain silicates like pyroxene and double chain silicates like amphibole.

Ring – Rings of tetrahedra are bonded in the basic compound such as tourmaline.

Sheet – Sheets of tetrahedra are bonded in the basic compound such as micas.

Framework – Three-dimensional frameworks of tetrahedra are bonded in the basic compound such as quartz and feldspar.

6.2 Classification of Rocks

Rocks are naturally occurring solid masses or aggregates of one or more minerals, organic material, and/or mineraloid matter such as volcanic glass. This is far less stringent a definition than that for minerals. There are three rock categories: igneous, sedimentary, and metamorphic but all rocks are classified by texture and composition. Texture is the size of grains and, in some cases, their orientation. Composition is either based on their mineral contents or by their bulk chemistry determined analytically. There are many features of rocks that impact the environmental controls on an area.

Igneous rocks are cooled or crystallized from melted rock, either magma if underground or lava if on the surface. It is hot underground and magma crystallizes slowly, producing mineral grains that are large enough to see. These are intrusive rocks forming plutons, which are bodies of igneous rock. It is cool at the surface and lava cools quickly as a result. Quick cooling does not give minerals time to grow or even crystallize in some cases and forms very fine-grained, extrusive rocks. The textural term for coarse-grained, plutonic rocks is phaneritic and it is aphanitic for fine-grained extrusive rocks. If the magma partly crystallized before it was extruded from a volcano, it will have a mix of grain sizes and be termed porphyritic.

The compositional terms for igneous rocks are mafic if they are rich in magnesium (Mg) and iron (Fe) silicates, felsic if they are rich in feldspar and quartz (silica), and intermediate if they are somewhere between the two. Table 6.1 shows the classification system for igneous rocks including major minerals and relative color. For example, granite is plutonic (coarse-grained), light gray or pink in color, and composed primarily of quartz and feldspar. Intermediate rocks have two rock types because there are more mafic types (Diorite, Andesite) and more felsic types (Granodiorite, Dacite) within the compositional range. There are no volcanic ultramafic rocks and the plutonic rocks are rare at the surface but compose the mantle.

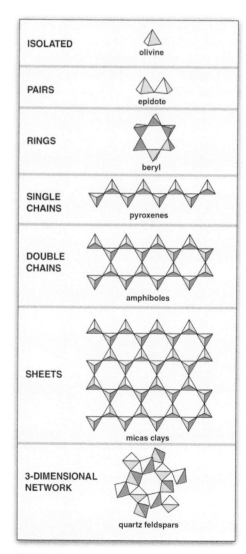

FIGURE 6.4 Diagram showing the types of silicates with the geometry of their silica tetrahedra.

TABLE 6.1 Classification of igneous rocks.

Composition	Ultramafic	Mafic	Intermediate	Felsic
Plutonic	Peridotite	Gabbro	Diorite, Granodiorite	Granite
Volcanic		Basalt	Andesite, Dacite	Rhyolite
Minerals	Olivine, Pyroxene	Plagioclase, Pyroxene, Olivine	Plagioclase, Hornblende, Biotite	Quartz, Feldspar
Color	Black	Black	Medium gray	Light gray/pink

Sedimentary rocks are produced by the chemical and/or physical erosion of existing rocks, transport of the material, deposition of the material and lithification into a rock. They are generally divided into two categories, clastic and precipitants. Clastic rocks are eroded into particles that are then physically transported and deposited before being lithified. These rocks are classified by grainsize as texture and grain type as composition. Texture is based on the grainsize of the particles (Table 6.2). As sediment, coarse material is gravel which forms conglomerate if lithified but it has a number of subdivisions which should be used in naming the rock. Otherwise, sand, silt, and clay are the particle names, though sand is typically subdivided into coarse, medium, and fine. The clay is lithified into shale.

Sandstone is compositionally classified as additional terms to grainsize (Table 6.3). This is a two-part system. If the rock has some clay, it is a wacke, or generally a graywacke, or an arenite if it has no clay. In addition, although it is mostly quartz, if it contains mostly feldspar in addition to quartz, it is feldspathic. If instead of feldspar it is composed mostly of rock fragments, it is termed lithic.

If the original erosion was chemical weathering or dissolution of the original rock, then the material is transported in solution as water. Precipitants are the rocks formed by the removal and reforming of this material as a solid either by precipitation or the removal from the fluid by plants and animals. There are numerous types including carbonates, evaporites, coal, and chert. The economically most important of these may be coal but the most voluminous are the carbonates, which are primarily limestones. Some of these are precipitated into rocks either directly to the sea floor or coating other materials but most result from removal of the elements from sea water by bacteria or animals. The rocks are generally composed of lime mud mixed with carbonate grains. The grains are fossils, rock fragments (intraclasts), coated sand grains called ooids, and fossilized animal droppings called pellets. These clast types are used to modify the rock names. There are several classification systems but the one described here is the Dunham classification system. Table 6.4 shows the system based on the proportion of grains to mud. If the grains are touching, it is considered grain supported but if not, it is mud supported.

Metamorphic rocks are any preexisting rock that has mineral, compositional and/or textural changes by changes in temperature, pressure, and/or fluids. Metamorphic rocks are largely chemical systems in which the mineral compounds react with each other and the fluids circulating through them. The minerals produced often reflect the degree of metamorphism as does the texture. Classification is by texture and composition like the others, but texture is both by grainsize as well as alignment and banding of minerals. Table 6.5 shows the textural classification of metamorphic rocks. Foliated means that platy minerals align in the rock that are generally not parallel to compositional layering. Slate is flat layered but in phyllite the layers have a sheen but otherwise the grains are too small to see. Schist is primarily aligned and contains parallel visible flat minerals. Gneiss has distinct compositional/color banding that is formed through metamorphism. Fine-grained metamorphic rock with no alignment is hornfels but there is no accepted convention for coarse grains and can include several names.

TABLE 6.2 Textural classification of particles and rocks.

Grains	Boulder	Cobble	Pebble	Granule			
Sediment		Gravel			Sand	Silt	Clay
Rock		Conglomerate			Sandstone	Siltstone	Shale
Size (mm)	>256	64–256	4–64	2–4	2–1/16	1/16–1/256	<1/256

TABLE 6.3 Compositional classification of sandstone.

Rock	Amounts		Modifier	
Arenite	<15% clay	Quartz	Feldspathic	Lithic
		>75% Quartz	<75% Quartz	<75% Quartz
Wacke	>15% clay		More feldspar	More rock fragments

TABLE 6.4 Dunham classification system for limestone.

	Mud supported		Grain supported	
Mud/Grains	<10% grains	>10% grains	>10% mud	<10% mud
Rock name	Mudstone	Wackestone	Packstone	Grainstone

TABLE 6.5 Textural classification for metamorphic rocks.

	No visible minerals		Visible minerals	
Foliated	Slate	Phyllite	Schist	Gneiss
Non-foliated		Hornfels	Granofels/skarn	

Composition of metamorphic rocks is much more complicated than with other rock types. The problem is that there are so many starting compositions that the metamorphic rock can vary greatly. If the starting rock or protolith was a limestone, the metamorphic rock will be marble. Sandstone metamorphoses to quartzite and granite, feldspathic sandstone, and rhyolite metamorphose to quartzofeldspathic gneiss. Shale, however, develops a sequence of mineral assemblages depending upon temperature and pressure. These assemblages include index minerals that are used to name the rock. In increasing metamorphic order they are: chlorite, biotite, garnet, staurolite, kyanite, sillimanite unless the pressure is low. In this case, there will be cordierite instead of staurolite and andalusite instead of kyanite.

Dirty limestone or marls metamorphose into calcsilicates composed of tremolite, K-spar, diopside, and other minerals that depend on increasing temperature and pressure but also the composition of the fluids. Ultramafic rocks metamorphose into minerals like talc, serpentine, brucite, and olivine but also depend on fluid compositions, so determining metamorphic conditions of these rocks is also complicated. Mafic igneous rocks progressively metamorphose to greenstones made of chlorite, amphibolites made of hornblende and granulites made of pyroxene. Under high-pressure conditions in a subduction zone, different minerals are formed that are relatively rare. These rocks are blueschists and eclogites.

6.3 Pollution by Rocks and Minerals

Fresh rocks and minerals contain elements and compounds that can be very damaging to the environment. However, they are not common and restricted to certain rocks and situations. In highly weathered rocks, most dangerous elements and compounds have been removed by many years of chemical reactions. Fresh material that is hazardous is common in mining, especially metal mining, and possible in tectonically active areas, elevated areas undergoing rapid erosion, areas of surface excavation, and possible drilling for underground resources.

Certainly, eruption and deposition of volcanic rocks can lead to extreme environmental destruction through air, water, and soil. Generally, igneous rocks are not environmentally dangerous once emplaced. However, certain conditions can lead to a concentration of potentially hazardous minerals or even production of potentially hazardous ground and surface waters. Heated fluids from active volcanoes can deposit pure sulfur (solfatara) and other minerals that could be dangerous. Within plutons, local concentrations of dangerous minerals are rare. Around plutons, however, deposits of lead, arsenic, and uranium minerals are possible. Further, areas around crystallizing and dewatering plutons can contain potentially hazardous fluids. All of the hot springs at Yellowstone National Park are the result of a pluton at depth. These fluids can be very acidic with pH below 1 in some areas and can carry high concentrations of ions of many types including those that could pose a health hazard.

Evaporites can have extreme compositions that can be damaging. Salt deposits will disrupt most ecological systems, killing most vegetation and microorganisms. Sulfate and sulfide deposits can produce acids that are also very damaging. Borate or borax deposits contain high levels of boron that are hazardous to environmental and human health. Barite deposits are rich in barium, which is also unhealthy in high concentrations. Most sedimentary rocks are relatively safe except under specific conditions. For example, the cementing fluids that help lithify the rock can be rich in dangerous compounds and elements including uranium. In some areas, the sediment can be eroded from sources that are rich in dangerous compounds such as arsenic, mercury, or lead. These can lead to local contamination of groundwater.

However, surface and near-surface water contain nowhere near the contaminants that deep groundwater, called formational fluids, can contain. With the higher temperature and pressure, waters can contain much higher ion concentrations and remove contaminants from the surrounding rock and sediment. Formational fluids can contain high concentrations of arsenic, mercury, sulfur compounds, and even radium and other radioactive compounds. They can also contain crude oil or gas condensate as well as with as many toxicants as they can hold. Fortunately, these dangerous fluids are sealed underground except in the special circumstances. One is through hot springs and seeps which are rare. Another is through drilling for hydrocarbons. Drilling and formational fluids from deep underground can be brought to the surface and dumped or stored. The threat from formational fluid at the surface increased greatly when the practice of fracking exploded across the United States and across the world.

CASE STUDY 6.1 Fracking in the United States

Surface Water Pollution

The United States was energy self-sufficient until the 1960s when cheap oil was increasingly imported from the Middle East. By the 1970s, more than one third of the oil was imported. In 1973, Middle Eastern exporters suddenly cut the supply, throwing the United States into crisis. Drivers had to wait two to three hours to fill up their car with gasoline, were rationed, and were limited to certain days. The situation did not resolve until the 1980s through increased exploration, but mainly by finding more stable sources for imports. The percentage of energy imports actually increased over the next few decades, placing the United States in an even more precarious situation. The situation, however, began to change dramatically in about 2005 with the expansion of fracking.

Hydrofracturing or fracking was developed in 1947 and instituted as an oil production technique in 1950. The problem with some oil and gas-bearing deposits is that the permeability is too low to flow the oil and gas to the well where it can be produced. To improve the permeability and flow, a hydrofracturing procedure is attempted. A fluid mixture of approximately 90% water, 9.5% ground quartz, and 0.5% chemicals

is pumped at extremely high pressure down a well that has a sealed top but openings in the well casing at the production depth. The high pressure of the fluid fractures the rock around the well bore and the quartz sand is driven up into the cracks, where it acts as a "proppant" that props open the cracks so they do not shut when the pressure is released. This allows the oil and gas to flow to the well at high rates (Figure 6.5).

The small amount of chemicals used in the wells are lubricants, antimicrobial agents, gelling compounds, corrosion preventers, viscosity maintainers, and possible antifreeze if the drilling is in the winter. Each hydraulic fractured well requires between 1.2 and 3.5 million gallons (4500 and 13 200 m³) of relatively clean freshwater. Deeper wells can require 5 million gallons (19 000 m³) and if wells require refracturing, single wells can involve the use of 3–8 million gallons (11 000–30 000 m³) of water to complete. The vast amount of water needed to frack a well can cause problems for areas with water shortage issues.

Many people worry about the 0.5% of added chemicals in the fracking fluid. However, the real danger is the process fluid. Once the fracking is complete, part of the fracking fluid and some formation fluid may come back up the well. Some wells are considered dry gas if 90% or more of the fracking fluid is absorbed by the production strata.

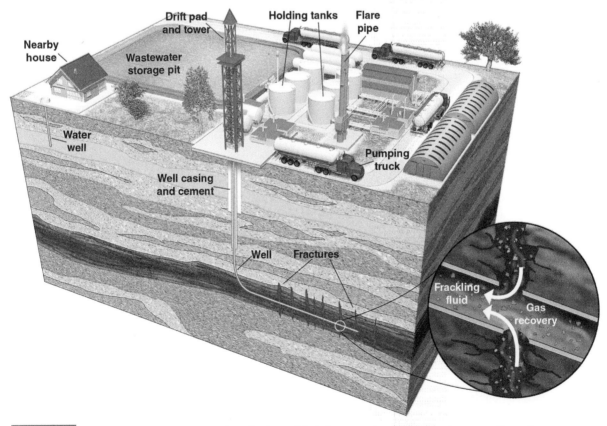

FIGURE 6.5 Diagram showing horizontal drilling, fracking of shale for gas and produce water lagoon on the surface.

However, the mix of the fracking fluid and formational fluids can be highly contaminated with heavy metals, sulfur compounds, barium, radium, and crude oil among other contaminants. This process water is dumped into onsite lagoons that are supposed to be lined. However, they are open at the top and so accessible to wildlife and susceptible to overflowing and spillage. This can pollute surface and groundwater and presents the greatest danger to the environment and human health.

Fracking was only used in special circumstances for most of its history. However, the big breakthrough came in 2005 not in advancing fracking techniques but in directional drilling. Previously, wells had to be drilled near vertically but now, wells could be drilled horizontally along producing strata (Figure 6.5). This combined with fracking opened a whole new area of exploration in shale. Shale is highly organic and can contain oil and especially natural gas. The problem is the very low permeability restricts flow. Further, drilling vertically

means that only the vertical thickness can be fracked, yielding non-economic quantities of gas. However, the new horizontal drilling technique allows the well to lie along the length of the shale unit, which is essentially infinite. The amount of fracked natural gas or shale gas that can be recovered from this method is immense. As a result, the United States quickly achieved energy independence by 2016.

One problem with all of this production was what to do with all of the toxic produce fluid. In some cases, even the drillers recognize the toxicity of the fluids and do not put them into lagoons. In Oklahoma, the decision was made to dispose of produce fluids in deep wells. The problem is that pumping fluid at high pressure into these wells produces earthquakes. Oklahoma went from one of the quieter states seismically to as active as California over a very short period. Oklahoma was forced to abandon the deep disposal practice and find an alternate solution.

Mining for any economic resources is extremely destructive to the environment. In addition to the heavy machinery, fuel and chemicals used in the operations, bringing fresh highly reactive minerals to the surface can release dangerous inorganic elements to the environment. These can include heavy metals such as lead, mercury, and arsenic as well as many other elements that are unhealthy in high concentrations such as selenium, boron, and sulfur.

CASE STUDY 6.2 Vermiculite Mountain Libby, Montana

Asbestos Pollution

The vermiculite deposit in Libby, Montana was discovered in 1919 by a hotel owner who was prospecting for gold (Figure 6.6a). He was searching in old mine workings where he noticed that the rocks near his carbide lantern sputtered and popped if he got too close. He sent a sample to the US Bureau of Mines, who informed him that the mineral was vermiculite and valuable. Vermiculite is an important industrial mineral used for insulation, sound deadener, soil conditioner, absorbant in packaging, and for fireproofing. Vermiculite is a mica that has such a wide range of uses because it exfoliates and pops if heated to 1472–2012 °F (800–1100 °C) for a few seconds (Figure 6.6b). Once popped it is so light that it can float on water, resembling Styrofoam, but it is chemically resistant and temperature-resistant.

The owner sold his hotel and became a full-time miner by 1924. He built a small plant to size and pop the vermiculite after it was mined. By 1925, the vermiculite was being shipped around the country for use as insulation and as an addition to roof sealant to improve durability and fireproofing. By 1926, the Libby workings was producing more than 100 tons (90.7 mt) of vermiculite per day.

The problem is that the vermiculite from Libby contains significant amounts of asbestos forms of the mineral group amphibole and, in particular, the mineral varieties tremolite and lesser winchite and richite, which are also asbestiform minerals (Figure 6.6c). Asbestos is composed of crystals that are long and fibrous with length to width ratios of 20:1 to 100:1 or greater.

When vermiculite from Libby was extracted, processed, transported, or used in manufacture, it produced asbestos-laden dust that wound up distributed throughout these areas.

Mining at the Vermiculite Mountain involved stripping off a thin layer of soil and digging out the vermiculite ore with regular excavating equipment. It was then loaded onto large trucks, and transported to a mill for processing. Later, an open-topped conveyor system was used to transport the ore from the mine and dumped into large piles at the mill near

town (Figure 6.6d). The ore was heated with steam until it popped and the finished product, called *Zonolite*, was packaged and shipped out. In some cases, the ore was shipped to other processing mills for processing.

In 1939, the company merged with another vermiculite mining company and the Zonolite Corporation was formed. In 1963, W. R. Grace purchased the Zonolite Corporation and expanded operations to become the world's largest producer of vermiculite, controlling more than 80% of the world market. This company developed into the world's largest specialty chemicals and materials operations, with over 6000 employees and $2.5 billion in annual sales. Between 1963 and 1990, sales of Zonolite earned over $180 million in after tax profits.

During the 1970s and 1980s, dangers from exposure to asbestos became apparent. In 1978, the US Environmental Protection Agency (EPA) began investigating the Libby operation when workers at a chemical fertilizer plant that used vermiculite in the products exhibited asbestos-related diseases. The Libby vermiculite was the suspected source of the asbestos and the EPA began to investigate asbestos-contaminated vermiculite. In 1982, the EPA sampled vermiculite at three major U. S. mines, including Libby, but the researchers were shifted from the investigation. Instead, the EPA was pressured into investigating asbestos in schools. They felt that asbestos-contaminated vermiculite was a less significant risk.

With the pressure off, Grace was able to cease operations at the Libby mine in 1990 without drawing attention. The price of vermiculite was dropping because of new, more profitable mines in Africa and Asia anyway. However, in 1992, a letter to the EPA complained about possible asbestos exposure from the destruction of a building at the inactive vermiculite mill. The complaint was sent to the Montana Department of Environmental Quality (DEQ), who determined that the asbestos was not removed from the building from before demolition nor had they taken precautions to prevent its airborne release. This violated the Clean Air Act and Grace was fined $500 000. A second letter in 1994, complained about blowing dust from the mill and an adjacent haul road having adverse health effects on Libby's residents. The Montana

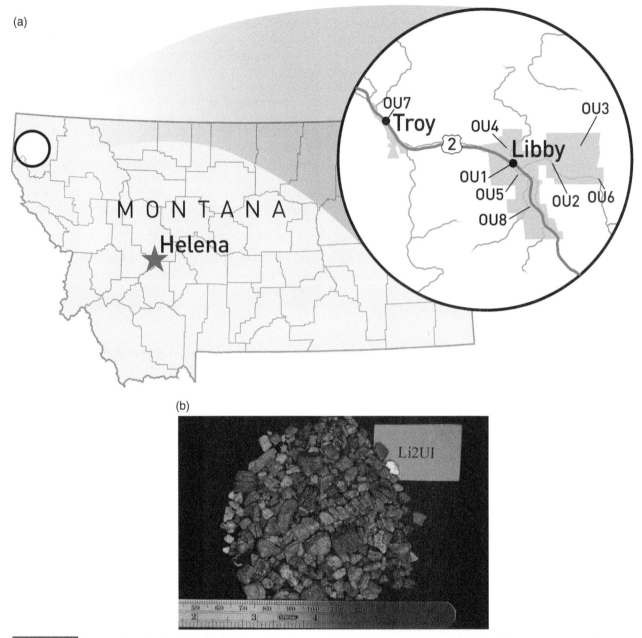

(a)

(b)

FIGURE 6.6 (a) Map showing the Superfund site remediation areas in Libby and surrounding towns as a blow-up on a map of the state of Montana. (b) Photo of vermiculite from the Libby, MT mine. *Source:* Courtesy of USGS. (c) SEM images of samples of asbestos, a-f are from the Libby mine. *Source:* Courtesy of USGS. (d) Aerial photograph of the Libby mine *Source:* Courtesy of US Environmental Protection Agency.

DEQ investigated this complaint as well by collecting samples from the mill and road. Asbestos was found but no action was taken because it was not commercial asbestos. Vermiculite was not mined for asbestos and therefore was exempt from regulation. In 1995, the citizens were informed that no action would be taken.

For the next five years, many of Libby's citizens grew sicker. In 1999, a Seattle newspaper published an article entitled "Uncivil Action – A Town Left To Die" that brought Libby to national attention. More than 500 Libby residents who were exposed to the asbestos-contaminated vermiculite from the Libby mine had either died or become seriously ill. Further, asbestos-bearing dust continued to be a major public health threat. Shortly after the newspaper article was released, the EPA and Montana Department of Health investigated Libby. They found that more than 30% of the collected samples contained asbestos. They also found

that from 1979 to 1998, the death rate in Libby from asbestos-related disease was 40–80 times higher than in other areas. In 2001, X-rays showed that 18% of more than 6700 Libby residents had asbestos-related lung damage. In all, the asbestos contamination resulted in the deaths of at least 200 people from lung disease and especially mesothelioma, a form of lung cancer directly related to asbestos exposure.

The EPA began cleanup of the Libby area in 2000. They removed and replaced asbestos-contaminated soil in schoolyards and playgrounds and removed asbestos dust from homes among other actions. These efforts continued through 2007 at a cost of over $180 million. Another $7 million was required to remediate the more than 170 sites around the country that received asbestos-contaminated vermiculite. The action was designated as an EPA Superfund cleanup project and it was among the most extensive Superfund cleanups on record. It was

(c)

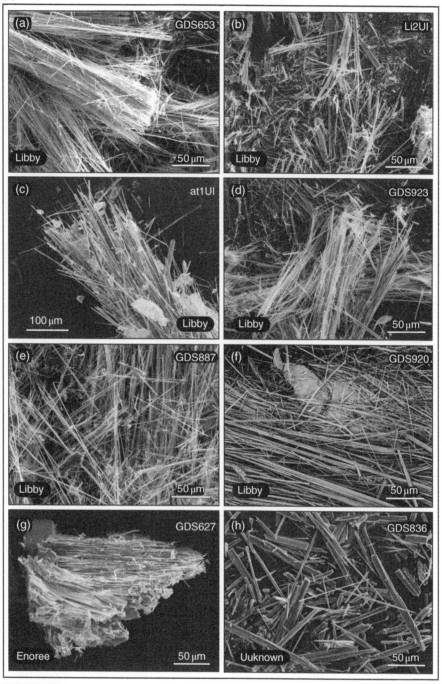

FIGURE 6.6 (Continued)

so extensive that the EPA declared Libby to be a Public Health Emergency, the first ever for the agency.

The Grace Company was a Fortune 100 company at the time. They refused to pay for the remediation because they felt that the EPA was doing far more than was necessary to remediate the property and town. The EPA sued Grace, which was ordered to reimburse EPA $50 million for the first part of the project. This began a series of asbestos-damage lawsuits against Grace from both workers and people who developed health issues from using Zonolite. The Grace company was forced to file for bankruptcy in 2001.

This was not the end of Grace's troubles. In 2005, the US Attorney General filed criminal charges against eight Grace executives, claiming that they knowingly released asbestos into the air, endangering miners, their families and residents of Libby. They were also charged with obstructing the EPA investigation. The trial began in 2006, with executives facing with up to 15 years in prison and the company facing fines of up to twice the profits it made from the Libby mine, estimated at $140 million. However, in 2008, the courts ordered Grace to provide $250 million for cleanup costs, the largest award in history, even though the total cost was more than $600 million. In 2009, the criminal charges were dropped against the defendants. It took until 2018 for the EPA's cleanup efforts to wind down. Grace's troubles, however, continue as on 16 February 2022, a single mine worker was awarded $36.5 million for damage to his lungs.

(d)

FIGURE 6.6 (Continued)

6.4 Environmental Remediation by Rocks, Minerals, and Soils

Certainly, rocks and other materials can cause pollution and it can be deadly. However, minerals, rocks, and soils have likely prevented more human illness and death than caused them. The main ways geological materials can be remedial is by filtering out unwanted contaminants from surface and groundwater, both chemical and biological, including actually killing off unwanted microorganisms. In addition, they can act as a buffer to certain chemical pollutants. The other way that they can peripherally help the environment is to act as a storage facility to keep certain pollutants sequestered. This, however, usually requires human intervention to make it work.

6.4.1 Physical/Chemical Filtering

The most prevalent removal of pollutants from groundwater is filtering through clastic sediments and sedimentary rocks. Clastic means eroded rock materials were transported as particles and deposited as sediments before being consolidated into rocks. The classification of the sediments is primarily based on their size, including boulder, cobbles, pebbles, and granules which are grouped as gravel (though the larger sizes are excluded from some definitions), sand, silt, and mud or clay. As lithified rocks, these grains produce conglomerates or breccias, sandstones, siltstones, and shales or mudstones, respectively. Sands are subdivided into coarse, medium, and fine grainsizes.

Clastic sediments and sedimentary rocks remove both pollutants and microorganisms by percolation and filtration. The grains are packed together in an aggregate with open pores around them (Figure 6.7). The pores constitute

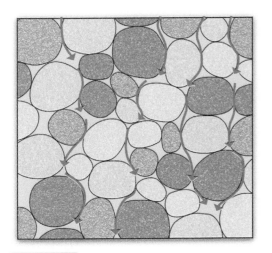

FIGURE 6.7 Diagram showing percolation of water through pores in sediments.

porosity that is expressed as a percentage of the rock or sediment. The pores are generally interconnected, allowing fluid to filter through. The interconnectivity of pores and property of fluid or gas flow through the material is the permeability. Consider a sponge. The amount of water it can sop up is the porosity but its ability to sop up or squeeze out the water is the permeability. Some materials can have porosity but no permeability, like bubble wrap. All of the cells/bubbles hold gas (air) but it cannot move unless the bubble is popped.

As water passes through the sediment or rock, it must wind its way around the grains. The outside of these grains or boundaries are typically highly chemically reactive depending upon the mineral type. They react with the infiltrating water and remove or exchange ions with them. This can remove chemical pollutants from the groundwater. The more the water interacts with grain surfaces, in general, the more chemical pollutant is removed. This means that finer-grained sediments and rocks have a higher potential to remove pollutants because the smaller the grains, the more the surface area available. Some pollutants are "sticky" as well. As they pass along grain boundaries, they can adhere to them without a real chemical reaction. The stickiness can be based on electrostatic charges, like static electricity, or surface tension of the pollutant.

The grains are commonly packed tightly together, especially in clastic rocks. Infiltrating water must squeeze between the narrow necks between grains. These necks are microscopic and so small that particles and even microorganisms cannot pass through. In this way, bacteria and other dangerous microorganisms are removed from infiltrating water thereby purifying it. If the sediments or grains in sedimentary rocks are not tightly packed, microorganisms can remain in the water. The other problem is if the rock contains fractures or other open passageways like caves. Water will take the quickest route, which is through the open passages. In these materials, the microorganisms will not be removed and there will be minimal contact with grain boundaries so the pollutants will not be removed either. Great care must be taken in these areas to safeguard groundwater quality.

6.4.2   Chemical Filtering

Some pure chemical filtering of surface and groundwater also takes place. The top soil horizon in any vegetated and especially forested area is the O-horizon, where O refers to organic. These horizons are composed almost exclusively of decaying or decayed leaf litter and are highly organic. The carbon in these layers is highly reactive and can act to fix most chemical pollutants and especially metals. Any infiltrating water is significantly purified as it passes through. These layers act like home water filters but for groundwater.

In areas lacking or with poorly developed O-horizons, the A and B-horizons in soil can also provide some filtering though not as complete as O-horizons. A-horizons can have some organic material which acts similar to the O-horizon. Otherwise, the clays in these layers can also react with infiltrating water to remove pollutants, though not as effectively.

There are many other mineral filters that operate in the surface and subsurface on a daily basis such as diatomaceous earth or dolomite. Zeolites are a group of about 40 naturally occurring minerals that have the ability to adsorb high amounts of impurities from the water, reduce bacteria, eliminate ammonia, and maintain an alkaline pH. They have a unique open structure that allows fluids to infiltrate but removes ions. There are numerous other minerals that filter chemicals and even those that reduce radiation exposure.

6.4.3   Chemical Buffering

Some soils, sediments, and rocks can buffer the chemical composition of infiltrating or even surface waters. These systems work similar to exchange columns like those found in home water softeners. Unwanted ions that occur naturally in the water are exchanged for less unwanted ions in the water softener. Many water softeners are designed to remove calcium and even iron ions by filtering the water through salt. This same process can occur in nature where percolating groundwater can pass through a salt deposit where it is softened similar to a home. The process can occur in the opposite direction as well. Calcium carbonate or calcite is the primary constituent of limestone and marble and it is highly alkaline. If acidic water passes through these rocks, the acid will be neutralized, thereby reducing its corrosiveness and danger to plants and animals. This is very useful in addressing issues of acid mine drainage from coal mines and sulfide mines. Many homeowners apply lime to their lawns to offset the acidity produced by decaying leaves. The carbonic acid produced by vegetation will slowly kill the grass.

Even surface water passing through a limestone terrane will be buffered by the rock. As the result of exchange with bedrock in the area, surface water which normally has a pH of about 5.2 as the result of equilibration with CO_2 in the air can have pH of 7–8 or even higher depending upon conditions. Low pH in surface water can damage aquatic life.

6.4.4   Antimicrobial Properties

Microbes in water and soil at the surface can be dangerous, especially to humans. The development of antibiotic drugs helped to lessen the impact of infections on the health and well-being of humans. However, with the overuse of antibiotic medications and the application of these drugs to common items like soap, many species of bacteria have become immune to the medications.

One such resistant bacteria is methicillin-resistant *Staphylococcus aureus* (MRSA), which has been referred to as a "super bug," threatening the lives of tens of thousands of Americans annually. This is becoming much more common and, as a result, mortality from untreatable infections is predicted to increase by more than 10-fold by 2050.

However, a new class of antibiotics has emerged from soils that currently may filter these dangerous bacteria from entering the water supply and in the future may be prescribed to fight infections in humans. These newly discovered antibiotics are called malacidins and are a class of calcium-rich antibiotic chemicals produced by existing soil bacteria in microbiomes. In addition, certain naturally occurring clay minerals have also been found to be powerful antibiotics. Reduced clays from hydrothermally altered volcanic deposits in Crater Lake, Oregon were found to be effective against *E. coli* and Staphylococcus. The clays are common but the oxidation state makes them antibiotic. These too are currently sterilizing groundwater and could be the basis of future medications.

CASE STUDY 6.3 Waste Isolation Pilot Plant, New Mexico

Radioactive Waste Storage in Natural Rock

The United States developed a repository for the disposal of transuranic (TRU) radioactive wastes in the Chihuahuan Desert 26 miles (42 km) east of Carlsbad, New Mexico (Figure 6.8a). TRU wastes include protective clothing, tools, equipment, soils, and sludge contaminated by radioactive materials at nuclear weapon facilities. The Waste Isolation Pilot Plant (WIPP) complex was developed for 50 years at a cost of billions of dollars to safely manage, store, and dispose of toxic, defense-related radioactive waste. The WIPP does not accept high-level radioactive waste or spent nuclear fuel from commercial power reactors. There is no plan for this much more dangerous waste now that the Yucca Mountain, Nevada nuclear waste repository is not being pursued. This leaves a big problem for the United States as the high-level waste stockpiles continue to increase with no solution in sight.

In 1957, the National Academy of Sciences recommended the disposal of radioactive wastes in salt beds or salt domes. These features have high resistance to radiation and salt quickly dissipates the heat produced by the radioactive decay without damage. Salt layers are generally impermeable and water cannot easily enter or move through salt because it saturates with salt and precipitates new salt crystals. It can seal the radioactive waste from the environment.

After extensive environmental site characterization, the US Congress established a radioactive repository in a salt formation in New Mexico, in 1980. This repository is designed for permanent disposal of nearly 6 million cubic feet (175 000 m³) of waste materials with levels of radioactivity low enough to be handled by workers using standard safety precautions. It can also accommodate 25 000 cubic feet (2323 m³) of waste materials with levels of radioactivity requiring noncontact or remote handling.

Construction of WIPP began in 1983 and continues today. This site contains no circulating groundwater, geological stability in terms of earthquakes and volcanoes, and can allow a safety buffer of rock. The Department of the Interior Bureau of Land Management already owned the land. New Mexico fought the development of WIPP, but in 1996 the US Congress withdrew the property from public use. TRU wastes are placed 2150 ft (655 m) below the surface in the 2000 ft (610 m) thick Salado salt formation. Rainfall is less than 10 in. (25.4 cm) per year, minimizing the water that could infiltrate the waste (Figure 6.8b).

Four vertical shafts are complete to the 2150 ft (655 m) disposal level, with seven disposal chambers of 300 ft (91.4 m) length, 33 ft (10 m) width, and 13 ft (4 m) height. In the end, the facility will include 56 rooms containing 850 000 55-gal (208.2-L), impact- and flame-resistant drums filled with TRU waste. WIPP became operational in March 1999 and disposal operations will be ongoing through 2070 (Figure 6.8c).

WIPP operations includes controversy and environmental risk. The long half-lives of TRU wastes mean that there will be radioactive risk for thousands of years. The WIPP was designed to isolate the waste from the environment for more than 10 000 years. Water has occurred in the rooms where TRU wastes are being placed. Water in the disposal layer is problematic because salt water will corrode the metal works and containers. The possible interaction of waste and water could produce pressurized gases and force radioactive waste out of the repository or even to the surface. There are about 30 TRU waste storage sites around the United States of various types and safety. Moving the waste to one centralized specially constructed and heavily monitored location will reduce overall public health and environmental risks from these dangerous materials.

(a)

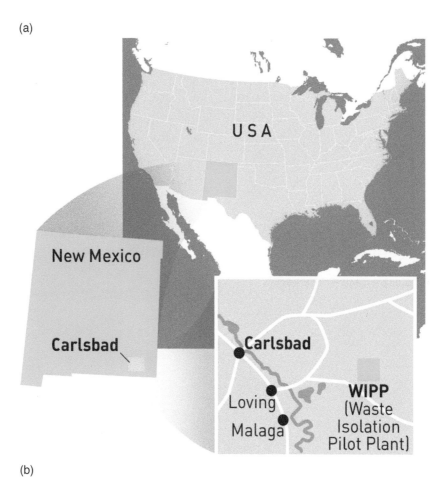

(b)

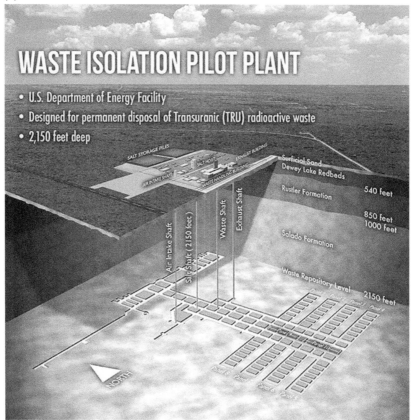

FIGURE 6.8   (a) Maps showing the location of WIPP in New Mexico and the United States. (b) Diagram showing the subsurface works of the WIPP. *Source:* Courtesy of USDOE. (c) Aerial photograph of the WIPP facility. *Source:* Courtesy of the USDOE.

(c)

FIGURE 6.8 (Continued)

CHAPTER 7

Natural Hazards and Pollution

CHAPTER OUTLINE

Words you should know:

Debris flow – A fast-moving downslope flow of water, rocks and mud.

Earthquake – Release of seismic waves from abrupt rupture and movement on a fault as a result of stress.

Hurricane – An intense storm originating in the tropics and having a sustained wind speed of more than 74 miles per hour (119 kph).

Magnitude – A scale to measure the power of an earthquake based on energy released.

Mass wasting – The downslope movement of material by gravity. Includes avalanches, landslides, mudflows, and debris flows, among others.

Mudflow – A rapid downhill flow of a mud river.

NATECH accidents – Natural hazard triggered technological accidents that release chemical pollutants into the environment.

Safir–Simpson Scale – The scale to classify the intensity of a hurricane from Category 1 (weakest) to 5 (strongest).

Tsunami – A huge ocean wave that is low and broad and moves at high speed in the open ocean but grows into a very tall destructive wave as it approaches the shore. Translates to "harbor wave" in Japanese.

Volcano – The release of lava, solid, and gases from an underground source through a vent to the surface.

7.1   Natural Disasters

Natural disasters cause such immense damage and loss of life on their own that the collateral environmental impacts pale in comparison. For that reason, they are commonly not reported or added as a footnote. However, they can cause damage, death, and hardships for the survivors that persist well after the media have moved onto other events. These impacts fall into two categories: those that are part of the natural disaster itself and those that are the result of damage to manmade structures.

The environmental impacts that are part of the natural disaster include encroachment by any pollutant it generates. This includes saltwater pushed or dropped onto dry land where it contaminates freshwater and soil; ash, aerosols, and dust that form air pollution; rubble and mud that covers soil and vegetation; and redirected or impounded surface water that floods the surface. Disease also commonly develops from these impacts or damage to public services like water and sewage disposal. The pollution that results from damage to manmade structures by the disaster largely involves chemical spills. These are termed "Natural hazard triggered technological accidents" or NATECH accidents by the chemical industry. They include damage to and the release of hazardous materials from a fixed position chemical plant as secondary consequences to a natural disaster. These can also include release of petroleum products from pipelines. These chemical pollutants are understood to be toxic and highly damaging to the environment.

An entire book is necessary to cover all of the details of pollution from natural disasters. The natural disasters capable of producing NATECH accidents include earthquakes, volcanoes, tsunamis, landslides, tornadoes, hurricanes, extraterrestrial impacts, and others, each requiring a separate chapter to describe the classifications, processes, and case studies. Therefore, this chapter will focus on a few types of natural disasters, both with accompanying pollution and accompanying NATECH accidents.

7.2   Volcanic Eruptions

Volcanoes can cause pollution of the surface of the Earth and the atmosphere unequaled by any single pollution event caused by humans. People are awed by the glowing lava and explosion of an eruption but these are insignificant in comparison with the danger from the gas and particulate emitted. These emissions are natural, but as a single event, they can be more damaging to the environment than anthropogenic pollution. They can cause radical climate change in a short period of time. The success of humans over the past 10 000 years as a species is partly the result of the lack of any large volcanic eruptions over the period.

Highly polluting aerosol compounds are emitted during a volcanic eruption. They occur in two forms: particulate and liquid. Volcanic eruptions always release aerosols into the troposphere where they directly affect humans, but eruptions can be so powerful that the aerosols are shot into the stratosphere and indirectly affect humans through climate impacts. In both cases, the effects are commonly devastating, depending upon the size of the volcano.

Explosive eruptions from the summit of a volcano can shoot an eruption column more than ten miles (16 km) vertically (Figure 7.1). Considering the amount of ash and other rock fragments it carries, this requires a massive amount of power. The reason that eruptions are so powerful is that volcanoes along magmatic arcs, like the Andes and Cascades, or volcanic island

arcs, like the Caribbean and Indonesia, always have water as part of the magma. When the magma is underground, it is under pressure. The pressure causes the water to remain liquid and mixed with the melted rock. This situation is just like carbon dioxide in soda or beer. It remains as part of the soda or beer in closed bottles. If the cap is on the bottle, even if the bottle is shaken, it does not foam. Removing the cap releases the pressure and the carbon dioxide changes from liquid to gas, which allows it to bubble, foam, or shoot out. Under standard indoor conditions, the gas will take up 22.4 times as much volume as it does as a liquid. When magma moves up the volcanic pipe in an eruption, the pressure is released and the water converts to gas in an instant. With the magma at 1768 °F (1000 °C) or more, the instantaneous volume increase is hundreds to thousands of times. The resulting explosion is more intense than our largest bombs.

The ejecta and aerosols can pollute the troposphere. One of the main differences between high-altitude and low-level material is the quantity and size. Gravity keeps the heavier volcanic particles closer to the surface, and they settle first. The larger fragments are bombs and blocks; lapilli are pebble-sized and ash is dust-sized. Ash can spread out and be transported great distances before settling. Nearer to the volcano, it may pile up and cause weak structures to collapse. It is dangerous to breathe at any concentration. Ash is mainly composed of fragments of volcanic glass and pumice. It can cause scarring of the respiratory system, resulting in silicosis and possible lung cancer. If aircraft fly through ash, it can abrade and destroy the engines, potentially causing them to crash.

There are a variety of chemicals emitted during volcanic eruptions. Chemicals include sulfur dioxide, carbon dioxide, carbon monoxide, numerous inorganic pollutants including lead, arsenic, chromium, cadmium, mercury, and even a few organic compounds such as methane and benzene. Most of these are emitted in such small amounts that they do not contribute to global pollution. Some chemicals, however, can be in quantities that can cause local to global environmental problems or even disasters.

Sulfur is the most common and abundant volcanic chemical pollutant. Sulfur is converted to sulfuric acid in the atmosphere, which contributes to acid precipitation. A large quantity of sulfur may be emitted in a single eruption and dissipate in weeks to months. Prolonged volcanic emissions, however, can produce substantial amounts of acid rain. Although industrial sulfur emissions are relatively low in Hawaii, there is an acid rain problem from the frequent volcanic emissions that is monitored for public health.

In rare cases, emissions of other dangerous chemicals may present a problem. The most common are light and volatile elements such as chlorine and boron. Among the best examples of dangerous volcanic emissions was the 1783 eruption of Laki in Iceland. This eruption emitted basalt lava, normally a relatively safe eruption. However, it was the largest of its kind on record, producing 3.5 cubic miles (14.7 km³) of lava, spreading out over 218 square miles (565 km²). An accompanying massive emission of fluorine gas settled on vegetation across Iceland. Fluorine poisoned the livestock that ate it killing hundreds of thousands of animals. An ensuing famine resulted in the deaths of more than 9000 residents of Iceland or 25% of the population at the time.

CASE STUDY 7.1 1783 Laki Fissure Eruptions

Poison Gas Emissions

The Laki (or Lakagígar) volcano commenced an eruption on 8 June 1783. It erupted from a fissure that is 16.8 miles (27 km) long including 130 craters. The eruption lasted for eight months and ended on 7 February 1784. The eruption was the largest single outpouring of lava in human history. It emitted 3.5 cubic miles (14.7 km³) of lava and 0.2 cubic miles (0.91 km³) of pyroclastic tephra (Figure 7.2a). It was the most voluminous tephra emission in Iceland in 250 years.

Laki erupted from fire fountains that reached heights of 2600 to 4600 ft (800–1400 m) (Figure 7.2b). Peak lava discharge is estimated at 176 573 to 211 888 cubic feet (5000–6600 m³) per second though some estimates place it up to 295 000 cubic feet (8600 m³) per second. The lava field covered 218 square miles (565 km²) and flowed as far as 15 miles (24 km) (Figure 7.2c). The flow is 10 miles (16 km) wide at its widest point and has a thickness of 100 ft (30 m) on average. It filled two valleys where the flow was dammed and, as a result, the thickness reached several hundred feet.

(a)

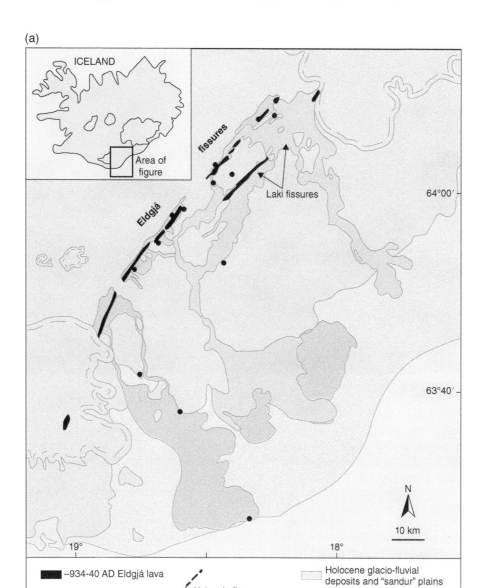

ICELAND

Area of figure

fissures

Laki fissures

Eldgjá

64°00′

63°40′

N

10 km

19°

18°

■ −934-40 AD Eldgjá lava

░ 1783-84 Laki lava

▰▰ Volcanic fissures

▒ Holocene glacio-fluvial deposits and "sandur" plains

▓ Holocene and Quaternary eruptives undifferentiated

(b)

FIGURE 7.2 (a) Geologic map showing the fissures and flows of the 1783 Laki eruption. Inset shows position of the area in Iceland. (b) Small volcanic cone in the Laki volcanic field. *Source:* Courtesy of USGS. (c) Fissure/rift in the Laki volcanic field. *Source:* Courtesy of USGS.

(c)

FIGURE 7.2 (Continued)

The catastrophic impact of the Laki eruption was the emission of aerosols. The eruption columns are estimated to have reached altitudes of 8.1 miles (13 km) during the first part of the eruption and stayed at 6.2 miles (10 km) high or more for the first three months. At this altitude, the eruption column reached into the jet stream which spread emissions across Europe and beyond. In addition to ash, Laki emitted an estimated 8 million tons (7.3 million mt) of hydrogen fluoride and 120 million tons (109 million mt) of sulfur dioxide. The sulfur was converted into sulfuric acid in the atmosphere and produced acid fog and rain.

The most devastating effects were in Iceland. Ash settled over the island and a dry fog known as "Laki haze" clouded the atmosphere for more than five months. Fluorine combined with water in the atmosphere, on plants, and on mucous membranes in humans and animals to form hydrofluoric acid, which is so strong that it can dissolve glass. Fallout of fluorine poisoned the grass in grazing areas, killing about 80% of the sheep, 50% of the cattle, and 50% of the horses in Iceland. By the end of 1785, 20–25% of the human population in Iceland (>10 000 people) had also perished in the fluorine poisoning and ensuing famine. Ash fallout extended into Great Britain, where the summer of 1783 was called "sand summer" as a result of the fallout. Laki haze also spread across Europe. By 30 June, Europe was covered by the haze, reaching as far as Lisbon, St. Petersburg, and Moscow. By mid-July, the haze reached the Altai Mountains in Central Asia, 4350 miles (7000 km) from the volcano. Later, Laki haze was seen over Alaska and China, which means the aerosol plume spread across the entire Northern Hemisphere from 35° north latitude to the North Pole.

The acid fallout from the haze was devastating to vegetation and human health. In Iceland, acid fallout burned leaves and produced sores on animals and humans. Crops failed in Iceland and there were low yields and crop failures throughout Europe. Fish kills caused by acid deposition were reported in Scotland. Inhaling the haze caused fatigue, shortness of breath, and heart palpitations in most people. It was very hard on the old, the young, and on people with respiratory ailments.

The weather was also affected by the Laki aerosol and ash. The month of July 1783 was unusually hot in western Europe but it was unusually cold during the rest of the summer and in many other areas. Rainfall in the Nile River valley was well below average and produced record low river levels. The Altai Mountains experienced harsh frosts in August. A severe drought occurred in India and throughout the Yangtze River valley in China. In general, the summer was unusually cold throughout China. In Japan, the excessive precipitation and cold late-summer temperatures resulted in widespread rice crop failures and the worst famine in Japanese history.

The winter of 1783–1784 was no better and included exceptionally long periods of sub-freezing temperatures. There was extremely cold weather in Great Britain, France, and Scandinavia. New England experienced its longest period of sub-zero temperatures on record and accumulation of snow in New Jersey was the highest on record. In the Chesapeake Bay in Maryland and Virginia, ice caused the longest closure of the harbors and channels on record. In New Orleans, the Mississippi River had ice rafts on 13–19 February 1784.

The combination of acid haze and precipitation and the severe weather in 1783–1784 is estimated to have killed more than 23 000 people in Great Britain and more than 16 000 in France. There was a severe famine in Egypt, China, and Japan, causing tens to hundreds of thousands of deaths. All totaled, estimates are that as many as 6 million people perished. The food shortage caused the economy in western Europe to falter, which led to widespread poverty and unrest. This situation persisted until 1788 and is likely a major factor in the French Revolution in 1789.

Summit eruptions come from the top of the volcano and can shoot thousands of tons of rock and aerosols into the stratosphere. Depending on the magma, these aerosols are mostly ash, water, and sulfur compounds. Sulfur dioxide is typically the primary of these stratospheric aerosols and makes the greatest impact. Aerosols can reflect sunlight back into space, which prevents it from reaching the Earth. This filtering effect causes global cooling. The most famous eruption that produced major temporary climate change was the eruption of Tambora in Indonesia. In April 1815, Tambora produced an eruption column that reached 28 miles (44 km) height. The eruption produced 36 cubic miles (150 km³) of particulate and injected 200 million tons (182 million mt) of sulfur dioxide into the stratosphere. The aerosols spread over the entire atmosphere and the reflected sunlight causing global cooling by 5 °F (3 °C). The year after the eruption still had most of the Northern Hemisphere experiencing cooler summer months. In North America and Europe, 1816 was known as "the year without a summer." Snow was on the ground through June in New England.

7.3 Earthquakes

In addition to the devastation inflicted on life and property, earthquakes can cause environmental impacts that last for years. In most cases, the environmental effects are minor in comparison to the damage and death from the main event and are largely overlooked. If anything, only water contamination and the resulting disease are ever reported.

Earthquakes are classified by magnitude, which translates to strength. Early efforts to measure the power of an earthquake were based upon damage to structures and the perception of witnesses. This measure is termed intensity rather than magnitude, and, at present, the Modified Mercalli Scale is the accepted method of measuring the intensity of earthquakes. It rates intensity from I to XII and reflects the magnitude as well as location of the earthquake in terms of geologic materials, building construction, and population density. Even though magnitude defines earthquake power, intensity producing damage at the surface, is the most important designation for purposes of pollution. Magnitude is the energy released in an earthquake through the instrumental measure of the waves as they pass through the Earth. Magnitude is a logarithmic scale based upon the measured amplitude of the strongest waves of an earthquake and based on the types of rock and soil they pass through. The moment magnitude scale uses ground acceleration as the waves pass through the Earth. Moment magnitude is the current standard for earthquake strength.

Earthquakes can cause huge clouds of dust or particulate to be raised into the air, creating air pollution. The earthquake in Lisbon, Portugal in 1755 is said to have raised so much dust that it blocked out the sun. As the waves pass through an arid region, the loose soil and silt at the surface is disturbed and thrown into the air. During the Assam, India 1897 earthquake, rocks on roads were said to vibrate "like peas on a drum," as the surface waves passed through. The collapsing of buildings also commonly raises significant dust. Dust like this can be seen when large buildings are demolished for urban renewal. A large dust cloud is emitted when they are blown up. These old buildings can contain asbestos, lead paint, lead solder, and fiber glass, making this dust cloud a health hazard.

Fire is a threat from earthquakes but not nearly as much as it was previously. In historic times, cooking was done on fires, lighting was from candles and oil lamps, and heating was from coal or wood stoves or furnaces. Seismic waves frequently toppled these sources, and old wood houses were quickly engulfed in flames. Buildings commonly topple into the street when destroyed. The resulting rubble slows fire engines and fire brigades, making the infernos even worse. Before hydrant systems, horses pulled tanks of water to the fire, and the small amount of water was poured on the fire. In many famous historical earthquakes, the main destruction was by fire rather than the seismic waves themselves. The Great Kanto earthquake of 1923 destroyed Tokyo, Japan, primarily by fire. More than 142 000 people were killed in this disaster, including 35 000 in a single fire storm. The smoke from such fires was loaded with particulate, PAHs, benzene, methyl ethyl ketone (MEK), and other dangerous chemicals, depending upon what was burning at the time.

Both direct surface movement and seismic waves can sever all forms of utilities during a major earthquake. The most obvious problem is the severing of natural gas lines. Gas leaks can occur throughout the affected city and not be stopped until the system is shut down. Methane can cause significant adverse health effects, particularly in the respiratory system, and even lead to death. It is also highly flammable, representing a great risk for explosion. In areas of oil production, this danger is even more pronounced if oil pipelines are severed. Spilled oil can contaminate surface water bodies and other low-lying features such as wetlands, lakes, or ponds and present a fire hazard, as well.

Typically, the most dire environmental problem resulting from an earthquake is the destruction of water supplies. Without plumbing, fresh water supplies quickly dwindle and available surface water becomes contaminated with bacteria. Disease commonly follows major disasters as a result of problems with water supplies. This contamination is usually exacerbated by damage to sewer lines. They rupture as readily as the water mains and spill raw sewage into surface water and low-lying topography. The release of untreated sewage can cause damage that takes years to repair. Well water can also be contaminated during earthquakes. The shaking from seismic waves drives groundwater upward in the soil during earthquakes causing liquefaction, where soil flows like a viscous liquid. The upward movement of groundwater can cause it to become turbid and unpotable because of the excessive suspended silt.

CASE STUDY 7.2 2010 Port-au-Prince

Cholera Outbreak After Haiti Earthquake

A devastating earthquake struck the Port-au-Prince area, Haiti on Tuesday, 12 January 2010 at 04:53:10 p.m. It had a strong magnitude of 7.0 and a focus of 8.1 miles (13 km) depth, with the epicenter a close 15 miles (25 km) southwest of the capital city. Surface waves produced ground shaking at a Modified Mercalli Intensity of IX (maximum XII) in Port-au-Prince and the suburbs. The earthquake was on the Enriquillo-Plantain Garden fault system which produced large historical earthquakes in 1860, 1770, and 1751 (Figure 7.3a). The fault system

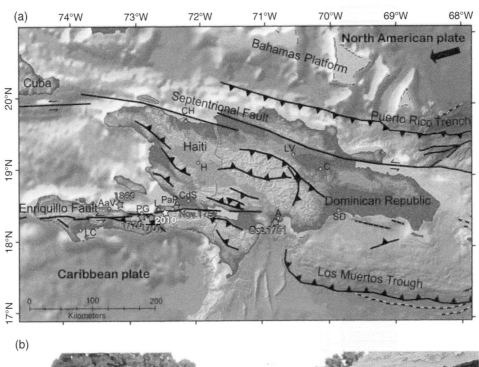

FIGURE 7.3 (a) Map of Hispaniola showing the location of major faults and the epicenters of major earthquakes including the 2010 earthquake. *Source:* Courtesy of the USGS. (b) Photograph of a collapsed hotel in the 2010 Port-au-Prince, Haiti earthquake. *Source:* Courtesy of USGS. (c) Photograph of a collapsed building in the 2010 Port-au-Prince, Haiti earthquake. *Source:* Courtesy of the USGS. (d) Map of Haiti showing the distribution of the number of cases of cholera by the dates shown. *Source:* Courtesy of US Department of Health & Human Services.

(c)

(d)

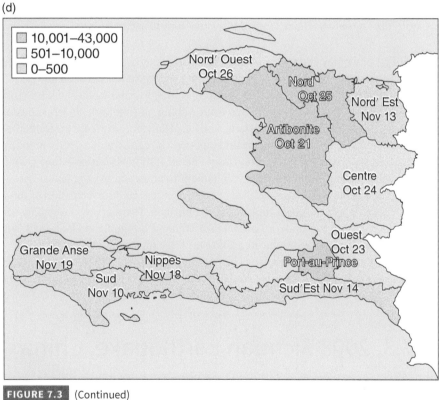

FIGURE 7.3 (Continued)

is a segment in the margin between the Caribbean tectonic plate and the North American plate. This transform boundary experiences left-lateral strike slip or motion and links the Caribbean arc and the Central American arc. The motion across the entire boundary is 0.8 in. (2.0 cm) per year. It produced numerous earthquakes such as the 1692 Port Royal and 1907 Kingston earthquakes in Jamaica and the 1972 Motagua earthquake in Guatemala that killed 23 000 people.

This devastating earthquake was the worst in more than 200 years and Haiti is not prepared for earthquakes (Figure 7.3b). It is the poorest country in the Western Hemisphere, with 80% of the population living in poverty and 50% in abject poverty. Haiti has substandard building practices; buildings barely survive normal conditions, much less shaking from a strong earthquake. The buildings collapsing caused most loss of life. Approximately 105 000 homes were destroyed and 208 000 damaged, mostly in Port-au-Prince (Figure 7.3c).

It is estimated that 3 000 000 people were impacted by the earthquake. The Haitian government official estimates are that 316 000 people perished, 300 000 were injured, and 1 000 000 were made homeless. This death toll is disputed by other international organizations that estimate the death toll, between 92 000 and 160 000. The total cost of the disaster was estimated at between $7.8 billion and $8.5 billion, based on a death toll of 200 000 to 250 000.

Many world leaders pledged generous humanitarian aid but delivering on this aid became a logistical nightmare. After the earthquake, humanitarian aid arrived in a trickle, electricity was non-existent, and telecommunications were intermittent. Ships were forced to remain offshore because of Haiti's damaged port and could not deliver their supplies. The airport was barely functional, with severely limited flights in and out of the country. All main roads and most secondary roads were impassable as a result of the debris and throngs of survivors.

The Haitian government developed an aggressive plan to begin reconstruction. They used this plan to raise $10 billion in pledges at a March 2010 donor conference. The rebuilding, however, was very slow and disorganized, with multiple accusations of misuse of funds.

The earthquake severely damaged the already poor public sanitation and public water systems in Haiti. Prior to the earthquake, only 27% of the country had basic sewerage, and 70% of the households had rudimentary toilets if at all. The cost of water from private companies and the lack of widespread distribution systems made clean water inaccessible for many. This means that drinking water for most residents was commonly impacted by raw sewage. These conditions produce regular outbreaks of infectious diseases such as typhoid, intestinal parasitosis, and bacterial dysentery but not cholera.

Nevertheless, on 21 October 2010, the first case of cholera in Haiti was confirmed in more than a century (Figure 7.3d). By

9 November there were 11 125 hospitalized cholera patients and 724 confirmed deaths. However, the outbreak did not appear to be the direct result of the earthquake but instead by contamination from infected United Nations peacekeepers. The United Nations denied the allegations but a video of UN soldiers trying to excavate a leaking pipe in the center of the outbreak area was posted online. As a result of the death of a young Haitian inside the Cap-Haitian UN base, a deadly riot broke out in the city of Cap-Haitian on 15 November 2010.

The riot was resolved after several days but in November 2011, the United Nations was served with a petition from 5000 victims demanding hundreds of millions of dollars in reparations over the outbreak. In February 2013, the United Nations invoked its right to immunity from lawsuits and the issue was dropped.

Meanwhile, by March 2011, the epidemic had killed 4672 people. By August 2012, the cholera outbreak caused 586 625 cases and 7490 deaths. By 21 November 2013, there had been 689 448 cases of cholera and 8448 deaths. By 2018, there had been a total of 812 586 cases of cholera and 9606 deaths. The epidemic apparently ended by 2019. However, in a rural community of 2500 studied residents, 18% reported a cholera diagnosis but 64% had antibodies against it. This indicated that the epidemic may have been far more extensive than previously thought.

All chemicals stored in surface and many underground tanks can potentially cause environmental disasters in an earthquake. Oil refineries or chemical plants are especially vulnerable. Refineries contain large amounts of oil and petroleum biproducts stored in large tanks connected by pipes. Refineries are commonly located on major water bodies, where crude oil can be offloaded from tankers. Major spills are more common at such locations even without earthquakes. Several coastal areas in northeastern New Jersey, on the Mississippi River near Baton Rouge, Louisiana, and near Houston, Texas, are profoundly polluted as a result. If an earthquake strikes these kinds of plants, spills from the tanks and pipes can be catastrophic. These areas also typically have major chemical industries that utilize refinery products and maintain large chemical supplies onsite in tanks, which can also rupture.

The oil industry commonly takes precautions to the potential damage to pipes and storage tanks. The trans-Alaskan oil pipeline crosses major active faults with documented large earthquakes and surface ruptures. In anticipation of this activity, designers have installed joints and surface mounts that guard against ruptures of the pipeline. There are also regularly spaced automatic shutoff valves to stop the flow in an emergency. A magnitude 7.9 earthquake occurred on the Denali fault on 3 November 2002. This fault crosses the Trans-Alaska Pipeline. The earthquake produced a 209 miles (335 km) long surface rupture with as much as 29 ft (8.7 m) of lateral offset. Shaking lasted 90 seconds, resulting in a series of landslides 16–24 miles (9.6–14.5 km) long. The ground shifted more than 14 ft (4.3 m) under the pipeline and shaking was intense. However, the pipeline did not rupture, proving the excellent engineering.

CASE STUDY 7.3 2008 Sichuan Earthquake, China

Chemical Spills from Earthquakes

A very powerful earthquake struck Sichuan, China on 12 May 2008. The epicenter was located 55 miles (90 km) northwest of Chengdu, the capital of Sichuan (Figure 7.4a). The depth of the focus was 11.8 miles (19 km). The earthquake had a strong magnitude of 7.9 and the maximum intensity of the shaking was XI of XII on the Modified Mercalli Scale. The shaking from surface waves lasted nearly 120 seconds. This earthquake caused devastation in surrounding counties and was felt in Beijing and Shanghai which are 932 miles (1500 km) and 1056 miles (1700 km) away, respectively.

Most of the devastation was caused by the intense shaking. The maximum shaking intensity on the Modified Mercalli Scale was XI in Wenchuan, VIII in Deyang and Mianyang, VII in Chengdu, VI in Luzhou and Xi'an, and V in Chongqing, Guozhen, Lanzhou, Leshan, Wu'an,

Xichang, and Ya'an (Figure 7.4b). This shaking caused catastrophic mass movement including landslides, avalanches, and rockfalls. They destroyed or badly damaged numerous mountain roads and railroads and buried buildings throughout the Beichuan-Wenchuan area. The massive debris and severe damage to roads eliminated access for search and rescue teams for several days (Figure 7.4c). A single landslide in Qingchuan buried more than 700 people. Landslides dammed up several rivers creating 34 "landslide" lakes. Huge amounts of water pooled behind these debris dams at very high rates. The government feared that the dams would fail behind the building water pressure. If they did, more than 700 000 people downstream of the dams were in peril. Chinese troops with heavy machinery reduced and removed the dams to drain the landslide lakes to control the release of water.

On 12 May, 50 000 Chinese army troops and police were dispatched to conduct disaster rescue and relief operations in Wenchuan County by the Chengdu Military Command. On 16 May, teams from Russia, Japan, South Korea, and others joined the rescue operations. The United States provided satellite images of earthquake-stricken areas and sent two US Air Force transport airplanes full of supplies, tents, and generators. There were 135 000 Chinese troops and medics involved in the rescue effort in 58 counties several days later. The aircraft involved in rescue and relief operations by air force, army, and civil aviation exceeded 150, making it China's largest non-combat airlift operation.

In all, nearly 2500 dams were damaged, more than 32 933 miles (53 000 km) of roads and 29 825 miles (48 000 km) of water pipelines were also damaged. More than 5.36 million buildings collapsed and 21 million buildings suffered damage in Sichuan and in parts of Chongqing, Gansu, Hubei, Shaanxi, and Yunnan. More than 80% of buildings collapsed in the cities of Beichuan, Dujiangyan, Wuolong, and Yingxiu. The total loss caused by the earthquake is estimated at US$86 billion.

The collapsing buildings led to catastrophic loss of human life. More than 45.5 million people in the 10 provinces and regions were impacted by the earthquake. At least 15 million people were ordered to evacuate their homes and at least 5 million were left homeless. The official death toll for the earthquake was 87 587 people. This massive number makes it the 16th deadliest earthquake in recorded history.

The other less reported impact was the damage to the environment from this earthquake in a NATECH accident. This highly industrialized area contains more than 100 chemical plants in the epicentral cities of Chengdu, Deyang, Mianyang, Guangyuan, Ya'an, Meishan, and Aba alone. Many of these were damaged or destroyed releasing toxic chemicals into the environment. In Shifang, 30 miles (50 km) east of the epicenter, the collapse of two chemical plants led to a spill of more than 88.2 tons (80 mt) of liquid ammonia. More than 100 plant workers were buried in the plant wreckage. The spill forced more than 6000 people to evacuate their homes. The spills were quickly brought under control but water sampling near the plants revealed increases in the levels of ammonia and nitrogen in groundwater. This water would require processing to be safe to drink.

The earthquake disaster area also contained China's main nuclear weapons research lab in the city of Mianyang as well as several secretive atomic sites. The earthquake damage resulted in 50 "hazardous sources of radiation" to be missing. The government reported that all had been located but there was some question whether this was true. The questions came when Chinese media inadvertently revealed that the communist regime had been hiding a toxic chemical spill. Apparently, ammonia was released by at least 5 facilities instead of two and they caused pollution of a river and damage to crops. There were also reports of sulfuric acid spills, a sulfur explosion and fire, and fires in phosphorus processing facilities. These spills caused a significant amount of damage to agricultural fields.

(a)

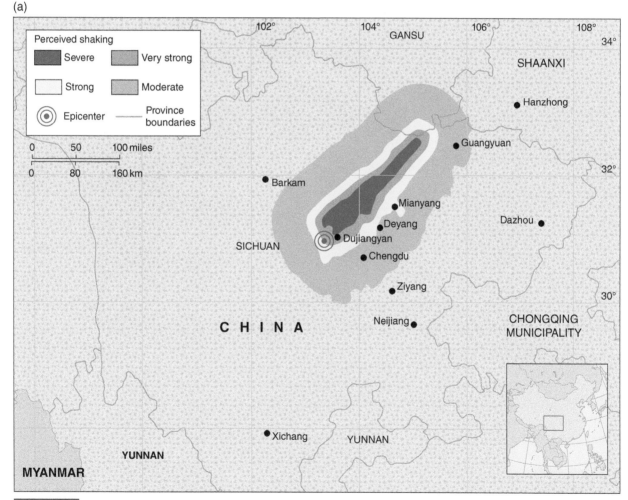

FIGURE 7.4 (a) Map of the epicenter and damage intensity of the 2008 Sichuan earthquake. Inset shows the location of the map in China. (b) Photo of damage done to buildings by the 2008 Sichuan earthquake. *Source:* Courtesy of USGS. (c) Photo of a rock slide crushing a plant in the 2008 Sichuan earthquake. *Source:* Courtesy of USGS.

(b)

(c)

FIGURE 7.4 (Continued)

Fortunately, earthquakes have not directly caused a disaster at a nuclear power plant. Nuclear power plants require large amounts of cooling water so are commonly located on rivers. Faulting breaks rocks into poorly consolidated material allowing them to be more easily removed and form a valley. Rivers follow valleys and can flow along faults. The US Nuclear Regulatory Commission requires a rigorous geological evaluation before a power plant permit is granted. Even after they are operational, close monitoring and reevaluation are common. The caution exercised by nuclear power plants and government regulators has resulted in no environmental disasters directly caused by earthquake induced damage of a nuclear power plant.

However, nuclear power plants have suffered direct earthquake damage and even leaks. The magnitude 6.8 earthquake in Niigata, Japan on 16 July 2007 damaged the Kashiwazaki Kariwa nuclear power plant, the largest nuclear power plant in the world. Some 315 gal (1192 l) of radioactive water spilled into the Sea of Japan and it took two hours to extinguish the transformer fires caused by ground shaking. There was no danger that the plant would experience a catastrophic meltdown, but Japan experiences much stronger and more damaging earthquakes.

The worst nuclear accident from an earthquake generated natural disaster also occurred in Japan. On Friday, 11 March 2011, a powerful earthquake struck offshore of northeast Honshu. The Tohoku earthquake had a magnitude of 9.0, the most powerful earthquake recorded in Japan and the fourth strongest recorded earthquake ever. It generated a large tsunami that damaged the Fukushima I and Fukushima II nuclear powerplants and the backup generators failed.

CASE STUDY 7.4 2011 Tohoku Earthquake

Tsunami-Generated Nuclear Accident

Large tsunamis can cause tremendous damage by themselves but they are also capable of causing significant collateral environmental damage that exacerbate the disaster. The Tohoku earthquake produced a devastating tsunami but it also generated the second worst nuclear power plant accident in history. The powerful earthquake struck the offshore area of northeast Honshu, Japan on Friday, 11 March 2011 (Figure 7.5a). The magnitude was a phenomenal 9.0, making it the fourth strongest earthquake ever recorded and the strongest earthquake ever in Japan. The epicenter was 80 miles (129 km) east of Sendai, and 231 miles (373 km) northeast of Tokyo Japan and the focus was at 19.9 miles (32 km) depth.

(a)

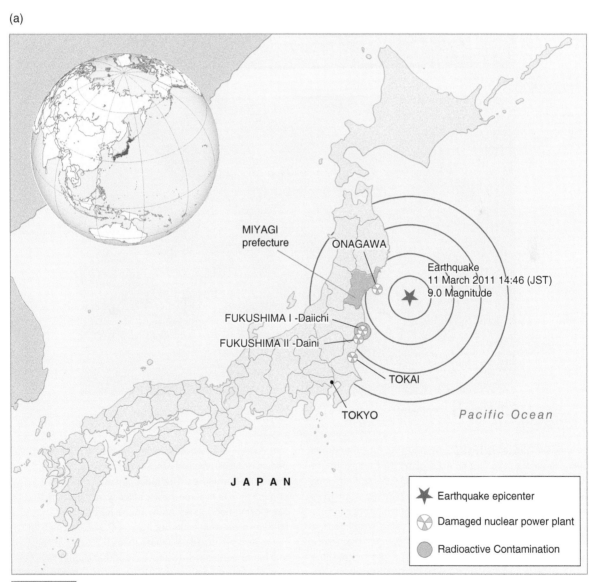

FIGURE 7.5 (a) Map of the 2011 Tohoku earthquake epicenter and damaged nuclear powerplants. Inset shows the location of the map on Earth. (b) Panoramic view of a Japanese coastal city destroyed by the Tohoku tsunami.
Source: Courtesy of NOAA. (c) Photo of the destroyed part of the Fukushima nuclear power plant resulting from the Tohoku tsunami *Source:* komaruko / Adobe Stock.

(b)

(c)

FIGURE 7.5 (Continued)

The earthquake was generated by the subduction zone that drives the Pacific Plate under the Japanese arc/Asian plate. The rupture surface on the fault that generated the earthquake was 186 miles (300 km) long and 93 miles (150 km) wide. It moved between 98–131 ft (30–40 m) at its maximum point. This movement caused a rapid uplift of the ocean floor between 8 and 26 ft (2.4–8 m) locally. The seafloor uplift also uplifted the Pacific Ocean water above it, producing a deadly tsunami that struck the coast of northeast Japan. The earthquake also caused the coastline to subside up to 2.0 ft (0.6 m) along a 250 miles (400 km) stretch.

The most destructive hazard of this huge earthquake was the tsunami which struck the coast with run-up heights as much as 127 ft (38.9 m) above mean sea level and it flooded as far as 6 miles (10 km) inland (Figure 7.5b). Ocean floor sensors detected the tsunami

immediately and broadcasted an alert for the entire northeast coast of Japan. However, the tsunami arrived very quickly after the earthquake; in just 10–30 minutes. The quick arrival did not give many residents enough time to evacuate or emergency officials time to set safeguards like closing flood gates. Many of the harbors and coastal areas have tsunami protection walls but the waves were too high and overtopped them. As a result, 181.5 square miles (470 km^2) of coast was flooded and many residents drowned in areas that should have been safe.

The sensors also sent the warning to the Pacific Tsunami Warning System (PTWS) Center in Hawaii, which, in turn, issued tsunami warnings for the entire Pacific Ocean rim. The tsunami caused damage in communities around the entire Pacific coast. In California and Oregon, waves as high as 8 ft (2.4 m) high damaged docks and boats and caused

$10 million in damage and a man in California was killed. In Chile, tsunami waves as much as 10 ft (3 m) high damaged 200 houses more than 11 000 miles (17 000 km) away from the earthquake epicenter.

Ground shaking from the earthquake surface waves lasted six minutes and locally produced ground acceleration of 2.99 g (29.33 m/s²). The intensity of this shaking produced damage at IX on the Modified Mercalli Scale along the coast but there were broad areas of VIII intensity throughout northeast Japan. There was also widespread liquefaction of the soil and landslides along hill slopes, causing great damage. Between these earthquake effects and the tsunami, there were 190 000 buildings either destroyed (45 700) or damaged (144 300), producing an estimated 25 million tons (23 million mt) of debris. They also caused power outages impacting 4.4 million households and water disruption to 1.5 million households.

The earthquake and tsunami left 15 901 people dead, 6157 injured, and 2529 people missing. The final death toll translates to about 18 500 people. The cost of this disaster is about $360 billion, making it the most expensive natural disaster in history.

In addition to the environmental damage caused directly by the earthquake and tsunami, another major environmental problem of global concern arose. Japan generates much its electricity from nuclear power plants and four of these plants are located on the northeast coast in the earthquake and tsunami impact area. Because Japan has a lot of earthquakes, the power plants were designed to withstand large earthquakes but none were designed to withstand a magnitude 9.

In addition, none of them were designed to withstand a tsunami. All four plants sustained some damage from the earthquake but the Fukushima I and Fukushima II plants were directly impacted by the tsunami. As a result of the flooding, the backup generators also failed. Fukushima I experienced a number of relatively large explosions of hydrogen gas that built up in the outer containment walls of the plant (Figure 7.5c). The explosions caused exposure of the reactor core and the leaking of radioactive materials into the environment. Cooling water in the containment vessel of the reactor reached radiation levels of 1000 times normal background radiation. Even the area around the plant reached radiation levels eight times the normal background.

The Japanese government immediately evacuated more than 200 000 residents from the area within a 12-mile (20 km) radius of the plant. The US government issued recommendations that American citizens should remain at least 50 miles (80 km) away from the damaged power plant. Fortunately for the residents, a weather system moved in and winds blew most of the airborne radioactive release away from land and over the Pacific Ocean. However, radioactive iodine, strontium, and cesium were found in soil and in tap water in areas around the damaged plant. As a result, Japan banned the consumption of food from the area and there was a general panic among residents. Even in California, residents stripped store shelves of potassium pills in fear that the radiation cloud would reach the United States. The Fukushima I event is second only to the Chernobyl meltdown as the worst civilian nuclear disaster to date.

Tsunamis, like that in the 2011 Tohoku earthquake, can produce other environmental disasters. Uplift of the seafloor or a submarine landslide abruptly changes the height of the water column and forms a temporary "step" on the ocean surface. Water cannot have abrupt elevation changes and a tsunami is a re-leveling of sea level. Tsunamis travel at speeds of 400–700 miles (644–1127 km) per hour in open ocean where they are less than 3 ft (0.9 m) high with a wavelength approximately 328 ft (100 m). Ships typically cannot detect them. Tsunami translates to "harbor wave" in Japanese, because the wave rises up in a harbor as the wave approaches the shore.

Tsunamis can drown people but the real danger is that the waves come ashore traveling at 50–60 miles (80–97 km) per hour compared to a typical wave speed of 5–10 miles (8–16 km) per hour. Depending upon height, they can damage coastal structures like chemical storage tanks and ships including tankers. The salt water can be pushed inland 1 mile (1.6 km) or more, contaminating surface water and groundwater supplies. The weight of the floodwater backs up sewer systems and raw sewage spills on the surface. Debris and decaying of human and animal corpses spreads disease. Environmental damage in Thailand and Indonesia from the 26 December 2004, Indian Ocean tsunami took more than a decade to recover.

7.4   Pollution from Hurricanes

When a hurricane makes landfall, the media covers the accompanying death and destruction from wind and flooding as headline stories. Considering the devastating impact on communities and residents, this priority is appropriate and necessary. Everything else about the effects of the storm is incidental. One of these incidental aspects is the environmental damage that it does. This damage can be natural or a result of the destruction of manmade structures. This damage is usually not reported in the same detail as the primary effects of the hurricane. The environmental damage, however, can produce significant health hazards and take several years to recover.

Hurricanes are large storms that form in tropical waters and develop into the most powerful meteorological event on Earth. A fully developed hurricane produces enough energy per day to supply energy needs of the United States for a year. These storms are called hurricanes in the Atlantic Ocean, the Caribbean Sea, the Gulf of Mexico, and the Pacific Ocean along the North American coast. They are called typhoons in the rest of the Pacific Ocean and cyclones in the Indian Ocean. All are produced in the same way and are equally powerful.

Hurricanes are not associated with major weather-producing systems such as fronts nor are they associated with permanent weather driving features such as the jet stream. Instead, they begin inauspiciously as a modest cluster of thunderstorms off the west coast of Africa in the eastern Atlantic Ocean. They can only develop within large stable air masses with little wind. They also only develop in areas where the water temperature on the surface of the ocean is 79 °F (25 °C) or greater. The humid air above this warm ocean is swept upward by the updrafts of the thunderstorms. As the air rises, the moisture in it condenses to form rain, and the phase change transfers latent heat into the storm. This energy added to the storm causes the pressure to decrease and the updraft to intensify. The updraft winds whip up the surface of the ocean into waves that expose more water surface, from which humidity forms. Increased humidity produces more condensation and more latent heat transfer, which powers up the storm.

The intensifying storm rotates counterclockwise in the Northern Hemisphere as the result of the Coriolis Effect from rotation of the Earth. Storms rotate clockwise in the Southern Hemisphere. As the storm grows, it develops a clear area at its center called the eye, which ranges from 20 to 40 miles (32–64 km) across. An eyewall surrounds the eye and contains the strongest winds and heaviest rains. It is the most dangerous part of the storm. Spiral rain bands of 5–10 miles (8–16 km) width and several hundred miles in length radiate outward from the eyewall. The rain bands form the outer part of the hurricane. Hurricanes are usually around 300 miles (480 km) across but can be from 200 to more than 500 miles (320–800 km). The areas in a hurricane with clouds and rain show where air is rising. Where there is no rain, the air is falling.

The path and speed that a hurricane will travel are very difficult to predict. They normally travel at 15–20 miles (24–32 km) per hour but can reach speeds of 60 miles (96 km) per hour. Hurricanes are actually fragile. If they encounter ocean with surface temperatures below 79 °F (85 °C), a front with high-level winds, wind shear, or they make landfall, they quickly degrade.

These storms are classified as tropical depressions, tropical storms, or hurricanes. Tropical depressions are organized systems of thunderstorms with surface circulation and maximum sustained of winds of 38 miles per hour (61 km/h). Tropical storms are organized systems of strong thunderstorms with surface circulation and maximum sustained winds of 39–73 miles (62–117 km) per hour. Hurricanes are intense tropical systems of strong thunderstorms with well-defined circulation and maximum sustained winds of 74 miles (118 km) per hour or more.

The Saffir–Simpson Damage Potential scale is the standard system for classifying hurricanes. This scale is primarily determined by the maximum sustained winds but has several factors that are also evaluated. Factors include height of the storm surge waves, inland penetration of ocean water and damage done to homes and other structures. There are five categories of hurricanes in this scale:

- *Category 1 hurricanes* have sustained winds of 74–95 miles per hour (119–153 kph). The storms damage unanchored mobile homes, shrubbery, and trees but not buildings.
- *Category 2 hurricanes* have sustained winds of 96–110 miles per hour (154–177 kph). They damage mobile homes, shrubbery, and trees, with some trees blown down. They also damage roofing material, doors, and windows as well as causing coastal flooding.
- *Category 3 hurricanes* have sustained winds of 111–130 miles per hour (178–209 kph). The storms cause structural damage to small homes and utility buildings as well as shrubbery and trees, which are typically defoliated and blown down. Mobile homes are destroyed. Coastal flooding spreads inland as far as 8 miles (13 km).
- *Category 4 hurricanes* have sustained winds of 131–155 miles per hour (210–249 kph). Storms destroy mobile homes and cause complete roof failure on small residences and extensive damage to doors and windows. Areas at less than 10 ft (3 m) above sea level may flood, requiring evacuation as far as 6 miles (10 km) inland.
- *Category 5 hurricanes* have sustained winds greater than 155 miles per hour (249 kph). The massive storms produce roof failure of houses and industrial buildings and some complete building failures. Small utility buildings are blown away. Flooding causes major damage to lower floors of all structures below 15 ft. (4.6 m) above sea level and less than 1500 ft (462 m) from the shoreline. All residential areas on low ground within 5–10 miles (8–16 km) of the shoreline are typically evacuated.

There are several severe hazards when hurricanes make landfall, including high winds, storm surge, heavy rains, and tornadoes, which typically lead to pollution and environmental damage.

- *High winds* Hurricanes have extreme wind speeds, especially around the eye from about 25 miles (40 km) to 150 miles (240 km) from the center. The counterclockwise rotation and forward movement results in high and low wind speed sides of the hurricane. In the Northern Hemisphere, the forward speed adds to the sustained wind speed on the right side of the storm, producing higher wind speeds than the rest of the storm. On the left side of the storm, the forward motion of the storm reduces the wind speed by that amount. For example, if a Category 3 hurricane with average wind speed of 125 miles (200 km) per hour moves northward at 45 miles (72 km) per hour, the right side has a wind speed of 170 miles (272 km) per hour and the left side has a wind speed of 80 miles (128 km) per hour. The right side is much more dangerous, with enhanced hazards.

    The high winds destroy buildings, storage facilities, manufacturing plants, power lines, telecommunications service of all types and can damage above-ground transmission and transportation lines. They can overturn trucks and railroad cars. It is this type of destruction that can cause chemical spills and related types of pollution on land.

    Winds can damage oil platforms and petroleum distribution facilities. These are commonly at sea or along the coast where hurricanes are strong. The damage is a combined effect with storm surge because it is driven primarily by the wind.

- *Storm surge* The forward motion of a hurricane combined with the high wind speed forms a dome of water on the surface of the ocean 50–100 miles (80–160 km) wide. The dome moves ahead of the hurricane and devastates coastal areas as a storm surge. Category 1 storms have a storm surge of 4–5 ft (1.2–1.5 m) height above mean sea level. Category 2 is 6–8 ft (1.8–2.5 m), Category 3 is 9–12 ft (2.8–3.7 m), and Category 4 is 13–18 ft (4–5.5 m) above mean sea level. Category 5 has a surge more than 18 ft (5.5 m) above normal.

  The height of the storm surge impacting the coast does not just depend on the wind speed, it also depends on the slope of the seafloor and the shape of the coastline. A shallow seafloor slope allows the storm surge waves to bottom out and rise up. This produces larger storm surges. Concave coastlines including harbors and bays focus incoming waves to increase their height and speed. If the shoreline is convex, waves will dissipate because their energy is spread out along a longer wave front. There are additional factors including tidal cycle and angle of incidence of the storm that also affect the storm surge. The glancing blow of a hurricane along its left (weaker) side while traveling quickly produces a far lower surge than a head-on landfall.

  Storm surge produces heavy damage to coastal structures. The distance storm surge penetrates inland depends on the topography of the coast. In flat coastal terrains, a storm surge can penetrate a half mile (0.8 km) inland or more. The waves demolish all but the strongest structures and flood everything with salt water. Utility transmission lines are severed and above-ground chemical storage facilities can be removed from their footings. Coastal oil refineries and production facilities can be badly damaged or destroyed. The saltwater stands in puddles for days, killing all but the hardiest of surviving plants and animals, and deposits salt onto the soil, making revegetation difficult for years. The stagnant sea water can breed disease and ruin local water supplies.

- *Heavy rain* The total rainfall of a hurricane depends on the storm and land conditions but usually exceeds 6–12 in (15–30 cm) and can be up to 40 in (102 cm). The main controlling factor on rainfall is the size and intensity of the storm. Larger storms with lower pressure usually include more rain. Rain is heavier in slower-moving storms because it rains for a longer period in a given location. Topography is also a major factor. Air that is driven up hill slopes cools, causing increased condensation and heavier rain. This is called orographic precipitation. Rugged areas suffer greatly if a hurricane moves inland. In Central America, westward moving storms move up the volcanic mountains of the west. In 1998, Hurricane Mitch dumped 40 in (102 cm) of rain over a few days, resulting in massive floods and mudslides that caused more than 11 000 deaths.

  Inland flooding is the primary cause of hurricane fatalities. It can cause heavy environmental damage as well such as overflowing of landfills, including Superfund sites, and wash out the waste. This is particularly true in areas where hurricanes are uncommon. The flood water results in high pressure on septic systems and sewers, which can cause them to back up and overflow onto the surface or even into houses. Flood damage to buildings, utility transmission lines, gas stations, chemical storage facilities, oil refineries, and other vulnerable structures can cause spills and leaks.

- *Tornadoes* The rain bands of a hurricane contain thunderstorms which can be severe. The severest thunderstorms can spawn tornadoes. Tornado threats are most common in the more intense hurricanes. These tornadoes mainly form within 150 miles (240 km) of the coastline because the hurricane intensity decreases as it moves inland.

  Tornadoes can cause disruption of utilities and damage to buildings, storage facilities, and most permanent structures depending upon the strength of the tornado. Pollutants are commonly discharged into the environment during a tornado. They can uproot trees that can upset local ecosystems for years. The amount of debris generated by tornadoes is great and litters the landscapes of impacted areas.

These catastrophic impacts from hurricanes cause the major environmental damage. In impacted areas, reports of releases to the US Coast Guard, can number in the dozens or more. Most accidental releases are small but together, they constitute a significant amount of environmental pollution. Hurricane Katrina in 2005 caused New Orleans to be covered in standing water of raw sewage and chemicals from releases. There were 25 chemical releases reported of 50 barrels or more during this storm. When Hurricane Ike struck the Houston area in 2008, it resulted in 51 pollution reports, with 15 requiring action.

CASE STUDY 7.5 2005 Hurricane Katrina

Toxic Soup in New Orleans, Louisiana

Hurricane Katrina was the costliest natural disaster in United States history. It was spawned in the Atlantic Ocean in August 2005 and strengthened from a tropical depression to a tropical storm on 23 August just before it struck south Florida as a Category 1 hurricane. It crossed into the Gulf of Mexico and the warm waters caused it to quickly strengthen to a Category 5 on 28 August. It reached its peak of power that day, with winds of 175 miles (280 km) per hour. Katrina turned northward from its westward track and headed to the east side of New Orleans, Louisiana (Figure 7.6a). This was the most dangerous track for New Orleans.

The stronger left side of the storm drove a massive storm surge northward onto the Mississippi Gulf Coast, causing extensive damage.

The westward winds on the leading edge of the storm drove this surge of Gulf of Mexico water westward through a narrow opening in a strip of land, into Lake Ponchartrain, Louisiana. This huge lake sits immediately north of New Orleans and measures 35 miles (56 km) long and 25 miles (40 km) wide. Earthen levees along the south shore of the lake were supposed to protect New Orleans from flooding. Hurricane Katrina passed over the east of the lake at Category 3 and the southward winds of the weaker left side drove a storm surge southward on the flooded lake that overtopped the levees and sent a wall of water southward, completely flooding the city. This surge breached the levees in 53 places and submerged 80% of New Orleans.

More than 1836 people were killed in Hurricane Katrina or the subsequent flooding, making it among the deadliest hurricanes in the United States. It caused upwards of $81.2 billion in damage, making it the costliest natural disaster in US history. New Orleans would take nearly a decade to recover despite extensive rebuilding efforts.

New Orleans was literally submerged in flood waters by Katrina (Figure 7.6b). These brackish flood waters were pumped back into Lake Pontchartrain using powerful pumps that operated 24 hours per day for 43 days. As they struck, flood waters overflowed the sewer systems and raw sewage poured out and mixed with them. Numerous homes, industrial plants, businesses, landfills, and cemeteries were flooded, sweeping toxic substances and diseases into the flood water, threatening public health. They contained a mix of raw sewage, toxic chemicals, heavy metals, bacteria, pesticides, and crude and refined oil, which fermented in the Louisiana sun. These dangerous waters were termed

(a)

(b)

FIGURE 7.6 (a) Satellite image of 2005 Hurricane Katrina just before it struck the Gulf Coast. *Source:* Courtesy of NASA. (b) Aerial photograph of downtown New Orleans flooded in toxic soup. *Source:* Courtesy of NOAA. (c) Close up aerial photograph of New Orleans flooded in toxic soup. *Source:* Courtesy of NOAA.

(c)

FIGURE 7.6   (Continued)

"toxic soup" and cleaned up by the state and military for as long as the next few years. The toxic soup was handled very carefully because of the potential for poisoning and disease (Figure 7.6c).

As a result of Hurricane Katrina, there were at least 25 spills of 50 barrels (2100 gal) or more of petroleum and petrochemicals, all requiring careful cleanup. Of these, six were major, five medium, and, in addition, there were more than 5000 minor oil and hazardous substance spills. Additional minor spills were identified for the next year or two. It is estimated that more than 9 million gallons (34.2 million l) of oil was released in the major spills alone.

Major crude oil spills from the petroleum industry required extensive treatment and repairs and made up the largest volume of the pollution. The Murphy Oil Corporation Refinery, Chalmette, LA released more than 819 000 gal (3.1 million l) of oil during the storm. Hundreds of homes adjacent to the facility were saturated with oil during the hurricane. Many of the homeowners agreed to a settlement from Murphy Oil, but many others began a class action lawsuit against the company. In Pilot Town, LA, more than 1.1 million gallons (4.2 million l) of crude oil leaked from an above-ground storage tank owned by the Shell Pipeline Company and into a tank dike and the surrounding area. In Nairn, LA, more than 136 290 gal (517 900 l) of crude oil spilled from a pipeline. The Chevron pipeline at Port Sulfur spilled approximately 1.4 million gallons (5.3 million l) of oil. There were other smaller spills that together constituted a considerable amount of oil.

7.5   Mass Movement

Mass wasting is the movement of material down a slope through the force of gravity. It is classified by the process of failure and transport. The main subdivision for classification of mass movements is based on the material. Rock failures have a system separate from the movement of soil and surficial materials.

Rock failures. The failure of solid rock on cliff faces and slopes is based on the way the rock falls, slides, or topples. The release of the rock is not part of the classification but might be the result of an earthquake or the freezing and thawing of water in cracks called frost wedging.

Soil and debris failures. The term "landslide" includes many subdivisions of soil and debris slope failures (Figure 7.7). The three factors in the classification of these failures are the proportion of water in the failing material, the speed of movement, and whether the material is coherent or fragmented. If the material has less than 20% water, the fastest failure is a sturzstrom and debris avalanche, which may begin as a rockfall. Avalanches contain rocks and debris and travel at speeds of 60 miles (96 km) per hour to more than 250 miles (400 km) per hour. Grain flows are made of similar materials but travel at speeds between 60 miles (96 km) per hour and 325 ft (100 m) per hour. Earth flows are true landslides in which coherent masses slide down a slope between 325 ft (100 m) per hour and 0.5 in (1 cm) per year. The process of creep is slower than this.

In a slurry flow, the moving material contains between 20 and 40% water and the speed is 250 miles (400 km) per hour. A mudflow is slower than this but faster than 0.63 miles (1 km) per hour. Debris flows are more cohesive than mudflows and move 0.63 miles (1 km) per hour to 0.5 in (1 cm) per year. Solifluction flows at less than 0.5 in (1 cm) per year.

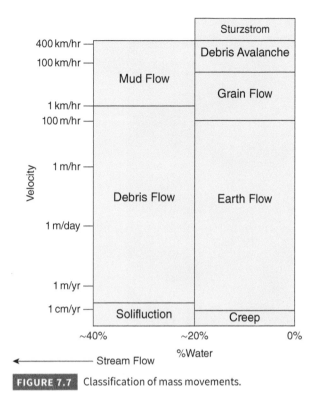

FIGURE 7.7 Classification of mass movements.

Several types of pollution are caused by mass movements either directly by slope failure or indirectly from structures destroyed by the failure. Direct pollution is largely from material flowing into a stream or other surface water body. The added material causes siltation of the river which can be muddy for days or even weeks. If a failure impacts a reservoir, it leaves the water undrinkable. Significant amounts of particulate air pollution (dust) can be generated by an avalanche. Debris flows and mudflows can also cover fertile soil with material that impacts its ability to grow.

The main direct pollution is from a mudflow, which can flow through towns, inundating everything. A thick flow can burst through walls of buildings and crush smaller buildings. The heavy weight of the mud can cause sewers to overflow and infiltrate surface water supplies, promoting disease. Rock failures and high-speed slope failures destroy most structures in their path including above-ground storage tanks, pipelines, rail lines and tanker cars, manufacturing plants, and power generating plants. Spills of chemicals are possible from such failure, which further spread the pollution by continued movement. Civil and geotechnical engineers attempt to anticipate these issues and plan accordingly.

Slope failures may also rupture underground structures. They can unearth pipelines and underground storage tanks if they are on the failing slopes. These structures can rupture or slide down the slope producing a chemical spill which is spread by the mass movement and flows down the slope. If a landfill is located on a hillslope, it can be disrupted by a mass movement causing a dispersed pollution. If oil or gas pipelines are ruptured, a leak can cause local pollution and a significant risk for fires and explosion.

CASE STUDY 7.6 1987 Reventador Landslides, Ecuador

Destruction of Oil Pipelines

The large Trans-Ecuadorian oil pipeline has been transporting as much as 360 000 barrels (42 927 m³) of crude oil per day since 1972. The pipeline transports oil from eastern Ecuador, 309 miles (498 km) across the rugged Andes Mountains to the port of Esmeraldas on the Pacific Ocean (Figure 7.8). Part of the pipeline passes along the steep-sloped El Reventador volcano, which is covered in ash. To pass over this terrain requires numerous pumping stations which increase the pressure in the pipeline depending upon the volume at the time.

The month of February 1987 was very rainy near Reventador, producing 23.6 in (60 cm) of precipitation. This soaked the volcanic ash of the volcano slopes, producing a thick layer of mud. On 5 March two strong earthquakes struck the area, the first with magnitude 6.1 at 8:54 p.m. and the second with magnitude 6.9 at 11:10 p.m. The epicenters were close together in the Napo Province about 46.6 miles (75 km) east-northeast of the capital city of Quito and about 15.5 miles (25 km) north of the Reventador volcano. Surface shaking produced a Modified Mercalli Scale intensity of IX, which qualifies as violent.

Damage directly from the earthquake was moderate but the real disaster resulted from the shaking loose of the steep water-soaked slopes. Debris avalanches, rock slides and falls, and debris, mud, and earth flows covered the east side of the Andes but especially on Reventador. Estimates of the volume of mud, rocks and debris from this mass wasting on the volcano range from 98 million to 144 million cubic yards (75–110 million m³), which left deposits up to 30 ft (9.1 m) deep. Many structures were destroyed by the mass wasting as was the only

highway from Quito to Ecuador's oil fields and northeastern rainforests. They also caused almost all of the deaths from this event which numbered at least 1000 and possibly 2000 people or more as 4000 people were reported missing.

The debris and mudflows also caused a significant NATECH accident. At the time of the mass movement the 24-in (60.6 cm) diameter Trans-Ecuadorian pipeline was transporting 250 000 barrels (29 810 m³) of oil per day at a pressure of 1400 pounds per square inch (96.5 bars). Debris flows completely destroyed a 6.5 mile (10.5 km) section of the pipeline spilling 20 000–30 000 barrels (2385–3577 m³) of oil and another 10 miles (16.1 km) was severely damaged by mudflows. The bridge bearing the pipeline across the Aquarico River was swept away. The Salado and other pumping stations were also badly damaged causing additional spills including a 4500-barrel (537 m³) spill from a storage tank at Salado. The footings that supported the pipeline were damaged, badly deforming another 5 miles (8 km) of pipeline.

In all, about 43.5 miles (70 km) of the Trans-Ecuadorian oil pipeline was damaged, with approximately 25 miles (40 km) requiring full reconstruction. This work took several months. This large, replaced segment makes the Reventador NATECH accident the single largest pipeline failure in history. In addition, as much as 100 000 barrels (11 924 m³) of crude oil were spilled in the accident. This spill badly polluted the Coca and Aguarico Rivers, killing millions of fish and destroying the environment all around the damage area. Oil even spread into other countries. It is estimated that the cost of the disaster was $1 billion in 1987 US dollars.

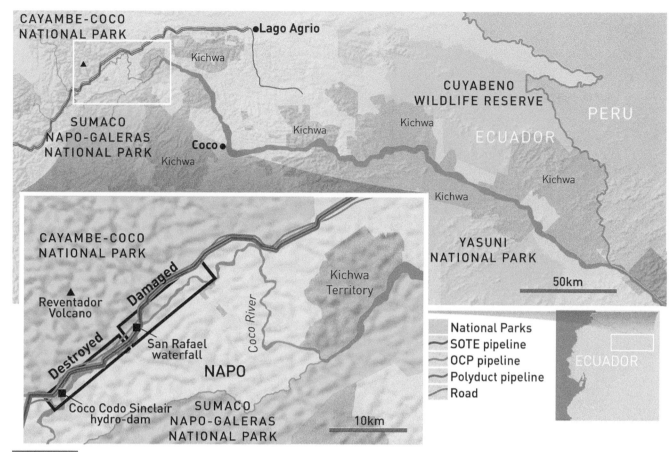

FIGURE 7.8 Map of the area of the 1987 damaged and destroyed pipeline in Ecuador (inset) showing the location in larger area maps.

Pollution of Groundwater

Polluted Earth: The Science of the Earth's Environment, First Edition. Alexander Gates.
© 2023 John Wiley & Sons, Inc. Published 2023 by John Wiley & Sons, Inc.
Companion website: www.wiley.com/go/gates/pollutedearth

Words you should know:

Artesian well – A well that is drilled into a pressurized aquifer so the water flows from it without pumping.

Aquifer – A soil, sediment, or rock that contains and transmits groundwater.

Aquifer recharge zone – The area where an aquifer reaches the surface and surface water infiltrates and fills it.

Capillary fringe – Thin area above the water table but containing some groundwater through capillary forces.

Cone of depression – Pumping wells remove enough water to draw down the water table in a cone-shaped depression.

Darcy's Law – A relationship relating groundwater flow to the hydraulic conductivity of the aquifer through which it flows.

Fractured rock aquifer – A body or section of bedrock that stores and transmits groundwater through intersecting open fractures.

Groundwater – Water that occurs in rock, sediment, or soil below the ground surface.

Hydraulic conductivity – The ability of groundwater to flow through an aquifer.

Karst aquifer – A body or section of limestone bedrock that stores and transmits groundwater through dissolved openings such as caves and caverns.

Perched aquifer – An isolated aquifer at a higher elevation than the main shallow aquifer in an area.

Phreatic/saturated zone – Rock, sediment, and/or soil below the water table that is constantly saturated with groundwater.

Pollutant plume – The underground spread of pollutant from a source in the groundwater system.

Vadose/unsaturated zone – Rock, sediment, and/or soil above the water table through which percolating water passes as it travels from the surface to the groundwater system but which is dry most of the time.

Water table – The top surface of groundwater.

8.1 The Groundwater System

In contrast to surface water which forms bodies on the surface of the Earth, groundwater saturates soil, sediment, or rock below the surface like an underground water body. The top surface of that body is the water table (Figure 8.1). The depth of the water table varies by location but also over time. In wetlands and some lakes and streams, the water table is at the surface and even above it whereas in dry areas it can be 1000 ft (320 m) deep or more. The depth also varies throughout the year in areas based on the amount

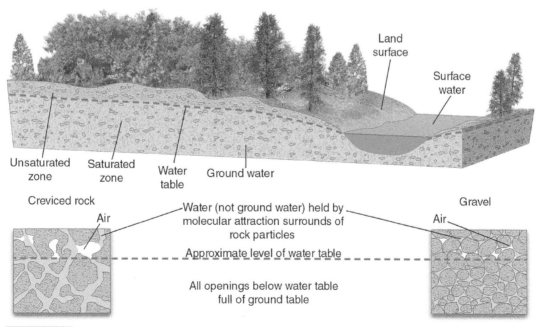

FIGURE 8.1   Diagram showing the water table with the saturated zone below and the unsaturated zone above and its relation to surface water. Insets show fractures (left) and gravel (right) both saturated with groundwater and in the fringe zone.

of precipitation and the thawing of ice and snow. The depth of the water table can therefore be seasonal and have longer-term variation reflecting droughts, wet periods, and large storms.

The rock, sediment, and/or soil below the water table is the phreatic or saturated zone (Figure 8.1). This zone is constantly saturated with groundwater in all of the pore spaces. Water can be held in some of the pore spaces just above the water table in the capillary fringe where capillary action allows water to fill some pore spaces. The rock, sediment, and soil above the water table is called the vadose or unsaturated zone. Water from rain, thawing snow and ice, or human activity passes through the unsaturated zone as it percolates to the water table. Otherwise, the pore space in the zone is filled with soil gas, which is mostly air but can include other gases depending on the area. These designations are only for the shallowest groundwater and those which are open to surface influence. Such shallow groundwater-bearing units are called unconfined aquifers.

An aquifer is a zone within or a layer of soil, sediment, or rock that can yield water in large enough quantities to be useful depending upon the need to be served. They can be at the surface or be at 1000 ft (320 m) or more in depth. Aquifers can be a few feet to hundreds of feet thick and serve small areas, states, or multiple states. If a single house is the sole user, even a small aquifer may be sufficient. Rock, soil, or sediment aquifers must have enough porosity to store water and good permeability so the water can flow (Figure 8.1). Water is most commonly pumped out of an aquifer from a well. The water that is pumped out must be quickly replaced by other water in the aquifer to prevent the well from going dry. In most cases, it is not quick enough and a cone of depression forms on the water table around the well (Figure 8.2). Once the well stops pumping, groundwater fills in the cone and it returns to normal. The best type of material for an aquifer is a large unlithified or weakly lithified beach sand-type deposit. The Ogallala Aquifer is composed of weakly consolidated river sand and gravel and supplies water to parts of South Dakota, Wyoming, Nebraska, Colorado, Kansas, Oklahoma, New Mexico, and Texas. It is the largest single aquifer in the United States.

Groundwater flows in the pore spaces between sediment, soil, or rock grains in an aquifer (Figure 8.1). This is the primary porosity. Infiltrating water travels around individual grains as it travels deeper into the Earth. The water chemically reacts with the surface of these grains, dissolving some chemical compounds and precipitating others. Pollutants and other ions are removed from the water, and other ions are added through this process. "Hard" water is produced by dissolving calcium carbonate (calcite or lime) and other compounds from the grains. Finer grains in the aquifer sediment have more surface area for groundwater to contact and facilitate chemical exchange. This filtering is one way purification takes place. The infiltrating water squeezes between tightly packed grains. The necks at grain contacts are so narrow that bacteria, viruses, and all solids in the infiltrating groundwater cannot pass through. They are filtered out by these passages. It is these natural filtering processes that purify groundwater by the earth into potable water for household use.

In contrast, an aquiclude has little to no permeability. As such, it restricts or prevents flow of groundwater, regardless of its porosity, and cannot be used for water production. Aquicludes are primarily composed of shale or clay, but they can be other rock and sediment types as well. An aquitard has minor permeability and can transmit water slowly, which is like a leaky aquiclude. They can produce water for small applications but they cannot be major water producers.

There are typically at least two aquifers in many areas. Wells in the unconfined or shallow aquifer were historically dug by hand and water was retrieved by means of a hand pump or bucket. Unfortunately, shallow unconfined aquifers are more easily contaminated from surface spills such as herbicides or pesticides from lawn care, or subsurface leakage of underground septic or oil tanks. People generally do not use water from the shallow aquifer in populated areas. Coastal areas are an exception to this because deep waters are saline and not potable.

Deeper aquifers are usually confined between aquiclude and/or aquitard layers and are drilled to produce water (Figure 8.3). The water in this aquifer has traveled a long distance from the surface, filtering through many rock and soil types. It contacts many mineral surfaces where contaminants can be removed and filters through tightly packed grains, thereby removing the bacteria and viruses. Water in deep confined aquifers is usually potable and provides abundant community and private well water. These confined aquifers, however, are not always clean. Pollutants that are denser than water (Dense Non-Aqueous Phase Liquids – DNAPLs) filter to these deeper aquifers more quickly than water. Most of these deep aquifers get their water from a significant distance away where the aquifer unit reaches the surface. These areas are called aquifer recharge zones (Figure 8.3). Surface pollutants

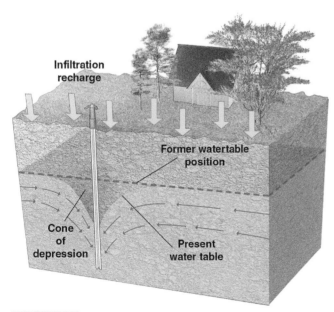

FIGURE 8.2 Block diagram showing a cone of depression on the water table and flow of groundwater around a pumping well.

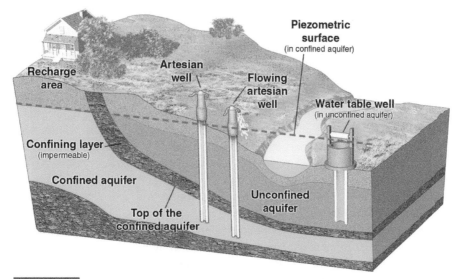

FIGURE 8.3 Block diagram showing the recharge area for an aquifer confined between aquicludes and water wells penetrating it. The piezometric or potentiometric surface show the height artesian wells and springs will shoot to.

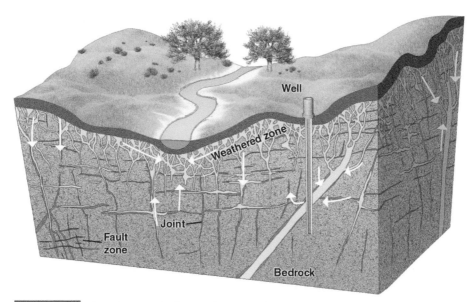

FIGURE 8.4 Block diagram of a fractured rock aquifer illustrating both the inability of them to purify water and the difficulty to drill a well that produces a sufficient supply.

there can slowly contaminate the aquifer, eventually making it unpotable. At that point, a still deeper aquifer must be found for production and care must be taken not to create a conduit to contaminate the deeper aquifer in the process of drilling wells.

There is no primary porosity and permeability in crystalline metamorphic and igneous rocks because the mineral grains are tightly knit, leaving no pore space. The only way these rocks can produce water is if they contain enough fractured rock to form aquifers (Figure 8.4). The fractures form interlocking networks that transmit water from the surface to depth but they are independent of rock type and difficult to predict. The fractures are called joints if they are relatively planar and parallel or subparallel to each other. Fractures and joints can be short or as long as tens to hundreds of feet and be simple cracks or have apertures of 0.25 in. (5 mm) or more. These joints and fractures flow groundwater like pipes in a house. Groundwater flowing through joints and fractures is fast and makes minimal contact with the rocks (Figure 8.4). This reduces chemical reactions and thus purification. The apertures of the joints and fractures are so large that microorganisms and particulate are not removed but migrate to depth unimpeded. As a result, groundwater in crystalline rocks is generally of lower quality and is sensitive to human activity both on the surface and underground. With high density of the joints and fractures, drilling for water in crystalline rocks potentially yields higher production. Zones of very high fracture density are fracture zones or fracture trends and yield very high production. In areas of low joint and fracture density, drilling may have to be very deep to encounter producible water. The inconsistent distribution of joints and fractures can result in one house producing water from a shallow well and the neighboring house having to drill a deep well for their water.

Many areas of the United States are underlain by crystalline rock and rely on fractured rock aquifers. Besides the Mesozoic basins, the entire Piedmont Province on the East Coast and the entire Green Mountains, Highlands Province, and Blue Ridge Mountains are underlain by crystalline rocks. This means that the cities of Boston, New York, Philadelphia, Baltimore, Washington, D.C., and Atlanta are underlain by crystalline rock with unconfined fractured rock aquifers, though not used for drinking water. The Northeast has glacial deposits overlying the crystalline rock which forms a complex double aquifer system. Crystalline rocks also underlie the entire Canadian shield including the part that juts into the United States across northern Michigan, Wisconsin, and Minnesota. It has glacial cover as well forming a similar double aquifer system. From the Rocky Mountain front westward, there is a complex interfingering of crystalline rocks and sediments. In the Basin and Range Province, many of the ranges are crystalline and many of the basins are filled with sediment. There are many other crystalline areas in North America and around the world. Virtually all older rocks are crystalline as are craton interiors.

The other common situation of large primarily unconfined aquifers is in areas underlain by limestone, which erodes to form karst aquifers (Figure 8.5). These geologic formations can yield usable to enormous quantities of groundwater. Karst aquifers form by acidic surface and groundwater infiltrating fractures in the limestone and dissolving the rock around them. With time, cave systems develop and transmit groundwater in huge underground rivers. Caves act like large pipes and as such are ineffective at purifying water. Virtually all surface water quickly drains through cracks and sinkholes into the caves below, making rivers few and small in karst areas. In areas with a shallow water table, sinkholes appear as small circular ponds and lakes. The water table is marked by the surface of these water bodies.

Karst systems are located where there is limestone, but they are especially prevalent in areas of adequate rainfall and extensive groundwater. Among the most famous karst aquifers is the Floridan Aquifer which underlies the entire state of Florida and parts of Georgia and Alabama. It is the site of a struggle between the agricultural and industrial growth and keeping drinking water clean. Karst aquifers cover a large area of the eastern central part of the United States from Vermont through Ohio, Illinois, and Michigan, southward through West Virginia, Kentucky, and Tennessee as well as Texas among other areas.

A perched aquifer is a lens of porous sediment within aquiclude and sits at a higher elevation than the main subsurface aquifer in an area (Figure 8.6). Perched aquifers form by unusual depositional circumstances and are common in glaciated areas. An example is a kame which is a ridge of sand and gravel deposited by melting glacial ice in glacial lakes. Kame sediments have high porosity and permeability and are encased within impermeable clays of the lake deposits. Kame deltas form in glacial lakes at various levels and also produce potential perched aquifers of sands and gravels within clay deposits. Bowl-shaped depressions in crystalline rocks may be filled with sediments that are cut off from the main groundwater system and serve as perched aquifers.

Perched aquifers contain small quantities of groundwater, which can quickly dry up during droughts or be drained by pumping wells. Their small size also means that they can be easily polluted. However, because they are isolated from the main aquifer system, the pollution cannot easily enter them.

FIGURE 8.5   Block diagram of a karst terrane showing that groundwater flows unfiltered through underground rivers in caves.

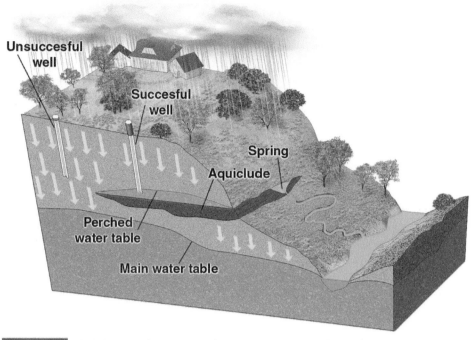

FIGURE 8.6 Block diagram showing groundwater in a perched aquifer.

8.2 Groundwater Movement

In most cases, groundwater moves. Because it percolates through soil, sediment, and rock, the movement is very slow, especially compared to surface water. The speed is a function of the properties of the soil, rock, and sediment it passes through and is largely a function of permeability. The term for this is hydraulic conductivity, which is the function of how easily groundwater passes through pore spaces or fractures in materials. The amount of groundwater movement through materials is approximately determined by Darcy's Law, which is:

$$Q = KA\, dh/dl \tag{8.1}$$

where K is the hydraulic conductivity, Q is the rate of discharge (the amount of water passing through the material over time), A is the area through which the water is passing, and dh/dl is the hydraulic gradient or slope on the water table, surface, or aquifer unit. Although listed in a single term K, the controls on hydraulic conductivity can be quite complicated and change from location to location as well as with time.

As shown by Darcy's Law, the drive on groundwater movement is the slope. In areas with any sort of topography, groundwater flows downhill just like surface water (Figure 8.7). In general, it flows directly downhill which means perpendicular to topographic contour lines. However, this is not always true. Lateral variations in soil or sediment can cause it to deviate significantly. A long deposit of gravel at an angle across a hillslope will capture most of the water flowing downhill and redirect it. This is because gravel is much more permeable than most soil and has a higher hydraulic conductivity. In areas where the groundwater resides in fractured rock aquifers, zones of concentrated fractures will capture the groundwater as well and redirect it. If the fractures are few and widely spaced, their general orientation can slightly redirect the flow.

Streams flowing down hillslopes can also impact the groundwater flow direction. If a stream is draining water into the groundwater system because it is at a higher elevation than the water table, it is an influent stream (Figure 8.8a). The draining pushes the groundwater away from the stream. If the water table is at a higher elevation than the stream, groundwater will be drawn toward the stream and flow into it. This is called an effluent stream (Figure 8.8b).

Where topography is flat, the speed and direction of groundwater flow are much more complex. In wetlands (also known as swamps, bogs, and marshes), the water visible at the surface is the top of the water table. With no other changes, this water and that in the connected aquifer can sit for long periods of time without appreciable movement. Droughts can lower the water in the wetlands or aquifer and drive some minor flow between them as can storms or flooding events. Otherwise, groundwater flow in flat terrains is driven by interaction with streams like on the slopes or governed by larger regional groundwater flow patterns.

Aquicludes can seal deeper aquifers if they bound them top and bottom. The pressure of gravity as the groundwater flows down gradient is therefore built up in the aquifer because it cannot escape. This makes the water pressure increase in the

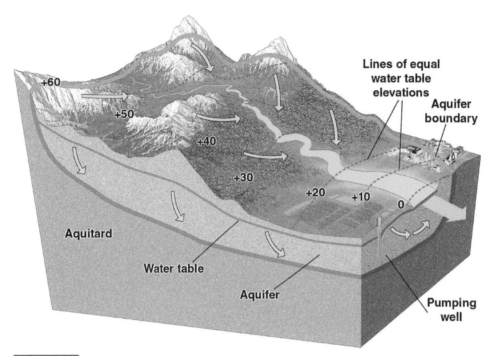

FIGURE 8.7 Block diagram showing the flow of groundwater down gradient (large arrows) and contour lines on the water table (dashed and numbered).

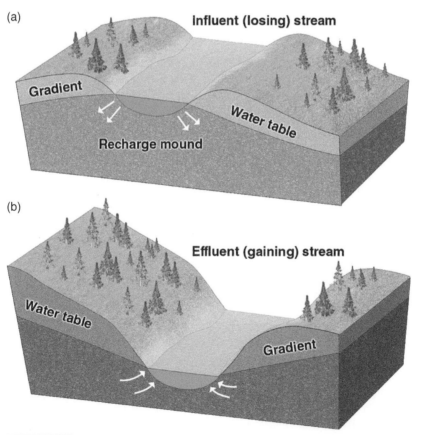

FIGURE 8.8 Block diagrams of streams showing the relation of groundwater to surface water for (a) influent streams and (b) effluent streams.

aquifer similar to a garden hose. This sealed and pressurized aquifer is called an artesian system (Figure 8.3). Wells drilled into an artesian system emit water without a pump and can shoot water tens of feet into the air to the piezometric or potentiometric surface, the height to which the water will rise. The pressure driving this rise is controlled by the weight and height of the water in the sealed aquifer known as the hydrostatic head. A great example of an artesian system is the Edwards Formation aquifer in

Texas. This aquifer is recharged in an elevated area and underlies the city of San Antonio at a much lower elevation. As a result, when San Antonio was first developed, the water would shoot up to 30 ft (9 m) into the air when a well was drilled. These were called artesian wells.

CASE STUDY 8.1 Edwards Aquifer, Texas

Artesian Groundwater System

The Edwards Aquifer of south-central Texas is among the most productive aquifers in the United States (Figure 8.9a). It is a limestone aquifer that serves the cities of Austin and San Antonio and provides water to more than 2 million people. This aquifer is contained within the Edwards Formation which is about 300–500 ft (92–154 m) thick. The aquifer area forms a 160-mile (256-km) long arc that is 5–40 miles (8–64 km) wide and extends through 13 Texas counties. This aquifer area forms three zones, the contributing zone, the recharge zone, and the artesian zone. The contributing zone is on the Edwards Plateau and covers about 4400 square miles (11 264 km²). From this

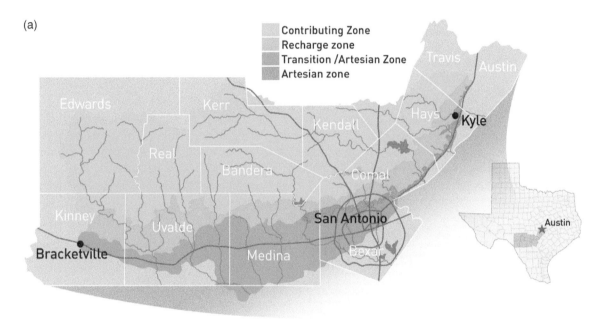

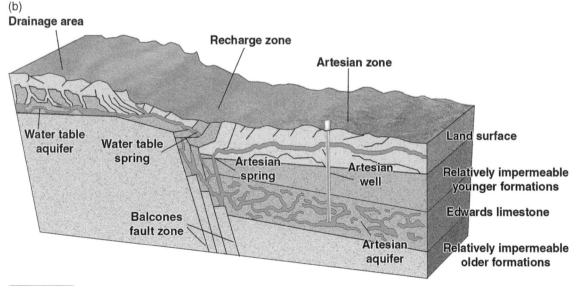

FIGURE 8.9 (a) Map showing the contributing, recharge, transition, and artesian zones of the Edwards Aquifer. Inset shows the location of the Edwards Aquifer in Texas. (b) Diagram showing a cross-section through the Edwards Aquifer from north (left) to south (right) with the zones shown.

higher elevation, the surface water moves downhill at a higher veloc- ity The aquifer recharge zone is downslope from the contributing zone and facilitated by the Balcones fault zone of fractured rocks, which allows infiltration of surface water into the groundwater sys- tem. Below this is the artesian zone where the spectacular flow rates occur (Figure 8.9b).

In the recharge or contributing area, the Edwards Aquifer con- tains numerous sinkholes and is unconfined. It is a shallow uncon- fined aquifer here, receiving water from surface precipitation and not under artesian pressure. A transition area exists between the recharge and artesian zones that has characteristics of each. The artesian zone occurs where the aquifer is confined between impermeable, aqui- clude layers. This allows the steep hillslope to increase the pressure in the aquifer.

The city of San Antonio started using well water from the Edwards Aquifer in 1891. The first well drilled into it resulted in water shooting 25 ft (7.7 m) into the air. The Edwards Aquifer quickly became the water supply for the city and by1896, there were 40 wells into the aquifer. As more wells were drilled, the pressure in the aquifer was soon reduced to a trickle. In other areas of Texas, however, it still forms springs in places. In 1991, one of the most productive wells in the world was drilled into Edwards Aquifer. The well yielded approximately 25 000 gal

per minute (95 000 l/m) without pumping. This is enough water to sup- ply the needs of small to mid-size towns and cities.

The karst features of Edwards Aquifer limestone produce the impressive flow rates. The groundwater is basically flowing through caves. Certain unique organisms have adapted to live in these caves. Where the limestone reaches the surface, there is interaction between surface and groundwater with spring ponds in some areas and disap- pearing streams in others. The springs from the aquifer contain sen- sitive ecological communities that include unique and endangered species. The organisms emerge from the caves where there is no light so many are blind and some have no eyes.

The water demand of the burgeoning population in the area and urban/suburban sprawl into aquifer recharge area have put the Edwards Aquifer system in peril. As a result, the Edwards Aquifer has become a prominent legislative battleground. Recent and ongoing droughts have threatened the ability of the aquifer to meet demand. Rationing has even been contemplated. Where industry and develop- ment are in the aquifer area, the quality of the water has been degraded by contamination with pesticides, fertilizer, bacteria, solvents, and heavy metals. These recharge area problems are causing more severe problems in the production areas. This heavy usage is drawing saline and unpotable water into producing wells.

Underground chemical reactions can also drive groundwater in a different direction than gravity dictates. If acidic water flow- ing underground encounters a limestone unit, it can cause the limestone to dissolve. The dissolution produces excessive amounts of carbon dioxide which, once pressure is released, turns to gas and greatly expands. This expansion increases the volume of groundwater and carbonates it so essentially seltzer water is emitted from springs. This is the mechanism that produces the famous Saratoga Springs, New York. Springs of carbonated water are still emitted as much as 10 ft (3 m) into the air even with all of the local usage. Each spring has a different flavor depending upon the rock it passes through.

8.3 Pollution of Groundwater

There are two sources of groundwater pollution: point source and non-point source (Figure 8.10). Non-point source pollution is from sources that are too spread out to be pinpointed such as salt and other chemicals washed off of roads, fertilizers and pesti- cides percolated from lawns and agriculture, runoff from buildings and industrial sites, and air pollution fallout. Point source pollu- tion is from identifiable sources such as industrial sites, landfills, leaking underground storage tanks (USTs), leaky sewers and septic systems, and injection wells among others.

Groundwater in unconfined aquifers is commonly polluted in urban and suburban areas. Only in pristine areas is groundwater in the shallow aquifer potable. Springs come from the shallow unconfined aquifer so although they are widely regarded as pure, in most cases, they are suspect. The widespread shallow aquifer pollution is from non-point sources, point sources, and interaction with polluted surface water. The main point sources are industrial and residential spills, overflowing sewers, leaky storage tanks, landfills, and road and industrial runoff. These generally degrade the quality of the water to an unpotable level but the groundwater quality can reach dangerous levels of toxicity around a point source.

The pollutants in the shallow aquifer occur in three types based on their physical properties. If they are water-soluble, they will mix with the groundwater and contaminate the entire phreatic zone and can be complicated. These can include most biologic pol- lutants but also many chemicals. If they float on the water table, they are light non-aqueous phase liquids (LNAPLs). These include most petroleum products such as oil, gasoline and jet fuel, and solvents such as benzene, toluene, and xylene, all of which are very toxic. However, their position on top of the water table makes them easier to recover. The final type are dense, non-aqueous phase liquids (DNAPLs) which sink through the groundwater column and are very difficult to recover. These compounds include chlorinated solvents, such as trichloroethylene (TCE), 1,1,1-trichloroethane (TCA) and carbon tetrachloride, coal tar, creosote polychlorinated biphenyl (PCBs), and heavy metals like mercury. These are also very dangerous pollutants.

Spills and leaks form a plume that extends from the point source to as far as it is carried (Figure 8.10). In addition to the complication of relation to the water column, although generally down gradient, the flow direction is strongly affected by the influence and effluence of other water bodies but also the surficial geology. Water will flow quickly through gravels and many sands but slowly through clays and loams which are mixed materials. The variations in these units can be both vertical as stacked layers but also lateral. Old stream beds can be made of sand and gravel but the flood plains around them will be clay-rich. Pollutants in these areas spread quickly in the old bed but around them the spread is slow or non-existent. If the soil or sediment sits on top of fractured crystalline bedrock the plume might spread one way in the shallow zone and completely differently in the bedrock.

CASE STUDY 8.2 Price's Pit, New Jersey

Pollution of a Sedimentary Aquifer

In 1968, Charles Price closed a 26-acre (10.4-ha) sand and gravel quarry that crossed from Egg Harbor Township into Pleasantville, New Jersey, 5 miles (8 km) east of Atlantic City (Figure 8.11a). It operated for just two years because the sand and gravel was mostly not in economic quantities. In 1970, Price obtained a permit to operate a sanitary landfill in the abandoned quarry that became known as Price's Pit.

The permit was for household trash and construction debris, but soon after it opened, Price's Pit started to accept chemical wastes for disposal despite the specific restriction on the permit prohibiting the disposal of chemicals. In 1972, the Price's Pit operators requested permission to accept chemical waste. This request was approved with the condition that they were not to accept liquid or soluble industrial waste, petrochemicals, waste oils, sewage sludge, or septic tank wastes. The restrictions were so comprehensive that they excluded all chemical and industrial wastes. By the time the permit was issued, Price's Pit had already disposed of more than 9 million gallons (34 million l) of chemical waste. These raw wastes had been dumped on the ground, commonly just poured directly from tanker trucks. In other cases, drums of hazardous chemicals were dumped onto the quarry floor and covered with trash. There are rumors of late-night dumping and abandoning of whole tankers in the pit because the contents were so hazardous.

After the new permit was issued, in November 1972, Price's Pit operators stopped accepting chemical waste and continued operating as a landfill until 1976. At this point, waste disposal operations were terminated and the landfill was covered with a few inches of sand. A gate was erected across the access road and the operators abandoned one of the largest buried chemical waste dumps in the United States.

After this, precipitation infiltrated the sand cover and percolated through the waste, rusting the containers until they leaked, dissolving other waste, and washing the waste into the groundwater. These dissolved contaminants then moved eastward by the groundwater flow. The polluted groundwater flowed in the Cohansey aquifer, a sand

that lies under most of the coastal plain of southern New Jersey. This aquifer provides drinking water for Atlantic City, which pumps approximately 10–15 million gallons (38–57 million l) per day from the aquifer. Of the 10 wells in Atlantic City's public water well field, four lie between the city and Price's Pit.

The contaminant plume migrated away from the pit and expanded, putting almost half of the drinking-water supply for 40 000 people at risk (Figure 8.11b). People in the 35 private homes along the northeast side of Price's Pit had shallow wells as their source of water. Many of the residents noted drastic degradation of their water quality shortly after the disposal chemical wastes began. In addition, residents reported their skin burning while showering and trees began dying.

The movement of the contaminant plume toward the Atlantic City well field, which was a half mile (0.8 km) east of Price's Pit, was discovered when these people reported the problems. The Atlantic County Health Department (ACHD) collected water samples from the affected wells. They found organic solvents and heavy metals at concentrations far above the limits and contacted the US EPA and NJ Department of Environmental Protection (NJDEP) for help. These agencies quickly implemented several short-term actions. They shut down or deepened the threatened Atlantic City wells or fitted them with water treatment systems. Bottled water was temporarily provided to residents until city water could be extended to the impacted homes. Price's Pit was identified as the source of the groundwater contamination and it was designated as a Superfund site in October 1981. This designation provided federal funding to remediate the site. The EPA, NJDEP, and Atlantic County negotiated an agreement with 50 companies that sent waste to Price's Pit to pay for most of the $17 million cleanup.

The long-term remedial plan included the installation of an 80-ft (24.4-m) deep underground containment wall that circled the site and controlled the spread of contaminants. Groundwater extraction wells were drilled into the Cohansey aquifer to east of the landfill to pump water out at 200 000 gal (756 000 l) per day. This contaminated groundwater was sent through a treatment plant. These extraction wells operated for 30 years with regular monitoring of homeowner wells

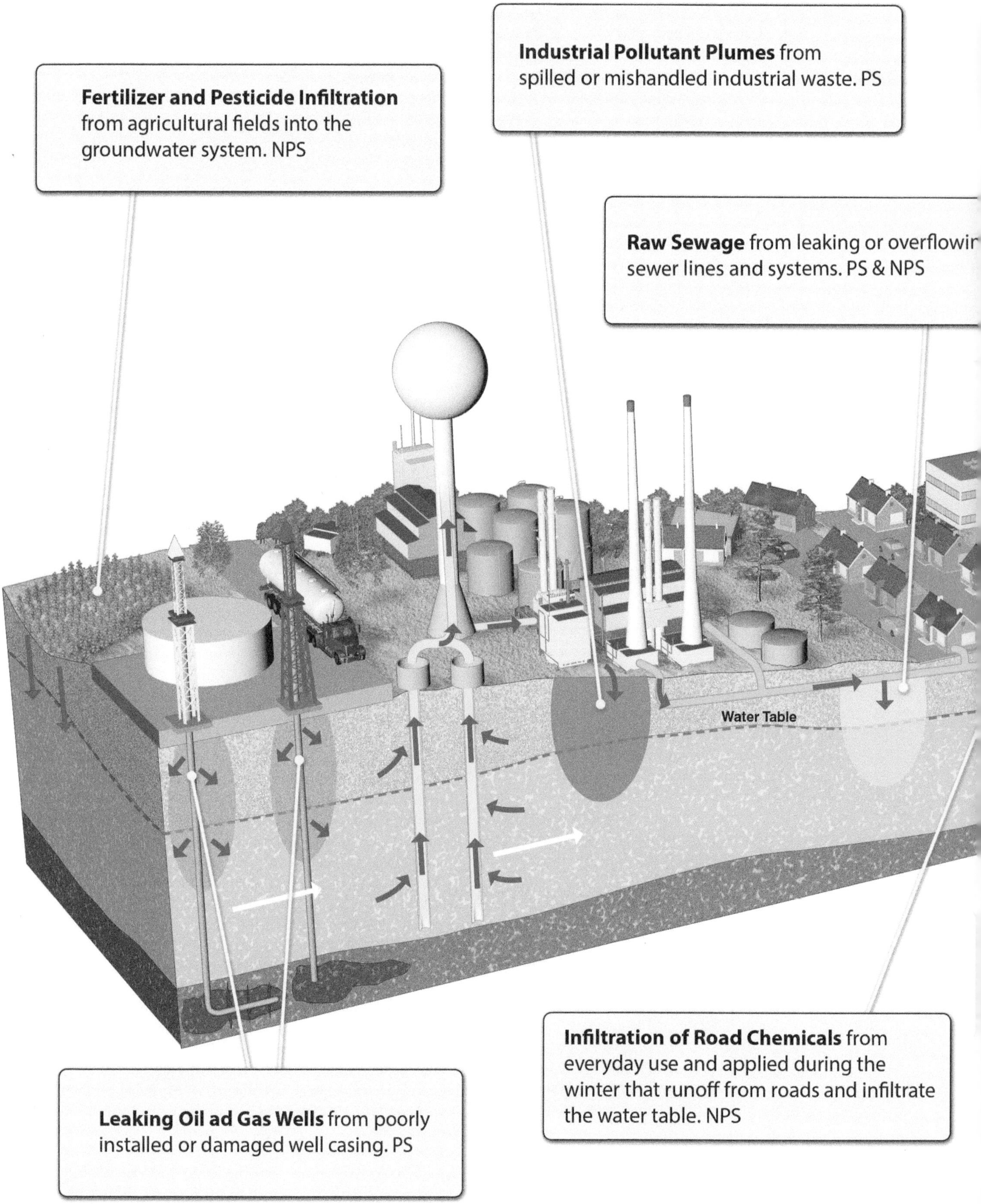

Fertilizer and Pesticide Infiltration from agricultural fields into the groundwater system. NPS

Industrial Pollutant Plumes from spilled or mishandled industrial waste. PS

Raw Sewage from leaking or overflowing sewer lines and systems. PS & NPS

Water Table

Infiltration of Road Chemicals from everyday use and applied during the winter that runoff from roads and infiltrate the water table. NPS

Leaking Oil ad Gas Wells from poorly installed or damaged well casing. PS

FIGURE 8.10   Diagram showing the major sources of groundwater pollution both point source and non-point source as described.

Unlined Industrial Waste Lagoons that leak industrial waste into the water table. PS

Leaking Sanitary Landfill that emits leachate into the soil and water table under it. PS

Infiltration of Surface Pollutant Spills through accidents or on purpose that reach the water table. PS

Leaking Underground Storage Tank (LUST) that releases pollutant into the water table. PS

Influent Polluted Streams that leak polluted water into the groundwater system. PS & NPS

Saltwater Incursion from the ocean encroaches on a freshwater aquifer and displaces the fresh water. NPS

and a series of monitoring wells around the site. The landfill itself is capped with heavy plastic under clay to control surface water infiltration. Although thorough, the onsite decontamination still leaves high concentrations of heavy metals in the water and it is pumped to the Atlantic County sewage plant for further treatment before it is released to the environment.

(a)

FIGURE 8.11 (a) Map of New Jersey showing major roads and the location of Pleasantville and Atlantic City. (b) Map of Prices Pit in Pleasantville showing the extent of the plume and the future extent of the pollume if untreated.

(b)

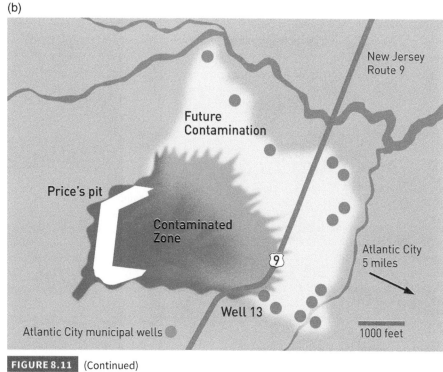

FIGURE 8.11   (Continued)

Mid-depth aquifers are generally safe from surface pollutants and provide good sources of drinking water from individual and community wells. Deep aquifers are less reliable. They can contain formational fluids. They are generally buffered by or in equilibrium with the chemistry of the rocks they reside in but at much higher pressure and elevated temperature relative to shallower groundwater. They can therefore contain significant amount of naturally occurring hazardous substances. They can contain high levels of heavy metals like arsenic and mercury, but even elements that are usually safe like iron can be at such high concentrations that they render the water unsafe. These fluids can contain crude oil and even radioactive elements like radium. They also tend to be saline. The main danger of the controversial process of fracking to produce natural gas is the dumping of the formational fluids (called produce waters) on the ground surface.

There are only a few situations that can make midlevel aquifers unsafe. If the deep groundwater system is open to the shallow water system like in deep karst terranes, it will be just as prone to problems. If the deep system of formational fluids extends to the surface or near-surface, it will not be usable for drinking water. This extension of deep system hazards to the shallow aquifer or surface is not unheard of. The La Brea tar pits in Los Angeles, California is a seep of crude oil to the surface. The shallow aquifer in much of Oklahoma has so much barium that it is not drinkable. The other hazards are if the recharge area of mid-depth aquifers is polluted, the pollutants will slowly contaminate it along its entire length. If an industry uses deep injection wells to dispose of waste, it can contaminate these aquifers. Deep injection as a method for disposal is expensive. Typically, if deep injection is used, the waste is particularly hazardous.

8.4   Pollutants in Groundwater

A number of pollutants can occur in groundwater. Most municipal drinking water in urban and suburban areas is supplied from surface water reservoirs. This water is highly treated and monitored and for good reason. Surface water is much more susceptible to problems than groundwater. It can carry diseases such as cholera and typhoid, parasites, and relatively undiluted and unfiltered pollutants. Problems in most municipal water supplies rarely occur but may require boiling or temporary use of bottled water. There are some contaminants that are not easily removed in the treatment process like pharmaceuticals and personal care products (PPCPs). These wind up in many surface water supplies including reservoirs. They are in very low concentrations but it is unclear how they impact human health.

Even though the shallow aquifer was the source of drinking water for much of human history, in urban and suburban areas, it can contain more dangerous pollutants than surface water. Leaking USTs release heating oil in residential areas, gasoline and its additives in many neighborhoods, and many pollutants in industrial sites. Other chemicals can be stored in USTs as well such as all kinds of fuels and some chlorinated solvents. Otherwise, all surface pollution filters into the shallow water aquifer with little to no filtering. In fractured rock and karst aquifers, surface water is delivered directly to groundwater. It can contain diseases such as cholera and many dangerous microorganisms such as giardia, *Escherichia* coli, and cryptosporidium even in rural areas. Even if the pollutants filter through soil and sediment before reaching the water table, they commonly are not removed. Many of these pollutants include industrial chlorinated solvents such as trichloroethene (TCE), tetrachloroethene (PCE), and 1,1,1 trichloroethane (TCA). These can be as common as in dry cleaners or only in processing plants.

CASE STUDY 8.3 Woburn Wells G and H, Massachusetts

Pollution of a Fractured Rock Aquifer

The lawsuit resulting from pollution in Woburn, Massachusetts was popularized in a book in 1995 and a movie in 1998 entitled *A Civil Action*. The city of Woburn is located just 10 miles (16 km) north of downtown Boston, Massachusetts (Figure 8.12a). Its convenient location makes Woburn ideal for industry including numerous tanneries. Woburn sits on the 6-mile (9.7-km) long Aberjona River, described as among the most heavily urbanized in the northeastern United States (Figure 8.12b). The largely channelized Aberjona extends from Reading through Woburn and into the Mystic Lakes. The river is of major economic importance as a supplier of fresh water that can transport goods to the major markets of Boston. By the mid-1860s, more than 20 tanneries and leather treatment facilities were operating in Woburn, using the river as both a water supply and a sewer. By the 1870s, the water quality of the river was so bad that the discharge of wastes into one of the major tributaries was banned. By 1911, all discharge of waste into the Aberjona River was outlawed.

Tanning of leather involves chemical processes using trivalent chromium and tannins and may include alum, syntans, formaldehyde, glutaraldehyde, or heavy oils in the process. The hides are soaked in progressively stronger concentrations of chemical tannins. The physical preparation and chemical treatment of hides generates large amounts of waste. The chemicals include the chlorinated organic solvent trichloroethene (TCE) in addition to ammonia and chromium. Tanneries use water in almost every step of their processing, especially high-volume or production-type operations.

In spite of attempts to improve the quality of Aberjona River, the city of Woburn as well as many local businesses continued to use it as both a water supply and to dispose of sanitary and industrial waste (Figure 8.12c). In the mid-1960s, Woburn designated a 245-acre (98 ha) tract of land near the intersection of two major highways, as an Industri-plex (Figure 8.12b). It was specially zoned for manufacturing and heavy industry. The intent was to reinvigorate the declining industry of the area. This site was progressively redeveloped through the 1970s and into the early 1980s as a center for paper, textile, pesticide (lead arsenate), and expanded leather-goods industries.

In addition, the city of Woburn drilled two additional water supply wells: Well G in 1964 and Well H in 1967. These wells pumped groundwater from the fractured rock aquifer beneath the Aberjona River Valley and yielded up to 2 million gallons (7.6 million l) per day. Tests for bacteria and pollutants indicated that the water from Wells G and H was of potable quality and it was piped into the municipal supply system without treatment. By the late 1970s, Wells G and H supplied about 30% of the city's water. However, in 1979, the local police discovered 200 drums of waste solvent in a vacant lot near Wells G and H while on routine patrol. They decided that if these drums were leaking, it could pose a public health hazard and they contacted health officials. Water samples were collected from the wells and they were found to contain unsafe levels of volatile organic compounds (VOCs), most notably tetrachloroethene (TCE) and tetrachloroethylene (PCE).

However, this was not the end of the problem. An official of the Massachusetts Department of Public Health noticed an increase in mortality rates from various cancers in Woburn and investigated whether they might be the result of pollution from the Industriplex. These findings were reported to the Centers for Disease Control (CDC). Then, a pediatric hematologist at Massachusetts General Hospital in Boston advised the CDC that he found six children with acute lymphocytic leukemia, within a six-block radius in Woburn. A local clergyman later informed the mayor and the local newspaper that he identified 10 cases of childhood leukemia in one part of Woburn, all within the past 15 years. In a retrospective analysis, the CDC determined that death rates in Woburn increased by 13% between 1969 and 1978.

In response, the city of Woburn arranged for an alternate water supply. The US Environmental Protection Agency (EPA) determined that the Industri-plex site and several surrounding sites were the source of the pollution and designated them as a Superfund site on 8 September 1983. The 330-acre (132 ha) site includes waste disposal areas and their adjacent wetlands as well as the Aberjona River. Contaminated runoff flowed through wetlands and accumulated in Aberjona River sediment, which contained PAHs and the heavy metals arsenic, chromium, and mercury. This pollution resulted from accidental spills and intentional disposal of waste materials on properties owned by six private corporations. These companies agreed to a negotiated settlement with EPA for $70 million toward remediation of the site.

Remediation has included removal of onsite waste materials including drums of solvent and PCBs, unprocessed or off-spec animal skins, and other debris. Excavation of more than 200 tons (180 mt) of soil was removed for offsite disposal, and other soil was treated in place by chemical oxidation and vapor extraction. Ponds and wetlands impacted by overland flow of surface runoff were dredged and restored whereas other impacted areas were capped to reduce additional contaminated runoff. Groundwater has been pumped and treated to reduce migration and recharge into the Aberjona River aquifer system. Remedial activities at this Superfund site are expected to continue for many years.

In 1982, eight Woburn families whose children died from leukemia filed a highly publicized lawsuit against companies responsible for the contamination of Wells G and H. Among the main issues of dispute was whether ingesting TCE and PCE at the concentrations in Woburn well water could have caused the leukemia. Unfortunately, regardless of the evidence, there was no absolute scientific consensus that the link could be made. In the end, the plaintiffs were awarded a modest settlement that could never make up for the pain and suffering from the loss of their children. This was the case described in the book and movie, *A Civil Action*.

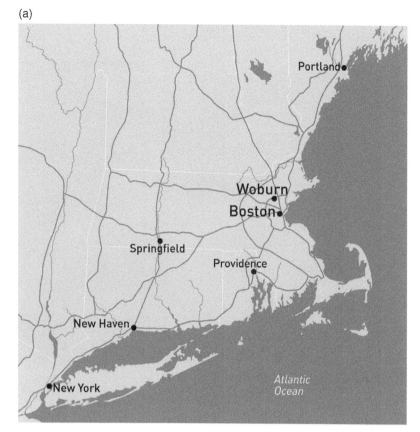

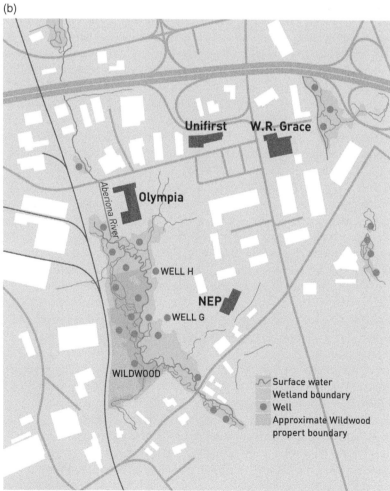

FIGURE 8.12 (a) Map of New England and major cities showing the location of Woburn. (b) Map of the Woburn industrial park showing the location of Wells G and H. *Source:* Courtesy of US Environmental Protection Agency. (c) Photo of a polluted stream near wells G and H *Source:* Daderot / Wikimedia Commons / Public Domain.

(c)

FIGURE 8.12 (Continued)

Groundwater can also contain inorganic pollutants including heavy metals such as arsenic, mercury, and chromium. Arsenic is used in many industrial applications including the manufacture of pesticide but it can also be naturally occurring at dangerous levels. The most notorious case of natural arsenic in the shallow aquifer is in Bangladesh where many residents suffer from arsenic poisoning, largely as a result of a UN effort to have residents stop using surface water for drinking and using shallow well water instead. However, there is a belt of elevated arsenic levels across the northern United States from New Jersey and New York across to Ohio but it even continues to Montana at lower levels. This belt apparently marks the termination of the ice sheet of the last ice age. In the areas of high concentrations, more than 50% of well water can test above the limits for arsenic in drinking water.

Liquid mercury is not that hazardous, but if it is methylated, it can be quite toxic, with one drop of dimethyl mercury being fatal. Even if in its less toxic state, methylated mercury can cause a variety of health issues largely involving organ failure. Like mercury, chromium is not toxic under most conditions, but if it is hexavalent (+6 charge), it is quite dangerous. It can cause damage to most mucosal tissue but more importantly, it is carcinogenic. Groundwater may also be contaminated with the heavy metal lead but lead in drinking water is more commonly the result of leaching from old pipes. The solder used in joining copper pipes together was largely lead before it was replaced with antimony in 1986. This was the case in the Flint, Michigan crisis. Lead causes brain damage and other organ failure. Other elements and compounds that are generally less dangerous can be very dangerous in concentrations above acceptable levels. These can include zinc, nickel, aluminum, sulfates, barium, and others. Some even less dangerous inorganic elements and compounds can be naturally occurring but in high concentrations in these aquifers such as calcium carbonate. These waters are referred to as "hard" and can clog pipes and appliances that use water such as water heaters, dishwashers, and washing machines.

Groundwater can also contain radioactive elements, both naturally occurring and as the result of manufacturing, that can be quite dangerous. Radium can occur in groundwater. It can leach out of rocks and sediment in areas that are rich in uranium. This is especially problematic in areas of uranium mining, where tailings dumped on the surface can weather at much higher rates. Radium was also used in painting the luminous parts of watch faces in the past. If radium-tainted waste was dumped on the surface, it can leach into the groundwater in many areas. This was the case in Orange, New Jersey, chronicled in the book *Radium Girls*. In addition to radium, its daughter, radon can also contaminate groundwater as a naturally occurring problem or through industrial processes. Water can hold far more radon than air and will release it to indoor air if agitated like in a shower, dishwasher, or washing machine. Indoor radon is the most dangerous natural pollutant and estimated to be responsible for 25 000 lung cancer deaths per year according to the EPA. It is unclear if drinking water containing radon is dangerous but it is released in the bloodstream and exhaled in breath. Other radioactive elements are also possible in groundwater both through natural processes and from industrial sources but their danger is not well established.

CASE STUDY 8.4 Pacific Gas and Electric, Hinkley, California

Heavy Metal Water Pollution

The Pacific Gas and Electric site in Hinkley, California (Figure 8.13a) became famous for its portrayal in the movie *Erin Brockovich*, released in 2000. This is the most successful movie about a pollution disaster and won an Academy Award for Julia Roberts as the leading actress. This movie brought attention to the dangers of chromium contamination. The heavy metal occurs in many valence states many of which are marginally dangerous. Hexavalent (+6) chromium, however, is highly hazardous to human health. Breathing hexavalent chromium irritates or damages the nose, throat, and lungs and produce ulcers on mucus membranes in these areas. It can also irritate or damage the eyes and skin. Even worse, however, is that hexavalent chromium is carcinogenic. It greatly increases risk of developing lung cancer and rates of lung cancer mortality.

Hexavalent chromium is artificially manufactured and is used in combination with sulfuric acid and water in industrial metal plating baths and in leather tanning operations. Other valence states of chromium can also be converted to the +6 state by certain microbial activity.

The small town of Hinkley, California is located approximately 120 miles (192 km) northeast of Los Angeles. In 1952, Pacific Gas & Electric (PG&E) constructed a pipeline to distribute natural gas throughout most of California. This natural gas pipeline contained a series of compressor stations spaced at about 300 miles (500 km). These compressors increased the pressure in the natural gas to ensure uninterrupted flow to customers. Hinkley is the site of a compressor station. The process of gas pressurization generates large amounts of heat. PG&E dissipated the heat from the compressors by circulating water through

(a)

FIGURE 8.13 (a) Map of California showing the Pacific Electric & Gas (PG&E) pipeline with a substation in Hinkley. (b) Maps showing the location of Hinkley, major roads and topography in the area (left) and the plumes of chromium tainted groundwater around the PG&E substation (right). (c) Poster for the movie *Erin Brockovich* about the impact of the PG&E chromium contamination *Source:* MT martaro / Flickr / Public Domain.

(b)

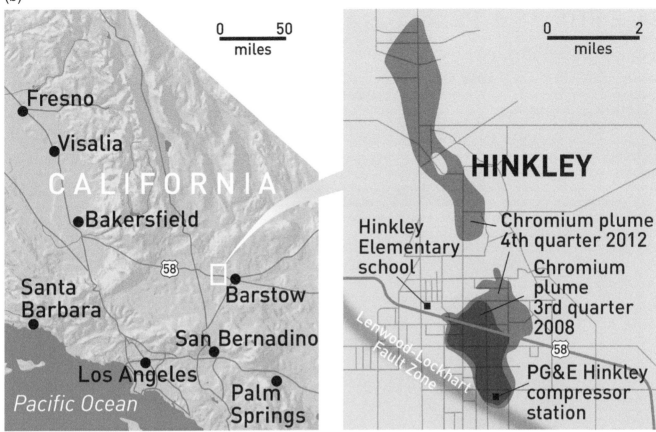

(c)

FIGURE 8.13 (Continued)

them. The heated water was circulated through a series of cooling towers and the cooled water recycled through the compressors.

As part of the routine equipment maintenance, rust inhibitor containing hexavalent chromium was added to the cooling water. The cooling water slowly became contaminated with small particles of grit and metal and the water was discharged into unlined pits in the bare soil when the concentrations reached a certain level. This practice continued until 1972, when new regulations required that the pits be lined. This lining reduced the infiltration contaminants into the soil and groundwater system.

Unfortunately, it was already too late as PG&E had discharged tens of thousands of gallons of hexavalent chromium contaminated water into the pits. It had already contaminated the groundwater system that the residents of Hinkley used for drinking, bathing, and cooking. PG&E had also used a spray aeration technique in which wastewater was jetted into the air to dispose of it. As the water evaporated, the solids became airborne as hexavalent contaminated dust and particulate that settled across the area.

In 1987, PG&E detected hexavalent chromium in several groundwater wells. The concentrations were reported to be 0.6 parts per million (ppm) which is above the federal limit of 0.1 ppm and the California limit of 0.05 ppm. However, they did not alert the residents of Hinkley that its water had been contaminated by hexavalent chromium

(Figure 8.13b). In response to the impact on their health and property values, 648 of the 1900 total residents of Hinkley established a class action suit against PG&E, in the early 1990s. Erin Brockovich was a legal assistant in the law firm handling their case.

The lawsuit claimed that the long-term exposure to hexavalent chromium in groundwater resulted in illnesses and disabilities among residents including uterine cancer, breast cancer, Hodgkin's disease, cancer of the brain, gastrointestinal cancer, miscarriages, nosebleeds, asthma, heart failure, damaged immune systems, and other health problems. The parties agreed to settle the case through binding arbitration instead of a public trial. The arbitration included medical and scientific experts from both sides vigorously presenting evidence pro and con. In contrast to the concentrations reported by PG&E, Ms. Brockovich produced a laboratory report of hexavalent chromium levels of 20 ppm in one well.

In 1997, PG&E settled the claims for $333 million, which was among the largest industrial toxic exposure settlements ever awarded in the United States. The plaintiff's lawyers collected a fee of $133 million, with Ms. Brockovich collecting a $2 million bonus (Figure 8.13c). After all expenses were paid, there was $188 million remaining which was distributed to the plaintiffs, based on the severity of their illnesses. The average payout was about $300 000 per individual.

8.5   What Can I Do?

Dumping any waste on the ground should be avoided. This can be simple littering or dumping of garbage because water percolating through it will leach components into the groundwater. Dumping leftover chemicals like paint thinner, gasoline, or motor oil on the ground will percolate directly into the groundwater system. The use of lawn chemicals like herbicides, pesticides, and chemical fertilizers should be avoided because they also leach into groundwater as do de-icing compounds. Water usage should also be limited because it impacts the supply. Watering lawns is commonly not needed to keep them from dying and it wastes water, reducing the aquifer and reservoir supplies. Baths rather than showers, multiple showers each day, running water using appliances when they are not full and letting water run excessively in sinks or hoses also reduce supply. Water saving appliances can also be chosen when replacing them.

Pollution of Rivers and Surface Waters

Words you should know:

Braided stream – river or stream in which the channels interweave with lozenge-shaped sand bars between them.

Drainage basin – the area bounded by hills or drainage divides that feed surface water into streams and rivers.

Effluent stream – a situation in which groundwater level is at a higher elevation than a stream and feeds into it.

Flood plain – the flat, low elevation area around a stream that floods when the river overflows its banks.

Influent stream – a situation in which groundwater level is at a lower elevation than a stream and feeds into it.

Lake – a large standing body of water of greater than 10 acres (4.0 ha) to 20 acres (8.0 ha).

Meandering stream – a stream or river with a highly sinuous course that are more common in areas of flat topography.

Natural levee – naturally occurring elevated strips of land adjacent to one or both sides of a river that help keep the water in the channel.

Pond – a smaller standing body of water typically less than 10 acres (4.0 ha) to 20 acres (8.0 ha).

Tributary – smaller stream that feeds into a larger stream or river.

Wetland – a low topography area of shallow standing water and/or saturated soil for most or all of the year. Also known as swamps, bogs, marshes, and fens among others.

9.1   Rivers and Streams

Freshwater occurs on the earth surface as rivers and streams if it is moving or lakes, ponds, and wetlands if it is stationary. Rivers and streams are bodies of surface water that flow in channels from higher to lower elevations. They typically contain water from surface runoff from precipitation and from the melting of snow and ice in colder areas and higher elevations. They can also contain base flow from groundwater that feeds the streams. The proportion of base flow and surface runoff is a function of the stream, season, climate, and other factors. Streams originate at their headwaters where tributaries coalesce to form larger streams (Figure 9.1). Streams are ranked based upon their position relative to the main river they form.

The area containing a system of coalescing streams is called a drainage basin and it is bounded by drainage divides around it (Figure 9.2). A watershed is a drainage basin with the groundwater included. The drainage pattern of the rivers and tributaries can reflect the underlying rock units and structures. In map view, a dendritic drainage pattern forms on horizontal strata and looks like the branches of a tree. In trellis drainage patterns, tributaries are primarily straight and parallel but that periodically contain right-angle turns. This pattern forms in areas of tilted strata (Figure 9.3). There are also several other less common patterns.

Water in streams is commonly in intimate contact with the groundwater system. Surface water from rain and melting ice infiltrates the soil to become groundwater. It passes through the normally dry rock, sediment, and soil of the vadose zone and into the phreatic zone which is constantly saturated with water. The top of the phreatic zone is the water table, the depth of which varies with the season and precipitation events. Like surface water, groundwater flows down slopes but in the subsurface and at much slower rates.

9.1.1   Influent or Effluent Streams

If water in a stream infiltrates the bed of the stream and feeds into the groundwater system the stream is termed influent (Figure 9.4a). Influent streams are common in dry areas where the water table is deep. Infiltrating water from the stream bed can travel a long distance downward to reach the water table. Surface water in these areas is the major mode of recharge for groundwater. The process has the effect of reducing the size and volume of the stream along its length.

FIGURE 9.1   Map-view diagram of tributaries joining together to form a major river. The numbers (and colors) show the rank of the stream.

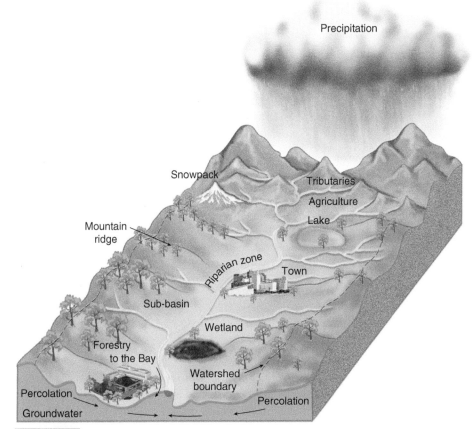

FIGURE 9.2 Block diagram of a drainage basin and watershed bounded by drainage divides.

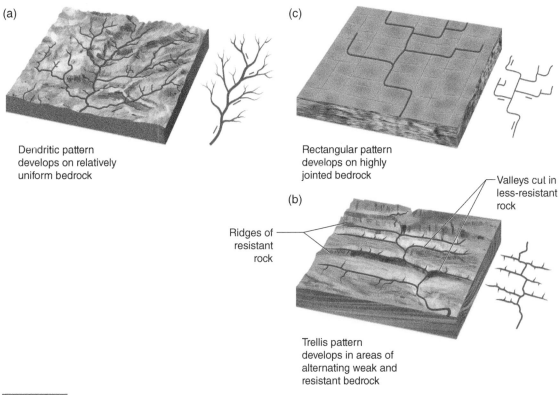

FIGURE 9.3 Drainage patterns for several river drainage systems based on the underlying bedrock: (a) dendritic pattern in horizontal strata, (b) trellis pattern in tilted strata, (c) rectangular pattern in jointed or faulted bedrock.

In contrast, if groundwater feeds into a stream from a spring or seep in the bed or side of the stream, the stream is termed effluent (Figure 9.4b). The water in the stream that originated from groundwater is the base flow and it typically comprises the major volume of the effluent stream. Base flow allows streams to flow even when it has not rained for a long time. The springs or seeps define the locations where the water table intersects the ground surface. This situation occurs with a shallow water table which requires regular precipitation in the watershed to keep the volume and elevation of groundwater high.

Whether a stream is influent or effluent can be a function of location and time. An individual stream can be influent during a dry season and effluent during a rainy season. In some streams, even a single major precipitation event can switch a stream from influent to effluent. Influence and effluence can also vary from location to location along a single stream. A stream may be effluent at its headwaters but influent farther downstream. This switch is typical in streams that flow from mountains and across dry areas such as in the southwestern United States. There are any combination of conditions of influence and effluence along the length of a stream.

9.1.1.1 River Classification

Streams and rivers are controlled by the relative ruggedness or "age" of the topography through which they flow, though age of the stream really has nothing to do with its condition. Immature or rugged topography is characterized by fast-flowing "young" streams within steep-sided valleys and canyons. In contrast, mature or soft topography is characterized by slow-flowing "old" streams that wind through broad, flat floodplains. There are also "middle-aged" streams which span the range between the two types. It is not uncommon for a single stream to have "young" characteristics in its headwaters, "middle-aged" features in the mid-section and "old" features near the mouth. There are two end member types of streams that are characteristic of these ages: young streams are braided streams and old streams are meanders.

(a)

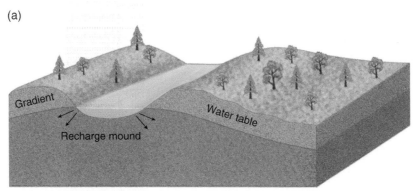

Influent (losing) stream

(b)

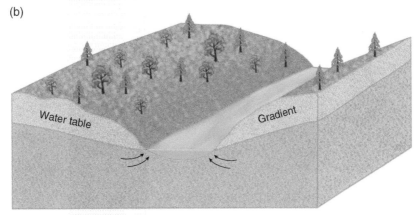

Effluent (gaining) stream

FIGURE 9.4   Block diagrams of streams and their relation to the groundwater table for (a) influent streams and (b) for effluent streams.

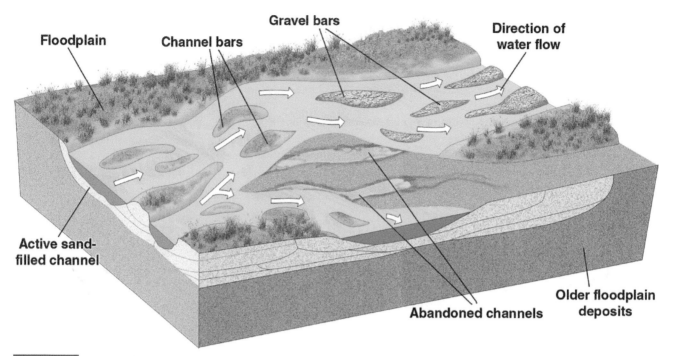

FIGURE 9.5 Diagram of a braided stream showing channels and channel bars during low stage.

Braided Streams

Braided streams are common in areas of rugged topography (Figure 9.5). They occur in stream valleys with steep walls, narrow flood plains, and relatively straight courses. The volume of water in the stream and consequently the river level varies greatly during the year. The steepness of the valley walls prevent lateral spreading of the stream during flood stage.

As indicated by the name, the channels in a braided stream form a braided or anastomosing pattern. These channels enclose lozenge-shaped channel bars composed of sand and gravel. During dry periods and especially over the summer, the channel bars form land and are surrounded by thin channels with gentle flow. Discharge, velocity, and sediment load are all low. The bars and channels tend to remain in the same place and vegetation may occur on the bars during this period.

During spring thaw or period of heavy rain, streams become raging torrents with high velocities, discharges, and sediment loads. The river levels rise and submerge the channel bars. All sizes of sediment are transported and channel bars are reshaped and migrate rapidly downstream. All vegetation and low velocity deposits on the channel bars are washed away by the fast flows.

Meandering Streams

Streams with highly sinuous courses are called meanders (Figure 9.6). The channels of the streams constantly migrate across the wide, flat flood plain that characterize meander belts. The stream flows straight and erodes the stream bank on the outer arc of curved channel, undercutting it and causing it to collapse. This is the cut bank and as it erodes the channel slowly migrates in the flow direction. On the opposite side or inner arc bank, the stream flows much slower. With the slower flow velocity, the stream can no longer carry sediments and they deposit on the point bar on the inner arc. The erosion of one bank and deposition on the facing bank results in a slow migration of the channel.

The channels can also shift quickly in the process of avulsion. This shift can be through neck cutoff or chute cutoff. Two loops in a channel meander belt can migrate toward each other. As channel loops come into contact, instead of traveling around the loop, the stream takes the shortest route and connects to two loops, abandoning the old channel loop. Water remains in the abandoned channel forming an oxbow lake along the side of the river. Eventually the lake fills with sediment and dries up. Chute cutoff is also a process in which channels are abandoned but it only occurs during floods. An overflow channel forms on the point

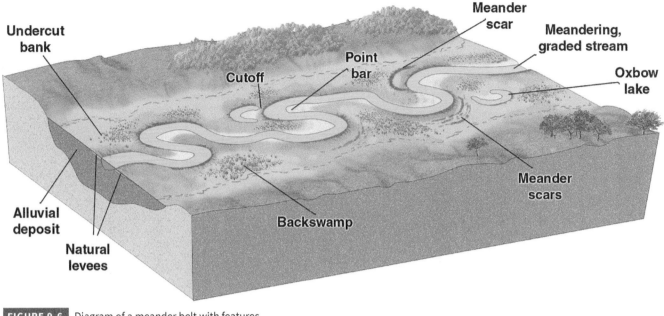

FIGURE 9.6   Diagram of a meander belt with features.

bar of the inside arc of the meander loop. This overflow channel or chute forms a shorter and straighter path for the river than the main channel. During floods, the stream overflows the banks and some of the flood water flows through the chutes. If the flood is powerful enough, erosion of the chute may be intense enough that the stream shifts into the chute and abandons the channel. An oxbow lake can also form by this process.

The topography around a meander belt is so flat that flooding is common. As a stream overflows the banks, water velocity slows and sediments deposit on the bank top. These deposits form narrow elongate mounds along the sides of the river called natural levees. They help keep the river in the channel and are usually structurally reinforced to reduce flooding. Flooding can be destructive in meander belts because the large flat flood plain allows the stream to spread across a wide area. As a result, these areas are extensively managed.

9.2   Pollution of Streams

Rivers in industrial areas are highly polluted, commonly supporting no aquatic life. Pollution can enter streams through several sources (Figure 9.7). Historically, rivers made a convenient place for waste disposal and pollutants were dumped directly into them. There were widespread cases of discharge pipes from public sewer systems, emptying raw sewage into rivers downstream of the city or town. The next city downstream had their drinking water source from the impacted river, further polluted it with their own waste, and discharged it downstream of the city. It has been estimated that by the time drinking water reached the city of New Orleans on the Mississippi River, it had been used six times.

Municipal wastewater discharge, dumping, and the spills and leaks from industrial plants along a river constitute most point source pollution in the water. Industrial plants are typically built along rivers because of the source of water for use in manufacturing and processing and because the rivers can be used for transporting products. Traditionally, these companies also discharged waste directly into the river, causing direct pollution of the water. In cases such as the Cuyahoga River in Cleveland, Ohio, there was such a high concentration of flammable chemicals in the water that the river periodically caught fire.

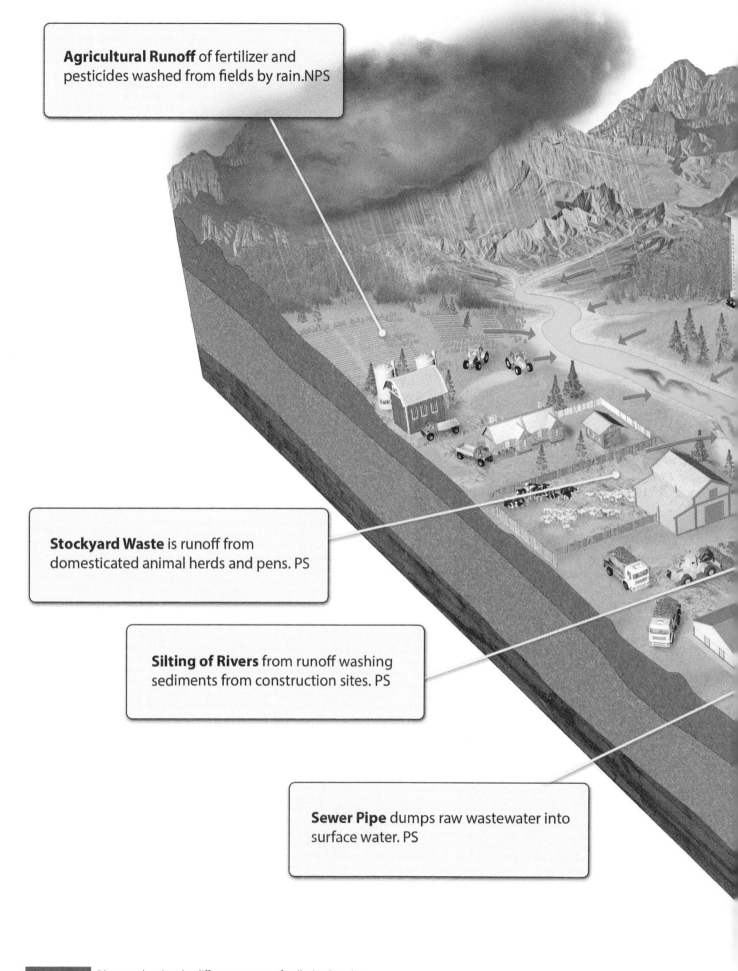

Agricultural Runoff of fertilizer and pesticides washed from fields by rain.NPS

Stockyard Waste is runoff from domesticated animal herds and pens. PS

Silting of Rivers from runoff washing sediments from construction sites. PS

Sewer Pipe dumps raw wastewater into surface water. PS

FIGURE 9.7 Diagram showing the different sources of pollution in a river.

River Fire - fuel, chemicals and garbage catches on fire and must be extinguished. PS

Urban Runoff from rain washing pollutants from urban areas into rivers. NPS

Industrial Spills are accidents at plants that release pollutants into rivers. PS

Pollutes Effluent Rivers receive base flow from polluted groundwater. NPS

Ship Accidents can spill pollutants into rivers. Tankers can spill high amounts of oil. PS

CASE STUDY 9.1 Cuyahoga River, Ohio

Waste Disposal into a River

The Cuyahoga River is the major watercourse through Cleveland, Ohio and it empties into Lake Erie (Figure 9.8a). To accommodate the demands of the area, by the early 1930s, the river was straightened, widened, and deepened by the U.S. Army Corps of Engineers to allow the passage of bigger ships into Cleveland. As a result of industrialization along the river, by 1936, parts were so polluted that it could not be used as cooling water or even shipment by barge (Figure 9.8b). Flammable pollutant concentrations increased to the level that the river actually caught fire. River fires began in the 1890s and early 1900s in a number of major cities but fires on the Cuyahoga River were more frequent and intense. Oil and debris in the river burned in 1868, 1883, and 1887. In 1916, sparks from a tugboat ignited oil leaking from a refinery, setting off massive explosions and a fire that killed five workers. As a result, the city passed an ordinance prohibiting the release of oil into the river by refineries but a fire occurred in the same area a decade later in 1922, and again in 1930. In 1936, a river fire burned for almost a week.

That year, a fire near a tugboat repair and maintenance facility burned for more than six hours, with flames five stories high and causing damage to a bridge, docks, piers, and barges. The fuel was the unused and waste oil dumped into the river and trash and debris thrown in by industries and municipalities (Figure 9.8c). The steel mills and oil refineries and terminals along the river dumped untreated wastewater, unrefined or unusable petroleum hydrocarbons, acids, solvents, and other organic chemicals into the water. It wasn't just industry that polluted the river. The city of Cleveland and the surrounding suburbs pumped raw sewage into the river for more than a century. Oily runoff from roads and fertilizer from agricultural areas also contributed to the pollution of the river.

As a result, the Cleveland City Fire Department became national experts in fighting river fires. They were equipped with fire-fighting boats and crews were regularly called to extinguish river fires. Prevention of river fires became an important function of the fire department and crews were specially trained to spot oil slicks and apply chemical dispersants to them.

After a fire in 1952, residents began to pressure the city of Cleveland to address Cuyahoga River pollution but a decade later in 1962, the river was still clogged with debris and heavily polluted. Only the main channel was kept open for barge and river traffic. In 1963, the city began an effort to remove debris from the river and several of the

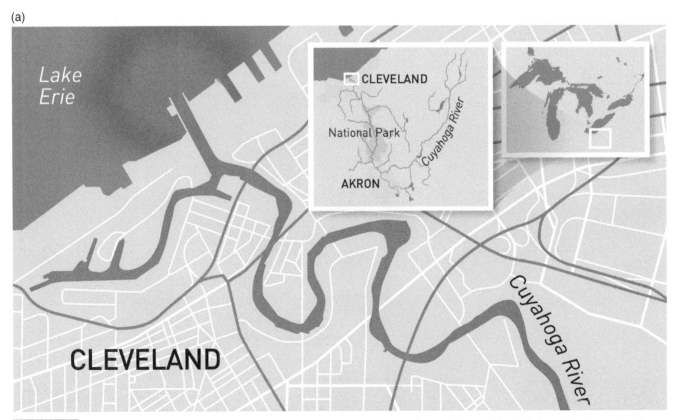

FIGURE 9.8 (a) Map showing the Cuyahoga River flowing through Cleveland, Ohio and into Lake Erie. (b) Aerial photo showing polluted Cuyahoga River water emptying into Lake Erie. *Source:* https://www.cleveland.com/metro/2018/11/cuyahoga-rivers-recovery-since-1969-fire-documented-in-new-ohio-epa-film-on-youtube.html. (c) Junk cars lined up along the banks of a Cuyahoga tributary. *Source:* Frank J./ Wikimedia Commons / Public Domain. (d) Chemicals and junk on fire in the Cuyahoga River. *Source:* locustsandhoney2005 / Flickr / Public Domain.

(b)

(c)

(d)

FIGURE 9.8 (Continued)

major point source polluters created the Cuyahoga River Basin Water Quality Committee. The Committee instituted a water quality surveillance program and began plans to reduce discharges to the river.

By the late 1960s, Cleveland began to lose population and its industrial base. In addition, national and local concern about the environment had become significant. In 1969, a $100 million bond issue was approved that had most of the funding earmarked for the cleanup and revitalization of the Cuyahoga River. These included major improvements to sewage treatment plants and aggressive debris removal and antidumping enforcement programs. Another initiative was launched in 1969 to institute methods to prevent oil from entering the river and to remove the oil that was there.

Despite these efforts, the river caught fire again on 22 June 1969. It was not a severe blaze and brought under control quickly by the fire department with only minimal damage (Figure 9.8d). However, it emerged as a seminal environmental event. It embarrassed the city of Cleveland, which wound up with the reputation as an environmental wasteland for many years. In response to the 1969 fire, the national media focused on Cleveland but because the fire had been extinguished before film crews could arrive, they used photos and newsreels of the much larger 1952 river fire. The Cuyahoga River fire, the Santa Barbara oil well blow-out, and several other environmental disasters at the time were the impetus for Federal action on environmental issues. In 1970, the U.S. Environmental Protection Agency was established, the Clean Air Act strengthened, and the first Earth Day was celebrated.

The Cuyahoga River is now one of 43 Areas of Concern in the Great Lakes Water Quality Agreement. This agreement by the United States and Canada attempts to repair the environmental damage done to the Great Lakes by the decades of industrial and municipal misuse. The Cuyahoga River plan addresses eutrophication, toxic substances including PCBs, and heavy metals in river water and sediment, bacterial contamination from sewer discharge, and over-development.

Non-point source pollution can be just as hazardous to river water quality as the point source pollution. This pollution is washed from its source by precipitation from the watershed and carried by surface runoff into a river. Anything that is applied to or spilled onto the surface can be swept into a river. In agricultural areas, runoff contains fertilizers and pesticides. In suburban areas, runoff from lawns, gardens, paved areas, and buildings is swept into storm drains which discharge into streams. This runoff typically contains pesticides, fertilizers, road salt, motor oil, antifreeze, paint residues, detergents, solvents, air pollution fallout, and other residential pollutants.

The interaction between surface water and groundwater explains that polluted groundwater may pollute a stream. A pollutant plume from a buried tank or chemical plant spreads away from the source in the direction of groundwater flow. The pollutant will enter an effluent stream with the groundwater. Leaky and overflowing septic systems are also potential sources. If the source of pollution is close to the stream, the stream will be more polluted because the groundwater will be less filtered. The addition of pollutant to surface water from groundwater is especially problematic in industrial areas. Efforts to clean and reduce pollution in the river, may be difficult until the groundwater pollution is addressed.

Polluted influent streams may also be more polluted through interaction with groundwater. The problem is that many pollutants float on water, especially if they are derived from hydrocarbons. Pollutants that are lighter than water are called light non-aqueous phase liquids or LNAPLs. Water in influent streams infiltrates the streambed which removes cleaner water and leaves LNAPLs in the stream water. In this case, the stream will be increasingly polluted along its length despite there being no apparent source for the pollution. This can only occur if the stream flow is slow so the LNAPL cannot mix in.

In addition to contaminating the river water, pollutants may be deposited in various sediments along a river. There are many factors that control where and how the pollutants will be deposited including density of the pollutant, chemical reactivity of the pollutant, chemistry of the river water, flow velocity and type of river (meander or braided) among others. This may control if the pollutant is deposited in the channel, sediment bar, or flood plain. For example, the Hudson River of New York had so much PCB dumped into it that a 100 mile (160 km) stretch of it is a single superfund site, making it the largest superfund site in the world.

CASE STUDY 9.2 Hudson River, New York

Pollution of River Channel Sediments

The General Electric Company (GE) recognized the value of a new class of dielectric fluids called polychlorinated biphenyls (PCBs) to produce capacitors for use in electric devices. By the late 1940s, GE had built two capacitor manufacturing facilities along the Hudson River of New York at Ft. Edward and Hudson Falls that extensively used PCBs. As part of their operations, these plants discharged PCB-laden wastewater into the river. The problem is that PCBs are bio-accumulating and biomagnifying in the environment. PCBs have been shown to cause liver and nerve damage and fetal injury in humans and they are probable carcinogens. The United States banned the production of PCBs in 1976 and planned their gradual phase-out and replacement. Most other countries passed similar bans.

The Hudson River has two sections, the Lower Hudson between the Battery in Manhattan (River Mile 0), and the Federal Dam near Troy at river mile 153 (Figure 9.9a). The Upper Hudson is a 40 mile (64.4 km) stretch of the River between Hudson Falls and the Federal Dam. These dams across the Lower Hudson had an environmental consequence

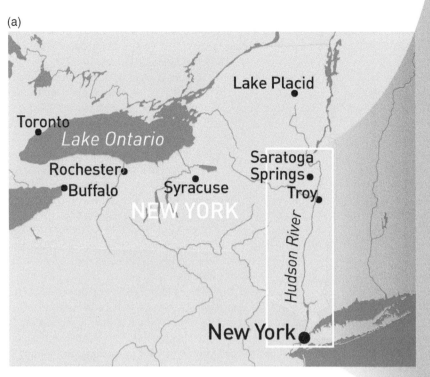
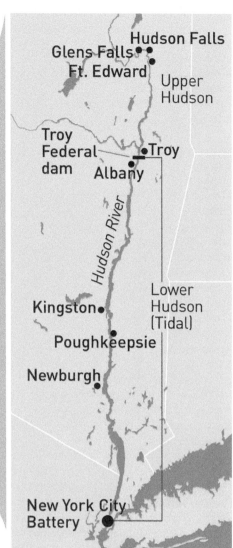

FIGURE 9.9 (a) Map showing the Hudson River from the upper portion near Troy to the Battery in New York City. Inset map of the northeastern United States. (b) Photo of dredging operation in the Hudson River by the EPA. *Source:* EPA.

with regard to PCBs. From post-World War II until 1977, the GE Ft. Edward and Hudson Falls capacitor manufacturing plants legally released wastewater containing PCBs into the Hudson River. PCBs are not very soluble, stable in the environment, and they are heavier than water. Once released, they rapidly accumulated downstream from the plants behind the dam at Ft. Edward.

In 1973, Niagara-Mohawk Power Company, removed the dam at Ft. Edward because it had become unstable and could no longer be properly maintained. Between July and October 1973, the dam was removed and the reservoir behind the dam drained. The new river channel formed, exposing the PCB-laden sediments that had deposited in the reservoir behind the dam. These contaminated sediments were in a 1.5-mile (2.4 km) segment of the Hudson River, upstream of the dam. The spring of 1974 was unusually wet in the area of the contaminated sediments, experiencing several floods. They swept the PCB-contaminated sediments downstream and subsequent flooding coupled with the tidal movements on the river spread the PCBs in the channel along its entire length.

PCB releases from the GE capacitor plants were discontinued in 1977. However, New York State Department of Environmental Conservation (NYSDEC), the U.S. Environmental Protection Agency (EPA), and GE estimated that some one million pounds (453 592 kg) of PCBs had been released into the Hudson River between 1947 and 1977. Much of the PCB-contaminated sediments was removed through dredging by GE or washed out to sea. However, the EPA estimates that 500 000 to 660 000 pounds (226 796 to 299 371 kg) of PCBs were still remaining in Hudson River sediments.

In response, in 1976, the NYSDEC banned commercial fishing for most of the Hudson River's main species. The New York State Department of Health (NYSDOH) advised that women of childbearing age and children under 15 should not eat fish from the Hudson River south of Hudson Falls and no one should eat any fish caught between Hudson Falls and the Federal Dam. In 2001, some of these restrictions were lifted but there are still recommended limits on consumption of fish from the river.

In 1984, 43 miles (69.2 km) of the Upper Hudson River site, was declared a superfund site. This site was later expanded to include the Lower Hudson River to the Battery in Manhattan, a distance of 200 miles (320 km). This makes the Hudson River one of the largest Superfund projects in the United States.

In September 1984, the USEPA declared that any attempt to remove the PCB-contaminated sediments would re-release them to the environment and cause more harm than benefit. The USEPA concluded that the PCB-contaminated sediments were being covered over by new sedimentation and thereby being isolated so they no longer posed a significant ecological or public health threat. NYSDEC did not agree and the NYSDEC and GE began remediation to reduce or prevent further migration of PCBs into the Hudson River.

In the mid-1970s, NYSDEC removed more than 770 000 cubic yards (588 707 m³) of PCB-contaminated sediment from the channel of the Champlain Canal near Ft. Edward. In addition, in mid-1990 and 1991, GE capped in situ 50–60 acres (20.2–24.3 ha) of remnant PCB deposits exposed on the banks and shoreline of the Hudson River. In 1997 and 1998, GE removed 1100 cubic yards (8410 m³) of PCB-contaminated sediment near the Ft. Edward plant to inspect the underlying bedrock for PCB seepage. In 1994 and 1995, GE excavated 3400 tons (3084 mt) of sediment containing 45 tons (40.8 mt) of PCBs from beneath the Ft. Edward plant. Fractures in the bedrock were then grouted and a groundwater collection system was installed. By 2001, more than 3000 gal (11 356 l) of PCB-contaminated water had been recovered. Some 230 groundwater monitoring and recovery wells were installed to protect the river from leakage.

As a result of these remedial activities, PCB levels in the Hudson River have been declining. Average PCB concentrations in sediments near Ft. Edward were greater than 50 ppm (ppm) in 1977 but were about 10 ppm in 1998. PCB concentrations in fish from the Upper Hudson River declined from 80 ppm in largemouth bass in the 1970s to 4 ppm in 1997 compared to the safe level for PCBs in fish of 2 ppm

In 2002, the EPA and NYSDEC decided to remove all PCB-contaminated sediment in the Upper Hudson River with concentrations above 3 g/m³. This involved removal of about 1.56 million cubic yards (1 192 706 m³) of sediments. In addition, 0.6 million cubic yards (458 733 m³) of sediment with PCB concentrations greater than 10 g/m² was removed from another part of the river. Another 0.5 million cubic yards (382 277 m³) of PCB removal also took place from "hot spots" with elevated levels.

The dredging project took place from 2009 to 2015 (Figure 9.9b). It occurred in two phases, the first of which occurred in 2009. Approximately, 283 000 cubic yards (216 000 m³) of contaminated sediment was removed from a six-mile (9.7 km) stretch in the Upper Hudson River. The second phase was from 2011 to 2015 in which approximately 2.5 million cubic yards (1.9 million m³) was dredged. Some of the dredged areas were repopulated with aquatic plants during 2015 and 2016.

Sedimentation of pollutants can also take place in other parts of a river valley. Suspended sediments settle as velocity of a stream slows. This relationship is explained by the Hjulstrom diagram (Figure 9.10). The fields of the diagram show the conditions of suspending versus depositing various sized grains under different velocity conditions. Higher velocity flow suspends sediments and lower velocity flow deposits them. The slope of the curve that separates these two fields shows that, in general, the higher the velocity of the stream, the coarser the sediments that can be suspended. With small particles such as clay and silt, the relationship breaks down. They require very low velocities to settle and stick together and once settled and attached, they require very high velocities to pry them back apart.

In a flood, the flow in the stream within the channel is fast, suspending particles of all sizes. However, as the water level rises, it overflows the banks and spills onto the flood plain. The velocity drops abruptly once on the flood plain without the force from the river or stream. The slower velocity allows the finer-grained sediments to be deposited on the flood plain. These fine-grained, chemically reactive sediments can capture a different type of pollutant than would be deposited in the channel.

Even if waste was dumped or buried along the shore of a river, flooding and runoff during precipitation and thawing events can wash pollution into the river. In the case of effluent streams, groundwater can also carry dissolved pollutants into the river through base flow. Clearly, industrial sites are capable of inflicting serious damage to the health of a river.

However, pollution of rivers is not only through spills of liquid waste or dissolution of pollutants out of soils. Pollution can also be in the form of solid or semi-solid waste as sludge. This kind of pollution can occur at disposal or construction sites where waste

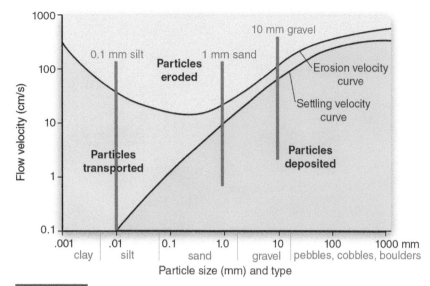

FIGURE 9.10 The Hjulstrom diagram. The positive slope shows that it takes faster velocity to suspend coarser particles (and all finer grains). The settling curve shows how slow the flow velocity must be for grains of a given size to settle to the surface. The erosion curve shows how fast the velocity must be to erode grains of a given size.

is illegally dumped into a river or slides into a river through mass wasting. However, a major solid waste threat to rivers is mines and mining operations. In addition to ore and usable products, mines generate a lot of waste. This can take the form of gangue or unusable minerals, rock with too little ore to be economic called tailings, and soils and sediments that were removed from the bedrock.

CASE STUDY 9.3 2015 and 2019 Minas Gerais Disasters, Brazil

Mine Waste Spills in Rivers

The worst mine waste disaster in the history of Brazil was quickly followed by a second similar disaster. The first disaster was the collapse of the Fundão dam in Mariana municipality, Minas Gerais State, Brazil, in November 2015 and the second was the failure of Dam B1 at the Córrego do Feijão mine in the city of Brumadinho, also in Minas Gerais State in 2019 (Figure 9.11a). The first was the worst environmental disaster and the second resulted in a large number of deaths. Both are in the same iron mining district with similar situations.

The mining operations generate tailings waste that is dumped into a lagoon outside the mine, contained by a dam. The tailings form sludge composed of wet rock and soil of uneconomic gangue minerals and materials in which the amount of ore is uneconomic and any other refuse in the operation. With highly active mines, such as the two in Brazil, the tailings waste can accumulate quickly and become quite massive. Even though iron mining does not generate dangerously toxic substances, the chemicals in the waste are still quite toxic to the environment.

The Fundão tailings dam at the Germano iron mine of the Samarco Mariana Mining Complex in Minas Gerais, Brazil collapsed on 5 November 2015. At 3:30 p.m., the Fundão dam began to leak (Figure 9.11b). In response, an emergency team attempted to mitigate the leak by reducing the volume of the lagoon. At 4:20 p.m., however, the dam failed, releasing 57.2 million cubic yards (43.7 million m³) of toxic sludge into the Santarém river valley and Doce River. This toxic

brown mudflow polluted the river and beaches all the way to the Atlantic Ocean. The town of Bento Rodrigues, 1.6 miles (2.5 km) down the river valley, was almost entirely flooded, killing 19 people and leaving it completely inaccessible by road, which prevented rescue efforts (Figure 9.11c). Other villages along the river valley suffered minor damage and some 600 people were evacuated. More than 1.6 million people in 39 municipalities across two states in the Doce River's watershed were impacted by the event. In the Barra Longa district of Minas Gerais, 77.9% of residents experienced health issues after the disaster, including headaches, coughing, leg pain, and allergic reactions.

The dam collapse and toxic flood is regarded as Brazil's worst environmental disaster. The toxic sludge was spread along 415 miles (668 km) of river valleys and even along 120 miles (200 km) of Atlantic Ocean coastline, within 17 days of the collapse. It spread down the Doce River in Minas Gerais, the largest river in the Southeast Atlantic. The sludge immediately destroyed 6.6 square miles (17 km²) of land including uprooting vegetation along 3.2 square miles (8.35 km²) of critically endangered Brazilian Atlantic riparian forest.

Chemically, the sludge contained arsenic, lead, cadmium, chromium, nickel, selenium, and manganese, all above maximum allowable levels, but sodium and amines were probably the greatest hazards. The sludge killed millions of freshwater fish; about 11 tons (10 mt) of dead fish were pulled from the river after the event. It degraded pristine riparian lands and polluted the Atlantic coastline in a vulnerable turtle-nesting area. The plume reached biodiversity conservation areas in the Atlantic Ocean, including four important

(a)

(b)

FIGURE 9.11 (a) Map showing the locations of the two mine flood disasters in Minas Gerais State, Brazil.
Source: (The Environment & Society Portal / CC BY 4.0). (b) Map showing the area impacted by the Fundao dam
disaster with inset showing the location in Brazil. (c) Photo of the town of Bento Rodrigues destroyed in the disaster.
Source: Senado Federal / Wikimedia Commons / CC BY 2.0. (d) Map showing the location of the Brumadinho dam
failure area and impacted rivers. Inset shows location in Brazil.

marine protected areas, threatening rare species of marine fauna. Besides chemical pollution, the sludge reduced oxygen availability in the stream water, increased its turbidity, interrupted reproductive cycles of many fish species, and altered entire ecological networks. It is estimated that the cost of the environmental impact exceeds $521 million (US) per year.

An estimated 26.2 million cubic yards (20 million m³) of sediment was deposited in a 74 mile (119.2 km) segment of the Gualaxo do Norte, Carmo, and Doce Rivers alone. This huge volume of sediment greatly altered the river morphology. The rivers actually changed from meandering types to braided stream types with multiple channels as a result of the massive sediment influx.

(c)

FIGURE 9.11   (Continued)

The owners of the mine and dam were prosecuted and government sanctions were imposed as a result of the disaster. In 2016, manslaughter and environmental damage charges were filed against 21 company executives. The case greatly expanded when a 2013 report was leaked, showing that these executives knew there were structural problems with the dam well in advance of the collapse.

Despite the death, cost, environmental impact, and legal ramifications of the 2015 mine waste dam collapse, a very similar collapse of the nearby Dam B1 at the Córrego do Feijão mine in the city of Brumadinho, Minas Gerais State, occurred on 25 January 2019, just over three years later (Figure 9.11d). The tailings dam collapse occurred at 12:28 p.m., sending a wave of toxic sludge through the break. The wave engulfed the mine's administrative area and cafeteria while hundreds of mine employees were eating lunch there. The wave proceeded through the small community of Vila Ferteco about 0.62 miles (1 km) from the mine. The mud flow reached the Paraopeba River at about 3:50 p.m. It is the region's main river, supplying water to about one third of the Greater Belo Horizonte region of Brazil.

This second tailings dam collapse released approximately 15.7 million cubic yards (12 million m³) of a similar type of toxic sludge as in the 2015 disaster. The sludge contaminated about 75 miles (120 km) of the Paraopeba river and impacted at least 10 riverside towns. It killed 270 people, primarily employees of the mine, making it much deadlier than the 2015 event. The wave also destroyed three locomotives, 132 wagons, two sections of railway bridge and about 330 ft (100 m) of railway track.

The toxic sludge released in the dam burst was composed of fine material of 30% sand and 70% silt-clay. Chemically, it was mainly iron, aluminum, manganese, and titanium but with significant trace amounts of toxic metals including uranium, cadmium, lead, arsenic, and mercury among others. Many elements were increased up to 10-fold above background in the river sediments, whereas cadmium, manganese, and phosphorous levels increased by at least 70-fold as a result. Water testing detected high levels of toxic metals in the river, including lead and chromium, within the first 12 miles (20 km) of the spill. In tests of river sediments at Retiro Baixo, 187.7 miles (302 km) downstream from the dam, only cadmium displayed marked enrichment, with all other metals showing only minor or no enrichment. This decrease shows the removal and dilution of metals with distance. The toxicity of the sludge impacts all levels of the food chain from primary producers such as algae to primary and secondary consumers such as microcrustaceans. It also causes a 20% increase in mortality rate in fish.

Just two days later, on 27 January at 5:30 a.m., sirens were again sounded at the mining operation because the mine's adjacent Dam VI appeared to be growing unstable. This was a process water reservoir and increasing water levels were detected. In anticipation of another dam burst, about 24 000 residents from the city of Brumadinho were evacuated, including the downtown area. The threat, however, never materialized.

The 2019 collapsed tailings dam was at a mine owned by the Brazilian mining giant, Vale. The company is the world's biggest producer of iron ore. Senior staff at Vale are facing murder charges over the disaster. The communities struck by the disaster received a $7 billion payout from Vale. It is estimated that there are at least 126 tailings dams in Brazil that are vulnerable to similar failure.

(d)

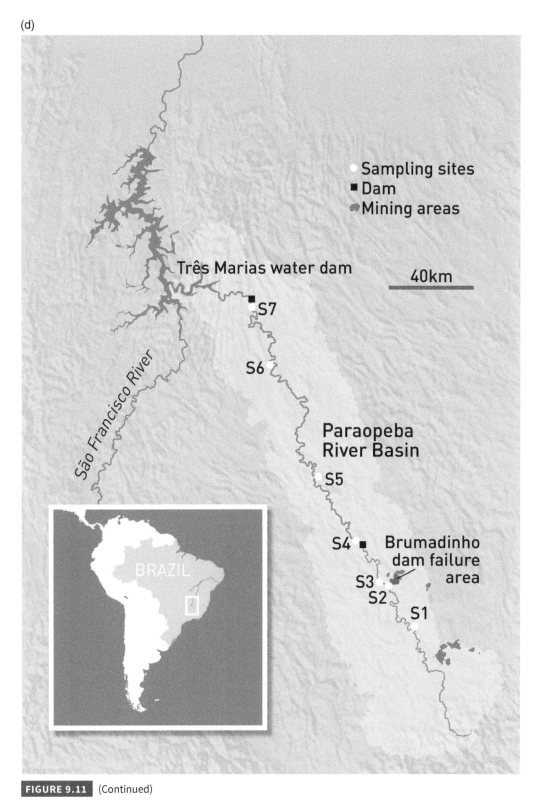

FIGURE 9.11 (Continued)

9.3 | Lakes, Ponds, and Wetlands

A lake is a relatively large inland body of stationary or slowly moving water. The difference between a pond and a lake is the surface area and depth. Sunlight is able to reach the bottom in all ponds but not necessarily in lakes. Some definitions require ponds to have a surface area of less than 10 acres (4.0 ha) but others set the limit at 20 acres (8.0 ha) and there are many ponds that greatly exceed this size. The precise dimension limits on lakes, ponds, and wetlands are not completely agreed upon. Wetlands typically contain grasses, trees, and/or shrubs.

Lakes contain about 0.009% of all free water, which is less than 0.4% of all fresh water on land. Inland seas and saline lakes contain about 0.0075% of free water. Freshwater lakes contain more than 98% of important surface waters that are available for use. About 80% of the 30 000 cubic miles (125 000 km³) of lake water occurs in some 40 major lakes. The largest of these major lakes are Lake Baikal, Russia, which contains about 5500 cubic miles (23 000 km³) of water, Lake Tanganyika, east Africa with about 4600 cubic miles (19 000 km³) of water, and Lake Superior, North America with 2900 cubic miles (12 000 km³) of water. The Great Lakes of North America contain about 6000 cubic miles (25 000 km³) of water in total and including the rest of the North American lakes larger than 2 cubic miles (10 cubic km³), contain about 25% of the total lake waters. About 70% of the total lake water is contained in the North American, African, and Asian continents.

CASE STUDY 9.4 Lake Uru Uru, Bolivian, Andes

"Made of Plastic" Lake

The Andes Mountains of South America are very high, relatively young mountains composed primarily of volcanic rocks intruded by large igneous plutons. The primitive composition of these igneous rocks makes them enriched in metals. As such, there are numerous mining districts throughout the range extracting base metals to be used in manufacturing. This brings large amounts of metals and metal-rich rich rocks to the surface where they can be weathered and become the source of pollution. However, even the surface water that is not impacted by mining can have elevated metal concentrations from the weathering of surface rocks. Lake Uru Uru of Bolivia not only has elevated metal levels from background and mining sources but additionally suffers from world record levels of bacterial and plastic waste.

There is a sizable area of surface water bodies in the Peruvian to Bolivian Andes. Lake Titicaca is the largest of these bodies as well as the largest freshwater lake in South America and the highest altitude large lake in the world (Figure 9.12a). It crosses between Peru and Bolivia and the surface elevation is 12 507 ft (3812 m). The surface of the lake is 118 miles (190 km) long by 50 miles (80 km) wide, yielding a surface area of 3232 square miles (8372 km²) with a maximum depth of 922 ft (281 m). Two other lakes occur in this high-altitude area, Lake Poopo, and the manmade Lake Uru Uru, both of which are in Bolivia. Lake Poopo received much attention because it has shrunk tremendously as a result of climate change and the loss of local alpine glaciers.

Lake Uru Uru is best known for excessive amounts of heavy metal contamination but all lakes have elevated levels of certain metals. The young volcanic rocks and their weathering products in the area produce excessive background levels of arsenic in surface and groundwater. For example, dissolved arsenic averages between 5.0 and 15 μg/l in Lake Titicaca in comparison to the World Health Organization limit for drinking water of 10 μg/l. As a result of this high background, some 4.5 million South Americans are constantly exposed to high levels of arsenic to the point of some Andean populations having adapted to arsenic toxicity.

However, Lake Uru Uru is much more polluted than the others. The lake surface is at 12 093 ft (3686 m) elevation with a surface area that varies between 46.3 and 135 square miles (120 and 350 km²) depending upon drought conditions. The average depth of the lake is a shallow 3.2 ft (1 m). The lake is a short 5 miles (8 km) south and along the Tagarete River from Oruro City. Oruro has about 220 000 inhabitants

and the city dumps sewage and garbage into the river which winds up in the lake.

Lake Uru Uru is characterized by an alkaline pH of 8.3 ± 0.6 and is continuously impacted by intense mining and smelting emissions primarily involving tin and antimony, but also silver, gold, copper, zinc, bismuth, tungsten, and including arsenic and mercury as biproducts. The Huanuni River transports products of acid mine drainage (AMD) to the lake from active mining operations. The Tagarete River transports smelting biproducts from the San Jose Mine and the Vinto smelting plants and dissolved material from surface water filtration through piles of at least 2.2 million tons (2 million mt) of mining waste and into the lake.

As a result, the concentrations of heavy metals including cadmium, lead, mercury, antimony, nickel, cobalt, chromium, zinc, copper, and arsenic in lake water are far in excess of standard limits for drinking water. Arsenic concentrations reach 78.5 μg/l in Lake Uru Uru, which is eightfold higher than in Lake Titicaca. In addition, the sulfur concentration in Uru Uru is 70% higher at 142 ± 0.9 mg/l. The Huanuni River is the main source of this arsenic and sulfur containing 84 ± 10 μg/l of arsenic and 376 ± 0.6 mg/l of sulfur.

The lake water has extremely high dissolved iron of about 111 mg/l. The downstream section of Lake Uru Uru has high dissolved copper of 45 ± 2 μg/l from mining input compared to the upper part which has <0.2 μg/l. Similarly, the sediment in the lake is enriched in arsenic and iron with 181 ± 3 μg/g of arsenic and 4.72 ± 0.11% of iron. It also contains high concentrations of zinc at 1811 mg/kg of sediments in addition to cadmium, lead, and nickel. In addition, there are numerous inputs of mercury from the surrounding mines. The mercury levels are 30 times higher in the lake than in unimpacted water. The proportion of the highly toxic methylated mercury in both sediment and the surface water of Lake Uru Uru is as high as 50% which is excessive. Periphyton aquatic plants act as sinks for the methylmercury. As a result, there is mercury accumulation as well as cadmium and lead in fish and in the entire aquatic food web of the lake.

The Tagarete River drains untreated wastewater with fecal pollution from Oruro city into the lake. This heavy nutrient input causes the hypersaline lake to be eutrophic most of the time. As a result, it is colonized by extremophile bacterial communities. Many of the bacterial colonies are hotbeds of antibiotic resistant gene proliferation as a result of this environmental pollution. Although the main source of this dangerous situation is fecal-rich wastewater containing antibiotic resistant bacteria, the high concentration of metals from mining waste could be contributing to the proliferation.

More recently, Lake Uru Uru became infamous for another type of pollution. Not only does Oruro city discharge sewage into the lake via the Tagaete River, they also dump huge amounts of garbage into the river. The lake has come to be known for the floating plastic, discarded bottles, containers, toys, and tires for several years (Figure 9.12b). However, in 2016, the Oruro area suffered a major drought that reduced Lake Uru Uru's water levels to around 25–30% of its capacity. This shrinkage of the surface area of the lake concentrated the floating waste to the point that observers could no longer see the water underneath. The situation became infamous when Reuters released a report on it entitled "Made of Plastic" which went viral on the internet. In response, local and even far traveled international volunteers descended on the hapless lake attempting to do the mammoth job of cleaning it up.

(a)

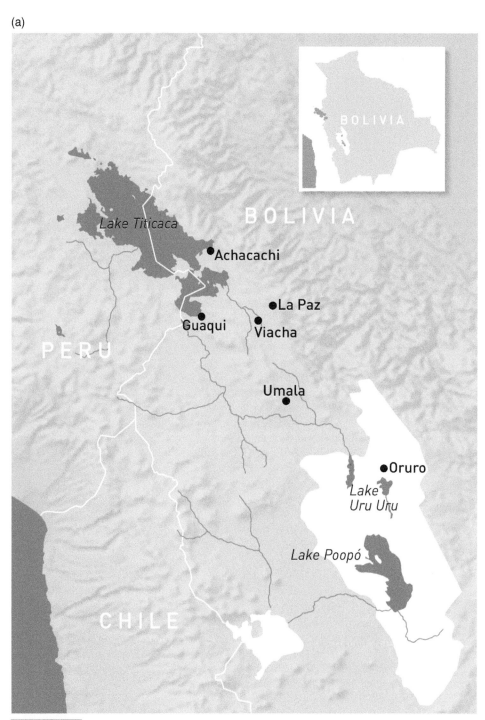

FIGURE 9.12 (a) Map showing the area of high-altitude lakes in the Peruvian-Bolivian Andes including Lake Titicaca and Lake Uru Uru. (b) Photo of Lake Uru Uru so covered with garbage that the water is not visible. *Source:* TWC Product and Technology LLC.

(b)

FIGURE 9.12 (Continued)

Wetlands are probably the most paradoxical of surface features. Otherwise referred to as swamps, bogs and marshes, inland wetlands were historically reviled as disease-, insect-, and snake-infested dangers to be avoided or destroyed. In reality, wetlands are beneficial to essential for human settlements for a number of functions including water storage, groundwater recharge, storm and flood protection and water purification through retention of nutrients, sediments, and pollutants.

As a result of the negative reputation, wetlands have been extensively filled with soil, sediment, and even waste to convert them into dry and usable land. More than 220 million acres (89 million ha) of wetlands are estimated to have originally existed across the continental United States in the 1600s. More than half of these wetlands were drained, filled, and converted to other uses (Figure 9.13). The major period of loss was from the mid-1950s to the mid-1970s when the urban and suburban spread peaked. The rate of loss has slowed since then thanks to laws preventing it in many states though it still has not stopped. It is estimated that

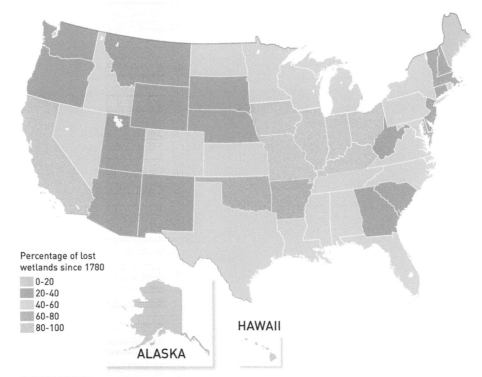

Percentage of lost
wetlands since 1780

- 0–20
- 20–40
- 40–60
- 60–80
- 80–100

HAWAII

ALASKA

FIGURE 9.13 Map showing the percent loss of wetlands in each state of the United States since 1780. *Source:* Doc Searls / Wikimedia Commons / CC BY-SA 2.0.

about 100 million acres (40.5 million ha) of wetlands remain in the continental United States with 22 states having lost 50% or more of their wetland acreage. There is still a loss of about 60 000 acres (24 281 ha) annually primarily in Louisiana, Mississippi, Arkansas, Florida, South Carolina, and North Carolina. Draining and reclamation of wetlands for agricultural purposes is a significant part of the loss but it is declining. Development remains the largest cause of current wetland loss.

The loss of wetlands on a global basis is no better and, in several cases, it is worse. It is estimated that the world has lost 50–71% of its wetlands since 1900. Long-term loss of wetlands is estimated at 54–57% but in some developed areas, it may be as high as 87% since 1700. The loss of wetlands accelerated to 3.7 times faster a rate during the 20th and early 21st centuries than in previous centuries. Losses have been much more extensive and faster for inland wetlands than those along the coasts. After massive losses through much of the twentieth century, the rate of wetland loss in Europe and North America has slowed in recent decades. The rate of loss, however, remains high in Asia and getting worse in Africa.

The Ramsar Convention on Wetlands was begun in 1971 to develop an accounting of threatened natural wetlands and the threats to them on a global basis. Table 9.1 shows the classification of impact factors. Some 55% of the Ramsar sites are affected by three or four of the impact factors. The most significant impact factors are pollution affecting 54% of the sites, biological resources use affecting 53%, natural system modification also affecting 53%, and agriculture and aquaculture affecting 42%.

There are 2303 Ramsar designated Sites on a worldwide basis. Europe has 1004 sites which is the largest number of any continent but the total area of them is just 6% of the total. Africa has 397 which is 17% of the total but they account for 48% of the total area at 272 million acres (110.0 million ha). South America has 146 sites, which is just 6% of the total but accounts for 17% of the total area at 98 million acres (39.6 million ha). The 368 sites in Asia constitute 16% of the total with a total area 68 million acres (27.5 million ha) or 12% of the total. North America has 13% of the total at 309 and 13% of the total area at 71.4 million acres (28.9 million ha). The nationally protected wetland areas listed on the Ramsar Convention are 92% in Africa, 7% in Asia, 9% in Europe, 21% in South America, and 9% in North America of the total wetland area on each continent.

Impact on the environment is the major subclass. The major continents of impact are South America where it is the issue for 75% of the sites and North America where it is 74%. The amount of sites where environment is the major subclass make up 60% in the other four continents. The environmental impact is from pollutants primarily from household sewage, urban wastewater, garbage and solid waste, agricultural and forestry runoff, and industrial runoff.

Even though every continent contains wetlands designated in the Ramsar Convention, the total area of them accounts for less than 19% of the global total. This means that there is no accounting of 81% of the wetland area at present. There could be much more wetland area that is under threat of damage or destruction. Many of these unaccounted areas are in remote locations and may be affected by natural impacts such as climate change. This is especially true for polar and subpolar and alpine areas.

TABLE 9.1 Ramsar impact classifications.

Class	Subclass	Subdivision
Human impact	Land area	Agriculture and aquaculture
		Natural system modifications
		Human settlements (non-agricultural)
		Transportation and service corridors
	Environment	Pollution
		Human intrusions and disturbance
		Energy production and mining
	Biodiversity	Biological resource use
		Invasive and other problematic species and genes
	Water resources	Water regulation
Natural impact	–	Climate change and severe weather
	–	Geological events

CASE STUDY 9.5 Hackensack Meadowlands, New Jersey

Severe Wetlands Pollution

There are unwanted, mosquito-filled wetlands around many major cities that were treated poorly and polluted. However, in addition to being the suspected final resting place of Union Boss Jimmy Hoffa, the worst impacted wetlands, by far, is the Hackensack Meadowlands or Meadowlands of northeastern New Jersey (Figure 9.14a). The Hackensack River is an estuary and the surrounding wetlands and mudflats formerly formed a glacial lake. Being in such an important location, early in the settlement of the area, residents attempted to convert it to usable land by infilling, canals, and dikes. This activity created silting of rivers and reduced freshwater input changing the chemistry of the water to more saline.

More serious pollution began in the later nineteenth century, when the raw sewage and industrial wastes that poured into the Passaic River, backed up into the southern portion of the Meadowlands. To address the ever-increasing waste in the Passaic River, Newark, and several other cities constructed a large sewer line to pump sewage directly into Newark Bay in the early 1900s. The sewer alleviated pollution in the river but Newark Bay became severely contaminated. The problem was that tides in Newark Bay drove polluted water back up the Hackensack River more than 21 miles (33.8 km) upstream.

Dredging of the river to deepen and widen it at about the same time, allowed even larger amounts of increasingly polluted tidal waters from the Newark Bay and New York Harbor into the Meadowlands.

Also at this time, artificial fill was being used to reclaim wetlands from all of eastern New Jersey including the Meadowlands. This fill was prepared by mixing urban garbage with clean fill composed of solid materials, such as rocks, gravel, ash, bricks, and concrete. Much of the garbage was being shipped in from New York City. One of these projects was the construction of Port Newark in 1914. This involved dredging a ship channel from Newark Bay into the Meadowlands. The dredged fill was mixed with garbage and ashes and dumped on the wetlands on the north side of the channel. This land was built a few feet above sea level, and docks and warehouses were built on it

Despite the swamps, pollution, and garbage-rich fill, in 1929, the Regional Plan Association designed plans to build a major city larger even than New York at the time that involved dredging and straightening the Hackensack River and filling the Meadowlands swamps with 200 million cubic yards (153 million m³) of dirt. It never got off the drawing board.

Instead, the Meadowlands became a dumping ground for more waste than anywhere else in the United States. Military waste was

(a)

FIGURE 9.14 (a) Aerial photo looking over the Hackensack Meadowlands and Hackensack River toward New York City. *Source:* Doc Searls / Wikimedia Commons / CC BY-SA 2.0. (b) Annotated satellite image of the western Hackensack Meadowlands showing the density of landfills. *Source:* Courtesy of the Meadowlands Environmental Research Institute (MERI).(c) Map of the Hackensack Meadowlands showing the locations of some EPA Superfund sites. Inset map shows the location of the Hackensack Meadowlands in New Jersey.

(b)

NJMC District Landfill
Approximate Landfill Acreage - 1379
◌ Landfill Leachate Sample Location

FIGURE 9.14 (Continued)

dumped in the Meadowlands during World War II, including rubble from the London blitz as well as chromium ore which had been used as ballast in ships. After the war, the Meadowlands was used for massive civilian waste disposal. The New Jersey Turnpike opened through the Meadowlands in 1952 which facilitated easier transport of waste into the area. Garbage was being deposited in open dumps and simply along roadsides. The expanding road system provided ever-increasing access to transport garbage by truck out of New York City and New Jersey communities and into the Meadowlands.

Uncontrolled open waste dumping continued throughout the 1950s and 1960s turning the Meadowlands into the world's largest landfill at that time. It is estimated that about 10 000 tons (9072 mt) of garbage was dumped in the Meadowlands each day. Larger dumps slowly accumulated into large hills, many of which were regularly ignited to save space and control wind-strewn debris (Figure 9.14b). Methane gas was produced from the decaying garbage and with other flammable debris caught fire and burned uncontrollably for years.

(c)

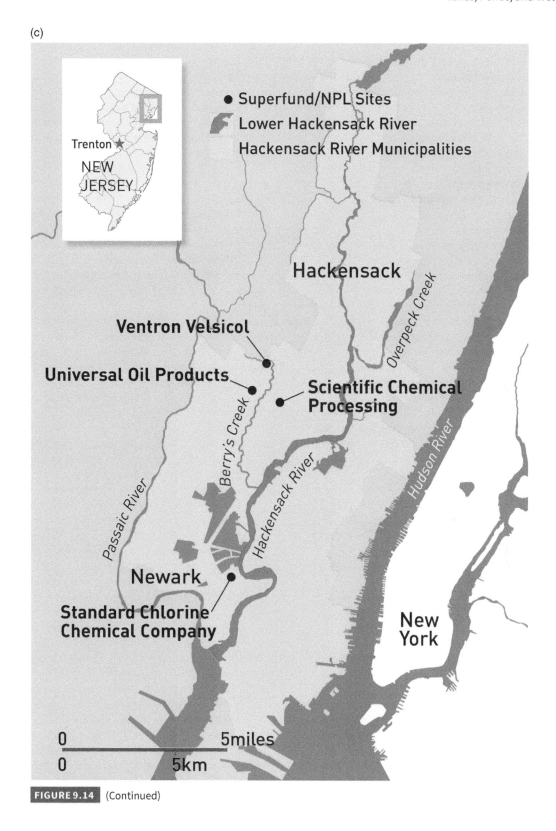

FIGURE 9.14 (Continued)

In 1957, New York City decided to stop providing garbage removal for companies. This required them to hire private garbage collectors some of which were associated with organized crime. These underworld enterprises eliminated the garbage dump costs by dumping collected waste at any unwatched location they could find as well as setting up toll booths to collect tipping fees for dumping by everyone else. The highway access and low population density of the northern Meadowlands made them the ideal dumping location. The explosive growth of the local garbage disposal industry greatly increased illegal dumping in the Meadowlands. Dozens of garbage haulers made continuous trips to an increasing number of illegal dumps attracting the attention of residents, newspapers, and finally law-enforcement officials who attempted to control the influx with marginal success.

During the peak of dumping activity in the 1960s, the Meadowlands received 40% of New Jersey's solid waste and 10 000 tons (9072 mt) per day from New York City. There were greater than 200 dumps, occupying 2500 acres (1012 ha) in the Meadowlands. The burning dumps filled the air with foul-smelling smoke accompanying the stench of decaying garbage, and millions of cubic feet of methane emitted from landfills. The odor sickened travelers on the New Jersey Turnpike and gave the state a bad reputation that would persist for decades. At one point, the smoke from burning garbage in the Meadowlands got so thick that it caused a massive traffic accident on the NJ Turnpike. Leachate from landfills, toxic chemicals, and raw sewage poured into the Hackensack River making fish kills and extensive dead zones commonplace.

The municipal waste and sewage was not the only thing being dumped into the Meadowlands at the time. Several companies were also depositing toxic industrial waste in several locations. These would later be designated as Superfund sites with clean up overseen by the US Environmental Protection Agency (EPA) (Figure 9.14c). The 40-acre Ventron/Velsicol Site operated as a mercury processing plant from 1929 to 1974. They removed mercury from lab equipment, batteries, and other devices and disposed of it into the Meadowlands. Some 160 tons (145 mt) of waste was buried on the site resulting in elemental mercury as high as 195 000 ppm (ppm) in on-site soils. It is said that there are layers of liquid mercury interlayered with soil. This property was designated as a Superfund site in 1984.

The Ventron/Velsicol operation also discharged waste directly into Berry's Creek and it spread several thousand feet downstream. It is estimated that the mercury-contaminated waste discharged into the creek between 1943 and 1974 reached a total of 268 tons (243 mt). Sediments from the creek have been measured as having between 1 and 2 g/kg (parts per thousand) of methyl mercury, which is the highest concentration of any fresh-water sediment in the world. Methyl mercury is a deadly poison and now contaminates all fish in the area through bioaccumulation. Even the air around the site has elevated mercury levels. Berry's Creek also contains chromium and PCB wastes at unsafe levels. It was designated as a Superfund site in 2018 with a cleanup price tag of $338 million.

The 42-acre (17 ha) Standard Chlorine property in the Meadowlands contained a chemical manufacturing plant used by a number of companies from 1916 to 1993. Processes included refinement of naphthalene, processing of liquid naphthalene, manufacture of lead-acid batteries and drain-cleaner and packing dichlorobenzene products. Soil, groundwater, and two waste lagoons contained high levels of dioxin, benzene, naphthalene, PCBs, and volatile organic compounds. The lagoons drained directly into the Hackensack River through a ditch. There were also tanks and drums of hazardous substances such as dioxin and asbestos. The property was designated as a Superfund site in September 2007.

The 20.2-acre (8.2 ha) Diamond Head Oil Refinery property in Kearny, NJ was an oil reprocessing plant that was operated by several companies, including PSC Resources, Inc., Ag-Met Oil Service, Inc., and Newtown Refining Corporation, between 1946 and 1979. The facility contained numerous aboveground storage tanks and several subsurface pits that were used to store oil waste. The waste was discharged onto adjacent wetlands creating a large "oil lake" because of the amount of waste pooling at its site. The property is polluted with LNAPL, chromium, dioxin, PCBs, lead, aldrin, thallium, and benzo[a]pyrene in both the soil and sediment. It is also a Superfund site.

The 6-acre (2.4 ha) Scientific Chemical Processing property was used as a processing facility for the recovery and disposal of various types of waste. When the facility was shut down in 1979–1980, 375 000 gal (1.42 million l) of hazardous waste were being stored on site in tanks, drums, and tank trailers. As a result of the operations, groundwater and soil were polluted with a variety of hazardous chemicals including PCBs, heavy metals, PAHs, and VOCs such as benzene, chloroform, and trichloroethylene. The Scientific Chemical Processing property was designated as a Superfund site on 1 September 1983.

Perhaps the worst polluted site in the Meadowlands area was actually on the Passaic River. The Diamond Alkali Superfund Site (DASS) includes portions of the Passaic River beginning at the lower Study Area (LPRSA), extending 17.4 miles (28 km) downstream and into the lower Hackensack River, and the Newark Bay. The facility began operations in the 1940s but in the 1950s and 1960s, the Diamond Alkali Company manufactured agricultural chemicals, including the herbicides used in the infamous defoliant "Agent Orange." Agent Orange was used extensively in the Vietnam War causing untold health issues for American soldiers as well as the Vietnamese. Agent orange contains the extremely toxic chemical, dioxin. The soil and groundwater at the site contained dioxin and other hazardous substances including PCBs, heavy metals, polycyclic aromatic hydrocarbons (PAHs) and pesticides. However, the real hazard was the dioxin in river sediments. Fish in the Passaic and Hackensack Rivers still have very high levels of bioaccumulated dioxin and cannot be consumed.

In 1969, the State of New Jersey finally put a halt to the uncontrolled pollution of the Meadowlands. At that point, more than 5000 tons (4535 mt) of garbage from 118 New Jersey municipalities was being dumped in the Meadowlands every day, six days per week. They identified 51 active landfills. In response, they established the Meadowlands Commission which created the district's first Master Plan, for the sanitary disposal of solid waste. They also planned the orderly development of the region, and the restoration of natural areas as much as possible. Old landfills were remediated, operational landfills were stringently regulated and 3500 acres (1416 ha) of environmentally sensitive and less impacted wetlands were protected. Despite these changes, sanitary landfill disposal of waste continued in the Meadowlands until 2020 and full remediation has quite a long way to go.

With the abundance of dangerous pollutants in surface waters, it is surprising that overabundance of fertilizer might be the greatest threat to water quality. The primary active ingredients of fertilizers are phosphates and nitrates. These compounds nourish plants allowing them to grow faster, larger, and stronger. This is beneficial for crops and ornamental plants. In water, however, they nourish algae and other aquatic plants, making them flourish even more than land plants. When the algae and aquatic plants die, they decompose and microorganisms consume them. The microorganisms also consume oxygen in the process. The reduction in oxygen produces hypoxic conditions in the water, which drives away or kills fish and other aquatic animals that depend on the oxygen in the water. This process of oxygen reduction is called eutrophication and nearly every surface water body on a global basis suffers from it. Eutrophication is largely driven by runoff from agricultural and recreational fields and ornamental plantings in urban and suburban areas. However, it can also be the result of runoff from grazing areas, feedlots and other animal facilities, leaky sewers and septic systems, and even air pollution fallout. It can also be naturally occurring from heavy rain events in vegetated areas.

Coastal areas in the oceans can also suffer from fertilizers carried by rivers. Rivers deliver water laden with fertilizers to the ocean. It was thought that oceans could absorb the excess nutrients but in the 1960s, a large zone of eutrophication was found in the Black Sea in the USSR. These areas were named "dead zones" and have appeared at the mouth of most major rivers worldwide. These dead zones make large areas of ocean unproductive and reduce our ability to produce enough food for the huge human population.

9.4   What Can I Do?

With surface water, even small efforts make a difference. The problem with surface water is that anything in your yard will wind up in runoff and pollute surface water. Dumping leftover chemicals like paint thinner, gasoline, or motor oil on the ground should be avoided. The use of lawn chemicals like herbicides, pesticides, and chemical fertilizers will wind up in runoff. They are very damaging to the environment but mean that the lawn will not be as green or weed-free. Overfertilizing will cause extensive eutrophication of surface water. De-icing compounds will also wind up in runoff so should be used sparingly. If your car leaks any fluids, like oil, anti-freeze, brake fluid, or gasoline, they too will wind up in runoff. Otherwise, any surface litter will find its way to water bodies and should be avoided. Volunteering for river cleanups in your town will further help the situation.

Pollution of Soil

CHAPTER OUTLINE

Polluted Earth: The Science of the Earth's Environment, First Edition. Alexander Gates.
© 2023 John Wiley & Sons, Inc. Published 2023 by John Wiley & Sons, Inc.
Companion website: www.wiley.com/go/gates/pollutedearth

Words you should know:

Alluvium – Young soils and sediments deposited on hillslopes by weathering and gravity.

Drift – Sediments deposited by glacial or related activity.

Fluvial deposits – Sediments deposited by rivers and streams.

Laterites – Deeply weathered soils rich in aluminum through extreme dissolution and common in tropical areas.

Mine spoils – Human-produced deposits of waste rock around mining and smelting sites.

Outwash – Sediments deposited by meltwaters beneath and in front of an advancing glacier, typically occurring in valleys.

Pedalfers – Soils that are enriched in aluminum and iron as a result of deep weathering in temperate areas and the removal of other elements by dissolution.

Pedocals – Soils rich in calcium that are common in arid and hot regions.

Soil horizons – Distinct layers developed in soils through weathering and vegetation including from the surface, O-, A-, B-, C-horizons with depth.

Soil structures – Shape features that typically form in various types of soils through drying, weathering, and erosion.

Till – Glacial sediments deposited by ice, typically occurring on slopes.

10.1 Soil Basics

Soil is among our most precious resources and yet it is generally regarded with disdain. Common sayings like "treated like dirt" illustrate this. Most of our food supply grows in soil or survives only because it eats food grown in the soil. Little life would be possible outside of the oceans without soil. Soil is defined as unconsolidated material lying on bedrock and formed by surficial processes. It is composed of variably degraded rock, organic detritus, microorganisms, and materials added by natural and artificial processes. There are classifications for soil by texture and chemistry, by structural development, and by regional associations.

One classification of soil composition is by the amount of clay, silt, and sand (Figure 10.1). The system uses a ternary graphical diagram that is subdivided into fields with basic names. The apices of the diagram are clay, silt, and sand and the fields with mixes of two or three are called loam. The soil name can be modified by the type of clay. There are several types of clays but it requires chemical analysis to identify them. Clays can adsorb chemicals so can be important sources of nutrients or dangerously polluted.

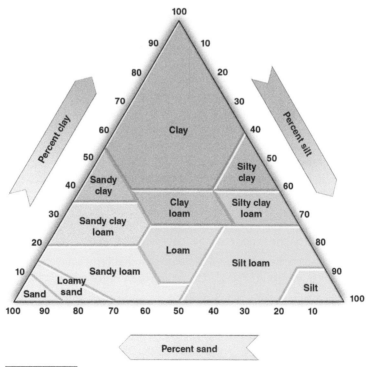

FIGURE 10.1 Ternary diagram of clay, sand, and silt for the classification of soil.

Horizons

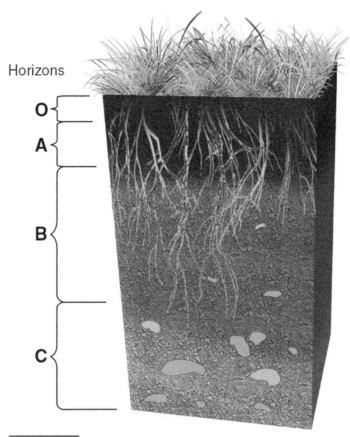

O {
A {
B {
C {

FIGURE 10.2 Diagram showing the dominant soil horizons in a soil profile.

Soils form profiles with characteristic horizons if they are formed from the weathering of the underlying bedrock (Figure 10.2). The O-horizon is at the surface, black in color, and completely organic, composed of decaying leaf litter. It is also called humus and is home to plants and animals. The A-horizon is below and generally termed topsoil. It is also dark-colored and organic-rich but contains clays and degraded rock. It also displays biologic activity. The underlying B-horizon is called subsoil, generally tan colored and clay-rich. It commonly contains variably degraded rock fragments and only the roots of trees and other large plants. Some animals burrow into the subsoil but far fewer than in the shallower horizons. The C-horizon is composed of bedrock fragments in various stages of decomposition.

The size of each horizon depends on the local environment. In arid areas, there is little to no vegetation to provide organic material and consequently the O-horizon and A-horizon are typically absent or poorly developed at best. Glaciers removed most of the existing soil during the last and previous ice ages. As a result, the O-horizon can lie directly on bedrock or on glacial sediments in northern areas, with thin or absent A-, B-, and C-horizons. In old, deeply weathered areas nearer to the tropics, the A-, B-, and C-horizons can be very thick, 50 ft (15 m) or more. In the southeastern United States, the soil contains the layering and structure of the parent rock but the original minerals have altered to clay and oxides. This soil is called saprolite.

There are five types of soil structures based on the shape they break into under natural conditions. Soils with no structure are massive as are many agricultural soils.

Platy soil structure breaks into flat, plate-like aggregates. The plates generally impede infiltration of water and downward growth of roots. Platy structure is in B- and C-horizons that have been flattened by animals or machinery.

Prismatic soil structure contains aggregates separated by flat to rounded vertical faces and prism shaped tops. They are common in B-horizons where there is wetting and drying, freezing and thawing, or strong infiltration of water or root growth.

Columnar soil structure is similar to prismatic but vertical cracks are deeper and the prismatic tops are more distinct. The structure gives soil a columned appearance.

Blocky soil structure has equidimensional aggregates with slightly curved cracks. They may be angular if the edges are straight. They are common in areas with swelling clays.

Granular soil structure has rounded aggregates of mixed sizes that do not fit tightly together. They drain and accumulate organic material well allowing them to be fertile and productive.

Soils are also classified by general compositional characteristics through long-term weathering. The most common are pedalfers, pedocals, and laterites in the original classification system but there are numerous others. In this system, "ped" and "sol" refers to soil, and the rest is based on chemistry or usage.

Pedalfers are soils rich in aluminum (Al) and iron (Fe). They are common in areas that are temperate with significant rainfall and vegetation. In these areas, decaying vegetation produces acid that dissolves most minerals, especially those rich in calcium and alkalines.

Pedocals are soils rich in calcium. They occur in hot, dry climates and produce hard pans on the surface that commonly contain caliche.

Laterites are soils rich in aluminum and some clay. They are common in rainy tropical areas where the weathering penetrates very deeply. Decaying vegetation produces acids that remove everything but the aluminum and minor iron. They develop bauxite deposits that are mined for aluminum.

The advanced system of soil classification in the United States is more comprehensive to account for the variations in composition and texture. This US Comprehensive Soil Classification System contains 12 major orders of soil based on environment and physical character. There are many suborders and 19 000 soil series. There are numerous other classifications in the world that are not included in the United States system. The major 12 soils include *Alfisol*, a young soil rich in clay and plant nutrients that forms in humid forests. "Alf" is aluminum and iron which makes it a pedalfer of the old system. *Andisol* is a soil that forms in volcanic ash from altered glass and unstable mineral fragments. *Aridosol* is a soil that forms in the desert and is equivalent to a pedocal. *Entisol* is very young soil, lacking in structure and showing characteristics of the parent material. These soils form on active slopes or in rivers. *Gelisol* is young soil that forms in permafrost so is frozen part of the year and soil profile development is poor. *Histosol* is black, highly organic-rich soil referred to as peat and muck. It is composed of partially decomposed vegetation and is therefore O-horizon. *Inceptisol* is a soil that forms quickly through alteration of rock and sediment in humid climates. *Mollisol* is a soft, dark, highly fertile soil that develops from long-term growth of grass. They have good A-horizons and high calcium and magnesium contents. *Oxisol* is a leached soil that is equivalent to a laterite. *Spondosol* is soil that develops in humid regions on a sandy base. The A-horizon is weathered organic material and the B-horizon is light-colored leached material. *Ultisol* is a highly leached red to yellow soil from extensive weathering. The "ulti" part is from "ultimate" because they form as the ultimate product of weathering. *Vertisol* is a soil has a high content of expanding clays that are common in arid areas with wet and dry seasons.

10.2   Soil Pollution

Soil has been a dumping ground for refuse throughout human history. Before the Industrial Revolution, mining and smelting operations caused problems but in most areas waste was biodegradable. This meant that the impacted area would largely recover through natural processes known as natural attenuation. Mining operations have always caused pollution. Mining wastes do not biodegrade, bring dangerous chemicals to the surface, and require hundreds to thousands of years to break down naturally. Smelting and forging of ore is even worse. Smelting splits the chemicals in ores to yield a target element, like a metal, and waste elements. The waste elements include unwanted materials like sulfur that could cause local acid rain and acid mine drainage and heavy metals like mercury, lead, chromium, and cadmium. All of these contaminants cause severe environmental problems. This disregard for soil only grew worse during the Industrial Revolution first when coal mining and usage caused widespread air, water, and soil pollution. Later, petroleum and biproducts, most of which are toxic, were added to soil contamination. Agricultural pollutants like pesticides and chemical fertilizers further degraded soils. Pollution of soil continued at a rapid pace throughout most of the twentieth century. It slowed in most developed countries in the later part but on a global basis, it has continued into the twenty-first century unabated.

Soil pollution can contribute to pollution of water and air. Soluble soil pollutants can be leached by infiltrating water into the groundwater system and contaminate drinking water. Some contaminants can be absorbed by plants and concentrate in the leaves, stems, and fruit. If contaminants are not soluble, erosion can transport them as particles into surface water and contaminate sediments. Some contaminants can evaporate from the soil and be carried by wind as particulate. These processes contribute to air pollution.

10.2.1   Heavy Metals

Heavy metals have a density greater than 6.0 g/ml but this is not strictly followed. Heavy metals include arsenic, cadmium, chromium, lead, mercury, and zinc, among others that can be dangerous to toxic. They pollute soil from mine waste, settle out or wash out of air pollution, and from direct dumping. The toxicity depends on the element but also the valence state. Dimethyl mercury is highly toxic in any quantity but non-methylated mercury is far less dangerous. Hexavalent chromium is much more dangerous than other forms. Many heavy metals occur naturally in rock and soil in very small quantities. If concentrated by natural or industrial processes, however, they are much more dangerous.

Mine wastes produce extreme but localized heavy metal pollution because the minerals containing them are unstable at the surface. The weathering of rock surfaces occurs over thousands to millions of years which breaks down these minerals. Mining brings unweathered and highly reactive minerals to the surface where they can degrade. In addition, mined rocks are crushed into small fragments which creates more surface area to weather. Chemical reactions release heavy metals from metallic deposits and into the environment. The result is that mine soils can be large sources of pollution.

CASE STUDY 10.1 Bunker Hill Complex, Idaho

Soil Pollution from Mining

The Bunker Hill Mine site, Idaho was designated as a Superfund site by the U. S. Environmental Protection Agency (EPA) in 1983 (Figure 10.3a). The property is among the largest superfund sites in the United States at 21 square miles (54.4 km²). Operations first opened in 1889 to mine lead and silver but they expanded and in 1917, a lead smelter was opened to process and refine ores. At peak production, the mine processing area was 6200 acres (2509 ha), and the mines were 1 mile (1.6 km) deep and included 125 miles (201.2 km) of underground roads or tunnels. The location of the operation also promoted this growth because it was convenient for transport of ore and refined metals and included a continuous water supply for ore processing, local agriculture, and drinking.

The processing of mined Bunker Hill lead ore was done in three steps (Figure 10.3b). First, the mined ore was crushed and some of the waste rock or gangue was removed by hand and/or mechanical sorting. Second, sintering was done in which lead ore was mixed mainly with limestone flux and heated to just below the melting point to allow chemical bonding. Third, the sintered ore was placed in a blast furnace where carbon in the flux allowed molten lead bullion to sink to the bottom of the container. The slag is the waste from this process that rose to the top of the molten mass and was skimmed off and discarded.

Bunker Hill produced enormous quantities of hazardous waste by-products in these processes. Following regulations at the time, mine tailings and dust-laden wastewater and fine solid waste were dumped into the nearby Coeur d'Alene River (Figure 10.3c). Slag was dumped on the site, allowing contaminated runoff and blowing contaminated dust. There was 100+ years of mining that took place, allowing lead to be emitted into the atmosphere for the entire time. The lead-bearing dust quickly settled to the soil in the vicinity of the mill, where it contaminated the soil and washed into surface and groundwater. The lead in the dust was a serious threat to the health of the mine and smelter workers as well as the residents. In addition, mine and smelter tailings were used to sand icy streets in the winter as well as backfill in new construction. Lead was dispersed throughout the Coeur d'Alene River basin and caused the deaths of large numbers of aquatic organisms, native fish, and waterfowl.

By the early 1900s, environmental conditions were terrible at the facility. In response, a baghouse was installed to filter particulate emissions from the smelter in 1923 in which fumes from lead refining passed through cloth traps. This solution was only partly effective, allowing lead-bearing particulate to be released to the environment from the smelter for 50 more years. In 1938, a tailings pond was installed in the outwash plain of the Coeur d'Alene River. The pond was built from low-grade ore and waste rock and designed to receive and hold process and mine water. It allowed particulate to settle prior to discharge into the Coeur d'Alene River. The tailings pond was only partly effective in controlling the releases of lead wastes from Bunker Hill.

Mining and processing of ore at the Bunker Hill complex rapidly expanded throughout the 1940s and 1950s. Workers and residents were reluctant to complain about the deteriorating environmental conditions because the mine provided secure, high-paying jobs. The economic health of the area depended on the mining companies to produce lead cheaply and quickly. The Bunker Hill smelter was the second largest in the United States by the 1970s. It produced one fifth of the lead in the world and more than 20% of the lead and zinc in the United States. The smelter released 6 million pounds (2.7 million kg) of lead into the atmosphere between 1965 and 1981. In 1973, a fire disabled the baghouse resulting in a sharp increase in lead emissions. Nothing was done to fix the situation because the price of lead was near an all-time high.

Public awareness increased about the danger of lead and, as a result, political pressure began to mount to monitor the health of people working or living near Bunker Hill. In the 1970s, the passage of the Federal Clean Air Act and Clean Water Act forced environmental improvements to be made. The company constructed a $1 million wastewater treatment plant at the tailings pond to treat the water before being discharged to the Coeur d'Alene River. The baghouse was repaired and modernized by 1975. However, the Bunker Hill company fought environmental requirements with public relations and lobbying campaigns. They also purchased all of the houses within a half mile (0.8 km) of the smelter, burned them, and replaced the contaminated soil with fresh topsoil. It was too little and too late and the Idaho Department of Health & Welfare began testing lead in air, soil, and vegetation in the area and found excessive concentrations of lead throughout the area.

In response, the company commissioned their own study, which included collecting and testing urine samples from children without informing the parents or obtaining their consent. The company refused to release these results. Finally, public pressure forced them to cosponsor and fund a blood lead level monitoring program on local children. In January 1975, the results released and 170 of the 172 children tested near the smelter had blood lead levels above 40 µg/dl, the hospitalization level. Emergency treatment is required at blood lead levels above 70 µg/dl, and 45 of the 172 children had levels above 80 µg/dl. One child had a level of 160 µg/dl, the highest lead concentration in blood ever recorded in the United States.

The company made several last-ditch efforts to control the environmental situation. By 1982, they had spent more than $20 million for pollution control, but still could not meet new EPA standards. The mine and smelter closed as a result and 2000 people lost their jobs. In 1983, cleanup of the area began as a Superfund site.

(a)

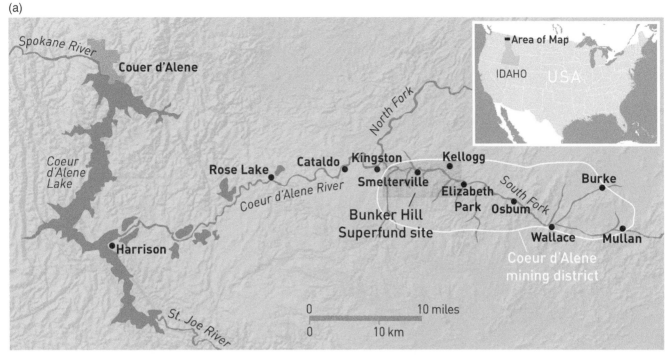

(b)

FIGURE 10.3 (a) Map of the Bunker Hill complex and Superfund site. Inset map shows the map location in Idaho with surrounding states. (b) Photo of the Bunker Hill mine and smelter complex. *Source:* Courtesy of the US Library of Congress. (c) Aerial photograph of the Bunker Hill complex. *Source:* Unknown source/ Public Domain.

It involved treatment or removal of thousands to millions of tons of tailings, waste rock, unprocessed ore, and lead-contaminated soils spread throughout the area. Early activity involved removal of lead-contaminated soil from 16 public parks, and 2000 private and commercial properties. The company declared bankruptcy in 1994, and tax funds had to be used for most of this work. A 32-acre (13-ha) mine waste and debris landfill was built at the smelter and closed and capped. More than 1.25 million cubic yards (1 million m³) of lead-contaminated soil were disposed of onsite. More than 30 000 cubic yards (22 937 m³) of contaminated sediment was removed from the Coeur d'Alene River. Although much remediation has been completed and many of the residents are out of danger, the Bunker Hill site is still considered as one of the worst cases of lead poisoning and environmental contamination in the world.

(c)

FIGURE 10.3 (Continued)

Dumping of industrial waste contributes heavy metals to the environment. The metals in industrial waste are typically in a hazardous state. Until the 1970s, all industry dumped waste solids and wastewater in dumps or lagoons on the grounds of the manufacturing facility. Berry's Creek, New Jersey, for example, includes a Superfund site with layers of liquid mercury interlayered with soil strata. There are examples of dumps and lagoons for most heavy metals.

Heavy metals can contaminate soil through fallout from air pollution or wash out by precipitation. Manufacturing facilities have released heavy metals into the air as dust or vapor. Metal processing facilities are the main sources of these. Incinerators released metals to the air in the fly ash from smokestack emissions for many years but filtering has eliminated most of them. Especially in the past, coal-burning power and industrial plants have released mercury, other metals and sulfur compounds carried in air and deposited on the surface. The concentration of metals are highest in soil closest to the source. It is also highest toward the direction of prevailing winds. Recent regulations on smokestack emissions of many pollutants require scrubbers and other emission reduction devices which have drastically reduced contamination. The legacy of old pollution still remains.

Heavy metals can remain in the soil for a long period of time. They are mostly weakly soluble or insoluble and are rarely converted into other less dangerous chemicals through reactions like organic pollutants other than by changing valence states. Some can undergo limited bioconversion from one valence state to another. Some plants such as tobacco uptake specific metals into the plant structure. Otherwise, they mainly remain where they are deposited until physically moved or diluted with clean soil.

10.2.2 Organic Pollutants

Organic contaminants are introduced to soils through leaks and spills from manufacturing, transportation and storage, air pollution fall-out and wash-out, spraying, and direct dumping. They tend to be less persistent than heavy metals and degrade in a few hours to as long as several decades. Many organics can biodegrade and alter because they contain carbon compounds that are used for cellular energy. Organic pollutants are far more abundant than inorganic pollutants to the point that essentially every person comes into contact with an organic pollutant every day.

Purposefully sprayed or applied organic compounds to soil are primarily for agricultural and landscaping purposes. These pollutants include pesticides, herbicides, and fertilizers. Most are soluble and readily leached into groundwater and occur in drinking-water supplies throughout the United States. Some compounds can be adsorbed to clay or organic particles and be chemically fixed as the most persistent organic compounds. Some pesticides are toxic to microorganisms in the soil and can only be broken down by chemical reactions. Some pesticides are systemic and absorbed into root systems of plants and distributed throughout them. Fertilizers and pesticides are washed from soil by surface runoff and transported to surface water systems where they can have devastating effects.

There are organic pollutants in the air from both point and non-point sources. The most abundant organic pollutants are from automobile exhaust, evaporation of volatile organic compounds (VOCs), and burning. Most organic pollutants remain in the air until they are degraded or are washed out by precipitation to the surface and soil. Particulate can settle or fall out to the surface by gravity and become part of the soil system.

Organic compounds commonly leak into soil. Most leaks are from underground storage tanks (USTs) and from gasoline stations and home heating oil tanks. The contamination passes through soil and into the groundwater system. The gasoline and oil plumes float on top of the water table, infiltrate soil as they move and leave petroleum behind, coating soil particles and filling pore spaces. Leaks also occur in pipelines and at manufacturing facilities.

Deliberate dumping of organic pollutants is very dangerous. Some of the most infamous superfund sites contain industrial organic pollutants. Two of the worst cases of organic soil pollution in the world are Love Canal, New York and Times Beach, Missouri. There are numerous others where organic pollution of soil is the important impact.

CASE STUDY 10.2 Times Beach Superfund Site, Missouri

Organic Soil Pollution

Times Beach, Missouri, was a small community 20 miles (32 km) west of Saint Louis on the flood plain of the Meramec River. It began in the 1930s as a vacation town and grew into a community of 1200 people by the 1970s (Figure 10.4a).

Between 1972 and 1976, the town of Times Beach and several businesses paid a local waste hauler to spray oil on unpaved roads to control dust. The waste hauler collected used oil from service stations, heating oil distributors, and machine shops among others, either as waste removal service or as a purchase. He sprayed the oil on unpaved, dirt roads, a common method of dust control.

The waste hauler was also hired by Northeast Pharmaceutical and Chemical Company to dispose of waste chemicals from its agricultural chemical manufacturing operation. Instead of disposing of this waste, the waste hauler mixed them with the waste oil and sprayed the mixture on the roads, horse racetracks, and farms around the Times Beach area and 26 other locations. As a result of this action, some 4000 to 5000 gal (11 356–18 927 l) of oil containing more than 300 parts per million (ppm) of the very hazardous chemical, dioxin, were distributed.

Dioxin was the main active ingredient in many of the pesticides and herbicides as well as in the waste and it is highly toxic. In 1974, indications of a major environmental problem in Times Beach emerged. There was an increase in the deaths of otherwise healthy horses at local stables. In all, 63 horses, 6 dogs, 12 cats, and 70 chickens died and 10 people were sickened. The common link among these deaths was they occurred in stables whose tracks, paddocks, and access roads had been oiled using the chemical waste. The waste hauler was confronted by the stable owners but he assured them that he only used motor oil in his spray and was not responsible for the death of the horses. Unconvinced, the stable owners contacted the Centers for Disease Prevention and Control (CDC) and in late 1979, after an extensive investigation by multiple agencies, Northeast Pharmaceutical and Chemical Company admitted that dioxin was in their waste.

The U.S. EPA launched a full investigation of the waste hauler's activities. In November 1982, the EPA notified the residents of Times Beach that the area was likely sprayed with waste oil contaminated with dioxin. They collected soil samples in Times Beach, even though the roads had been paved. In early December 1982, soil containing dioxin concentrations 300 times greater than the maximum allowable were found in the town. During this project, a freak rainstorm caused the worst flood in Times Beach history spreading dioxin-contaminated soil and mud throughout the town (Figure 10.4b). As the flood water subsided, the EPA confirmed the widespread dangerous levels of dioxin, 100 times maximum. In response, EPA officials advised residents to abandon Times Beach.

Panic spread throughout the community and surrounding areas, and many illnesses, miscarriages, and animal deaths were attributed to the dioxin. The Missouri state police and National Guard were called to blockade roads and to prevent people from entering or exiting the town. Dioxin had been spread throughout the area impacted by floodwaters and covered the entire 0.8 square miles (2.1 km²) area of Times Beach. Missouri public health officials were concerned that fleeing residents could track the contamination to other areas of Missouri.

In 1983, the EPA attempted a $33 million buyout of the entire town of Times Beach including 437 houses, 364 mobile homes, and 45 businesses. Residents rejected this and the second offer by EPA because they were less than market value. It took several years for an agreement to be reached. Having lived in Times Beach, however, was considered a stigma as evacuees were often shunned because many people were afraid that effects of dioxin exposure could be contagious.

The waste disposer denied wrongdoing and claimed that Northeast Pharmaceutical and Chemical Company had not informed him that the waste he received was hazardous. This conflict of testimony kept both defendants from being convicted of a crime. However, in 1983, both were sued by 183 Times Beach residents for $1.8 billion in damages. As a result, Northeast Pharmaceutical and Chemical Company declared bankruptcy the following year. Its insurers settled the lawsuit for $19 million.

The cleanup implemented by the EPA was radical, highly criticized and never repeated on such a large scale (Figure 10.4c). Between March 1996 and June 1997, well more than 10 years after the incident, about 240 000 tons (217 724 mt) of soil and every building, and structure was burned in a high-temperature incinerator that the EPA constructed near the town. The incinerator was a 75-ft (23-m)-long rotary kiln that was lined with a 9-in. (22.9 cm)-thick layer of high-temperature, acid resistant, insulation. The kiln processed about 40 tons (36.3 mt) of waste per hour, at a temperature of 1250 °F (677 °C). These temperatures dissociated the chemical bonds in the organic compounds and they were reduced to elemental components. To ensure that dioxin was destroyed, the exhaust gas was sent to a secondary combustion chamber that operated at a higher temperature of 1750 °F (954 °C). The dioxin removal efficiency of the system is estimated at 99.999% and the cost of remediation was approximately $140 million.

(a)

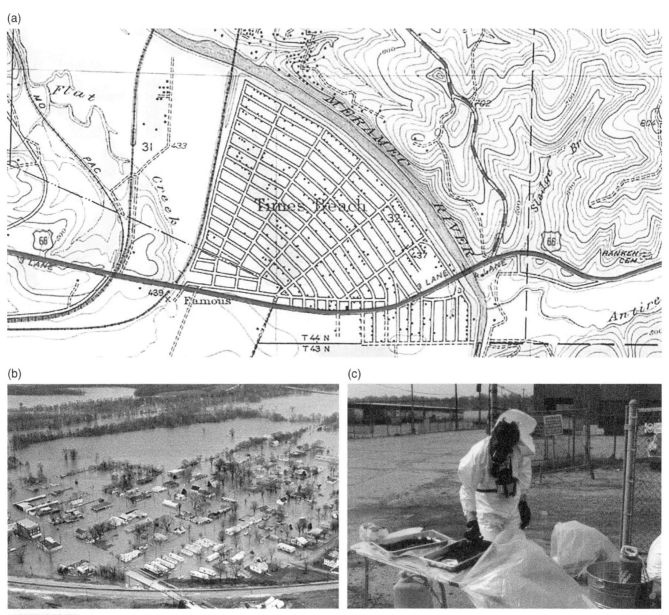

(b)   (c)

FIGURE 10.4   (a) Topographic/land use map of Times Beach before the disaster. The lines are roads and the small dots are houses (numbering in the hundreds), showing the size of the town. Map source, US Geological Survey. (b) Aerial photograph of Times Beach at the peak of the flood. Courtesy of the US EPA. (c) Photo of EPA personnel working to remove contaminated materials. *Source:* Courtesy of USEPA.

Soil pollution and potential sterilization is a real danger for the survival of the human race. Soil produces most food we consume either directly or indirectly in the feed for livestock. The more humans attempt to continually increase productivity from agricultural lands, the more likely there will be a catastrophic collapse of the system. Soil mismanagement in the past has restricted productivity of agriculture such as the Dust Bowl of the American Midwest in addition to numerous cases of desertification. There have been numerous cases of soil pollution that left large areas ecologically dead. Mining and smelting operations around Norilsk, Russia, Sudbury, Canada, Ducktown, Tennessee, and Palmertown, Pennsylvania are examples of such disasters. Fortunately, society realizes the danger of such situations and, in most cases, has taken action to prevent them from continuing.

10.3   Fluvial and Alluvial Sediments and Soils

Fluvial and alluvial sediments and soils are sometimes interchanged, but in strict definition, fluvial refers to rivers and streams and alluvial refers to slope deposits. In mountainous arid regions, alluvial fans form at the base of slopes. These are large fan-shaped geomorphic features with an apex that sits at the mouth of a large canyon or valley. Water feeds down this valley and onto the fan during storms and spring thaws but may be dry and inactive most of the year. The sediments on the fan are supplied and transported in one of two ways, by river flow or by debris flow. The fans are classified by whether they are debris flow or river flow dominant. Rivers deposit mainly sand with some gravel whereas debris flows transport up to boulder-sized sediments and commonly have their coarser sediments at the top of the layer or bed. In most other deposits, the coarsest sediment is at the bottom of the bed.

Alluvial fans grade into alluvial plains which are composed of fine sands and silts interlayered with coarser river deposits. They too are only actively deposited during wet seasons and lie dry and inactive most of the year. They occur in arid areas so contain little vegetation or organic material. The alluvial plains can lead into fluvial environments with braided streams or playa lakes depending upon the topography of the area.

Playa lakes occur in the center of basins. They too are filled with water during wet seasons or thaws and evaporate and shrink the rest of the year. They tend to have sandy shores but the extensive evaporation and lack of additional water makes them briney. In addition to salt, these lakes can carry other important elements and compounds like potassium, bromine, lithium, borates, and sulfates. They can create economic deposits of these important resources. Playa lakes can also deposit limestone depending upon the chemistry of the water.

Alluvial systems are very sensitive ecologically with landscapes and ecosystems taking long periods of time to develop. However, the lack of water, relative instability, hot climate, and distance from population centers typically makes them sparsely inhabited. Pollution of these sites is commonly imposed both by individuals and military operations but it is not usually reported because of the low impact on humans.

Fluvial systems are discussed separately with regard to pollution of the water. However, they deposit sediments and develop characteristic soils as well. Rivers are generally divided into two end members, braided and meandering. Braided streams or rivers are characteristic of rugged topography and relatively dry conditions. They are fast-flowing, relatively straight, confined to steep, narrow valleys, and have small flood plains, if any. Most of the year, they appear as a series of anastomosing channels around lozenge-shaped channel bars composed of sediment. Soils may develop on the channel bars and vegetation may take hold because very little sediment moves in this condition. During high water periods, the river level is higher on the valley walls and the channel bars are completely submerged and move downstream. The fast-moving water rips up all soils, clays and vegetation that may have formed in the quiet period and large amounts of all sizes of sediment are transported.

Meander belts form in areas of flat topography (Figure 10.5). They have single, highly sinuous channels and very wide floodplains. The water level in the channels remains full throughout the year but floods occasionally. The river constantly erodes the outer arcs of the loops or cut banks and deposits new sediments and soils on the inner arcs or point bars. These are dynamic systems with constant shifting. The channels contain the water and coarse sediments but during floods, river water covers the

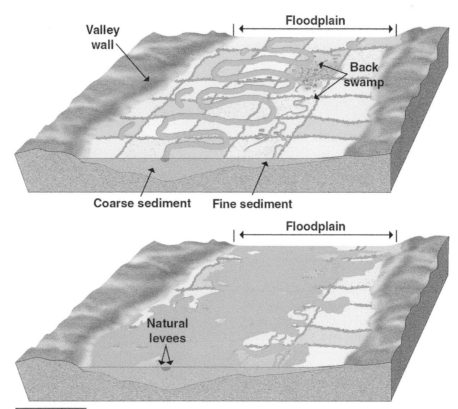

FIGURE 10.5 Diagrams showing a meandering river (top) and the same river during a flooding event with the river water covering the floodplain (bottom).

floodplain. Sediment is deposited on the floodplain during the flood but as the water recedes, large puddles of turbid floodwater remain and slowly dry. They leave behind clay and silt deposits across this broad area.

Most rivers fall somewhere between braided streams and meanders. Depending upon the topography, these rivers tend to be straighter than meanders but still retain significant floodplains. There is less channel migration but floods can be even more frequent with more extensive flood plain deposits.

Certain types of pollutants can be more prevalent in certain types of deposits. For example, heavy metals are more commonly deposited in channels because their density makes them deposit quickly. On the other hand, some chemicals tend to bind to clays in the process of adsorption and be spread across floodplains.

CASE STUDY 10.3 Tittabawassee River Contamination, Michigan

Soil and River Sediment Pollution

In 1879, the Dow company began in the town of Midland, Michigan, to extract bromine from the groundwater brines in parts of Midland, Manistee, Muskegon, Wayne, and Saint Clair Counties. Bromine is an important chemical that was used in pesticides, fumigants, dyes, water purification, and film. It is heavy, volatile, mobile, and hazardous. It can also be used as a bleaching compound, but care must be exercised because the vapors irritate the eyes and throat.

The Dow Corporation sold potassium bromide and bleach but grew to one of the major providers of chemicals, plastics, and agricultural products. Despite its growth, Dow never abandoned its Michigan facilities. The headquarters of Dow are still in Midland and it operates a 1900-acre (769-ha) plant in town.

In the early 1900s, Dow constructed a number of lagoons on a 600-acre (243-ha) tract of land in Midland to manage its liquid waste. Wastewater from the plant was discharged into the lagoons and the solids settled to the bottoms. The liquid was then released into the adjacent Tittabawassee River. Later, an advanced wastewater treatment plant was constructed to process liquid wastes before being released.

The Tittabawassee River flows south, where it joins the Shiawassee River to form the Saginaw River near Midland (Figure 10.6a). The Saginaw River enters Saginaw Bay off Lake Huron. During floods, Dow released more liquid waste from the lagoons into the river. These and other wastewater releases into the river had significant impacts on the environment. In 1986, a flood overwhelmed the wastewater treatment plant, submerged part of the manufacturing plant, and washed untreated chemical wastes and contaminated soil from the lagoons into the Tittabawassee River (Figure 10.6b). This flood contaminated a 20-mile (32-km) stretch of river downstream of plant with dioxins and furans.

Even short-term exposure to dioxins and furans result in a variety of health problems, including liver damage, immune system damage, the skin disorder chloracne, and the development of several types of cancers. Dioxin is the active ingredient in Agent Orange which caused terrible health issues for Vietnam War veterans. Both dioxins and furans are chemically stable in the environment and commonly move up the food chain, concentrating in animal fat. The primary human exposure to dioxins and furans is through meat, milk products, and fish.

Dioxins and furans are concentrated in channel and floodplain sediments of the Tittabawassee River downstream of the Dow plant in Midland (Figure 10.6c). During a wetland restoration project, soils in a field in the Tittabawassee River flood plain were sampled and analyzed. The dioxins and furans were at concentrations as high as 2200 parts per thousand (ppt), which is 25 times higher than maximum allowable levels. In response, the Michigan Department of Environmental Quality (MIDEQ) conducted a survey and found that dioxin and furan levels in sediment samples increased nearer to the Dow plant. They were dramatically lower upstream of the facility at less than 10 ppt. Soil samples outside the floodplain had concentrations of dioxins and furans at 6 ppt.

The survey also found that downstream of the Dow plant, dioxin in channel and floodplain sediments ranged from 39 ppt to more than 7200 ppt. Although the tainted sediments were thicker in the channels, the dioxin adhered the clays in the floodplain making them more prone to high levels of contamination. The maximum permissible level of dioxin in soil in residential areas is 90 ppt in Michigan. The Michigan law requires the release of hazardous substances to be immediately remediated at the polluters expense. Dow immediately began implementing several actions to remediate the dioxins and furans that had been released from the Midland facility. First, Dow mapped the area where dioxin levels exceeded 90 ppt. This area was slated for remedial action.

Dow worked with MIDEQ and local health officials to educate residents how to reduce dioxin exposure. Health advisories were issued for people against eating fish from the Tittabawassee River downstream of Midland. Bottom feeders such as carp, catfish, or white bass that feed on the contaminated sediments were strongly urged not to be eaten, and children and pregnant women were warned not eat any fish from the river more than once per month. Health officials further urged residents to avoid consuming game that grazed on vegetation in floodplain sediments downstream of the plant as well. They also advised landowners not to disturb floodplain sediments with any construction or landscaping projects. An orderly and organized removal of tainted soils and sediments then proceeded.

Dow also implemented measures to reduce exposure for children, the elderly and people with impaired immune systems because they have the greatest risk from dioxin. The cleanup was slow and several government officials, community leaders, and environmental activists, complained about delays. In 2005, 2000 landowners of property in the Tittabawassee River floodplain filed a class action lawsuit against Dow for loss of property values resulting from the dioxin contamination.

(a)

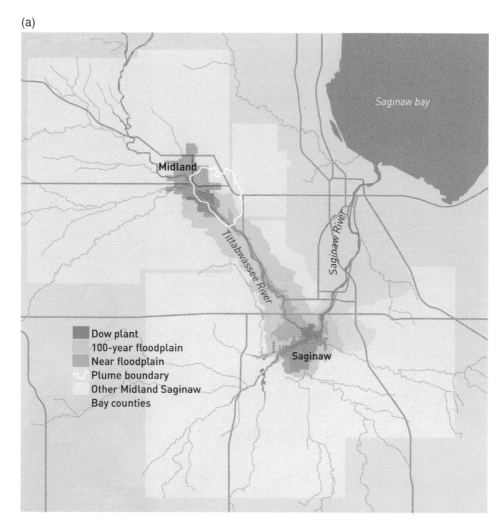

(b)

FIGURE 10.6 (a) Map showing the position of Midland, the Tittabawassee River and floodplain, the Saginaw River, and Saginaw Bay in Michigan. (b) Aerial photograph of the Dow Corporation plant in Midland during the peak of flooding. *Source:* https://www.ourmidland.com/news/article/Midland-County-releases-flooding-update-15283917.php#photo-19435209. (Michigan State Police) / Accessed on December 24, 2022. (Michigan State Police). (c) Map showing the dioxin concentrations in floodplain sediments around Midland and the Tittabawassee River.

(c)

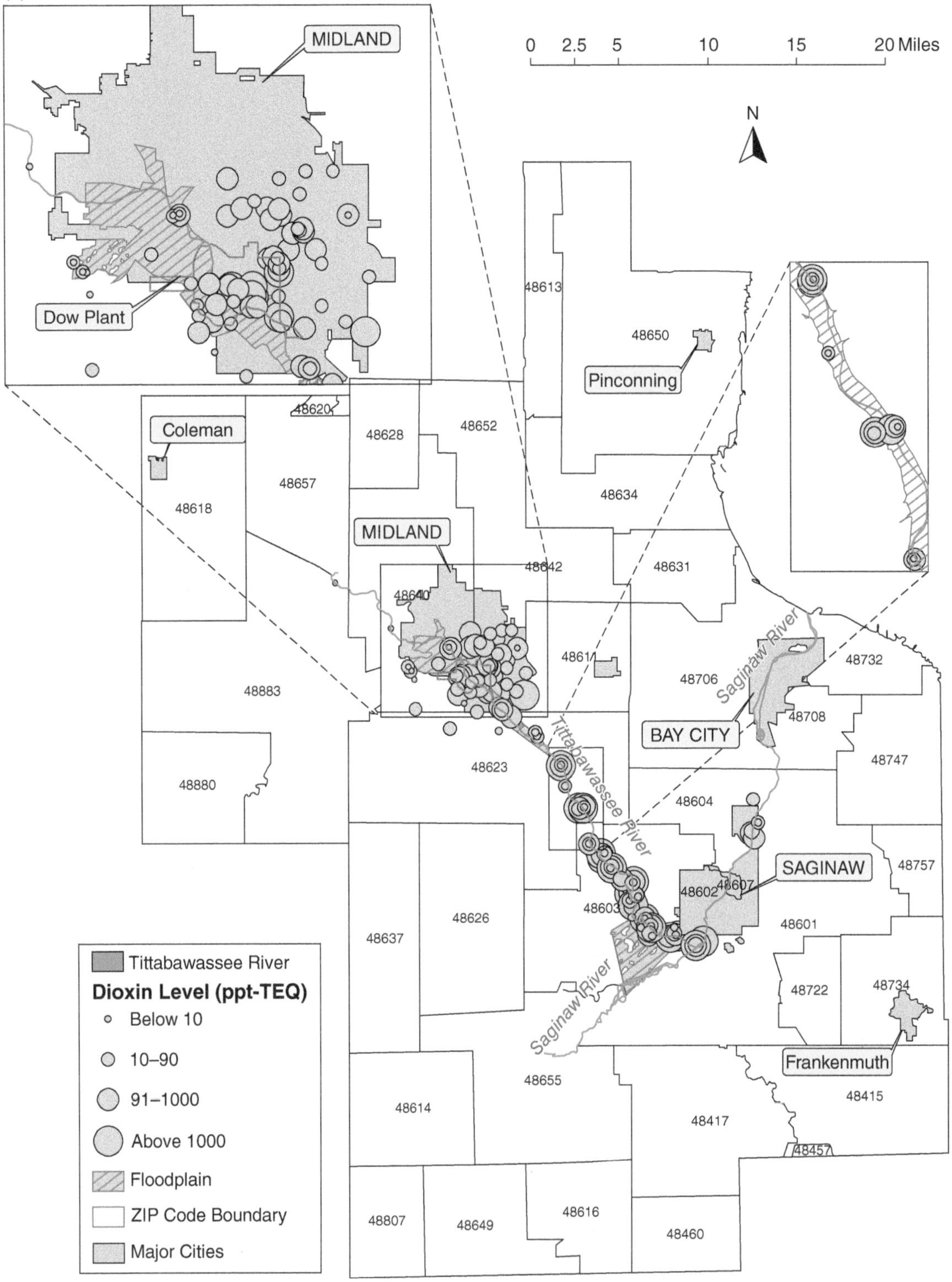

FIGURE 10.6 (Continued)

10.4 | Glacial Deposits and Pollution

Drift is the general name for glacial deposits (Figure 10.7). Till is a glacial sediment carried and deposited by ice and outwash is sediment carried by meltwater. Ice carries all sizes of sediment and, for this reason, till is very poorly sorted with mixed sizes from clay to boulders. In contrast, outwash sediments are deposited by water and are better sorted, with more like-sized sediments. Upland till consists of sediments on hillslopes and hilltops. There are a few deposit types from ice.

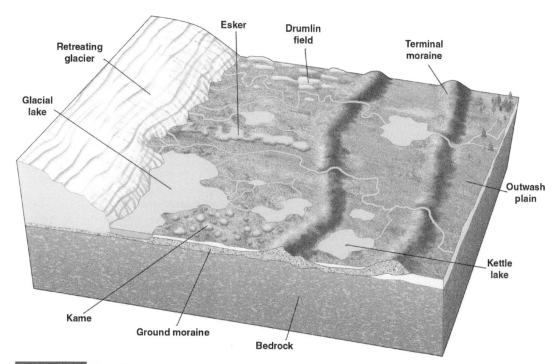

FIGURE 10.7   Block diagram showing the sediments and glacial geomorphic features in front of a retreating glacier.

Glacial erratics are solitary large boulders generally perched on bedrock. Typically, erratics are the only glacial deposits on bare bedrock surfaces. Erratics, however, can be encased in till deposits. Glacial ice carries all sizes of rocks by lifting them off the ground. As ice melts, sediment is released in the deposit. In rare cases, large rocks are dislodged from the ice by contact with bedrock. However, it is commonly dumped with other sediments. Erratics mainly sit on hills as a single or group of boulders by erosion of all other till by glacial outwash or later weathering.

A moraine is a large deposit of till that forms along the side or at the end of the glacier. Glaciers slowly move away from origination areas and toward warmer areas. The ambient temperature and consequently melting increases. At some point, the melting balances the forward movement which determines the end of the glacier. At that point, the entrained rock and soil deposits as the ice melts leaving a large deposit of till. The time that the end of the glacier stays in one spot determines the size of the moraine deposit. The largest moraine forms at the farthest extent of the glacier because the termination of the glacier remains there the longest. This is the terminal moraine.

As the ice age ends and the glacier retreats, it recedes quickly for a distance and then stabilizes for a time. Each time it stalls in retreat, a recessional moraine accumulates at the front of the glacier. Recessional moraines are much smaller than terminal moraines. These moraines form parallel to the end of the glacier. Moraine deposits disrupt local surface water drainage creating extensive wetlands and glacial lakes.

Glaciers are melted by the sun heating the ice surface and by the pressure at the base from the weight of the glacier. Subglacial meltwater is generated by infiltration of surface melt water through cracks in the glacier. Some of this water may also flow through tunnels in the ice at multiple levels in the glacier. All water exits the glacier and combines at its terminus, yielding large quantities of fresh water and producing outwash deposits.

Tunnels form at the base of the glacier that carry meltwater carrying sediments. As the water in the tunnel slows, it deposits the sediments. After the glacier is removed, these deposits in the tunnels are preserved as eskers. They are long narrow ridges up to 30 ft (9.2 m) high and miles long. The sediments in eskers are outwash deposits and are characterized by high porosity and permeability as a result.

There is an enormous amount of meltwater emitted by a glacier and coupled with the disrupted drainage produced by the moraines, glacial lakes as much as tens of miles across and hundreds of feet deep can form. The frequent ice cover on these lakes eliminates surface input and waves that could suspend the sediment. As a result, sedimentation is restricted to clays and silt most of the time. In the summer when the ice melts, coarser sediments can be deposited from higher volumes of meltwater and increased agitation by waves. The deposits of finely interlayered clay and silt are called varves. These deposits have low porosity and permeability. In some cases, the lakes contain extensive vegetation even forming swamps. These deposits are organic-rich and black in color, locally forming peat bogs.

Blocks of ice calve off into the glacial lake and form ice rafts. As they melt, they release sediment into the lake. Cobbles and boulders drop into the lake and the varve deposits, commonly deforming and disrupting the fine layering. These are called dropstones and the main distinguishing characteristic of a glacial lake in contrast to other lake deposits.

In many cases, the glacier forms part of the dam that forms the glacial lake. When the glacier retreats during warming, this dam may break and allow the entire lake to drain at once. This draining produces floods that are among the most catastrophic in the history of the Earth. The rushing flood waters scour huge canyons and erode large amounts of sediment in a matter of days. This is the mechanism that produced the catastrophic flood that channeled scablands in western Washington. Glacial Lake Missoula extended across Washington, Idaho, and Montana. It was 200 miles across (322 km) and 2000 ft (610 m) deep, containing more than 500 cubic miles (2 billion cubic decameters) of water. This is more water than Lake Erie and Lake Ontario combined. When the largest of the ice dams on the lake burst, water flooded out at 10 times the flow of all the rivers of the world combined. This flood stripped away hundreds of feet of soil and cut deep canyons or "coulees" into underlying bedrock which formed the Grand Coulee. The entire lake drained into the Pacific Ocean in about 48 hours. The flood removed 50 cubic miles (208 million cubic decameters) of rock and sediment, left piles of gravel 30 stories high, created giant dunes three-stories high, and scattered 200-ton (181 metric ton) boulders from the Rockies to the Willamette Valley.

Some of the calved icebergs carried large pockets of sediment into the glacial lakes as kame deposits. The pockets formed as potholes in the ice or part of tunnels in the glacier. Otherwise, moraine sediment also deposited on the ice. Icebergs rafted these sediments into the lake and then melted. The sediment deposited on the lake bottom in a jumble pile on the varved clays typically deforming and disrupting them. These porous and permeable deposits can form perched aquifers.

Glacial deposits are varied enough that a single summary of their potential for pollution is impossible. Outwash deposits are stratified and the best sorted glacial deposits which gives them the best porosity and permeability. They form shallow aquifers in valleys and lowlands where they are 100 ft (30 m) thick or thicker. These sediments are unconsolidated and transmit water. The high permeability means that groundwater is generally not purified in outwash deposits so can be polluted by chemical and sewage spills and leaks.

Glacial till is commonly thin and poorly sorted and does not form an acceptable aquifer. The thorough mix of grain sizes limits porosity and permeability. The freshly ground rock and soils are chemically reactive and the high clay content can filter water effectively. For this reason, till can purify groundwater. It can contain decomposed vegetation that produces methane gas within the groundwater. In some cases, this methane entered homes with tap water and accumulated in the indoor air. Methane displaces indoor air in the home and will suffocate humans. Early settlers hand dug water wells, and methane in them was fatal on occasion. Methane is flammable and has blown up homes if inadvertently ignited. Glacial deposits from swamps and lakes, can also produce methane.

Clay deposits in glacial lakes have low porosity and permeability. They can isolate hazardous materials and isolate them from the groundwater system. In Love Canal, New York, toxic substances in a chemical dump were relatively safe in the glacial clay deposit but the town was built on top of it penetrating the protective clay cap causing the disaster.

CASE STUDY 10.4   Usinsk Oil Spill, Russia

Petroleum Pollution of Arctic Soil

Russia has a huge reserve of petroleum and natural gas in sedimentary deposits of the Timan-Pechora Basin in the far north near the Barents Sea. The city of Usinsk is at the center of the major Kharyaga oil field in this economically important area (Figure 10.8a). The city lies immediately south of the Arctic Circle and 1000 miles (1609 km) northeast of Moscow. It could become a major petroleum center but the weather in the area is brutal. The average temperature in this cold area from 1961 to 1990 was 26 °F (−3.3 °C), with the average temperature of the warmest month at 58.6 °F (14.8 °C). The snow cover lasted an average of 210 days per year. As a result, the city includes just a few thousand oil field workers among a population of about 40 000 people.

The oil in Usinsk has been pumped into an old, 29-in. (73.7-cm) diameter, 11.2-mile-(18-km)-long pipeline that feeds into larger pipelines (Figure 10.8b). This oil is piped to refineries in more moderate climates farther south in Russia. There are no facilities to capture the natural gas produced with the oil, so it is burned in giant flares that are visible from satellites. They burn about as much natural gas as is used in Denver, Colorado annually.

This precarious situation led Usinsk to become the site of one of the largest land-based oil spills in history. In February 1994, four sections of the Usinsk pipeline began leaking. KomiNeft Oil was the company responsible for the repair and maintenance of the pipeline. In October 1994, they decided to capture the oil from the leaks and pump it into a hastily constructed, soil enclosed lagoon, rather than stopping the flow to make repairs. This soil enclosure created a huge oil lake that held more than 1.5 million barrels or almost 60 million gallons (227 million l). The leaks in the pipes were later blamed on the adding of river water to the oil to improve flow. The water caused the metal pipeline to rust, especially near fittings and connections.

The hastily constructed lagoon walls began to crumble as the Russian winter weather began. On 1 October, part of the lagoon wall collapsed and oil rushed onto the ecologically fragile Russian tundra (Figure 10.8c). Oil flowed over the landscape and into local streams and drainage channels. It eventually covered more than 0.3 square miles (70 ha) of the area. Once released into the Arctic climate of northern Russia, the oil was resistant to both evaporation and biological breakdown because of the low temperatures. In addition, the frozen ground slowed infiltration and oil readily flowed great distances over the landscape.

(a)

(b)

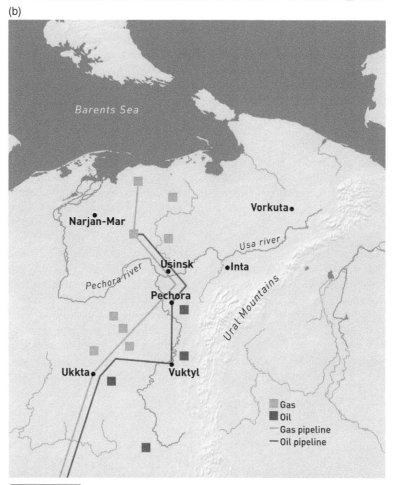

FIGURE 10.8 (a) Map of the location of Usinsk and the Komi Republic in Russia and the surrounding countries and Barents Sea. Scale bar = 250 miles. (b) Map showing the area around Usinsk including the location of oil wells and pipelines. (c) Photo of oil spilled on the Arctic tundra in Usinsk. *Source:* Greenpeace Russia / Wikimedia Commons / CC BY 3.0.

(c)

FIGURE 10.8 (Continued)

The soils in the Usinsk region are primarily Quaternary glaciofluvial and glaciolacustrine sediments. The glaciofluvial sediments include coarse sands and gravels. The glaciolacustrine sediments include extensive peat deposits and peat horizons. There are also recent fluvial sediments covering the glacial sediments. The area contains extensive wetlands and is considered to be in discontinuous permafrost though the true permafrost line has retreated northward with climate change.

After breaching the lagoon walls, the oil flooded the Kova and Usa Rivers, which are tributaries to the Pechora River. This river is a major salmon fishery and also serves as a habitat for many plants and animals. The Pechora River flows into the Pechorskoye Sea, which transitions into the Barents Sea.

KomiNeft was pumping some 4000 tons (3269 mt) of oil per day through the pipeline at the time of the spill. The flow was allowed to continue unabated for several weeks before local officials convinced company officials to shut down the pipeline to undertake cleanup operations. The cold weather thickened and slowed the oil which facilitated containment and recovery. The hasty cleanup consisted of bulldozers and earth-moving equipment scraping the oil contaminated soil off the ground and disposing of it. This environmentally unsound method had the side effect of destroying all vegetation. In the tundra, it takes many years to for vegetation to take hold. Oil that was difficult to reach was simply set on fire. This further destroyed vegetation and the fragile habitat. KomiNeft claimed that this approach was necessary to prevent a worse disaster in the spring, when warmer temperatures made the oil more mobile.

Although KomiNeft claims low amounts for the spill and environmentalists claim a catastrophic volume, the volume of the spill is most likely about 500 000 barrels (100 000 mt, 20 million gallons). The spring thaw and rains spread the oil across an additional 8 square miles (2100 ha) of tundra, most of which is grazing land for reindeer herds. KomiNeft was fined the equivalent of $600 000 by the Russian government for the 1994 spill and they were ordered to begin cleanup operations immediately. KomiNeft repaired the pipeline with international help from Australian and American teams and attempted recovery of oil spilled from hundreds of leaks along this short 11.2 miles (18 km) length. KomiNeft was purchased by LUKoil in 1999. This was not the first oil spill from this pipeline. Other spills occurred in 1988 with 20 000 tons (18 144 mt) of oil released and in 1992 when two leaks each released more than 30 000 tons (27 216 mt) of oil.

The 1994 spill did not end the spills in the area but Russia does not have a good record in terms of oil spills. An energy statistics bureau reported that there were 11 709 pipeline breaks in Russia in 2014. In comparison, Canada had five pipeline accidents that involved injury and 133 natural gas and oil pipelines leaks in 2014. Major spills the in the Usinsk area in 2013, 2014, and 2015 released many hundreds of tons of oil. As a result, the north-flowing Russian tundra rivers, including the Pechora carry 551 000 tons (500 000 mt) of oil into the Arctic Ocean every year.

The residents in the villages of the Usinsk area register complaints that the many oil spills have badly degraded the quality of their drinking water, contaminated the fish and reindeer they depend on for food, and have caused numerous chronic health conditions. Statistics from a local hospital in a small town on the Pechora River in 2010 show rises in most illnesses. Nervous system diseases in adults, increased from 26 in 1995 to 70 in 2009. In patients under 18 years old, these diseases increased dramatically from 72 to 254.

10.5 | What Can I Do?

With nearly 8 billion people on the planet, even small efforts make a difference. Dumping any waste on the ground should be avoided. This can be simple littering or dumping leftover chemicals like paint thinner, gasoline, or motor oil on the ground, but all should be avoided. Those are easy to avoid. More challenging are avoiding the use of lawn chemicals like herbicides, pesticides, and chemical fertilizers. They are very damaging to the environment but mean that the lawn will not be as green or weed-free. Mowing leaves into the lawn and adding lime improves productivity naturally and reduces the leaf waste. Recycling and reusing where possible reduces the need for natural resource acquisition which destroys soil.

CHAPTER 11

Ecosystem Pollution

CHAPTER OUTLINE

Words you should know:

Bioaccumulation – The intake and concentration of a persistent substance in the tissues of a living organism.

Biomagnification – The increase in the concentration of a pollutant at higher levels in the food web through predation.

Ecology – The science of the relation of organisms to each other and to their physical environment.

Ecosystem – A community of interacting organisms and their physical environment.

Habitat destruction – The processes that cause a natural habitat to no longer be able to support the native species.

Habitat fragmentation – The partitioning of a native landscape into unconnected fragments separated by impacted areas that do not support native species.

Invasive species – Plant, animals, insects, or other life introduced to a new area and ecosystem.

Landscape – Geographic areas of diverse interacting patches or ecosystems, both natural and human-impacted.

Light pollution – Nighttime artificial light that is inappropriate and damaging.

Patches – The area fragments of native species and landscape resulting from habitat fragmentation.

11.1 | Ecology and Ecosystems

Pollution does not only involve unwanted chemicals and microorganisms. Addition of larger biota to a harmonious system of life or even removal of components of that system can be as damaging and disrupting as chemical pollutants. The harmony and disruption of the system falls under the umbrella of ecology. Ecology is the science of the interrelations among organisms, including humans, and their environment. It determines the connections among plants, animals, climate, geology, and geography. Understanding ecology requires at least a full course. For the purposes of this book, the importance of ecology is that it demonstrates the benefits of ecosystems and how people can better use natural resources to preserve the environment.

Ecology includes the study of organisms, population, communities, and ecosystems. Of these, the most applicable is ecosystems. There are three general types of ecosystems including freshwater or aquatic, marine, and terrestrial. Within these categories are individual ecosystems based on organisms and their physical habitat including chemicals, bedrock, and soil. The main foci of ecosystem ecology are the processes and mechanisms that maintain the ecosystems. These include primary productivity, decomposition, and interactions. They function to maintain sustainable foraging, fiber, fuel, and provision of water for all living organisms. These processes allow systems to adjust to climate, disturbances, and management. Understanding the processes allows the modeling of global environmental problems, such as CO_2 production, global warming, and degradation of surface water.

Ecosystems can be laterally extensive and the boundaries are commonly nebulous and may vary in location and type with time. The organisms within ecosystems depend on the biological and physical processes of the ecosystem to function properly. Disruptions can have far-reaching and complex effects on the organisms that depend on the processes. This dependence varies with the degree the organism is tied to them. Adjacent ecosystems typically interact and are commonly interdependent for community structure and the processes that maintain productivity and biodiversity of the ecosystem. These interactions can be complex.

Basically, all of the living organisms in a particular area develop and evolve into a system that is in basic equilibrium with the physical conditions of the area. The development of life is controlled by the rock, soil, water, and air as well as the weather in the area. A mountain can have rocky surfaces, coarse sediments, thin soil, thin air, and cold temperatures. The life that develops on it is limited by these conditions. If the area is in a desert, by the ocean, around a river, in a rainforest, or in the tundra, these physical constraints are different and biota will be limited or encouraged accordingly. Biota also develops an equilibrium within itself among plants, herbivores, carnivores, scavengers, insects, and even microorganisms in terms of food or energy production and consumption. This equilibrium changes with season and may have long-term changes. If production changes, then consumption will change in response.

Ecosystems are in short-term equilibrium but include minor disruptions and even significant disruptions. These might be heavy rains, dry periods, cold periods, and others. During these times, the system may be in temporary disequilibrium while the biota shift to accommodate the changes. Fixed location biota like plants may die and mobile life might move to more appropriate areas or die as well. In favorable conditions, life flourishes and biota from other ecosystems might move to it. The equilibrium shifts to accommodate newly introduced species just as it shifts if species leave the area. Extreme conditions can be natural and can destroy the entire ecosystem. Wildfires, severe storms like a tornado, severe floods, or even beaver dams that flood the area may cause a complete replacement of the ecosystem, at least temporarily.

Ecosystems can also be mildly to profoundly disrupted by human intervention. If disruptions are minor, equilibrium may be returned quickly. Most, however, are profound and permanent in human time scale. By adding humans, the science shifts toward systems ecology. This study of ecosystems incorporates mathematical modeling, computation, and systems theory. Systems ecology takes a holistic view of interactions within and between biological and geological systems and includes the human dimension. Human industrial activity becomes a fundamental part of ecosystems.

Most of the chemical pollution events in this book are profound and permanent for one to several centuries. However, there are ecosystem changes and overlap between chemical and biological processes that can be equally profound and require significant amounts of time to correct, if they can be corrected. These might be considered biological pollutants and they are almost exclusively the result of human input. These include invasive species, habitat destruction and fragmentation, light pollution, and bioaccumulation and biomagnification.

11.2   Basic Ecology Concepts

Biodiversity is short for biological diversity and is the variation of microorganisms, plants, fungi, and animals in a given area. Biodiversity includes the wide variation of species of organisms on Earth. Some variation occurs within species, such as in shapes, sizes, and colors of the flowers within a single species of a plant.

Community ecology is the interaction among species in a community including distribution, structure, abundance, demography, and interactions among the coexisting populations. It also considers non-biologic factors that influence species distributions or interactions. For example, desert plant communities are different from tropical rainforest communities as the result of precipitation. Humans can affect the community structure as the result of habitat disturbance, such as the introduction of invasive species.

Within a community, each species occupies a specific niche. This niche determines how the species interacts with the environment and its role in the community. The number of niches in a community controls the number of species. If two species share the same niche, one species will outcompete the other. The more niches filled in a community, the higher the biodiversity.

A guild is a group of species in a community that compete for the same resources. Closely related species in a community are commonly in the same guild because traits can be inherited through descent from a common ancestor. However, guilds do not exclusively include closely related species. Carnivores, omnivores, and herbivores are basic guilds. More specific guilds could be vertebrates that hunt ground-dwelling insects and might include birds and mammals. Flowering plants with the same pollinator form a guild.

The trophic level of a species is its position in the food chain or web. The bottom of the food web contains primary producers. These organisms provide energy through photosynthesis or chemosynthesis. Plants are the most common primary producers. The next level is herbivores that feed on the plants for energy so are primary consumers. Herbivores are hunted and consumed by omnivores or carnivores, making them secondary and tertiary consumers. The additional levels of the trophic scale are smaller omnivores or carnivores being eaten by larger ones. The apex predator is the top species in the food web, which is not consumed by any other predator.

11.3   Invasive Species

Invasive species are among the main threats to native wild animals and plants. About 42% of both threatened and endangered species are the result of invasive species. An invasive species can be any type of organism, including plants, insects, fish, bacteria, amphibians, and fungi among others, that is alien to an ecosystem. They can be damaging to the environment, the economy, and/or human health. Invasive species can include any form of life that grows, spreads, and reproduces quickly. Invasive mammalian predators are probably the most damaging group of invasive species to biodiversity. These invasive predators are responsible for at least 87 bird, 45 mammal, and 10 reptile major extinctions, accounting for more than 58% of extinctions worldwide. In addition, 23 of the critically endangered species are considered as possibly extinct. Invasive mammal predators are currently endangering an additional 596 species for extinction.

Invasive species can be introduced naturally. Land bridges can allow non-native species to cross from one landmass to another. Land bridges can be built through plate tectonics or by climate change. For example, life in North America evolved to develop placental mammals but marsupial mammals in South America (Figure 11.1). Central America developed by plate tectonics to connect North and South America. This allowed placentals to migrate to South America and marsupials to migrate to North America. Opossums and armadillos are North American species that came from South America and there are dozens of mammal species in South America whose ancestors are from North America. During ice ages, glacial ice locks up so much water on land in continental

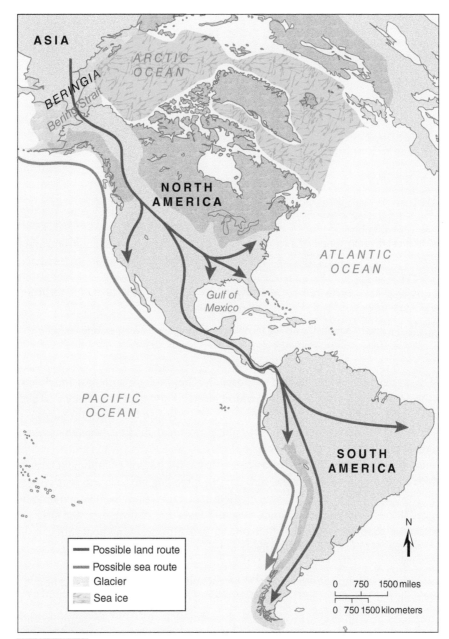

FIGURE 11.1 Map of North and South America showing migration paths of land animals that crossed from Asia via a land bridge.

glaciers that sea level can drop 200 ft (60 m) or more. All parts of the seafloor less than 200 ft (60 m) deep become land. If this land connects continents or islands, it becomes a land bridge, allowing land animals to migrate across. This is how humans crossed from Asia to North America among other invasive animals.

Even if land masses get close to each other through plate tectonics, certain biota can cross between them. These include some birds, short distance swimming or flying animals and plants, insects, and other biota that can be transported by the wind. There are documented examples of seeds, insects, and eggs of restricted marine life that were transported across ocean basins on pumice rafts produced during volcanic eruptions. Marine life can also be invasive by natural processes. Rifting can open seaways between two oceans, allowing marine life to migrate between them. These natural processes are slow and gradual.

The problem is that humans have overwhelmed all but a few land masses and ocean basins with invasive species at a much faster rate than ecosystems can adjust (Figure 11.2). The problem began by accident. Stowaway species like rats, insects, and barnacles on ship hulls, escaped animals that were kept for food, and grains that were spilled by mistake wound up in unintended locations during the age of exploration. Depending upon the success of the invasive species, these accidental introductions could have devastating effects on native species. Once emigrating humans, the most dangerous invasive species, settled in a new area,

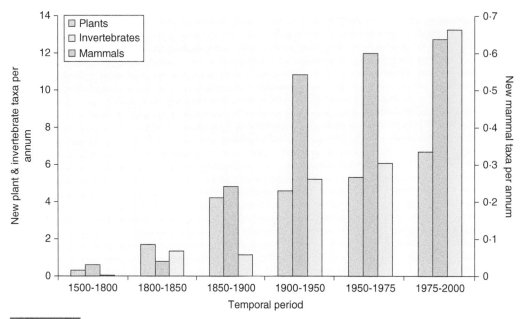

FIGURE 11.2 Bar graph showing the annual increase of invasive species in the world per year.

they brought many invasive species both by mistake and purposefully. They brought both pets and domesticated animals for food and work that damaged the environment while under the control of their human owners. Many also escaped into the wild. They competed for resources with native species as well as preying on them. They also facilitated the success of other invasive species. Some invasive species were introduced to develop economic ventures like meat, fur, fruit, vegetables, fibers, and lumber. They were also introduced just to make immigrants feel more at home.

Economic ventures increased the introduction of invasive species to the natural environment largely by mistake. Plantations introduced non-native crops that spread to native areas through the spreading of seeds and root systems. Animals also escaped from enclosures to inhabit the natural areas as well. Some invasive species were not even intended to be transported to another area. Insects were introduced through importation of wood, fruit, and vegetables both as eggs and larvae and as adults. Seeds were also contained in other imported goods that then took hold and spread. Stowaways on boats are invasive species including rats, insects, birds, and many microbes. Stowaways on the outside of boats include barnacles, snails, insects that bored into the hull and other invertebrates.

Some invasive species are especially aggressive and have caused terrible damage to the native environments. In terms of animals, birds, insects, amphibians and reptiles, aggressive predatory mammals have done the most to threaten native species and biodiversity. The worst threats are cats, rodents, dogs and pigs, with mongoose and stoats also locally damaging through predation, competition, disease transmission, and facilitation with other invasive species. The impacts of extinctions of native species by invasive predators can cascade through and significantly damage entire ecosystems. For example, feral cats and red foxes caused the decline or extinction of more than two thirds of digging mammal species in Australia. The absence of digging mammals reduced topsoil disturbance and led to barren landscapes with little organic matter and low rates of seed germination.

Invasive species are now everywhere. In the northeastern United States, invasive plants include garlic mustard, burdock, purple loosestrife, and multiflora rose among others. Invasive birds include starlings, house sparrows, and rock pigeons among others and invasive insects include gypsy moths and Japanese beetles among others. There are more than 800 invasive plant species in Texas and California has more than 1000. The originally area-restricted New Zealand mud snail can now be found throughout Europe, Asia, Australia, the Middle East, and the American West. The extensive global trade and travel have accelerated invasive species to an extreme. For example, more than 10000 species are being moved around the globe daily just in the ballast water of oil supertankers.

There are 473 birds species (32% of total) threatened by invasive species and disease. The majority of at-risk species are from predatory invasive species. Rats and cats are threatening 239 and 204 of these species, respectively. The problem is extreme on oceanic islands, which contain 74% of the species threatened by invasives. For example, the Galápagos Islands has more than 750 invasive plant species and 44 invasive vertebrate species and they are damaging the natural biodiversity. Of the 41 bird species on the Galápagos, 21 are threatened and of these 90% are threatened by invasive species.

CASE STUDY 11.1 New Zealand Flightless Birds

Invasive Species Disaster

New Zealand is probably the best example of the damage that invasive species can cause. It is also the best example of methods to address it. The reason that invasive species were so devastating to the natural ecology of the islands is that New Zealand evolved as a landmass far from other landmasses (Figure 11.3). About 85 million years ago, the land that developed into New Zealand split away from the supercontinent Gondwana and evolved in relative isolation.

Very few mammals developed or inhabited New Zealand. They only have a few species of native bats and several ocean-dwelling mammals such as dolphins, seals, and whales. The native plants and animals therefore evolved without predatory mammals. This absence allowed birds that previously inhabited trees for safety from predators, to inhabit the ground. With time, some evolved into ground-dwelling species and lost their ability to fly.

The lack of devastating predatory mammals allowed New Zealand to evolve many unique birds, plants, and invertebrates. One example was the moa, a large ground-dwelling bird with nine species. The largest was the South Island giant moa, which weighed as much as 500 lbs (227 kg) and, with its neck outstretched, could reach a height of 12 ft (3.7 m). Moas fed on native plants without fear of all but one predator, the Haast's eagle, which was the world's largest eagle and had a diet of moa. There were many other ground-dwelling birds on New Zealand such as kiwis and kakapos among others. In fact, when the first Polynesian explorers arrived, there were 245 bird species and a quarter of them were flightless.

The first human settlers were the Maori who arrived in about 1350 CE. They brought Pacific rats or kiore with them, intending to eat them (Figure 11.3a). However, the great abundance of easily hunted birds distracted the Maori and the quickly multiplying and spreading rats also invaded the islands. The Maori hunted down and killed all of the moas and Haast's eagle within a century or two. The rats fed on weta, young tuatara, and the eggs of many ground-nesting birds. Between the rats and Maori, several species of New Zealand's native ducks, flightless rails, and two species of flightless geese among others were gone within a century.

The Maori and kiore were mild problems compared with those brought by Europeans. The first European in New Zealand was the Dutch explorer Tasman in 1642 but the real European invasion began in 1769 with the arrival of James Cook, who claimed the islands for Great Britain and brought residents. With them came ship rats, which further decimated the native bird population. The transplanted Europeans brought brushtail opossums from Australia in 1830 to start a fur industry. Later came Norway rats from ships, which were equally devastating. Once a conflict between the British settlers and Maori in the late 1860s was resolved, large-scale immigration of British settlers began. They imported invasive species of red deer, fallow deer, white-tailed deer, sika deer, tahr, chamois, moose, elk, hedgehogs, wallabies, turkeys, pheasants, partridges, quail, mallards, house sparrows, blackbirds, brown trout, Atlantic salmon, herring, whitefish, and carp either by mistake as castaways or purposefully to make the islands seem more like Great Britain. In total, 34 exotic mammal species were introduced in New Zealand as were hundreds of exotic plant species (Figure 11.3b).

A big mistake was the importation of European rabbits. They multiplied so quickly that they overran the countryside within a few decades forcing out native ground birds. The infestation got so bad that in 1876 the New Zealand government passed an act to destroy rabbits. This had no impact, so in the 1880s stoats, weasels, and ferrets were imported and released into the wild to remove the rabbits. However, the stoats also targeted the flightless birds further reducing their population. As a result of these efforts, nearly 25% of native birds including at least 59 species are extinct and two thirds of the animals in cities are exotic. In addition, one endemic bat, one fish, at least a dozen invertebrates and 10 plants also were driven to extinction. About 30% of the remaining native birds are threatened with extinction. It would take 50 million years of evolution to recover the diversity of birds lost through human colonization.

By the 1970s, New Zealanders had become very concerned about the havoc wreaked by invasive predators. In response, the country has become the world leader in dealing with them. They established the Zealandia sanctuary in 1999 to protect and proliferate the remaining native species. To date, 18 species of native wildlife were reintroduced into the wild, six of which were absent from New Zealand for more than a century. For example, the large native Kākā parrot increased from a population of six to more than 800.

In addition, New Zealand began massive efforts to eradicate the invasive predatory mammals. The Database of Island Invasive Species Eradications (DIISE) tracks worldwide attempted eradications of vertebrate invasive predators on islands. By 2016, of the 1376 attempted eradications, 424 were conducted by New Zealand. The country began an ambitious project called the "Apollo program" to eliminate rats, stoats, and opossums by 2050. The New Zealand Department of Conservation conducted a widespread aerial drop of the toxin "1080." New Zealand has about one-tenth of 1% of the total land surface and is using 80% of the 1080. Residents were encouraged to trap and eliminate rats and stoats. Baited tags that identify nibbling predators and information brochures are widely distributed.

The programs are having positive results. Invasive predators have been completely eradicated on 117 of the New Zealand's offshore islands. The two large islands are more difficult but there has been headway made there as well. These efforts are not just important to New Zealand. Invasive predatory species are the leading cause of extinctions on ocean islands. Since 1500, about 80% of the recorded extinctions are on islands. In addition, 40% of all endangered species in the world live on or rely on ocean islands. New Zealand's successful methods have already been replicated around the world and, as a result, invasive predators have been removed from about 1000 offshore islands.

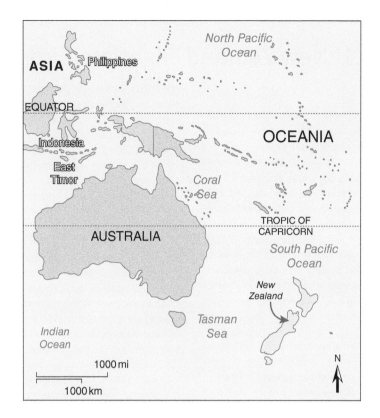

(a)

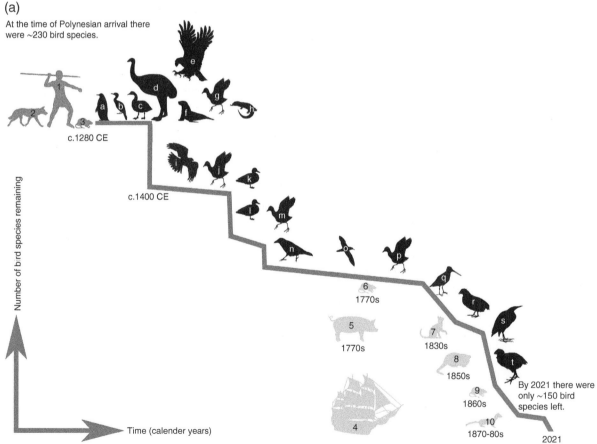

FIGURE 11.3 Map showing the location of New Zealand in the southwest Pacific Ocean. (a) Diagrammatic graph of the decline of bird species in New Zealand with the invasive species times and types shown. *Source:* Courtesy of Greig K and Rawlence NJ (2021). (b) Graph showing the increase of invasive plant species in New Zealand over the past two centuries.

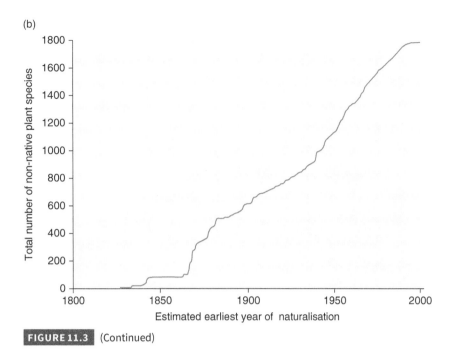

(b)

FIGURE 11.3   (Continued)

11.4 │ Impact on Landscapes

Landscapes are geographic areas characterized by interacting ecosystems including natural terrestrial and aquatic systems such as forests, grasslands, and lakes and human-dominated environments like agricultural and urban areas. Landscape ecology is the science of relationships among ecological processes and ecosystems. It is the synergetic result of biodiversity and geodiversity and also includes human impacts on landscape diversity.

The landscape develops from the interaction among ecosystems, in terms of intensity, frequency, and spatial scale. Components of scale include structure, function, and composition. Composition includes the number of patch types in a landscape and their abundance. For example, the amount of forest, wetland, and forest edge and the density of roads can be aspects of landscape composition. Patches are the basic unit of landscapes. A patch is an area that differs from its surroundings and that can fluctuate over time. Patches have definite shapes and configurations and are characterized by variables like number of trees, number of tree species, height of trees, and other similar factors, for example.

Landscape patches have boundaries between them which can be sharp or fuzzy. A boundary is made of edges of adjacent ecosystems. The edge is the area around the perimeter of an ecosystem. Edge areas of adjacent patches can cause environmental differences between the interior and edge of the patch. Edge effects can include distinctive species compositions and abundances. For example, if a landscape is composed of contrasting patches such as a forest adjacent to a grassland, the edge is the location where the two abut. If the landscape is continuous, such as a forest grading into open woodland, the edge location is considered fuzzy. It is determined by exceeding a certain threshold, such as where the tree cover is below 35%.

Connectivity is how connected or continuous a corridor of landscape types is. For example, a forested landscape with few gaps in forest cover has a higher connectivity. Corridors are strips of a specific type of landscape differing from that of the land on both sides. A network is an interconnected system of corridors. In contrast, a mosaic describes the pattern of patches, corridors, and matrix that form a landscape.

One type of boundary is an ecotone, which is a transitional zone between two communities. Ecotones can be natural, like a lakeshore, or human-made like a cleared agricultural field from a forest. Examples of ecotones include forest to marshland transitions, forest to grassland transitions, or land-water interfaces such as riparian zones in forests. In contrast, an ecocline is a gradual and continuous landscape boundary shown by a change in environmental conditions of an ecosystem. Ecoclines can explain the diversity and distribution of organisms within a landscape because certain organisms survive better under specific conditions which can change laterally. They contain mixed communities that are more environmentally stable than in ecotones.

11.4.1   Habitat Fragmentation and Destruction

Fragmentation is breaking a habitat, ecosystem, or land-use type into smaller areas (Figure 11.4). Habitat fragmentation is the development of breaks in an organism's habitat, causing fragmentation of the population and ecosystem decay. Habitat fragmentation can be the result of geological processes like a volcanic eruption or by human activity such as land conversion. In this process,

FIGURE 11.4 Illustration of the impact of ecosystem fragmentation on the amount of edge versus unimpacted habitat and number and amount and types of vertebrates versus unfragmented.

large and contiguous habitats are divided into smaller, isolated patches of habitats. Habitat fragmentation is an invasive threat to biodiversity, affecting large number of species.

Although habitat fragmentation can be natural through wildfires, flooding, or volcanoes, it is most frequently caused by humans clearing native plants for activities like agriculture, development, urbanization, and creation of reservoirs. Once-continuous habitats are divided into small areas through fragmentation. Clearing divides fragments into small islands of relict native habitats that are isolated from each other by cropland, pasture, pavement, or even barren land. For example, the wheat belt of New South Wales, Australia involved the removal of 90% of the native vegetation, resulting in extreme habitat fragmentation. With such clearing fragments transition from connected to disconnected. Habitat loss occurs through the process of habitat fragmentation and is regarded as the greatest threat to most species.

Forest fragmentation reduces food availability and habitat area for animals and thereby divides species populations. This makes the animals more susceptible to predators and more likely to interbreed. Forest fragmentation is one of the direst threats to forest biodiversity, especially in tropical areas. The individual forest fragments or patches are commonly unable to support viable species populations, especially large vertebrate animals. As a result, even moderate-sized patches can result in local extinction of species because they may not be able to support a viable population. In addition, edge effects on the patches may be large if they border abruptly different ecosystems like developed areas. The mixed conditions in the outer areas of the patch can reduce the interior amount of true native forest habitat. It is estimated that more than 70% of the remaining forest stands in the world are within 0.6 miles (1 km) of a forest edge.

The effects of fragmentation on the vegetation and animals in a forest patch depends on both the patch size and its isolation. The degree of isolation is a function of the distance to a similar patch and the contrast of the patch with the areas surrounding it. If an area is cleared and then reforested, the increased diversity of its vegetation lessens the isolation of the forest fragments in the area. On the other hand, if forests are permanently converted to agricultural fields, pastures, or developed areas, the other forest fragments become highly isolated. Repeated clearing breaks forest into permanently isolated patches. In this case, the cleared area exceeds a certain critical level, the degree of which depends on the climate. The landscapes are then either connected or disconnected.

Habitat fragmentation damages the ability of species to adapt to changing environments. It reduces biodiversity by reducing the amount of suitable habitat, mates, and food available for organisms. Plants and immobile organisms are disproportionately impacted by habitat fragmentation because they cannot respond quickly or escape. As such, fragmentation is an important cause of extinction. In unfragmented landscapes, a single declining population can be supplemented by immigration from nearby populations. The greater the distance between fragments, the less likely this will happen, depending upon the species.

Edge effects can also impact biodiversity by causing changes in light, temperature, and wind that alter the ecology within the fragment as well as around it. Fires are more common in these areas because the changes cause humidity to drop and temperature and wind to increase. Invasive and pest species establish themselves more easily in disturbed environments, and the ability of predatory domestic animals to penetrate these areas increases. Different species inhabit the edge of a fragment than in the interior which reduces the population sizes. A 10% continuous remnant native habitat generally results in a 50% loss in biodiversity.

The ever-increasing human demand for wood, pulp, paper, and other forest resources continuously reduces the forests. It is estimated that between 3.5 and 15 billion trees are harvested each year globally. The United States harvests some 900 million trees annually. Many uses cannot be easily reduced but the waste of paper can be. About 40% of tree harvesting is used to produce paper and paper products. About 27 000 trees are cut down each day just to make toilet paper.

CASE STUDY 11.2 Monarch Butterfly Population Collapse

Habitat Destruction and Extinction

It is reported that in the 1850s there were so many monarch butterflies (Figure 11.5a) migrating through the Mississippi River Valley that the clouds of them darkened the midday sky. It is also reported that in California so many monarchs gathered that tree branches broke under their weight. It is estimated that there were once billions of monarch butterflies and they were ubiquitous throughout the United States during the summer. They were considered the standard in American butterflies through the 1970s and even the 1980s.

The monarch butterfly is famous for its multi-generational 3000-miles migrations south to Mexico in the fall and north as far as Canada in the spring for the eastern population (Figure 11.5b). The eastern monarchs include all butterflies east of the Rocky Mountains and account for 99% of the North American monarch population. The western monarch population is exclusively west of the Rockies and migrates from as far north as Canada during the spring and summer to southern California each fall.

Monarch butterflies lay their eggs on the underside of milkweed plant leaves (Figure 11.5c). Eggs hatch in four days, producing larvae or a caterpillar which feeds on the milkweed to store fat to be used in the non-feeding pupal stage of the butterfly. The caterpillar passes through five stages, separated by molting growing larger with each. After two weeks, the caterpillar spins a silk cocoon hanging upside down and transforms into a butterfly. In 10 days, the butterfly emerges from the pupa and lives for 4–6 weeks and mates. The butterflies from the late-summer generation survive for six to seven months and migrate from Canada and the northern United States southward into Mexico. This migration begins in September and October and the return trip begins in March. They arrive at their summer habitat by July.

The problem is that the North American monarch butterfly population is in freefall and they could go extinct. From 1996 to 2020 alone, the eastern monarch butterfly population decreased by 88%, from 383 million to about 45 million. Overall, the eastern monarch population declined by greater than 80% during the past two decades (Figure 11.5d). The population of western monarchs that winters in the central and southern California coast decreased to 1914 butterflies in 2021 (Figure 11.5e). This is a 99.9% decrease since the 1980s, when as many as 10 million butterflies overwintered in the area.

The reasons for this precipitous decline are complex and manifold. Monarchs are threatened by pesticides, global climate change, urban and suburban sprawl, and illegal logging of the forests where they migrate for the winter but habitat destruction is a primary factor. Monarchs require milkweed as an integral part of their life cycle. The problem is that milkweed rapidly disappeared in the late 1990s because farmers planted genetically modified, herbicide-tolerant corn and soybean crops in the Upper Midwestern United States, where more eastern monarchs were produced per acre than anywhere else in North America. As a result, farmers could apply herbicide in higher quantities, which decimated the milkweed population. Billions of milkweed plants were killed through this herbicide use. Monarchs lost at least 167 million acres of breeding ground to the herbicide spraying but also the extensive development over the past several decades. In addition to herbicide, butterflies are also threatened by neonicotinoid insecticides, fungicides, and other chemicals that are toxic to caterpillars. Much of this happened through suburban sprawl. The development destroys milkweed plants through clearing for building as does the landscaping which uses the herbicides and insecticides in abundance in addition to planting non-native plant species.

The eastern monarchs winter in the Oyamel fir forests in the mountains of central Mexico. They pack onto the fir trees in such clusters so dense that it is difficult to count them. The wintering area has been greatly reduced over the past two decades. In 1996–1997, the wintering area covered 44.5 acres (18 ha) compared with the 2019–2020 season when it covered just 7 acres (2.83 ha). It has been estimated that 15 acres (6 ha) are the minimum required for monarchs to survive. Habitat destruction may be the ultimate cause of the loss of the monarchs.

The western monarch migration has collapsed to an even greater degree. The reason for this is habitat destruction, like in the east and south, but also because of invasive species. Over the past 30 years, the Sacramento Valley, the largest urban area in the Central Valley of California, has been massively developed. This urbanization has removed a huge number of the milkweed plants in the area as well as removing part of the wintering grounds for the monarchs. This urban growth also provides a major source of contaminants like pesticides and herbicides which further reduces both the butterflies and milkweed.

Two unexpected factors are also reducing the western migrating monarch population. First, is that warmer winters, as the result of climate change, are reducing the need to migrate. Many areas that were too cold to winter are now suitable. The second issue is that, although residents are not partial to milkweed, they are planting invasive tropical milkweed in abundance. Tropical milkweed is a problem in temperate areas because it survives later in the season and does not die in winter. When it grows later in the season, it can confuse monarchs into breeding at a time when they should be migrating. Eventually, because of these factors, monarchs abandon migrating and become non-migrating butterflies, which is now occurring in California.

A second problem with the tropical milkweed is that a protozoan parasite can be carried by monarchs and be deposited on leaves of plants. Native milkweeds die back after blooming and the parasites die as well, so that the following summer's monarchs feed on parasite-free

(a)

(b)

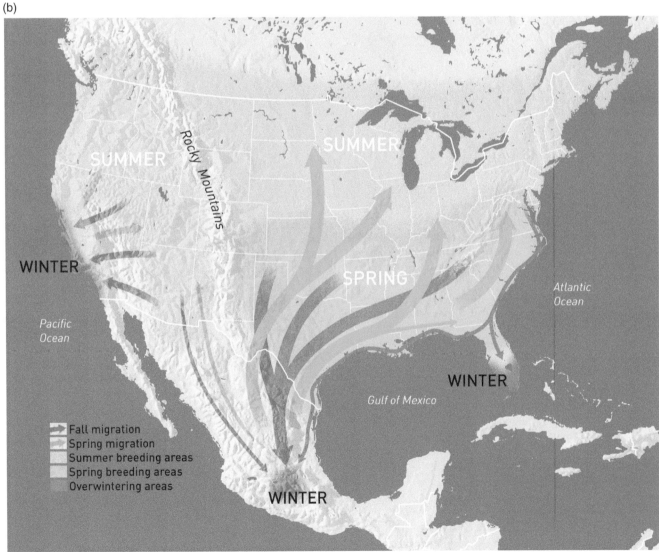

FIGURE 11.5 (a) Photograph of monarch butterflies. *Source:* Courtesy of US Fish & Wildlife. (b) Map of North America showing the migration paths of monarch butterflies and their locations during each season. (c) Photograph of butterfly milkweed. *Source:* Courtesy of the US Department of Agriculture. (d) Graph showing the decrease in area occupied by overwintering eastern monarch butterflies with time. (e) Graph showing the decrease in population of the western monarch butterflies with time.

(c)

(d)

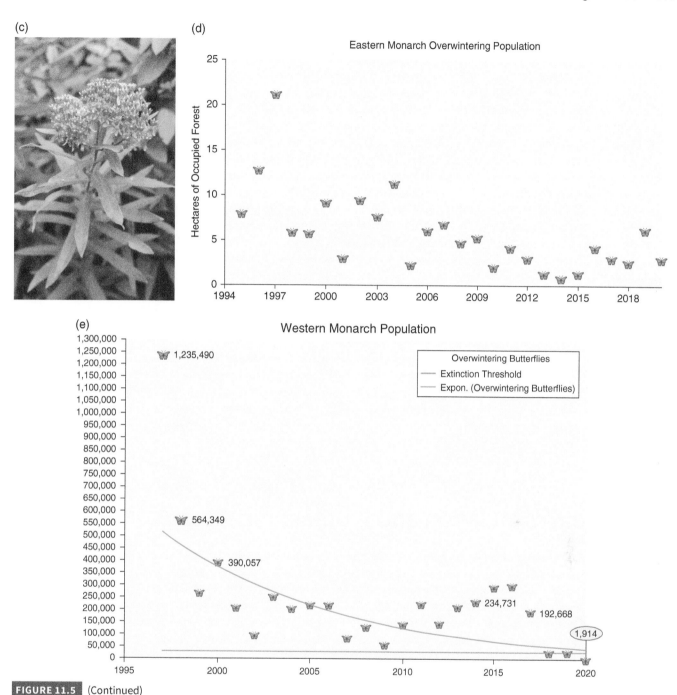

FIGURE 11.5 (Continued)

foliage. However, tropical milkweed remains green through winter which allows parasites to remain on the plant. This means monarch caterpillars feeding on the infected plant can be exposed to the parasites.

As a result of this radical decline, the US Fish and Wildlife Service placed monarchs on the list to be protected under the Endangered Species Act. It is estimated that there is an 80% probability of extinction of eastern monarchs within 50 years and a 96–100% probability of extinction for western monarchs. In Canada, monarchs are also slated for listing as endangered species and in Mexico they are classified as a species of special concern.

11.5 | Light Pollution

Most habitats in all but a few remaining native areas are broken into ever more distant and smaller patches. The edge areas are becoming ever larger because the contrast with the surrounding area is so profound, especially in developed areas. One of the most penetrating edge effects into the interior of the patch is light pollution. Light pollution is the excessive use of artificial outdoor light. It disrupts the natural wildlife patterns and even human sleep. There are four types of light pollution including glare or excessive brightness, skyglow or general brightening of the night sky, light trespass or unintended or needed light, and clutter or confusing and excessive light sources. Urban areas are hundreds to thousands of times brighter than they were 200 years ago as a result. This light severely impacts all species in a natural habitat and contribute to habitat destruction.

Plants, insects, animals, and other natural life depend on a daily cycle of light and dark to control life activity including nourishment, reproduction, sleep, and protection from predators. Light pollution disrupts this cycle in many habitat edge areas impacting all natural species in complex ways. For example, birds that migrate and hunt at night use moonlight and starlight to navigate. Light pollution commonly causes them to lose direction and fly toward urban and suburban areas. Millions of birds collide with illuminated buildings every year. Migrating birds follow schedules seasonal based on temperatures and lighting. Artificial lighting can cause them to migrate too early or too late and they can nest, hunt, and reproduce at poor times.

Many insects are drawn to light and are confused by artificial lights. Some predators use this attraction to hunt them more effectively but others are starved when the insects are not where they are expected. This confusion impacts food webs. Insect populations have been declining dramatically in the past several decades. In turn, this negatively impacts species that rely on insects for food. Bats are especially impacted by both the artificial light and shifting insect patterns as are insect hunting birds.

Plants require a period of darkness for proper metabolism. Plants respond to specific day and night cycles through the production of phytochrome hormone. Artificial night lighting alters flowering cycles, onset of fall dormancy, shoot growth, leaf growth, and seed germination. It may induce or suppress flowering in some species. Some fruits ripen in the darkness. Artificial night lighting can damage crops which depend on darkness for part of their production cycle. Strawberries require darkness to bloom and produce fruit and lettuce and spinach also require darkness. Trees depend on light duration to time their preparation for flowering, fruit production, and fall dormancy. Light pollution upsets these cycles and trees around streetlights can have delayed dormancy by more than a month. These trees can freeze with full leaves and sap in their branches. They are more subject to damage and it can kill the tree over time or in extreme conditions.

Light pollution also impacts amphibians. They detect light levels 100 times dimmer than humans. Exposure to strong artificial lighting impacts their hormones, skin coloration, temperature regulation, and reproduction. Sea turtles may live in the ocean but they hatch at night on beaches. The newborn hatchlings find their way to the ocean by the direction of the bright horizon. Light pollution confuses and diverts them away from the ocean. Along Florida beaches, millions of sea turtle hatchlings die because they cannot find the ocean. Most zooplankton swim to the near-surface levels of ponds, lakes, and oceans at night but remain at the bottom of during day to avoid predators. Light pollution causes them to remain at lower levels at night where they are not consumed by night predators. This removes one of the lowest levels of the foodweb.

11.6 | Bioaccumulation and Biomagnification

Bioaccumulation is the consumption and concentration of persistent substances within the tissues of living organisms. Bioaccumulation is an important aspect of the development and growth of all living organisms. It is a process within ecosystems with regard to the food web or food chain. Animals consume and plants uptake chemicals that will not readily break down under normal conditions and they are stored in various locations within the organism. In animals, persistent chemicals tend to be stored in fat but they can accumulate in certain organs as well. In humans, the process stores vitamins, trace minerals, and essential fats and amino acids in specific tissues and organs for use in metabolic processes. The mechanisms that accumulate nutrients and vitamins in the human body, also act to store persistent contaminants like the pesticide DDT.

DDT's persistence in naturally occurring bioaccumulation makes it a deadly environmental contaminant. Once on plants or in the soil, DDT resists natural processes that would degrade it like chemical interaction with other compounds, and biodegradation. DDT, like many other chlorinated pesticides, is a hydrophobic, lipophilic chemical that can remain in the environment for a long time with no change. Once consumed, these chemicals tend to move out of solution and into the fat cells of an organism. Lipophilic chemicals are not water-soluble and are not easily flushed from the system. Once a persistent organic pollutant (POP) like DDT and other pesticides is stored in body fat, it is not removed, and does not degrade. As additional contaminated fat is added, the amount of POP increases or bioaccumulates. When the fat reserves are finally used, much of the POPs are remobilized into the organism.

Bioaccumulation occurs at different rates in each species and even in individuals in the same species. Animals that are larger or accumulate fat will bioaccumulate at higher rates than short-lived and small species. An old, large fish will have bioaccumulated more POP than a small, young fish from the same lake. POPs enter organisms through biological uptake, or exposure mechanisms. These mechanisms include inhalation of vapor and dust, ingestion of contaminated plants, or absorption through skin or gills. Once inside an organism, POPs biomagnify with each step of the food chain or food web. Biomagnification is the accumulation of a pollutant in organisms at higher concentrations than are present in its food or surrounding environment.

When a substance enters the environment it can be taken up by organisms at the bottom of the food web or chain in an ecosystem where it can bioaccumulate (Figure 11.6). These organisms can be producers like plants, fungi, or invertebrates such as insects on land or small aquatic animals. These organisms are consumed by predators that are next up in the food chain. The food chain concept is older and simpler because it envisions a single line of consumption from producers and simpler life to more

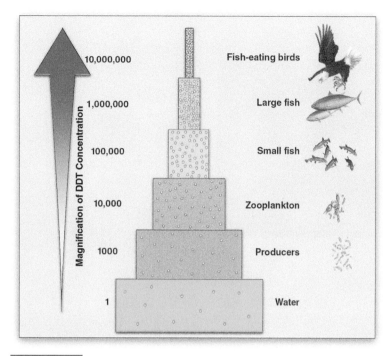

FIGURE 11.6   Illustration of the biomagnification of DDT from water through the aquatic food chain or web and to fish-eating birds.

aggressive predators. However, it is not so simple because there are many organisms consumed by each predator in each step with conditional and competing predators. Mapping all these interactions forms a web rather than a chain.

In each step upward in the web, the predator consumes prey from lower in the web. Each of the prey has bioaccumulated a contaminant that entered the ecosystem, and therefore has higher than background levels. The predator slowly also bioaccumulates the contaminant at even higher levels. This is the process of biomagnification and it can lead to damaging concentrations in animals at the top of the food web. This provides another process to advance pollutants to unsafe levels. For example, it was found that DDT biomagnifies from concentrations at 10 ppm in the soil, to more than 140 ppm in earthworms and 400 ppm in robins.

Bioaccumulation and biomagnification of pesticides and other industrial chemicals were not considered during their development and use. However, in 1962, Rachel Carson carefully documented how DDT and other organochlorinated pesticides entered and worked their way up the food chain in her book *Silent Spring*. As a result, DDT was banned for sale in the United States in 1972. There are many other elements and compounds that can be biomagnified through the food web, both organic and inorganic and both in fat deposits and specific organs. The filtering organs such as the liver and kidneys are especially susceptible to poisoning but the bones can store inorganic elements like lead.

Like DDT, all POPs are difficult to degrade and are mobile in the environment. The EPA recognizes 12 POPs including aldrin, chlordane, DDT, dieldrin, endrin, heptachlor, toxaphene, Mirex, HCB (hexachlorobenzene), PCB (polychlorinated biphenyls), dioxin, and furans. These pollutants can be either a solid or vapor, depending on temperature. Once released to the environment, if the temperature is warm, POPs like PCBs or toxaphene evaporate or are carried on dust particles, possibly by the wind. If the temperature cools or the dust settles, the POP could be thousands of miles from where it was originally released. With temperature and wind changes, settling and remobilizing results in worldwide distribution of POPs. POPs are commonly detected in areas where they have never been used. DDT occurs in Arctic ice and PCBs are in the tissues of deep ocean Atlantic cod.

CASE STUDY 11.3 American Bald Eagle

Near Extinction Through Biomagnification

When the bald eagle was adopted as the national symbol of the United States in 1782, there were approximately 100 000 nesting pairs of eagles in the United States, including Alaska (Figure 11.7a). In the early eighteenth century, the bald eagle population is estimated to have been 300 000–500 000. They inhabited every state except Hawaii.

However, they were not welcome in most places. Bald eagles were regularly shot and killed by poisoning because they were deemed a threat to livestock and salmon. Bounties were even offered for eagle carcasses. In 1917, the Alaskan government posted a 50 cent bounty per bald eagle killed, and later a dollar. This resulted in at least 120 000 confirmed eagle deaths. By 1940, the demise of bald eagles was so extensive that the US Congress passed the Bald Eagle Protection Act,

(a)

(b)

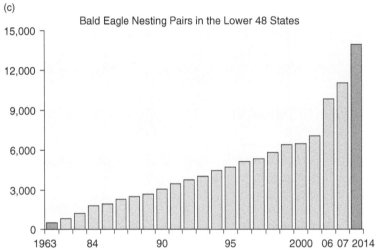

(c)

FIGURE 11.7 (a) Photograph of a bald eagle in flight. *Source:* Courtesy of US National Park Service. (b) Graph of nesting eagle pairs in Florida showing the impact of the introduction of DDT use as a pesticide. (c) Bar graph showing the increase in bald eagle population after the ban on DDT and through repopulation efforts.

outlawing the killing and even disturbing of eagles. The act further forbade possession of all eagle parts, including feathers, eggs, and nests.

The plight of eagles intersected with the story of the infamous pesticide dichloror-diphenyl-trichloro-methylmethane or DDT at that point. That story began with malaria-carrying mosquitoes, which are documented as problems for people since about 2700 BCE. In 1897, British doctor, Sir Ronald Ross, showed that mosquitoes transmitted malaria from person to person and received the Nobel Prize in Medicine in 1902 as a result. DDT was first synthesized by Othmar Zeidler in 1874 but its pesticide ability was not realized until 1939 by Dr. Paul Müller. He was also awarded a Nobel Prize for Physiology and Medicine in 1948 for this discovery. This began the mass application of DDT to control mosquito populations worldwide helping to save tens to hundreds of thousands of lives directly, and tens of thousands of lives through malaria eradication in many areas.

DDT is an inexpensive, broad-spectrum pesticide that protects people from deadly mosquito-borne illnesses. It was first widely used in World War II to kill mosquitoes. However, in 1945, after the war, it was adopted for widespread agricultural use to control many insects that damaged crops like tobacco budworms, Colorado beetles, flies, potato beetles as well as managing Dutch Elm disease. Farmers applied DDT

in agricultural areas as a liquid or powder or mixed with oil-based dispersants using airplanes and helicopters. In urban and suburban areas, DDT was sprayed on walls and ceilings of houses and public buildings. In protected areas, it remains on surfaces for six months or more. It was also sprayed from trucks on most city streets and applied directly to people using handheld applicators. These nationwide activities spread DDT throughout the environment on a massive scale.

The agricultural DDT was carried by surface-water runoff to rivers and lakes, where it entered the water cycle. DDT breaks down in only a few days in sunlight in the air. In soil, however, DDT has a half-life of 15 years. A half-life is the time it takes for one half of the mass of the contaminant to break down to other compounds. This means that DDT can persist in soil for many decades. This persistence gives small organisms time to accumulate it, leading to high concentrations. The processes involved in the concentrating of DDT are bioaccumulation and biomagnification. Bioaccumulation reflects the ease with which an organism encounters and uptakes a chemical. In contrast, biomagnification involves concentrating DDT through the food chain.

The long-term effect of this overuse of DDT impacted the populations of animals, birds, and fish in particular, which began to rapidly decrease. Many species of birds of prey dropped markedly during the

1950s and 1960s (Figure 11.7b). By the 1960s, some biologists began to believe that DDT was the culprit. DDT entered the bald eagle food chain through aquatic producers, which were then eaten by fish which, in turn, were eaten by increasingly larger and more predatory fish which are the main food source for bald eagles. The DDT accumulates in the eagle as more fish are eaten through biomagnification, which results in 10 times the DDT in eagles as in the fish.

The factors making bald eagles susceptible to biomagnification are that they are large predators, live at least 25 years, are unable to reproduce until 5 years old, and their diet is mostly fish. That means they consume a lot of DDT through fish and it has time to accumulate over their long lifetime. Osprey, peregrine falcons, brown pelicans, and many other birds of prey are similarly at risk. The high amount of DDT metabolites found in eagles was the first evidence of cause. The complete exposure pathway from DDT application to the birds was further evidence. However, laboratory testing of DDT in adult eagles showed no increase in mortality from exposure.

As a result, it was difficult to explain that by 1963 there were just 412 documented breeding pairs of bald eagles in the United States. It took extensive work to determine that the environmental metabolite of DDT, dichlorodiphenyldi-chloroethylene (DDE), was the pathway to damaging the bald eagle and other birds of prey. Field observations revealed reproductive failure of bald eagles and other birds of prey due to thin eggshells. Laboratory experiments confirmed these observations that DDE causes thinning of eagle eggshells. DDE made it difficult for the birds to absorb calcium and the low levels made their eggshells thinner than normal. This resulted in female eagles crushing their eggs when they sat on them. Many eggs were broken before they could hatch. Field studies confirmed that exposure to DDE were sufficient to cause the drop in populations.

When bald eagles were listed as an endangered species in 1967, there were only 417 recorded breeding pairs in the lower 48 states. Many people believed that eagles would soon be extinct. Federal protection stopped the further demise of eagles but they did not recover. It took the ban on the widespread use of DDT in the United States in 1972 as proposed by the newly formed EPA and recommended by Rachel Carson, nine years earlier, for progress to be made. Due to the ban, successful breeding programs, reintroduction of eagles into the wild, the continued protection through the Endangered Species Act, the bald eagle population steadily climbed (Figure 11.7c). The improvement was so impressive that in August 1995, the eagle was removed from endangered and added to threatened status under the Endangered Species Act. By 1996, there were more than 5000 pairs of eagles counted. In 2009, there were 72 000 bald eagles in the United States, with about 30 000 breeding pairs. In 2007, the bald eagle was removed from federal protection.

In 2022, 40 years after the banning of DDT, there were nearly 15 000 nesting pairs of bald eagles in the lower 48 states and 20 000 pairs in Alaska, though some sources place the bald eagle population at 316 000. Other birds benefitted from these laws as well. In 1975, there were only 39 breeding pairs of peregrine falcons in the lower 48 states and only in the west. Peregrines in the eastern United States had been eradicated by DDT poisoning. By 1996, there were 993 pairs in the lower 48 states, representing a 20-fold increase. Ospreys increased from 8000 breeding pairs nationwide in 1981 to 14 246 pairs in 1994. Brown pelicans saw similar increases and Canada also recorded a great resurgence in birds of prey as well.

11.7 What Can I Do?

Suburban residents love well-manicured properties with exotic plants. A new trend in planting native species is much more sustainable and helps native birds, animals, and insects among others. Mowing plants down also does not promote the rest of the biota. Bushes promote bird success and should be planted, as should trees where safe and possible. They should not be removed unless necessary. Use of pesticides kills native insects, microbes, and other life forms reduces biodiversity both in the biota that is directly killed or the animals that prey on them. Herbicides also reduce native plants and plant biodiversity in addition to reducing food for native animal species. Pet cats should not be allowed to roam free and feral cats should be reduced. Outdoor poisons for animals can kill native species.

On a national and global scale, reducing waste is the best way to address the impacts on ecosystems. Reducing the wasting of wood is the first step. Using rags and sponges instead of paper towels, plates, and glasses instead of paper plates and cups, stopping printing out papers, reading the news online, receiving e-bills and other mail will all make an impact.

The other problem that can help is reducing food waste. It is estimated that between 30% and 35% of all food is wasted in the United States and Europe and the vast majority is disposed of in landfills. Much native land must be cultivated to meet the food demands, which reduces wilderness and native species. Agriculture is almost always non-native. Reducing food waste reduces pressure on natural ecosystems. Further, the more refined foods require more agricultural products so are more impactful. Addressing these issues on a large scale can make a significant impact.

Finally, driving automobiles and motorcycles through the forests for recreation contributes to fragmentation. The noise also frightens the animals and disrupts their activities.

Reference

Greig K and Rawlence N.J. (2021). The Contribution of Kurī (Polynesian Dog) to the Ecological Impacts of the Human Settlement of Aotearoa New Zealand. *Front. Ecol. Evol,* 9:757988. doi: 10.3389/fevo.2021.757988.

Ocean Pollution

CHAPTER OUTLINE

Polluted Earth: The Science of the Earth's Environment, First Edition. Alexander Gates.
Companion website: www.wiley.com/go/gates/pollutedearth

Words you should know:

Barrier islands – Long, narrow islands of sand along coastlines.

Bathymetry – The seafloor topography beneath the oceans.

Beaches – Strips of sand and possible gravel along coastlines.

Delta – Land built into the ocean at the mouth of a river.

Estuary – A river that has been flooded upstream by the ocean.

Fetch – The width of ocean surface across which wind acts to produce waves.

Longshore currents – Ocean currents that flow parallel to the shoreline.

Tides – Movement of water toward and away from the shore as the result of the gravity of the Moon and Sun.

Waves – Wind-driven height changes of the ocean surface whose size reflects the size of the basin and strength of the wind.

12.1 Ocean Basics

Ocean health is crucial for the survival of the planet. Oceans cover about 71% of the surface of the Earth (Figure 12.1) and contain 97% of all water on the planet. Although we focus on life on land, the oceans actually contain 99% of the habitable space or volume of the planet. At least 50% and likely closer to 80% of all life on Earth lives in the oceans, with 226 000 species of life counted so far. This number will increase because humans have more than 80% of ocean volume yet to explore. Further, more than one third of the human population, about 2.4 billion people, lives within 60 miles (100 km) of a coast. The oceans are one of the primary controllers of atmospheric processes. They strongly influence all weather on the planet by providing moisture, modulating temperatures and controlling the proliferation of most storms. Ocean plants produce about half of the oxygen in the atmosphere from photosynthesis. Oceans also absorb about 50% of the carbon dioxide produced by humans through burning fossil fuels for energy.

Oceans sit in basins that are floored by ocean crust everywhere except along the coastlines because ocean crust is denser and thinner than continental crust. As a result, gravity causes it to lie at a lower elevation. Water goes to the lowest elevation also as a result of gravity which is why ocean crust is flooded with water. The bathymetry of ocean floors has a characteristic geometry depending on whether the margin is tectonically active or not. If the coast is not a plate margin, it is called a passive margin. The East Coast of North America is a good example of a passive margin. The area closest to the coast is the continental shelf which is flooded continental crust (Figure 12.2). Depending on the continent, the shelf can be as wide as several hundred miles at a very shallow slope of 0.5°. Storms impact this seafloor because the waves are so large that they penetrate deep into the water.

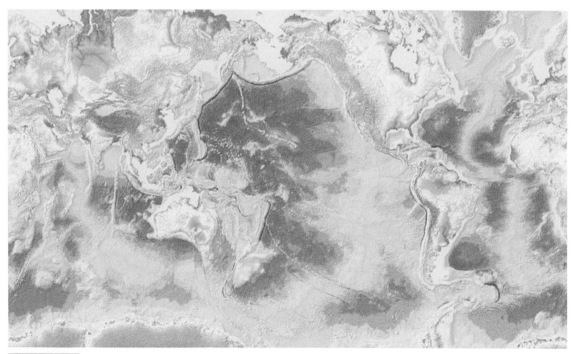

FIGURE 12.1 Map of the ocean floor bathymetry and continental topography. *Source:* Courtesy of NOAA.

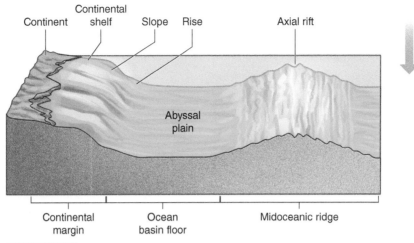

FIGURE 12.2 Block diagram of a typical ocean basin with a passive margin on the continent.

The continental slope occurs oceanward of the shelf and is relatively narrow. It is underlain by transitional continent-ocean crust and has a steeper slope up to 2°. The next feature oceanward is the rise which is also narrow and forms at the base of the slope on ocean crust. The rise transitions into the abyssal plain which is usually the widest part of ocean basins. It is flat and floored by ocean crust. An undersea mountain range marks the middle of the ocean basin. This range is called the mid-ocean ridge, the most rugged part of the ocean floor, but it is not necessarily located in the middle of the ocean because it depends on the plate tectonics.

Where there is a subduction zone in a convergent plate boundary, the slope transitions directly from the shore to the slope and then into an oceanic trench. Trenches are the deepest parts of the ocean and occur all around the Pacific Ocean and in some parts of the Atlantic and Indian Oceans. The abyssal plain is directly oceanward of the trench, and the distance to the mid-ocean ridge is variable if it is not absent.

Ocean meets the land at a coastline. Coastlines can have a number of geometries that can amplify or subdue the energy of incoming waves. Concave shorelines funnel the wave energy and increase wave height whereas convex shorelines dissipate the wave energy, reducing the height. Smooth shorelines produce waves that are generally the same height whereas jagged shorelines funnel or dissipate wave height from place to place.

Many shorelines contain beaches, bands of sand, and possibly gravel. Most sand beaches have dunes behind them which protect the coast from strong waves though sand is easily removed by them. During sea-level rise, the ocean can spill behind the beach forming a lagoon or bay landward of the beach (Figure 12.3). The elevated strip of beach remains an island aligned along the coast. These are barrier islands and form a semi-continuous band along the Atlantic coast of the United States from New York to Florida.

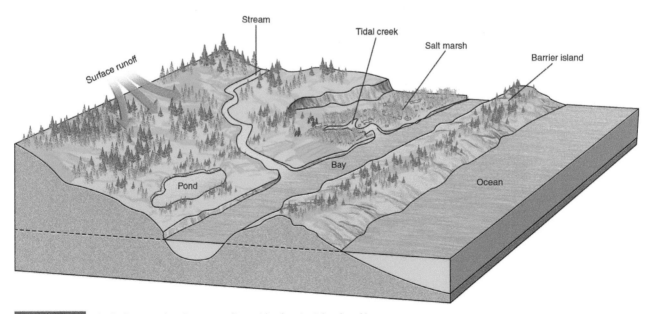

FIGURE 12.3 Block diagram showing a coastline with a barrier island and lagoon.

FIGURE 12.4   Satellite image of the end of the Mississippi River delta in the Gulf of Mexico. *Source:* Courtesy of NASA.

If the ocean has low wave and tidal energy, new land is built into the ocean by river deposits. This land is a delta and the best example is built by the Mississippi River into the Gulf of Mexico (Figure 12.4). There are several types of deltas depending on the amount the wave and tide energy and the ruggedness of the coast. During sea-level rise, water floods back up the river removing the land along the banks. This is an estuary as are most rivers along the mid-Atlantic and southeast coast of the United States fall. Low-lying areas along the coast that flood during high tide are tidal flats.

12.2 Ocean Processes

12.2.1 Tides

Tides are the rise and fall of the surface elevation of any water body. Tides are caused by the gravity of the Moon with input from the Sun acting on oceans or lakes. Ocean tides generally have two high tides and two low tides per day, each pair separated by 12 hours. High tides occur when the Moon is overhead or over the opposite side of the Earth whereas low tides occur when the Moon is on the horizon (Figure 12.5). If the high tide is at midnight, then the low tide will be at 6 a.m., the next high tide will be at noon and the next low tide will be at 6 p.m.

The height difference between the high and low tide is the tidal range. The distance the water floods upstream in a river during high tide is the tidal reach (Figure 12.6). The height of high tide and low tide varies and therefore the tidal reach varies. This variability is caused by the gravitational influence of the Sun. If the Sun, Moon, and Earth align with the Sun on either side of the Earth high tides are at their highest point and low tides are at their lowest point. The reason for this is the gravity of the Sun and Moon are additive in this geometry. This is spring tide and it is usually around the full Moon (Figure 12.7). If the Sun and Earth are at right angles to the Moon and Earth, it is neap tide. Neap tides have lower high tides and higher low tides than normal. Neap tides occur around half Moon. The distance between the Earth and Moon and Earth and Sun has a very small impact compared to the alignment.

The main factor in the height of the tides, the range or reach at a location is the shape of the ocean basin. Small, shallow ocean basins and lakes have a small tidal range and reach and may only have one tidal cycle per day. Large, deep basins typically have larger tides. However, the size and bathymetry of basins and particularly the shape of the coasts have a great impact on tidal range and reach. In the Bay of Fundy, Nova Scotia, Canada, the tidal range is an amazing 32 ft (9.85 m). This is because there is deep water off of Nova Scotia and therefore more water available. This and the funneling make large tides in Bay of Fundy.

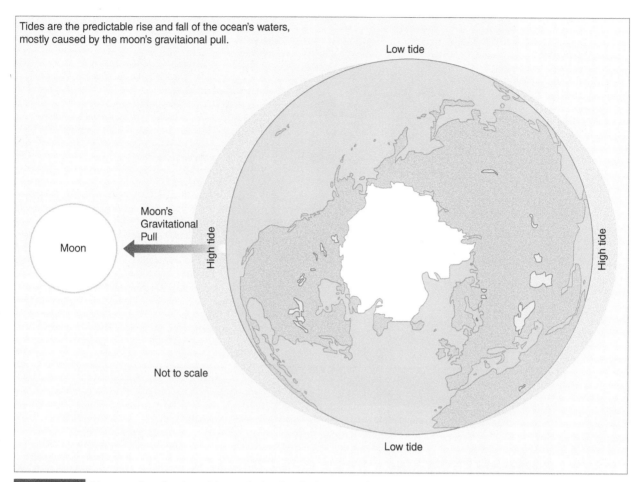

FIGURE 12.5 Diagram of a polar view of the Earth showing the locations of high and low tides relative to the Moon.

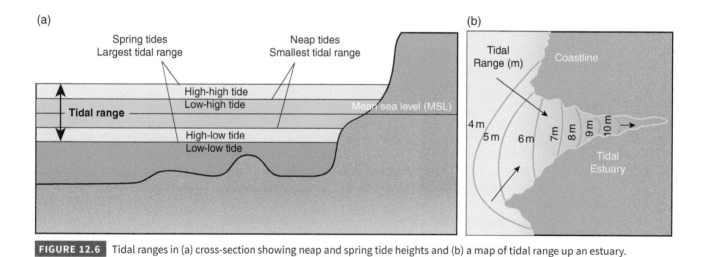

FIGURE 12.6 Tidal ranges in (a) cross-section showing neap and spring tide heights and (b) a map of tidal range up an estuary.

12.2.2 Waves

Waves occur all around an ocean basin but they are most prominent at the shore. Waves travel across the surface of water bodies driven by winds. The distance that the wind acts across the surface of the water governs the height of the waves produced. This distance of the interaction is the fetch. Wind speed and duration also determines the wave heights. Ocean waves have the same parameters of waves in physics (Figure 12.8). The crest is the peak of the wave, the trough is the bottom, and the wavelength

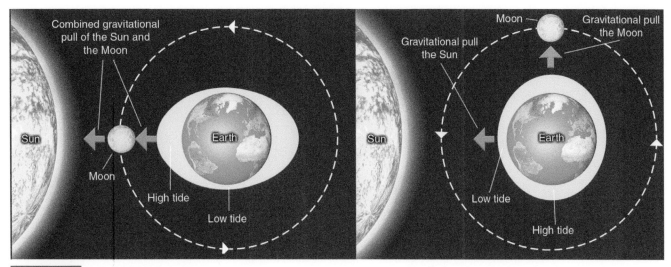

FIGURE 12.7 Polar view of the Earth relative to the Sun and Moon showing (left) spring tide arrangement with high and low tides and (right) neap tide arrangement with high and low tides.

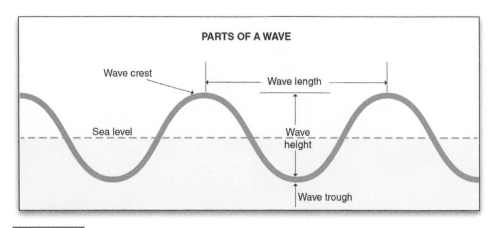

FIGURE 12.8 Cross-section showing parts of a wave.

is the distance between peaks or troughs. The wave height is the elevation difference between trough and crest and amplitude is one half the height. The wave speed is expressed by its period, the time it takes for similar points on adjacent waves to pass a reference point.

On the basin scale, water position remains essentially the same regardless of the waves. The water particles move in a circular or orbital motion as a wave moves through the open water (Figure 12.9). As the wave approaches the shore, the water is first pulled back before being lifted, then it falls and finally returns to its original position as the wave passes through. The motion of the water particles forms orbitals which are larger the nearer to the surface. The depth in the water to which the orbitals penetrate is one half of the wavelength and is termed the wave base. The water does not move below the wave base. The depth of wave base moves up and down with the tides. It is deeper at low tide and shallower at high tide. The variation in depth also has a bigger range during spring tide and a smaller range during neap tide. Wave base even varies seasonally with gentler summer waves penetrating much shallower than larger winter waves. Storm waves have a deeper wave base.

Wave base makes contact with the seafloor as the wave approaches the shore (Figure 12.10). The wave drags on the seafloor, slowing its forward movement. It slows more at the base than nearer to the surface, causing the wave to tilt forward. It also rises up with respect to sea level as it approaches the shore. Waves crowd together as the result of this friction and slowing as they approach the shore. The water orbitals flatten and the wave becomes more asymmetric as well. The waves fall over in the breaker zone and the forward movement of the wave pushes the water up the shoreface in the swash zone. The dragging of the ocean waves on the seafloor moves sediments there. Gentler waves have shallower wave bases and have less energy to suspend sediment. Sand deposits

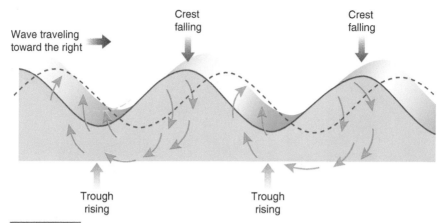

Wave traveling toward the right →

Crest falling

Crest falling

Trough rising

Trough rising

FIGURE 12.9 Cross-section of water particle motion as a wave passes through. The solid line shows the current position of a wave, the dashed line shows the future position of the wave and the orange arrows show the motion of the water with this movement.

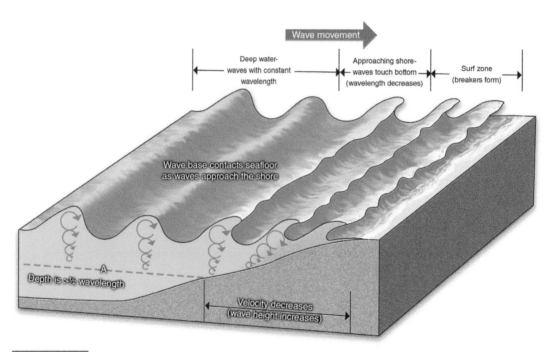

Wave movement →

Deep water- waves with constant wavelength

Approaching shore- waves touch bottom (wavelength decreases)

Surf zone (breakers form)

Wave base contacts seafloor as waves approach the shore

A
Depth is >½ wavelength

Velocity decreases (wave height increases)

FIGURE 12.10 Block diagram of waves and their water particle orbitals as they approach the shore and break.

closer to the shore and, as a result, beaches widen in summer. Waves are more energetic and have a deeper wave base in winter. They scour beach and near-shore sand and suspend it. Energy is too high for deposition up to a few hundred yards offshore. Sand is moved from the beach to offshore sand bars in winter. As a result, there is a cycle of sand being moved to the beach in summer and removed to the offshore in the winter.

In many areas, waves approach the shore at an angle. As a result of this angle, the approach angle pushes water parallel to the shore (Figure 12.11). This water movement is the longshore current, and the angle of approach and wave speed controls it. Stronger longshore currents are produced by more oblique waves. The leading edge of oblique waves drags more on the seafloor and slows more than the part of the wave in deeper water. The difference in speed along the wave crest causes it to bend as it approaches the shore. This is called wave refraction. The bending of waves can be quite complex as they approach jagged and

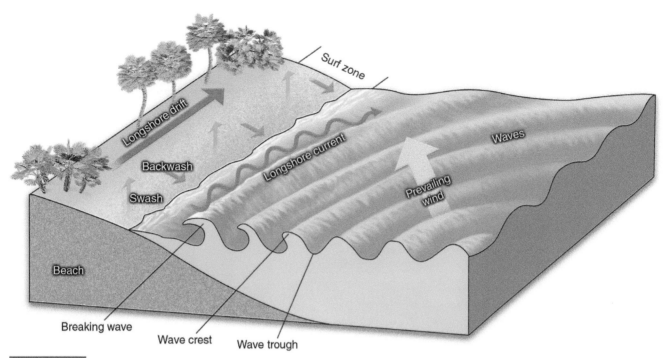

FIGURE 12.11 Block diagram showing oblique approach of waves to the shoreline and generation of longshore current and littoral drift.

uneven shorelines. In many shorelines, longshore currents move toward each other forming a build-up of water. This build-up escapes seaward in a dangerous rip current.

Longshore currents transport sediment along coastlines in a process called littoral drift. The sediment particles are carried toward and away from the beach with each wave in addition to moving laterally. In this process, transport of sand is along the beach which has been called a river of sand. Littoral drift can extend the entire length of barrier islands. The growth can cross a channel or shipping lane. If the shoreline angle varies or there is an obstruction, the sand is piled up in one place and starved from another. This also happens if residents build jetties, piers, and breakwaters. The flowing sand is blocked by jetties and builds up on one side whereas it is removed from the downflow side of the jetty, thereby producing a sand deficit. This gives shorelines a scalloped appearance. Breakwaters reduce the power of waves reaching the shore, allowing sand to be deposited behind them.

12.3 Pollution of the Coast

Many types of pollution impact the shore areas. There are current laws that prohibit discharge of raw sewage into the ocean in the United States but before 1972 and in much of the world, it still occurs (Figure 12.12). Sewage contains bacteria and nutrients that feed the bacteria that could be hazardous to marine life and humans. The bacteria can cause sickness and disease directly to people and animals from the water and indirectly to those who consume seafood from infected areas. Filter feeders and especially shellfish, are especially sensitive to bacteria. The most common disease in shellfish is hepatitis A.

Increased nutrients along the coast from sewers or runoff from agricultural or residential areas can cause dangerous algal blooms. Red tide is an increasingly common dangerous algal bloom on the Gulf Coast and East Coast of the United States. It is a condition where dinoflagellate algae quickly multiply, turning ocean water red brown. Depending upon the algae, mild to strong toxins can cause respiratory problems by airborne transport, in people who do not even enter the water. In some cases, the algae produce strong neurotoxins that can damage or kill marine mammals, either directly or if they consume contaminated seafood. These toxins are dangerous and potentially fatal to humans if not treated.

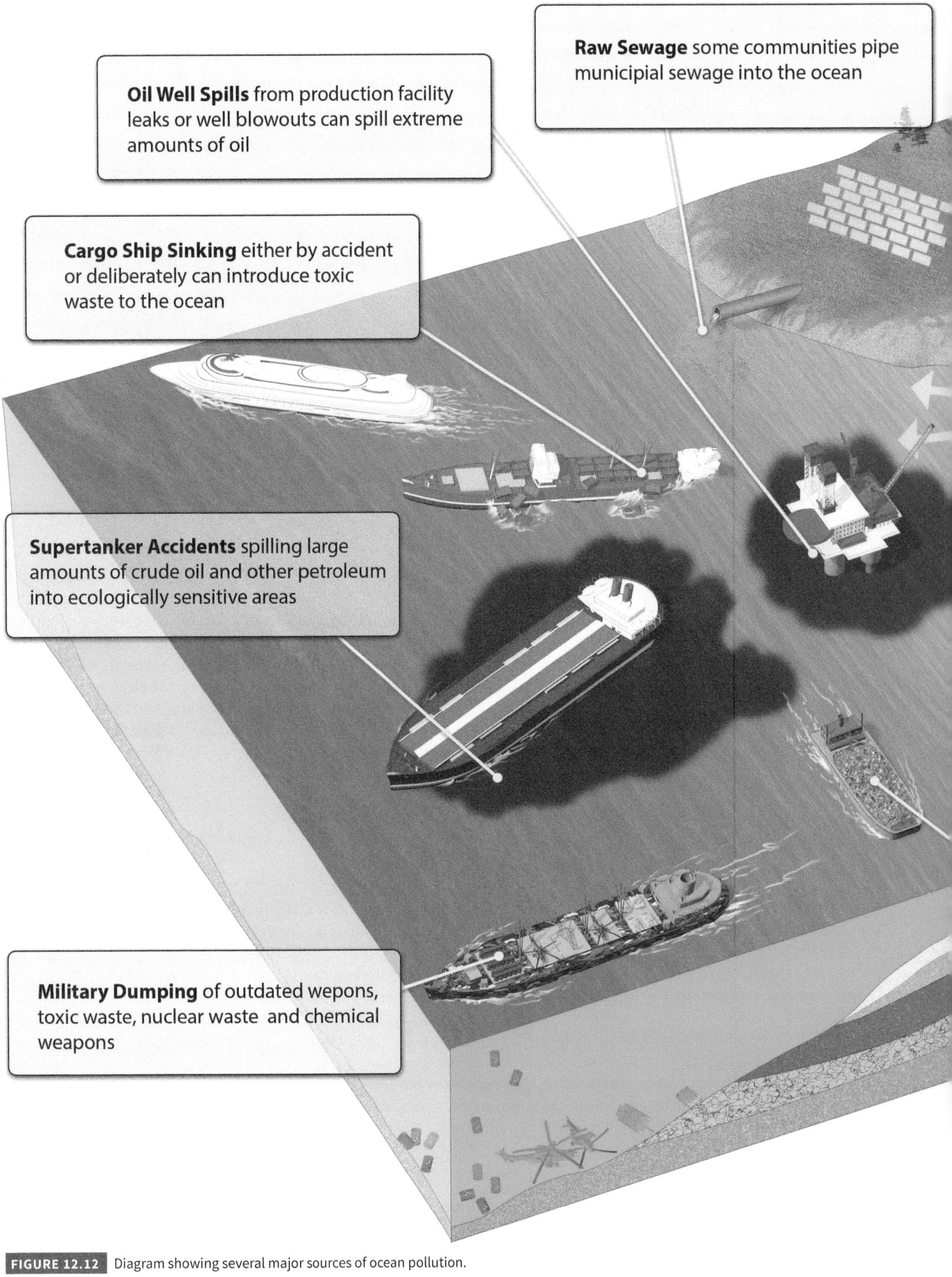

Oil Well Spills from production facility leaks or well blowouts can spill extreme amounts of oil

Raw Sewage some communities pipe municipial sewage into the ocean

Cargo Ship Sinking either by accident or deliberately can introduce toxic waste to the ocean

Supertanker Accidents spilling large amounts of crude oil and other petroleum into ecologically sensitive areas

Military Dumping of outdated wepons, toxic waste, nuclear waste and chemical weapons

FIGURE 12.12 Diagram showing several major sources of ocean pollution.

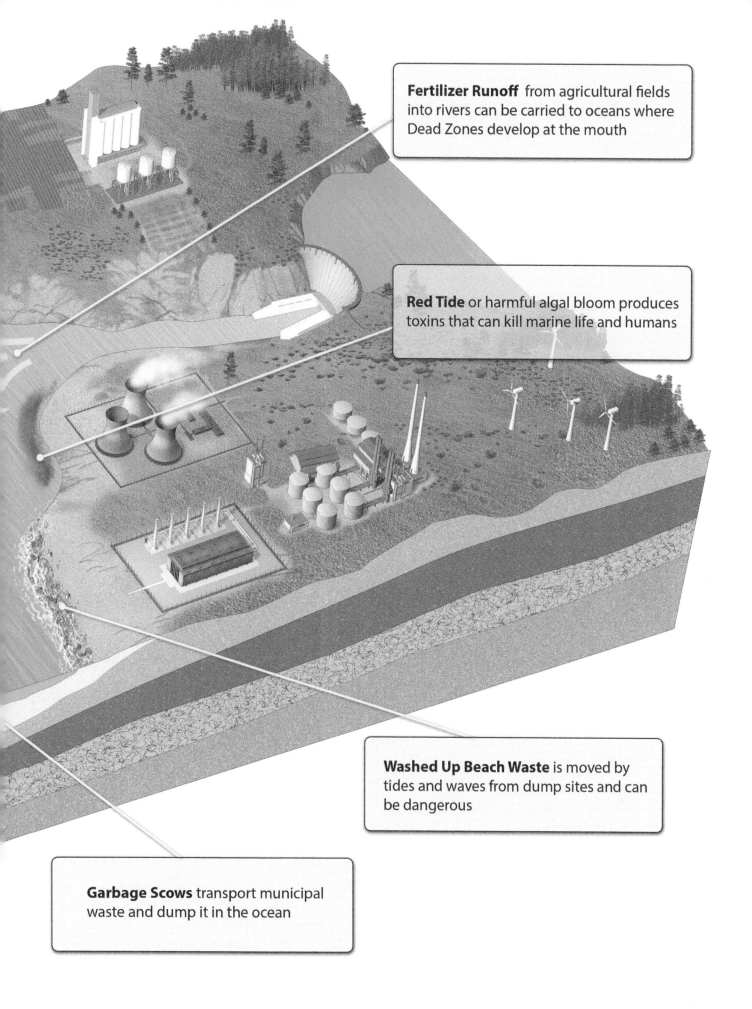

Fertilizer Runoff from agricultural fields into rivers can be carried to oceans where Dead Zones develop at the mouth

Red Tide or harmful algal bloom produces toxins that can kill marine life and humans

Washed Up Beach Waste is moved by tides and waves from dump sites and can be dangerous

Garbage Scows transport municipal waste and dump it in the ocean

Beach pollution also includes washed-up solid waste including medical waste. Typically, this waste includes household and construction waste. However, in the late 1980s, beaches had to be closed because used hypodermic needles, sample containers, gloves, and other medical waste washed up onto the beaches in New York and New Jersey.

Oil, fuel, and chemical spills can cause long-term damage to beaches. They are the most devastating type of beach pollution. Oil spills destroy the local ecosystem by poisoning and overwhelming it. Spills are mostly from tanker accidents, storage and transfer facilities, and from leaking oil wells. Crude oil, fuel, and most refined chemicals are toxic to marine life. Small amounts of fuel can spill from tanks of ocean vessels, but spills from accidents during transfer of fuel or leaking storage facilities are more common.

CASE STUDY 12.1 Torrey Canyon Scilly Islands, UK

Coastal Oil Tanker Accident

The *Torrey Canyon* supertanker was the 13th largest ship in commercial use (120 000 tons, or 108 862 mt) and almost 1000 ft (305 m) long and more than 125 ft (38 m) wide. It had a capacity of more than 75 000 tons (66 964 mt) of oil. The growing demand for oil in the 1960s caused a bottleneck in getting oil from fields to refineries. There were few overland pipelines and supertankers were the solution. The problem was the risk of moving such large quantities of toxic material near ecologically sensitive coastal settings.

The supertanker *Torrey Canyon* loaded 30 000 000 gal (114 million l) of crude oil in Kuwait on 17 February 1967. After a month at sea, the ship had gone around the Cape of Good Hope passed the Canary Islands and, by 17 March reached the Isles of Scilly off the English coast. The final destination was a BP refinery in Milford Haven in southeastern England.

Normally, a ship the size of *Torrey Canyon* would sail east of the Scilly Isles to Milford Haven. Otherwise, it would have to navigate a narrow (20-miles, or 32-km wide) gap between a spit of land jutting into the passage and the rocky coast of the Scillys. The Scilly Islands are a popular vacation spot and tourist attraction. The archipelago contains 55 islands and about 90 large half-submerged rocks. The islands are small, with the largest only 3 miles (4.8 km) across. The cliffs and submerged rocks circling the islands have been known to pose serious threats to shipping.

As the ship neared its destination, a navigational error resulted in it entering the dangerous passageway between the jutting land and the Scilly Isles (Figure 12.13). The captain realized the error and yet decided to risk the trip because if they did not reach Milford Haven by the next day, they would have to wait six days for a high enough tide to enter the shipping channel. The *Torrey Canyon* therefore continued at almost full speed, or about 20 miles per hour (32 kph).

The captain and crew were not worried because the seas were calm, the weather clear, and the ship was a modern and reliable vessel. However, a second navigational error was made by a junior officer on his first trip, who plotted the ship location well clear of the Seven Stones hazard when it was actually only 3 miles (4.8 km) from them. The captain discovered the error and ordered emergency course

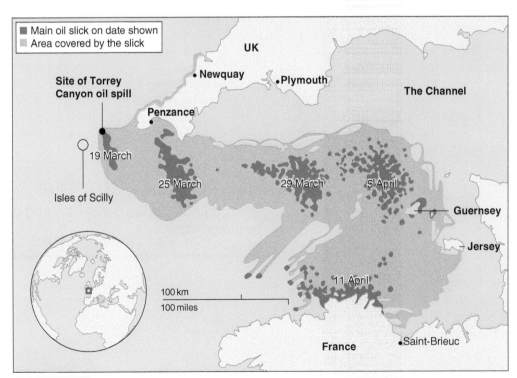

FIGURE 12.13 Map showing the movement and timing of the oil slick produced by the Torrey Canyon oil tanker accident.

changes. The ship, however, did not respond because the helmsman forgot to disengage the automatic steering. As a result, on 18 March 1967, the *Torrey Canyon* struck the Seven Stones hazard, which tore a 17-ft (5.2-m) long gash across six oil storage compartments. The entire cargo of 36-million gallons (136 million l) of crude oil spilled into these ecologically sensitive waters over the next 12 days.

This was the first spill accident involving a supertanker, so there were no protocols for addressing it. The crew was evacuated and a salvage company tried to refloat the tanker by removing it from the hazard. Salvage crews also tried to transfer the remaining oil to another tanker. These recovery efforts failed and resulted in deaths of workers so were abandoned. The Royal Navy attempted to bomb the *Torrey Canyon* to set the spreading oil on fire using napalm. Fighter jets dropped about 100 bombs on the tanker, many of which missed it and caused other ecological damage. Eventually, they were successful in blasting the tanker off the hazard and sinking it. Some floating oil was also burned by the bombing.

The oil slick from the spill spread to cover 270 square miles (700 km²). It contaminated 120 miles (192 km) of English coastline and 50 miles (81 km) of French coast. To clean the beaches, 1400 British soldiers used a mix of straw and a hay-like weed and recovered about 1 million gallons (3.8 million l) of oil. The French used chalk mixed with stearic acid, to treat the oil on their beaches. More than 3000 French troops treated 5 million gallons (18.9 million l) of oil. Both of these efforts took many weeks to complete.

To clean the slick at sea, skimmers pumps collected floating oil and 40 ships released 10 000 tons (9072 mt) of detergents to disperse the oil. The detergents broke up the floating oil fairly well but were extremely toxic to marine life. Studies later indicated that the dispersants were more ecological devastating than the oil. Marginally impacted coastal rocks and beaches recovered in five to eight years. Areas impacted by dispersant took 9–10 years to recover to their original ecological state. Areas where dispersants were heavily and repeatedly applied have yet to fully recover.

The spill also killed about 200 000 birds. Oiled birds picked up from beaches had a 1% survival rate. Volunteers washed these birds with detergent to remove the oil but it also removed the natural oils that allow seabirds to repel water and retain body heat. As the birds preened, they consumed detergent, which poisoned them.

12.4 Pollution of the Continental Shelf

The continental shelves are the most productive areas of the oceans. The depth of the water is shallow enough that the photic zone through which sunlight penetrates reaches the seafloor. Animals and plants that depend on sunlight must live within the photic zone. In the deep ocean, they can only live at the top of the water as floaters and swimmers. On the shelves, plants and animals can live throughout the water column and on the seafloor. Ocean-bearing natural resources such as oil and gas, minerals, fish, and other seafood are primarily from the shelves. These resources on the shelves are essential for human survival. However, overfishing and pollution has so damaged the shelves that they likely cannot recover without major intervention.

Human impacts on ocean shelf ecology has been devastating (Figure 12.12). In addition to fishing so extensively that it caused the extinction of many species, humans have used the shelves as dumping grounds for waste, munitions, and unwanted debris. Garbage scows loaded with refuse have been towed offshore and dumped in the ocean for many decades. Millions of tons of garbage and construction debris are now on the ocean floor. The military has dumped ordinance, weaponized gas, and even nuclear waste on the shelves. Beaches in Europe and eastern United States are closed or restricted because live ammunition has floated onto them. The military and military contractors in Europe, the former Soviet Union, and the United States have dumped untold numbers of barrels of nerve agent with no records of disposal locations. Countries have also tested bombs and rockets over the ocean, causing phosphorous pollution.

Accidental and sometimes intentional release of refined and unrefined petroleum products has also degraded the shelves. Oil to be delivered to refineries is often transported into ports in tankers. Oil tankers rarely leak in the open ocean, most accidents and spills are in shallow water near ports. Oil can be degraded by bacteria but it is very toxic to fish and shore birds. The oil spilled in oceans by accidents is staggering, and it devastates the ecology of the shelves. The average amount of oil spilled is 42.7 million gallons (162 million l) per year but megaspills can be 100 to 300 million gallons (380 million to 1140 million l). Years with megaspills in decreasing order of volume are 1979, 1991, 1983, 1978, and 1980. In addition to transportation accidents, oil drilling activities also cause leaks and spills. Most oil is produced from the continental shelves and leaks from drilling and production facilities are common.

Tankers were built larger to carry larger amounts of oil, increasing the impacts of accidents. The first supertankers, were built in the late 1960s, and the first accident with a supertanker was the Torrey Canyon, as described. Eleven years later on 16 March 1978, the Amoco Cadiz crashed in the English Channel and spilled 66.4 million gallons (210 000 mt) of oil. Strong winds and a high tide spread the spill along 180 miles (300 km) of French coast. At least 3000 birds died, and fishing and oyster farming suffered greatly.

The United States has also experienced tanker accidents. The March 1989 Exxon Valdez accident is the most famous but the volume of the spill was much smaller than others. The tanker hit a rocky shoreline in Prince William Sound that tore a gash in the hull and spilled 35 000 tons (31 818 mt) of oil. The remoteness of area, the strong tides, rugged coastline, and severe weather greatly exacerbated the impact of the spill. The oil covered about 900 square miles (2300 km²) and caused extensive damage to the fragile ecosystem. By 1992, 50% of the spill had degraded naturally, 20% evaporated, 13% settled into the sediments, and just 14% was cleaned up.

The ability to clean up oil spills or to determine the best method is still poor. Only 8–15% of oil spills are successfully cleaned up. The best methods for containment and removal of oil slicks are mechanical such as oil booms and oil skimmers. Booms are floating inflated rubber tubes that surround the spill and limit spreading. The most commonly used skimmer is a catamaran with a conveyor belt collector. The oil is gathered by the belt and it is carried to storage tanks. Skimmers are only effective in protected areas if the weather is good.

Onshore cleanup methods are very damaging to the environment. The spill is devastating to marine life and birds. Oil flattens feathers and reduces its resistance to cold. Oil clogs filter feeders and fish gills slowly suffocating them. Cleanup teams apply dispersants and detergents that commonly kill more wildlife than the oil. Hot water is sprayed onto beaches with hoses at high pressures (100 psi or 6.8 bars), streaming barnacles, snails, and other animals off of the rocks. It causes erosion of the beach and forces oil deeply into the sediments where it cannot evaporate. Another method is spreading hay on the beach which soaks up the oil. These procedures involve heavy equipment, hoses, and many people further damaging the area.

CASE STUDY 12.2 Deepwater Horizon Oil Spill, Gulf of Mexico

Deep Water Oil Spill

On 20 April 2010, the semi-submersible drilling rig, Deepwater Horizon, had an uncontrolled and catastrophic pressure release or "blowout" while drilling a well in the Gulf of Mexico, 40 miles (64 km) south of the Louisiana coast (Figure 12.14a). The explosion and fire killed 11 people, injured 17, and was the worst offshore oil spill in history. The drill rig was operated by Transocean Ltd., one of the largest oil drilling companies in the world, for BP. Over the following three months, as much as 2.5 million gallons (9.5 million l) of oil spilled into the Gulf of Mexico daily. This blowout is regarded as the most severe environmental disaster to occur in the United States.

The northern Gulf of Mexico has been the greatest source of oil and gas for the United States. Total hydrocarbons are an estimated 4300 billion barrels (684 trillion l) of petroleum and 170 000 billion cubic yards (130 000 billion m³) of natural gas in the offshore gulf. More than 60 000 people are employed for exploration and production operations. It still produces about 5% of the natural gas and 16–17% of oil in the United States.

In the spring of 2010, a new offshore drilling rig, named *Deepwater Horizon*, was moved into position. The *Deepwater Horizon* was drilling an exploratory oil well in a promising area on the Gulf Coast's continental slope. In 2008, BP (formerly British Petroleum) leased the right to drill for oil in the area code-named the "Macondo prospect." BP estimated that the Macondo prospect could hold as much as 50 million barrels (8 billion l) of recoverable oil and significant amounts of natural gas.

Once the *Deepwater Horizon* was in position, it was stabilized to begin drilling. Computer-controlled propellers and thrusters were designed to stabilize the rig over its borehole during drilling. *Deepwater Horizon* operated in water depths to 8000 ft (2.4 km) and it drilled as deep as 30 000 ft (9.1 km) into the sea floor. The Gulf of Mexico has an average water depth of approximately 5300 ft (1615 m).

Drilling began with steel pipe, called casing, being lowered to the sea floor and hammered several hundred feet into the sediment. The casing is 36–72 in. (0.9–1.8 m) in diameter depending on the total depth of the well. The drill string of 30–45 ft (9.1–13.7 m) long sections of six-inch (15.2 cm) steel pipe was connected together and attached to a diamond drill bit at the end. The drill string was lowered into the top of the wellhead one mile (1.6 km) below the water's surface. A powerful engine drove a turntable on the platform that turned the drill string, driving it deeper into the seafloor. Sections of drill pipe were added as the well progressed deeper. During drilling, a mixture of water, barium, and clay, called drilling fluid or mud, is circulated through the drill pipes in the borehole. Drilling mud cools the drill bit and flushes out ground up sediment, called cuttings, to the top of the well. The mud balances the pressures in the rock that is being drilled through. Oil and gas are under tremendous pressure underground and can rise up through the borehole, even explosively, if encountered unexpectedly. As the borehole is drilled deeper, the drill string is withdrawn and another length of 22–36 in. (55.9–90 cm) casing is linked and set into place using quick-setting cement.

Once casing is set into place, a blowout preventor (BOP) is attached to the top of the casing. The BOP is a sophisticated set of valves, and backup systems, that shut off the well to prevent high-pressure escape of oil and gas from the well. It prevents uncontrolled release of oil or natural gas called a blowout. These are extremely dangerous and have killed many oilfield and offshore platform workers.

On 20 April 2010, the Macondo prospect well was near completion, drilling at 18 000 ft (5.5 km) depth. The cementing of casing was finished and the installation of a production pipe to bring the oil and natural gas to the surface was in planning. Drilling was five weeks behind schedule. At 9:45, the BOP catastrophically failed and methane rushed out of the well and surrounded the rig. Eleven minutes later the platform exploded in flames (Figure 12.14b). The 126 crew members abandoned the rig in lifeboats to a service ship. Helicopters transported 17 injured survivors to hospitals. The US Coast Guard could not find 11 missing crewmembers, despite a three-day search. They are presumed dead.

The *Deepwater Horizon* burned for nearly 40 hours and then sank. At the time of the disaster, BP and Transocean executives were touring the rig but were not injured. BP tried to downplay the possibility of major ecological damage from the blowout. They informed the press and Coast Guard that little to no oil was leaking from the wellhead. However, using BP data, the Coast Guard calculated that oil was spilling from the uncapped well at about 1000 barrels (159 000 l) per day. This number was a gross underestimate because by early July, experts estimated that 1–2.5 million barrels (159–398 million l) of oil was spilling into the gulf from the well daily. The reason for the accident appears to have been improper cementation of casing.

The response to the disaster was slow and disorganized. Booms to contain the spill were delayed and skimming was spotty and ineffective. It took several weeks for the Coast Guard to establish a unified response. Under the oversight of the Coast Guard, BP implemented a comprehensive approach to the *Deepwater Horizon* disaster. Techniques used included capping the well with two separate bells to contain the flow, both of which failed. Mud and cement were pumped into the well to flood the wellhead but that failed as well (Figure 12.14c). Finally, a relief well was drilled into the main well to relieve the pressure and slow the flow. This allowed the well to be repaired and

(a)

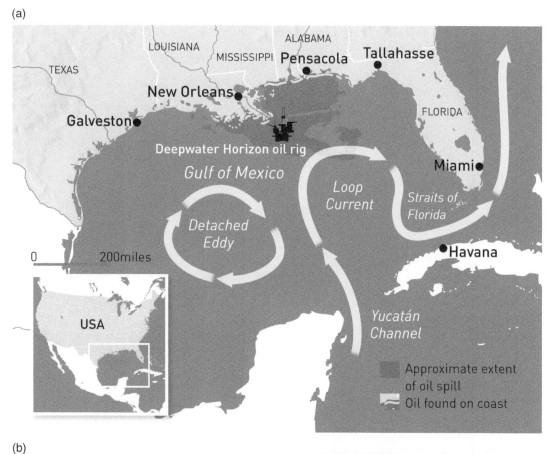

(b)

FIGURE 12.14   (a) Map of the oil spill, slick, and impacted beaches from the Deepwater Horizon accident. Inset map shows the location of the area in North America. (b) Photo of pumper tugboats spraying water on the burning Deepwater Horizon oil drilling rig. Courtesy of the US Coast Guard/Marine Photobank. (c) Diagram showing efforts to stop the flow of oil from the damaged Deepwater Horizon well.

(c)

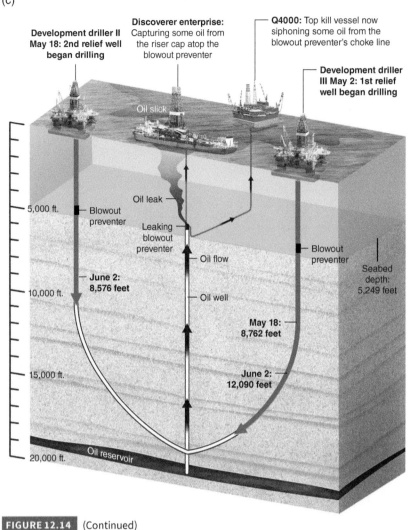

FIGURE 12.14 (Continued)

capped, finally sealing it on 19 September 2010. However, the spill had already resulted in an estimated discharge of 4.9 million barrels (210 million gal; 780 000 m³).

Even before the well was controlled, a program to protect Gulf Coast beaches, wetlands, and estuaries from the spreading oil had begun. This effort included using skimmer ships to collect oil, booms to restrict the spread, controlled burns of oil and gas around the well site and spraying of 1 840 000 gal (7000 m³) of oil dispersant. This slowed the spread significantly but the spill had spread for months and still caused extensive environmental damage. Even cleanup activities such as spreading chemical dispersant had adverse effects. More than 2450 tons (2223 mt) of oil and tar was removed from Louisiana

beaches in 2013 alone. Oil floated ashore as far away as the Florida Panhandle and Tampa Bay, where oil and dispersant stained the sand. Gulf Coast dolphins and other marine life died in record numbers and newborn dolphins died at six times the normal rate. Tuna and amberjack that were exposed to oil developed deformities of the heart and other organs.

The EPA temporarily banned BP from new contracts with the United States. In September 2014, BP was found primarily responsible for the oil spill as a result of negligence and reckless conduct. In April 2016, the company was ordered to pay $20.8 billion in fines, making it the costliest corporate settlement in United States history. By 2018, the cleanup costs and fines had cost BP more than $65 billion.

A lot of pollution is delivered by rivers to the shelves. Agricultural runoff in rivers is one source. These waters contain pesticides that can also be toxic to marine invertebrates, as well as fertilizers. Previously, the ocean was considered to be able to absorb all human waste, but this is not true. The fertilizers that rivers deliver to the ocean produces "dead zones" with hypoxic conditions. The fertilizer stimulates massive algal blooms and bacteria feed on the dying algae. The bacteria deplete oxygen in the water. The low dissolved oxygen level kills the sessile animals and forces the mobile animals to flee. The largest dead zone is in the Gulf of Mexico but many other dead zones are appearing and expanding.

Another source of pollution is air pollution fallout and washout through precipitation. Industry and transportation in coastal areas produces large amounts of air pollutants that are swept out to sea by winds. Considering the number of motor vehicles in coastal areas, the amount of fallout into the ocean is significant. The particulate and sulfur from coal burning and other industrial sources is also significant. These problems are especially rampant along coastal United States, Europe, and the east coast of China.

12.5 # 12.5 Ocean Dumping

Humans have also used the ocean as a dumping ground for many decades to centuries. Ocean dumping is the discarding of waste from a barge or ship on the ocean floor (Figure 12.12). During the 1970s and 1980s alone, the United States and other countries intentionally dumped an estimated 25 million tons (22.7 million mt) of industrial waste. The seemingly limitless expanses of ocean, remote from people, seemed the perfect place to dump sewage, scrap metals, mine tailings, radioactive waste, pesticides, among other waste. There was also no scientific information on impact on ecosystems that could exist on the deep seafloor.

CASE STUDY 12.3 Beaufort Dyke, Irish Sea, Scotland

Ocean Dumping

Beaufort's Dyke is a 31-mile (50-km) long and 2.5-mile (4 km) wide submarine trench in the North Channel of the Irish Sea, between Northern Ireland and Scotland. It is offshore of the Scottish port of Cairnryan, where thousands of British troops were based both during and after World War II. From this port, obsolete and excess munitions were loaded onto barges and dumped into Beaufort's Dyke in 984 ft (300 m) of water depth (Figure 12.15a). The great depth, closeness to the coast, and that it did not serve as a fishery made it the most appropriate place.

Ocean dumping of surplus munitions was a common practice because it was considered the most safe, efficient, and cost-effective method to dispose of these dangerous materials. After the world wars and until 1976, about 2 million tons (1.8 million mt) of torpedoes, rockets, bombs, grenades, bullets, and explosives were dumped from barges into Beaufort Dyke (Figure 12.15b). They also dumped canisters of sarin and tabun nerve gas, 120 000 tons (109 000 mt) of mustard and phosgene gas, cyanide, 330 tons (300 mt) of arsenic compounds, anthrax, phosphorus bombs, and radioactive waste containing cesium 137 and radium 226. Most of the explosives are at depths greater than 328 ft (100 m). However, some ships dump the munitions much closer to shore rather than make the long trip.

The Irish Sea is a commercial waterway linking Scotland and England to Ireland. The vast majority of goods are moved by ship between these areas. The Irish Sea also contains pipelines, electrical, telephone, and other cables that provide power, gas, and communications.

The stability and safety of the munitions in Beaufort Dyke became an issue when reports of underwater explosions in the area began in 1966. Seismograph records documented almost 50 explosions and concluded that it was very likely that many more undetected detonations had taken place. In 1969, the crew of a fishing trawler hauled up canisters of Eperite, a type of mustard gas or blister agent, in their nets. They had been bottom fishing well outside the 12-miles (19.2-km) exclusion zone surrounding Beaufort Dyke. Two of the crew members were so heavily exposed that their hair and skin began to peel and their urine burned their legs.

In 1995, British Gas began construction of a Scottish-Northern Ireland Pipeline using a 24-in. (61-cm) diameter submerged pipe across the Irish Sea. Submarine cables and pipelines are encased in layers of protective material in a flexible, metal membrane. To lay the pipeline, a barge loaded with spools of pipe is stationed on land at the starting point and the seaward end is attached to a marine plow. The plow is towed behind a ship and digs an 8 ft (2.5 m) trench in the seabed while feeding the pipe into the trench. The soft sediments fill or collapse into the

(a)

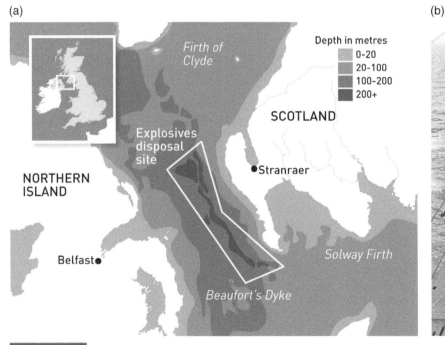

(b)

FIGURE 12.15 (a) Map showing the location of the Beaufort Dyke between Scotland and Northern Ireland. (b) Photo of sailors dumping military waste into Beaufort's Dyke. *Source:* History Department at the University of Saskatchewan / CC BY-ND 4.0.

trench around the pipeline, protecting it from ship anchors and fishing tackle. The placement of rocks or concrete mattresses and other cable protection techniques are commonly used closer to shorelines but open or unprotected cable laying is done in deeper ocean water.

British Gas installed the pipeline following the environmental and routing requirements and regulations. One of the restrictions was to avoid the Beaufort Dyke. Shortly after the pipeline was installed, approximately 4500 munitions, mostly phosphorous flares, started washing up on the Irish and Scottish coastlines. It became common for the beaches to be lit at night by burning flares. A child and an adult were seriously injured by picking up the metal tubes. The Royal Navy concluded that pipeline construction disturbed a pile of munitions that had been dumped outside of the Beaufort Dyke. Demolition teams cleared the beaches of these flares.

The attitude toward dumping changed as deep ocean exploration began in the early 1970s. In 1982, Woods Hole Oceanographic Institute, identified 800 species of sea life at depths of 6000 ft (1.8 km) or more off the coast of New Jersey and Delaware. This biodiversity was even greater to the south in warmer waters. It is estimated that more than 10 million species may be present on each ½ square mile (1.3 km²) of ocean floor, at depths of 3000 ft (0.9 km) or more. Ocean dumping can disrupt these fragile, deep sea ecosystems, significantly damaging marine life and ultimately damaging fisheries.

Trash, industrial waste, and sewage sludge can contain pesticides and heavy metals like lead, copper, and cadmium that can wind up in the food chain. Through bioaccumulation of these contaminants, fish and sea mammals can be poisoned, reducing their reproduction and/or making them inedible. Bacteria, viruses, and disease-causing organisms disposed of at sea have led to fish being infected with illnesses such as black gill, or bacterial gill disease. The infection reduces the respiratory efficiency of gills, eventually killing the fish.

Restrictions on ocean dumping began in 1972 in the United States. Permits were finally required for ocean disposal even for activities of minimal environmental impact, such as burials at sea and disposal of vessels. Special permits were required to dispose of waste into the ocean or to transport waste from the United States for ocean dumping. Material used for biological warfare or that was radioactive could not be disposed of at sea at all. In 1988, a new act added sewage sludge, industrial waste, and medical waste to the list of banned items. The only exceptions were to protect the health and safety of a crew, or because of war or national emergency. As a result of these regulations, the amount of sewage sludge dumping was reduced from 18 million tons (16.3 million mt) in 1980 to about 12 million tons (10.9 million mt) for all of the 1990s. The rest of the world also reduced legal ocean dumping at this time. For example, in November 1993, more than 80 countries participated in the "London Dumping Convention," which banned the disposal radioactive waste from ships.

Discarded waste materials into the ocean are classified as black list, gray list, or white list. Black list includes chemicals like organohalogens, cadmium, mercury, plastic, petroleum products, highly radioactive substances, and chemical or biological warfare agents. Gray list includes arsenic, lead, copper, zinc, organosilicon-contaminated liquids, cyanide, fluoride, pesticides, acids and bases, certain metals, such as chromium, beryllium, nickel compounds, and vanadium, scrap metal, and low-level radioactive waste, among others. White list items include all materials not included on the black or gray lists. They can be safely discarded if it is not done in vulnerable areas, such as coral reefs.

CASE STUDY 12.4 "Ecomafia" in Italy and Somalia

Criminal Ocean Dumping

The institution of laws restricting and against ocean dumping began a new practice of criminal ocean dumping. As the entire coastline cannot be monitored, it was relatively easy to load and ship waste of all types without being detected. As a result, there is no way to know how much criminal ocean dumping was or is taking place. However, there were a few pieces of evidence emerged over the years that crime was occurring.

The ship *Rosso* washed ashore in December 1990 near Amantea, Italy and the police deemed it a failed attempt to sink it (Figure 12.16). No cargo was found but there was evidence that it was unloaded and buried on land. In 1994, an Italian television journalist and her cameraman were shot and killed near Mogadishu, Somalia while following the trail of a hazardous waste dumping ring. In 1995, a parliamentary commission concluded the potential existence of national and international radioactive waste trafficking that includes both business and criminal elements throughout Europe. It also reported threats against investigators.

The big break in the investigation came in 2005 when an Italian newspaper published an interview with a former member of the 'Ndrangheta, a criminal organization from Calabria, Italy. He reported that they had been involved in offshore dumping of radioactive waste since 1979. In the 1980s and 1990s, as many as 45 ships had been loaded with toxic and radioactive waste, sailed offshore of the Italian coast and sunk. He claimed to have personally sunk a ship with 120 barrels of toxic and radioactive waste in 1992 for $162 720. The article led to several widespread investigations into toxic and radioactive waste disposal rackets, several involving otherwise legitimate disposal companies. The term "ecomafia" was coined for these criminal groups.

The trafficking was reported to reach high levels in companies and even in an Italian government-sponsored research and development agency. Reportedly, they paid the criminal organization to dispose of 600 drums of toxic and radioactive waste from Italy, Switzerland, France, Germany, and the United States. The waste was shipped to Somalia where it was buried on land after bribing local politicians. This practice is reportedly to have occurred throughout

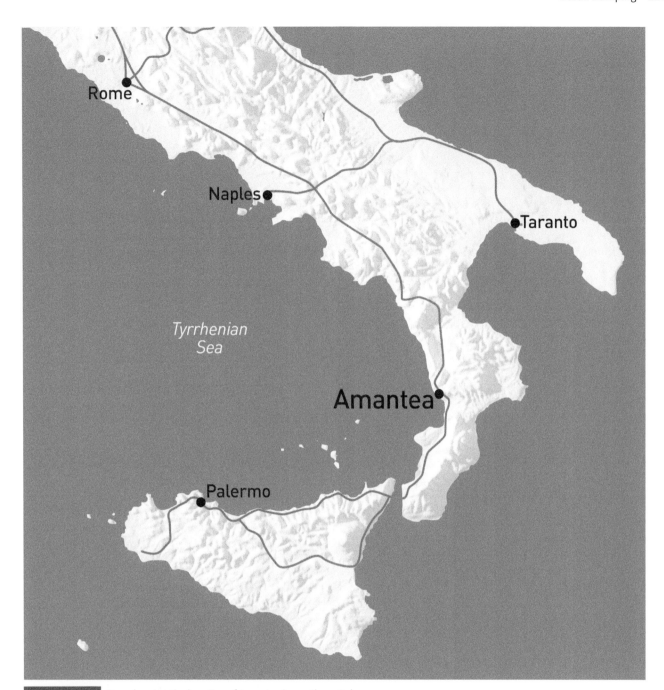

FIGURE 12.16 Map showing the location of Amantea in southern Italy.

the 1980s and much of the 1990s and included other countries such as Kenya and Zaire. The Prime Minister of Italy was even suggested as having been complicit.

The 2004 Indian Ocean tsunami is believed to have unearthed illegally dumped toxic and nuclear waste in northern Somalia, including heavy metals such as lead, cadmium, and mercury. There were other locations that also tested positive for hazardous waste and residents were becoming ill from it.

As a result of the report, in April 2007, Calabrian authorities halted fishing in the water off Cetraro because there are high levels of heavy metals in the sediments. In the area around Amantea, there was also an increase in cancer deaths between 1992 and 2001. Hospitalizations for some malignancies have risen in recent years. In 2009, a major environmental group released their Ecomafia Report, in which they estimated that the total missing toxic waste would pile into a mountain 10 171 ft (3100 m) high. It further concluded that Italy had produced 33 million tons (30 million mt) more industrial waste than was legally disposed of in 2008 alone. The problem is that there is no way to prove any of these claims or to find the missing waste with current technology.

Weather and Air Pollution

CHAPTER OUTLINE

Polluted Earth: The Science of the Earth's Environment, First Edition. Alexander Gates.
© 2023 John Wiley & Sons, Inc. Published 2023 by John Wiley & Sons, Inc.
Companion website: www.wiley.com/go/gates/pollutedearth

Words you should know:

Acid precipitation – Precipitation and fallout of acid and acidic rain and ice from air pollution to the surface.

Air masses – A large package of air in the atmosphere that takes on the physical characteristics of the area where it develops.

Climate – The long-term weather patterns of an area.

Coriolis Effect – Deflection of objects and sense of rotation generated by the rotating of the Earth and the resulting relative speed by latitude.

Criteria air pollutants – Air pollutants of greatest EPA concern including carbon monoxide, ground-level ozone, lead, nitrogen dioxide, particulate matter, and sulfur dioxide.

Fronts – The boundaries between two air masses.

Particulate – Solid particles and liquid droplets that are suspended in air. PM10 are inhalable particles that are 10–2.5 micrometers. PM2.5 are very small inhalable particles that are 2.5 micrometers or less.

Precipitation – The condensation of water vapor into droplets and the falling of the water to the earth in any state.

Primary air pollutants – Air pollutants that are directly released from their source without modification by chemical reactions in the atmosphere.

Secondary air pollutants: Air pollutants formed in the atmosphere through chemical reactions among primary air pollutants.

Smog – A mixed term from smoke and fog: any low-hanging pollution of the air but primarily applied to air pollution of urban areas, especially involving build-up of ground-level ozone.

Temperature inversion – An unusual situation where a layer in the atmosphere reverses temperature gradient sealing the lower atmosphere and allowing pollutants to accumulate.

13.1 Weather Basics

The weather conditions and patterns are primarily controlled by the following factors: vertical temperature gradient, spatial temperature distribution, moisture capacity of air, and the rotation of the Earth. Humidity is the amount of moisture contained as vapor in air. Warmer air holds more moisture than cooler air. If warm moist air is cooled enough, the moisture condenses into water droplets or ice if it is cold. Condensation can be illustrated by a cold drink on a humid summer day. The outside of the glass develops beads of sweat (water) as the moisture in the warm moist air condenses on the cold glass. The fogging of eyeglasses when a person walks from the cold outdoors into a warm room reflects the same process.

In the troposphere, temperature decreases with increasing elevation (Figure 13.1). That is why mountains can be snowcapped while the lower portions are not. Based upon warm air's ability to hold more moisture, there is also less moisture in the air at higher elevations. Temperatures are generally distributed geographically on the Earth as well. The incoming solar radiation strikes the Earth at right angles near the equator so is concentrated in comparison to the poles where the angle of incidence is much lower. This is why the temperature is warmer near the equator and colder near the poles. The humidity follows this temperature distribution as well.

The barometric or air pressure is the weight of the air above you, held to the Earth by gravity. It exerts about 14 lbs per square inch, 30 in. of mercury, or 1 bar of pressure on everyone at sea level. The air pressure also varies by elevation. People are used to the pressure and do not notice it unless they change elevation. If a person travels from sea level to the mountains, they notice that the air is "thin" and may have difficulty breathing. Movement of air impacts the pressure. Air that is sinking toward the surface increases the force of the air and produces high pressure. In contrast, rising air decreases the force producing low pressure. The upward movement of air in low pressure carries warmer, moister air from low elevations to high elevations where it is colder. This causes the moisture in the air to condense and precipitation occurs, which is why it rains in low pressure systems. In high pressure, the downward movement of air carries colder, drier air to lower altitudes where it is warmer and can hold more moisture, and clear skies prevail.

The Coriolis Effect (Figure 13.2) is the other major factor controlling weather and results from the Earth rotating. From a fixed vantage point in space, the speed of rotation depends upon position on the planet. A person at the equator travels around the widest part of the Earth or 24 000 miles (38 624 km) in one 24-hour day, which is a velocity of 1000 mph (1609 kph). On the other hand, a person would just spin around one full turn in 24 hours at the pole, which is essentially 0 mph. Locations on the Earth closer to the equator always move faster than those poleward. This difference deflects southward-moving airborne objects to the west and causes weather systems, high or low pressure, to rotate.

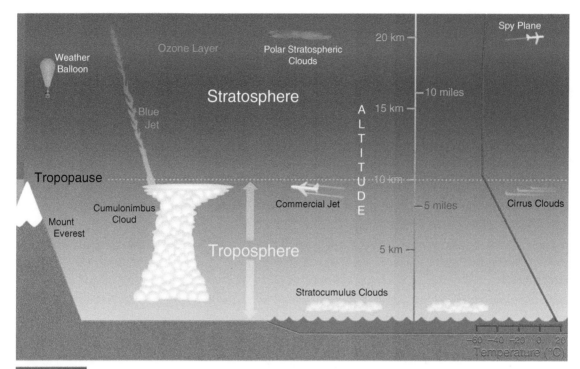

FIGURE 13.1 Diagram showing the decrease in temperature with height through the troposphere.

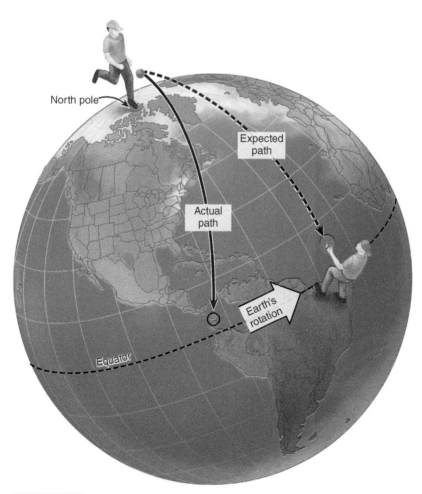

FIGURE 13.2 Diagram showing the straight path of a thrown ball with a non-rotating Earth and the curved path of a thrown ball with a rotating Earth, illustrating the Coriolis Effect.

13.2 Climate Systems of the Earth

These factors not only control local weather but also large-scale weather patterns known as climate. The main drivers of climate are the rising of heated and consequently less dense air at the equator and the sinking of cooler and denser air at the poles. This creates a conveyor belt effect, known as a convection cell, in which the rising air at the equator moves poleward at high altitude and the sinking air at the poles moves toward the equator along the surface (Figure 13.3).

This simple convection system is modified by the Coriolis Effect, which twists the flow and breaks the single convection cell into three separate convection cells per hemisphere called the Hadley, Ferrel (Mid-latitude), and Polar cells poleward (Figure 13.4). The

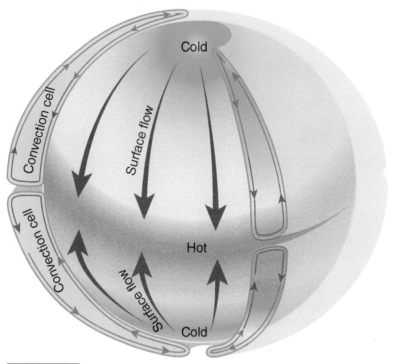

FIGURE 13.3 Diagram showing the simplified circulation of the atmosphere in a non-rotating Earth with a single convection cell from equator to pole.

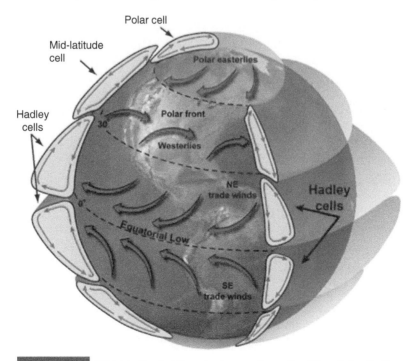

FIGURE 13.4 Diagram showing the actual convection cells of the rotating Earth with the Hadley cells closest to the equator and the major wind patterns shown.

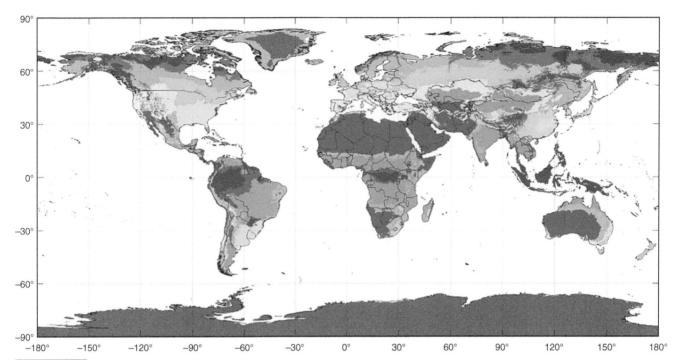

FIGURE 13.5 Map of the Earth showing the major climatic belts. Note that tropical rainforests occur around the equator and the deserts occur at 30° north and south latitude. *Source:* Peel et al. (2007).

rising air at the equator forms low pressure and precipitation, producing the band of tropical rainforests around the world (Figure 13.5). The Hadley cells drive dry air downward to the surface at 30° north and south latitudes, producing high pressure with clear and calm weather. As a result, the deserts of the world, such as the Sahara and Arabian deserts, are concentrated in these latitudes. The next poleward cells yield rising air at 60° north and south latitudes and produce the temperate rainforests like those in the northwestern North America and western Europe. The air falls at the poles, producing a desert. It is not intuitive that the poles are deserts because of the ice and snow, but these accumulations cover thousands of years. The twisting of the convection cells by the Coriolis Effect makes wind patterns along the surface more east–west than north–south, as would be expected in simple circulation (Figure 13.4). The sinking air from high-pressure slides along the surface as wind and lifts back upward in low pressure. These prevailing surface winds are the easterlies, westerlies, and trade winds depending upon location, though small systems also produce surface winds.

In addition to the major climates, local climates and microclimates also develop based on surface geography. Oceans modulate the local weather conditions and especially if they include strong currents. The Gulf Stream is a major ocean current that moves warm water from the Gulf of Mexico to Western Europe (Figure 13.6). It makes these areas warm to temperate conditions even though they should be much cooler based on their latitude. Oceans can also generate local storms that can increase the precipitation above that dictated by the climate belts. In contrast, an abundance of landmasses can reduce the amount of available water to evaporate into the atmosphere, making areas that should be rainy much drier than that dictated by climate model. For example, Siberia is within the northern-temperate rainforest zone but dry because there is so much landmass. Geography can also change the climate. In the Northwestern United States, moist air from the Pacific Ocean is driven up the west side of the Cascade Mountains (Figure 13.7) which is equivalent to lifting it in a low-pressure system. This produces heavy precipitation know as orographic precipitation.

Glaciation is an extreme climate type that results from the accumulation of ice on a long-term basis. The precipitation falls as snow and is compacted into glacial ice. There are two types of glaciation, continental and alpine, both of which involve the movement of ice bodies. Both types entrain rock, soil, and other debris, and form distinctive deposits. However, alpine is restricted to the mountains, and continental is located on relatively flat land in polar regions. It only extends away from those areas during ice ages.

13.3   Air Masses and the Fronts that Separate Them

A mass or package of air takes on the characteristics of an area if it remains for a sufficient period. For example, if an air mass develops over land in Canada it will be cold and dry most of the year. If it develops over the Gulf of Mexico, it will be warm and moist. Over the North Pacific or North Atlantic, it will be cold and moist most of the year. These air masses then move and interact with air masses from other areas. The boundary where two of these air masses meet is a front. There are cold fronts, warm fronts, and occluded fronts depending upon the types colliding of air masses.

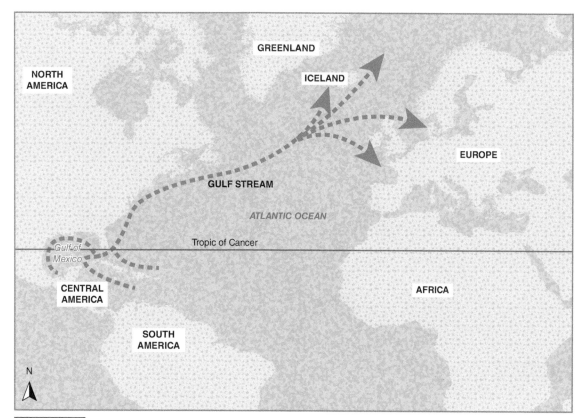

FIGURE 13.6 Map of the North Atlantic Ocean basin showing the Gulf Stream ocean current.

FIGURE 13.7 Diagram showing the orographic effect of moist rising up the front side of a mountain causing precipitation and sinking down the back side of the mountain as dry air.

13.3.1 Cold Front

The passage of a cold front is marked by the replacement of warm air with cold air in an area. Cold air masses largely move from northwest to southeast in North America or inland from the Pacific Ocean and less commonly inland from the Atlantic Ocean. Cold air masses from the north and northwest are drier than the warm air masses they contact. Cold air is denser than warm air and is driven under the warm air at the front (Figure 13.8). The front is steeply sloped, which makes the weather at the cold front intense and short-lived.

Typical cold fronts are marked by the appearance of thick cumulus clouds on a calm and humid day. They are most common in the summer. The thick, tall clouds build up quickly, with abrupt heavy rain and strong wind (Figure 13.8). Thunderstorms are common and tornadoes are possible along these fronts in extreme situations. The intense weather passes quickly and skies clear

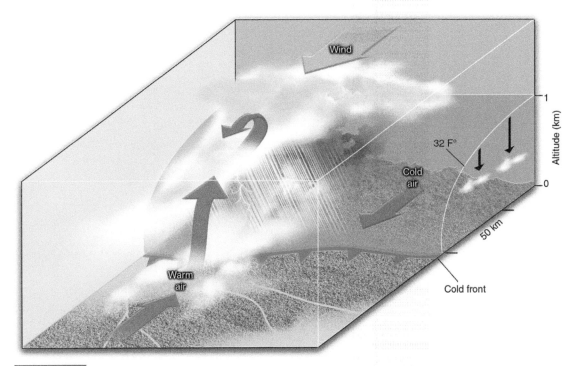

FIGURE 13.8 Block diagram of a cold front showing the front moving from back (right) to front (left).

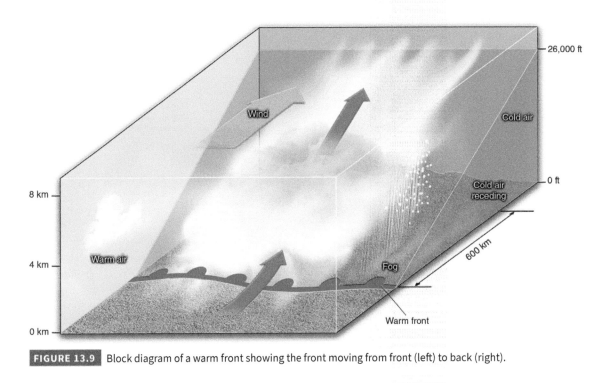

FIGURE 13.9 Block diagram of a warm front showing the front moving from front (left) to back (right).

once the front is through. The humidity and the temperature drop abruptly, though the wind may remain for a few days. In cold, wet air masses from oceans, there is less of a drop in humidity after passage of the front and fog may develop with cold, wet conditions.

13.3.2 Warm Front

Warm air replaces cold air during the passage of a warm front. If a warm air mass moves northward from the Gulf of Mexico into the United States, it is moist. If the mass moves northward from the southwestern United States, it is dry. Warm air is less dense than cold air, so it rises above the cold air at the ground surface (Figure 13.9). The leading edge of the warm air mass forms a wedge above the trailing edge of the cold air. This low angle causes the passage of the warm front to take longer than a cold front.

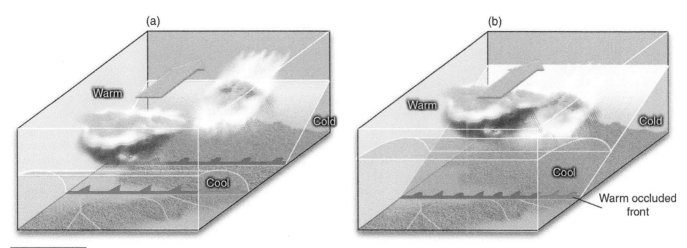

FIGURE 13.10 Block diagrams of (a) a cold front moving from front (left) to back (right) and overtaking a warm front to create (b) an occluded front.

The passage of a warm front with a humid air mass generally starts with a thin, high-level haze. Sun dogs or partial rainbows on either side of the sun can occur during the day and a halo may occur around a full moon at night. After several hours, thin, high-stratus clouds slowly build in thickness and lower altitude. Rain starts as drizzle but slowly intensifies to steady rain that lasts up to several days, clearing slowly.

13.3.3 Occluded Front

A rapidly advancing cold front can overtake a slower leading warm front. If these two fronts overlap, they form an occluded front. In this case, the steep-faced cold front overrides the shallow, wedge-shaped warm front (Figure 13.10). Dense, cold air sinks to the ground surface, forcing the wedge of lighter warm air above. At the surface, the result is that only the two cold air masses are felt. The temperature remains relatively unchanged with the passage of the front.

Occluded fronts produce clouds and precipitation that reflect both cold and warm fronts. As the front approaches, weather resembles a warm front with slow building of clouds and rain. The weather may then intensify when the steep-sloped cold front passes before the rain ends. Occluded fronts tend to move slowly and inclement conditions can persist for several days.

13.4 │ Air Pollutants

Input of chemical, aerosol, or particulate in the air that is not a significant component of everyday conditions is air pollution. This includes a vast numbers of pollutant types and chemicals. There are usually vast amounts of contaminants in the air but it is rare that they reach unhealthy proportions in most of the world. Ever since the start of the Industrial Revolution, the amounts and varieties of air pollutants have increased dramatically until after 1970 when the Clean Air Act was signed into law in the United States. Through wind that disperses pollutants and precipitation that washes it from air, the vigorous weather systems of the Earth can disperse considerable amounts of air pollutants and despite the number of seriously polluted urban areas, there would be many more air pollution disasters without vigorous weather.

CASE STUDY 13.1 1980 Chemical Control Corporation

Weather Prevents a Disaster

Chemical Control Corporation was a middleman company that removed hazardous waste from manufacturing companies and safely disposed of it. Between 1970 and 1978, the company removed wastes for processing and/or storage prior to shipping it to disposal sites. The facility was on a 2-acre (0.8 ha) lot in Elizabeth, New Jersey

(Figure 13.11a). The company accepted a variety of wastes including acids, arsenic, bases, cyanide, solvents, PCBs, compressed gas, biological waste, and pesticides among others. The owners were effective at receiving waste but not at processing it. The facility was cited for numerous violations for improper storage of waste and for dumping unprocessed waste in various places around Elizabeth. The situation became so bad that the State of New Jersey closed down the company in January 1979.

Over the next 15 months, the New Jersey Department of Environmental Protection (NJDEP) dealt with the situation. They drained and disposed of hazardous liquids from the seven above-ground storage tanks and identified and processed 10 000 of the more than 60 000 drums of waste in the warehouses and stacked around the property, some in piles six drums high (Figure 13.11b). This occurred before the US Environmental Protection Agency (EPA) Superfund program, so New Jersey was forced to remediate the site on their own. By April 1980, most of the loose solids and liquids that were not in containers had been removed as had much of the radioactive waste, explosive liquids, and infectious wastes but the company did not have a permit for much of it and NJDEP was preparing charges against the owners

Late in the evening on 21 April 1980, however, a fire spread quickly from a warehouse on the property that held more than 20 000 drums of waste (Figure 13.11c). Emergency crews arrived, to find flames shooting hundreds of feet into the air and drums filled with hazardous substances exploding and launching others into the air. Firefighters without respiratory protection were overcome by fumes and nearly 70 people had to be hospitalized with symptoms from difficulty breathing to skin rashes.

Thousands of residents of Elizabeth and nearby Staten Island, NY, were awakened by the explosions and rushed outside to see brightly glowing red, green, and yellow flames depending upon the chemical that was burning at the time. While the firefighters and emergency crews worked to control the four-alarm blaze, officials from NJDEP and New York Police Emergency Control Board worked late into the night to

(a)

(b)

FIGURE 13.11 (a) Map of the Elizabeth, NJ area. (b) Newspaper clipping showing the stacked drums of waste at Chemical Control Corporation property *Source:* Newspapers.com. (c) Photo of the massive fire caused by arson at the Chemical Control Corporation property. *Source:* Courtesy of US EPA.

(c)

Chemical
Wastes
Exploding!

FIGURE 13.11   (Continued)

assemble evacuation plans for the residents of Elizabeth and New York City Borough of Staten Island. If the cloud kept moving east, they might even need to evacuate parts of Brooklyn with its 5 million residents. Several hundred thousand people were being scheduled to be removed from hospitals, nursing homes, apartments, and houses when, luckily, the wind shifted and pushed the mushroom cloud of contaminants that had formed above the fire southeastward and out to sea.

Although disastrous, the situation could have been worse. In addition to the help from the weather, several months earlier, NJDEP had removed the more hazardous chemicals, including 500 pounds (226.8 kg) of TNT and numerous drums of benzene, picric acid, and radioactive waste. This action lessened the severity of the fire and reduced the toxicity of the smoke and ash. In addition, the fire was so intense that it incinerated many chemicals, which made them much less hazardous. However, the smoke and ash covered 15 square miles (38.9 km^2), forcing the closing of schools in Elizabeth, parts of Staten Island, and several other nearby towns. Residents were urged to remain indoors for several days after the fire, not to let children play outdoors, and to wash down cars and lawn furniture to remove potentially contaminated ash.

As a result of the fire, cleanup efforts accelerated dramatically, making this the largest emergency response action ever conducted by NJDEP. Within six months, the firefighting equipment was decontaminated and the remaining 49 000 drums and 250 000 gal (946 353 l)

of liquid chemicals were removed. Within 18 months of the fire, the drums that fell or been launched into the Elizabeth River, purposefully by the owners or as a result of the fire, were removed. The warehouses were demolished, and sediments of Elizabeth River, contaminated by runoff and firefighting operations, were dredged. It took five years to collect, test, and dispose of more than 200 fire-charred gas cylinders. An earthen berm was constructed to isolate the site from the Elizabeth River, and a chain-link fence was installed.

The cleanup cost $26 million US in 1980 dollars. About 80% of this was recovered from the 200 companies that generated the waste. Most of them had already paid to have it properly disposed of by Chemical Control Corporation. The EPA assumed responsibility for the cleanup in 1981, after NJDEP nominated it for the Superfund program. The cleanup was completed in 1993, but the EPA periodically monitors the site to inspect the effectiveness of the remediation. This monitoring is estimated to cost $60 000 per year until 2040. Criminal charges were filed against the owner of the company and a waste transporter who diverted truckloads of waste to the site, instead of processing it. They were found guilty and sentenced to prison. The owner was a reputed member of the Genovese organized crime family. Newspaper reports and testimony by one of the owner's associates claimed that members of this family started the fire in retaliation for the state takeover of the property and loss of revenue.

The 1970 Clean Air Act identified the six "criteria" air pollutants most in need of reduction including lead, nitrogen oxide, carbon monoxide, ozone, particulate, and sulfur dioxide. Each of them can have minor natural sources but most are from human activities. The worst danger to human health of these pollutants is ground-level ozone which is about exclusively from human activity. Since 1970, most of these pollutants decreased across the United States. In 1990, the US Congress revised the Clean Air Act, making many locations that had achieved attainment back into non-attainment. There were also amendments emphasizing cost-effective air pollution reduction methods. In 2007, carbon dioxide was added as air pollutant because it is a greenhouse gas.

The quality of air in the United States has greatly improved since the 1970 Clean Air Act. Europe, Japan, and other industrialized nations also have cleaner air as the result of government legislation. China had 16 of the 20 most polluted cities in the world in 2010 but through extensive cleanup efforts, China was down to having just one in 2020. Even with these efforts, the global air quality continues to decline. India has become quite polluted, with 22 of the 30 most polluted cites in 2020. Until all countries curtail their pollutants, conditions will not improve.

13.4.1 Primary Versus Secondary Air Pollutants

Air pollutants are classified as primary or secondary pollutants. Primary air pollutants are released directly from a source without modification or chemical reaction with other pollutants. The criteria pollutants are mostly primary air pollutants but ozone is mostly secondary. Particulate is mostly primary. Lead is always primary and carbon dioxide and sulfur dioxide are mostly primary. Nitrogen oxides are either.

Secondary air pollutants are produced by chemical reactions among primary air pollutants in the atmosphere. The reactions are largely photochemical, meaning that they are facilitated by sunlight. As such, secondary pollutants are more common in the summer when the incoming sunlight is stronger and the temperature is higher. They are also more common in warmer climates. The most common secondary air pollutant is ozone. It is produced by a photochemical reaction between nitrogen oxides and volatile organic compounds (VOCs). There are other reactions driven by sunlight that can also produce dangerous pollutants. Some photochemical reactions produce free radical ions that degrade other dangerous air pollutants. There are also some reactions that do not require strong sunlight, like the reaction of sulfur dioxide to sulfuric acid. These involve reaction with moisture in the atmosphere. Smog is composed primarily of secondary air pollutants. Most of the pollution in the southern California cities as well as Denver, Colorado is primarily secondary air pollutants.

CASE STUDY 13.2 Poza Rica, Mexico

Primary Pollutant Disaster

The town of Poza Rica is a center for the oil and gas refining industry in Mexico. Poza Rica sits on a valley floor surrounded by hills rising about 330 ft (100 m) (Figure 13.12). In 1950, most of the 22 000 residents lived in split bamboo houses with large windows and doors to increase air flow for cooling. The refineries and associated sulfur recovery plants were operated by Pemex, the Mexican government oil company. In November 1950, production activities were being expanded and a new

sulfur recovery plant was opened to remove hydrogen sulfide (H_2S) from a natural gas refinery. The gas contained about 3% H_2S. The sulfur recovery plant had two units, each processing 60 000 000 cubic feet (1 699 011 m³) of natural gas per day. However, facilities for the final step in the recovery process had not been completed. In this step, concentrated H_2S gas (16%) would be sent to a reactor to recover sulfur. Instead, H_2S-rich gas was piped to a tower where it was mixed with gas and ignited in a flare 30 ft (9.1 m) high.

FIGURE 13.12 Map of the Poza Rica area, Mexico and the location of the Pemex refineries.

H_2S is a dangerous asphyxiant. It is colorless, flammable and explosive, and heavier than air, allowing it to accumulate in low-lying or poorly ventilated areas. It has a smell of rotten eggs at concentrations as low as 0.5 parts per billion (ppb) but it deadens ability to smell within minutes. This leads exposed people to believe that the gas has dissipated when it has not. In industrial accidents, H_2S gradually displaces the air, killing the workers by suffocation.

H_2S decreases the oxygen-carrying capacity of blood. Low concentrations of H_2S (<300 ppm) cause fatigue, loss of appetite, and headaches after long periods of exposure. Concentrations of 700–800 ppm are usually fatal within five minutes. At 1000+ ppm, exposed people collapse after inhaling just one breath. The EPA established maximum exposure level for H_2S at 0.0014 ppm.

About 90% of H_2S in the atmosphere is from natural sources. The most common industrial source of H_2S is petroleum and natural gas production. It can occur in oil and natural gas from a few parts per million to as high as 30%. H_2S causes metal corrosion and fatigue. As the metal vessels and pipes of a refinery are exposed to H_2S, they become brittle and lose strength and flexibility, eventually failing. H_2S can also react with the steel parts, forming iron sulfides that plug pipes and foul valves. Oil refineries remove H_2S from oil and gas and convert it to elemental sulfur and burn the hydrogen.

The gas flare at the new Poza Rica plant had problems. Amine solution, which helped remove H_2S from the gas, regularly overflowed and extinguished the pilot light of the flare or clogged the gas lines. In response, flow rates were reduced to 400 000 cubic feet (1 132 674 m³) per day, which solved the problem. However, pressure on production staff to increase the plant output was intense. In response, at 2:00 a.m. on 24 November 1950, the facility's operator increased flow rates to the maximum output of 60 000 000 cubic feet (1 699 011 m³) per day. For 90 minutes, the increase was tolerated but at 3:30 a.m., the gas flow to the flare became erratic.

At 4:40 a.m., the supervisor began emergency shutdown procedures, but gas had already escaped. The amine solution overflowed and extinguished the flare's pilot light. For 20 minutes, between 4:50 a.m. and 5:10 a.m., 800 000 cubic feet (23 305 m³) of high H_2S gas escaped from the unlit flare vent. In Poza Rica, the air was still and a thin mist filled the valley, reducing atmospheric mixing and forcing the H_2S gas to move laterally and settle over a residential neighborhood. Exposure concentrations were calculated at 1000–2000 ppm H_2S.

In a home 400 ft (121.9 m) from the release, a father woke to the H_2S odor, grabbed his infant daughter from her crib and stumbled down the street before collapsing. He and the baby survived but his wife and three other children died in their sleep. Residents formed rescue squads, dragging seemingly lifeless neighbors from their homes and rushing them by taxi or truck to the hospital. In all, 22 people died from H_2S exposure and 320 were hospitalized. More than 50% of the pets and livestock in the area were also killed. A mass funeral was held a few days later for the victims.

13.4.2 Particulate

The 1970 Clean Air Act named particulate as one of the criteria air pollutants. At that point, it was regarded as among the least hazardous. With continuing research, particulate has emerged as one of the more dangerous and has more stringent regulations. Particulate is estimated to be responsible for up to 65 000 deaths per year in the United States from respiratory effects and as many as 200 000 deaths per year in Europe. This is a 17% increase in mortality risk.

Particulate is a general term for fine solid and liquid particles that are suspended in air. They include hundreds of elements and compounds, some of which may be hazardous including heavy metals and asbestos. Particulate is categorized based upon size. Total particulate matter (TPM) is all suspended matter between 100 and 0.1 μm. Within this group, PM_{10} classification includes particles less than 10 μm and $PM_{2.5}$ are less than 2.5 μm. The reason for these subcategories is medical: coarse PM_{10} particles are respirable but fine $PM_{2.5}$ particles can be absorbed directly into the bloodstream through the lungs. Primary or direct particles are primarily PM_{10} or larger and emitted directly from their source as solids. Secondary or indirect particles are primarily $PM_{2.5}$ and formed in the atmosphere through photochemical reactions. Particulates are from both natural and anthropogenic sources. Much of the fine particulate is from chemical reactions among sulfates from power plants, nitrates from vehicles and industrial emissions and ammonia from feed lots. A study of fine particulate in Ontario, Canada found that it was 30–50% organic and elemental carbon from burning or soot, 30–40% sulfate, 10–20% nitrates, and 3–10% soil, but the composition varies widely depending upon season and location.

Particulate of 50–100 μm tends to settle out close to the source or be washed out by precipitation. Coarse and fine particulate can be transported over great distances in suspension by wind before settling. This transport can produce acidic lakes and streams and alter the nutrient balance in coastal waters and lakes depending upon the composition of the particulate. It can deplete nutrients in soils and damage sensitive forests and agricultural crops. It can also destroy the balance in ecological systems. Fine particulate can cause reduced visibility and haze in urban, suburban, and rural areas.

More than 99% of inhaled particulate is exhaled or trapped in the upper respiratory tract and then exhaled. If it is not expelled, coarse particulate can lodge in lung tissue and slow oxygen exchange, causing shortness of breath. Exacerbated by irritation and swelling of airways, this can cause life-threatening asthma and COPD. Fine particulate can be absorbed directly into the bloodstream through the lungs. This situation is especially dangerous to people with heart or respiratory disease. Children and the elderly may be more sensitive to particulate. Short-term exposure to particulate can cause coughing, wheezing, irritation, and labored breathing depending upon the composition of the particulate. Long-term exposure to particulate increases the risk of lung diseases such as emphysema, pulmonary fibrosis, and lung cancer. It also damages the immune system and contributes to premature death. It has been estimated that chronic exposure to particulate shortens a person's lifespan by as much as two years.

13.4.3 Natural Versus Anthropogenic Sources

Surprisingly, much of our air pollution is from natural sources. The primary sources are windblown dust, fires, volcanoes, biological pollutants, and crossover pollutants, but there are numerous others from windblown salt spray to methane releases from the ocean floor. Strong winds passing over open sand or soil lift tons of fine material into the air, forming the dust storms and sandstorms that are common in arid regions. Inhalation of this dust, a particulate, can cause significant health problems.

Woodland wildfires can be natural, caused by lightning in dry conditions, or be started by humans. Wildfires produce enormous amounts of smoke composed of both fine and coarse particulate. The thickness of the smoke reflects the amount of particulate and it can turn the sky black for many miles around the fire. Particulate can persist for tens to thousands of miles away from the fire depending upon the size and intensity of the fire and the wind conditions. Typical chemical pollutants released by a fire are carbon monoxide, carbon dioxide, nitrogen oxides, and organic compounds. These organic compounds can be hazardous depending on the type of trees.

Large volcanic eruptions can emit more air pollutants in a shorter time than any other air pollution source. The main pollutant is particulate which is composed of pulverized rock and ash. The ash covers the ground for many hundreds of miles downwind of the volcano and may circulate in the stratosphere for years. Volcanoes also emit a significant amount of gas. These gases include carbon dioxide, methane, water, sulfur oxides, and minor amounts of many other compounds like fluorine, chlorine, argon, boron and lithium compounds, and metals. The gases are in such low quantity relative to the ash that they are usually overlooked.

Air pollutants produced by human activity are numerous and diverse. Anthropogenic sources include motor vehicles, construction, agriculture, unpaved roads, stone crushing, brickworks, cement works, quarrying, incineration, chemical factories, smokestack emissions, smelters, open burning, residential heating, and tobacco smoke among others. There are a number of industrial organic compounds used as solvents, plastics and chemical intermediaries, and heavy metals and heavy metal compounds that are mining-related and industrial.

Crude oil is a mixture of numerous chemicals with many uses. These chemicals are separated from each other in a refinery. The main component to a refinery is a distillation tower in which the oil is heated, vaporized, and condensed by density before further processing. This processing along with storage and transportation releases immense amounts of VOCs. Any time these chemicals are exposed to air, they evaporate and their fumes are released to the atmosphere.

13.5 Stagnation and Temperature Inversion

The degree of air pollution in an area is dependent upon: source, transport from source, and build up through stagnation. Stagnation is the result of light winds that restrict horizontal dispersion, stable conditions in the lower atmosphere that restrict vertical movement, and no precipitation to wash away pollution. These conditions are common in a persistent or slow-moving high-pressure system. The stagnation index is a measure of these conditions and it can result in high concentrations of pollutants and especially ground-level ozone. There is a stagnation index reported for the United States based upon number of days of stagnation (Figure 13.13)

Temperature inversion can result in exceptional stagnation that contributes to virtually all air pollution disasters (Figure 13.14). Normally, temperature decreases with elevation above the surface in the troposphere. However, in a temperature inversion, temperature increases with elevation for a short distance above the surface before returning to the normal gradient. An inversion occurs in very stable weather with no wind or precipitation. The Sun heats the Earth's surface during the day, which, in turn, heats the air above it. This is common in cities because of the blacktop, pavement, brick, cement, and glass buildings. The heated air can rise to high elevations. At night, the city cools off but air is a good thermal insulator and remains warm. As the surface cools, the air near it cools, making it denser and holding the pollutants near the ground. Topography can increase this effect when cold air flows down hills and concentrates in valleys. In either case, these inversions last on the order of hours before the sun heats the surface enough to re-establish a normal thermal gradient. However, in some cases, weather system created inversions can last for days.

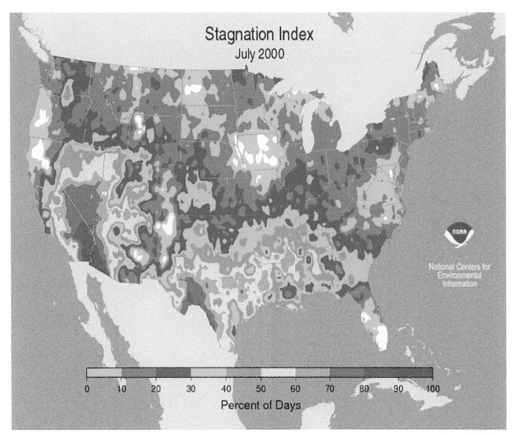

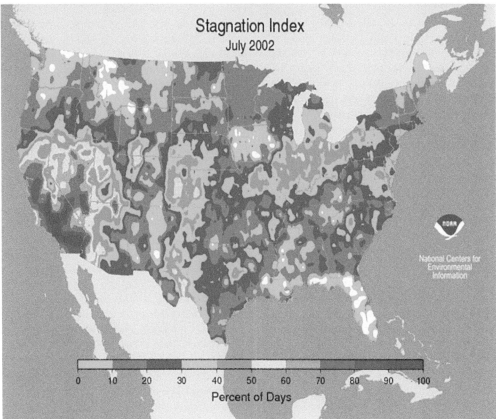

FIGURE 13.13 Maps showing stagnation index in the United States. During the summer of 2002 (bottom), conditions from Missouri east to New Jersey and down through South Carolina were conducive to air stagnation. Compared to the summer of 2000 (top), the percentage of days with air stagnation conditions nearly doubled. *Source:* Courtesy of NOAA.

FIGURE 13.14   Diagram showing a typical temperature inversion and the air pollution being sealed at the surface.

CASE STUDY 13.3 The Great London "Killer" Fog

Temperature Inversion Pollution

Coal was used to heat homes, cook meals, and fuel industrial furnaces and boilers in the 1950s in London, UK (Figure 13.15a). It produced thick smoke that contributed to the frequent thick fogs. In 1905, the term "smog" was derived by combining the words "smoke" from burning coal and "fog" to describe the dense pollution. This is "London type" or "classical" smog, a primary air pollutant as compared to "Los Angeles" smog which is caused by secondary air pollutants.

If coal is burned at a low temperature and without enough air like in a stove or fireplace, it gives off smoke, showing incomplete burning. Included in the smoke is hot gas and carbonaceous particulate or soot. London has historically been filled with smoke and covered with soot. For the normal healthy adult, short-term exposure to soot is a minor annoyance. It can lead to coughing, throat irritation, and temporary lung problems. On the other hand, if children, the elderly, or people with heart or lung diseases are exposed to particulate, they may be admitted to hospital emergency rooms and risk premature death.

The pollution was so bad that in 1300, the king banned the burning of coal because of the soot and odor it produced, under penalty of death. The law was rescinded when England's forests began to disappear. London is especially susceptible to pollution from coal burning because the weather is cold and damp for a lot of the year. Particulate stays suspended for long periods under these conditions and becomes concentrated in the atmosphere by attaching to the water vapor.

During the night of Friday, 5 December 1952, a layer of warm, moist air over London marked the start of an unusually long-lasting temperature inversion. With little wind, particulate emitted by coal burning began to accumulate in the humid air. The fog thickened and visibility dropped to a few feet (Figure 13.15b). For the next 114 hours over 5 days, London was continuously encased in smog so thick that it was difficult to tell night from day.

On Saturday, 6 December 1952, Londoners awoke to darkness. The increasing pollution began to cause serious health problems, especially to "at risk" residents, if they went outside. Hospitals were quickly overrun with patients as a result. Ambulances stopped being sent on calls because of the poor visibility and chaos in the streets (Figure 13.15c). As a result, thousands of ailing Londoners walked through the smog to the hospitals. That day, more than 500 people died in London from smog-related complications. Cattle in the city markets fell over gasping for air and died. Their carcasses were discarded before they could be butchered and sold.

By Sunday, 7 December visibility had decreased to 1 ft (30 cm) or less. Cars were abandoned on the roads, making it impossible to travel. The darkness forced cancelation of midday concerts. Smog infiltrated all homes and buildings, no matter how airtight. A theater performance was canceled when indoor smog became so thick that patrons and performers could not breathe. In some low-lying areas, visibility was so poor that people could not see their own feet. The infiltrating smog even caused adverse health effects for people who remained indoors. Patients arrived at the hospitals with blue lips from lack of oxygen. More than 750 of these patients died of suffocation.

Monday, 8 December and Tuesday, 9 December were equally perilous. Travel was impossible as the smog continued to worsen. The death toll exceeded 900 on each of these two days. Then, late on 9 December, the wind increased and swept out the killer smog. However, the daily death toll continued in the hundreds for several weeks until just before Christmas. Mortality from bronchitis and pneumonia increased more than sevenfold as a result of the smog.

Health officials estimated that the smog caused the premature death of about 4000 people. A later study estimated that smog caused a long-term death toll closer to 12 000 people through December 1952. The most susceptible people were those suffering from respiratory or cardiovascular illness. They were the majority of the deaths but healthy people also died.

Sulfur dioxide and particulate were estimated at 0.6 ppm and 1400 μg/m^3, respectively, during the disaster. The EPA limits sulfur

(a)

(b)

FIGURE 13.15 (a) Map of London with inset showing the location in the United Kingdom. (b) Photograph showing the smog covering Big Ben and other classic areas of London. *Source:* PeskyMonkey/Adobe Stock. (c) Photo of the impact of the smog on everyday life in London even during the day. *Source:* Courtesy of NASA.

(c)

FIGURE 13.15 (Continued)

dioxide (SO_2) levels to less than 0.03 ppm and particulate levels to less than 260 μg/m³. Estimates place the pollution trapped by the inversion at 1100 tons (1000 mt) of particulate, 2200 tons (2000 mt) of carbon dioxide, 154 tons (140 mt) of hydrochloric acid, 15.4 tons (14 mt) of fluorine, and 407 tons (370 mt) of SO_2 that were converted to 880 tons (800 mt) of sulfuric acid.

London experienced similar smog events in 1813, 1873, and 1891, but the 1952 event spurred serious epidemiological research into the effects of air pollution. The British Parliament slowly became convinced by the mortality data and in combination with strong pressure, they passed the 1956 United Kingdom Clean Air Act. This legislation placed tighter controls on the release of air pollution and established strict monitoring standards. However, in 1956, despite the new legislation, another killer fog struck London, killing an estimated 1000 people. The causes were the same, particulate and SO_2 trapped by a temperature inversion.

Another pollution condition is called subsidence inversion (Figure 13.16). This is a geographically controlled situation that occurs in cities in valleys or basins under large high-pressure systems. The high-pressure forces air down into a valley or basin and the steep topography prevents air from escaping laterally. As long as this situation persists, the air stagnates in the topographic enclosure. The air cannot circulate because of a subsidence inversion, it stagnates, and pollutants build to hazardous levels. If the weather does not change, an air pollution disaster can result.

One of the worst areas for subsidence inversions is Los Angeles, California. The Pacific high-pressure ridge is a semipermanent atmospheric feature off the California coast. In addition to the sinking air from the high pressure, the cool air from the Pacific Ocean significantly cools the air at lower elevations, causing a substantial inversion. Further, the San Gabriel Mountains to the northeast block the movement of the weather systems out of the Los Angeles basin for days to weeks. As a result, Los Angeles is smoggy and in a temperature inversion about 80% of the period between June and August. Denver, Colorado has a similarly unfortunate mix of topography, making it extremely susceptible to smog as well.

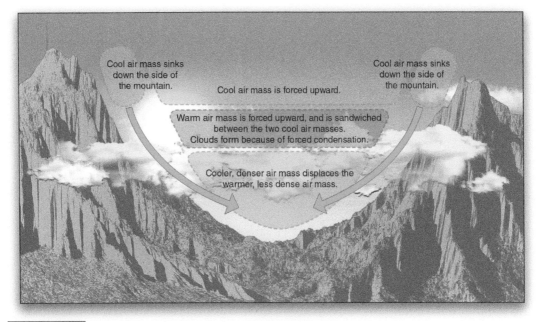

FIGURE 13.16 Diagram showing the effects of a subsidence inversion.

CASE STUDY 13.4 Donora Killer Fog

Subsidence Inversion

Donora, Pennsylvania was an industrial town with a population of about 14 000 in the 1940s, 40 miles (64.5 km) southwest of Pittsburgh, Pennsylvania (Figure 13.17a). Donora contained two of the largest metal working mills in the United States, including Donora Zinc Works and the American Steel & Wire. The four miles (6.4 km) long industrial belt was on the west side of a horseshoe bend in the Monongahela River, just upwind of Donora. The plants employed 6500 people who forged iron and zinc ores into steel, nails, wire, fencing, and other finished goods.

The one mile (1.6 km) wide Monongahela River valley was carved into the limestone bedrock that formed the surrounding cliffs rising 500 ft (152.4 m) above the river. Companies built plants in this area because it was near Pennsylvania's coal mines. In addition, using the Monongahela River, coal and other raw materials could be delivered, and finished goods could be shipped out. By 1916, the plants were continuously spewing thick smoke and there was a continuous line of freight-loaded river barges at this industrial center. This level of activity continued for 40 years.

The mill production rate made Donora the largest producer of nails in the world. Coal was shipped in at 40 barge-loads per day. This coal stoked the furnaces and vats where the iron and zinc were processed. The waste sulfur oxides, zinc, fluoride, other metals, coal ash and soot were discharged through 150-ft (45.7 m) smokestacks above the factories. In the 1940s, there were no air pollution control devices which meant that residents of Donora, and all other major industrial cities, endured foul smells and dead vegetation.

Impacts of this air pollution were common in Donora. From the 1920s, mill owners reimbursed claims for crop and livestock damage. From 1918, the zinc plant compensated Donora residents for health costs from exposure to plant emissions. However, the five-day pollution disaster beginning on 26 October 1948 made earlier events pale in comparison (Figure 13.17b). In evening of 25 October a high-pressure system blanketed the town with cool, dry air that slid down the valley walls forming a subsidence temperature inversion. This created a seal to the plant emissions. The cliffs surrounding Donora were so much higher than the smokestacks that the noxious smoke quickly built up in the ground-level air and reached dangerous levels (Figure 13.17c). The effects of this build-up were immediately noticed by the Donora residents.

The sealed air first turned yellow and then gray, leaving Donora covered in a dark haze. Nine elderly residents of Donora were found asphyxiated in their homes. They died from inhaling sulfuric acid from sulfur dioxide (SO_2) emissions that reacted with water vapor in the air. The next day, the death toll had increased to 18. The Saturday afternoon high-school football game was played in smog so dense that spectators could not to see the players on the field. By late-afternoon, visibility was reduced to 1 ft (0.3 m), making it unsafe to drive.

Doctors advised people with respiratory ailments to evacuate, but so many residents attempted to flee that the traffic jams made it impossible. By Sunday, the funeral home ran out of caskets. Firefighters went to homes giving people struggling to breathe, oxygen from their air packs. The zinc and steel mills kept operating and adding to the air pollution despite the disaster. Finally, on Sunday morning, 31 October, five days into the disaster, the plants were shut down. By then, 20 residents were dead and the community center was acting as a makeshift morgue. More than half the residents of Donora were suffering from headaches and stomach ailments.

Late Sunday, the inversion lifted and rain washed the pollution out of the air, allowing the mills to reopen. Fifty additional residents died of respiratory issues within one month of the disaster. Concentrations of SO$_2$ in air are estimated to have been between 1500 and 5500 μg/m^3 during the disaster. Current limits on SO$_2$ concentrations are 80 μg/m^3.

As a result of the Donora disaster, Pennsylvania passed the first air pollution law in the United States in 1955.

(a)

(b)

FIGURE 13.17 (a) Map of the Donora, PA area. Inset shows the location of Donora in Pennsylvania. (b) Photo of the thick smog covering Donora. *Source:* The National Library of Medicine. (c) Photo of the smog-producing factories on the banks of the Monongahela River. *Source:* Courtesy of The Library of Congress.

(c)

FIGURE 13.17   (Continued)

13.6   Acid Precipitation

Acid precipitation is the result of SO_2 and nitrogen oxides (NO_x) air pollutants that react with water and oxygen and produce sulfuric and nitric acids. The acids are precipitated to the surface by rain or snow or they can fallout as dry precipitation. There can also be acid fog. A small amount of the SO_2 and NO_x in acid precipitation is from natural sources such as volcanoes and wildfires, but it is primarily from fossil fuels. Burning of fossil fuels to generate electricity previously produced about 67% of the SO_2 and 25% of the NO_x in the atmosphere but because of increased use of natural gas rather than coal coupled with advanced filtering, between 1995 and 2021, annual emissions of SO_2 decreased by 92% and NO_x decreased by 87%. The rest is from motor vehicles, manufacturing, oil refineries, and other industries. However, with the extreme reduction of SO_2 and NO_x, most of the industrialized countries are well within the limits.

Normal precipitation is slightly acidic with pH of 5.6 because carbon dioxide (CO_2) in the atmosphere forms carbonic acid in it. Acid precipitation has a pH between 4.2 and 4.4. and is especially damaging to lakes and marshes, where it is harmful to plants and most wildlife. The acidity is directly damaging, but as it percolates through soil it can also leach and transport damaging metals to streams and lakes. In surface water with pH of 5, most fish eggs will not hatch. Below 5, adult fish can die and, as a result, lakes that might be productive can have no fish. Even if the fish can tolerate acidic water, the food might not. For example, frogs can survive pH of 4 but the mayflies they eat do not survive below 5.5. Trees and other vegetation are also damaged or destroyed by acidic conditions.

In the 1950s to late 1970s, acid precipitation was a great concern to society. Many major cities had so much acid precipitation that health effects were rampant. People regularly endured burning eyes and damage to their respiratory systems. The sulfuric and nitric acid in the atmosphere form fine particles that are inhaled, causing heart attacks and decreased lung function especially in people with asthma or COPD. Atmospheric nitric and sulfuric acid also damage surfaces of statues, buildings, and other structures. The acid corrodes metal and causes paint and stone to rapidly deteriorate. This became a serious issue for tourist areas in Europe, where antiquities began to degrade at an alarming rate.

However, acid precipitation is most importantly a story of how pollution problems can be overcome if society takes steps to address it. By the 1970s, acid precipitation was so bad that people in Japanese cities had to wear surgical masks when they went outside, traffic was restricted in Rome, Italy and Athens, Greece to slow the degradation of ancient treasures and 90% of high-altitude wilderness lakes in the Adirondacks Mountains, New York were dead. With awareness campaigns and public concern over the impact, public pressure on companies and the government resulted in regulations to address the problem and the formerly very serious issue of acid precipitation has greatly abated.

Case Study 13.5 Sudbury Canada Mines and Smelters

Extreme Acid Precipitation

There are numerous cases of areas severely damaged or destroyed by acid precipitation like Palmerton, Pennsylvania and Ducktown, Tennessee but perhaps the best example of acid precipitation and ecological collapse occurred in one of Canada's largest metropolitan areas. The 1000 square miles (2590 km²) Sudbury, Ontario area is 200 miles (322 km) north of the City of Toronto (Figure 13.18a). It appears that about 1.9 billion years ago, a large meteorite struck this area. The heat and pressure of the impact melted the bedrock and mineral-carrying fluids seeped into fractures. The minerals were deposited, forming one of the world's largest deposits of nickel and copper. About 60% of underground hard rock mining in Canada is within 300 miles (482.8 km) of Sudbury. It contains more than 3200 miles (5150 km) of mine tunnels.

In 1885, a surveyor discovered the rich nickel deposits and copper deposits were found soon after. Mining began immediately and it continues today. The copper and nickel occur in the minerals pentlandite and pyrrhotite, which are also rich in sulfur which must be removed to recover the valuable metals. Removing the sulfur started with the ore being cooked in open pits. A mixture of ore and wood was burned for months before the ore was placed in a furnace for processing. The heat and air converted sulfur to SO_2 during cooking, which escaped to the atmosphere. This practice was used to remove sulfur from ore until regulations prohibited it.

Open-air cooking was replaced by ore processing at high temperatures in tanks inside buildings. SO_2 was still released to the atmosphere but out of tall smokestacks that widely distributed it. In 1965, 2 500 000 tons (2.3 million mt) of SO_2 was released to the atmosphere

(a)

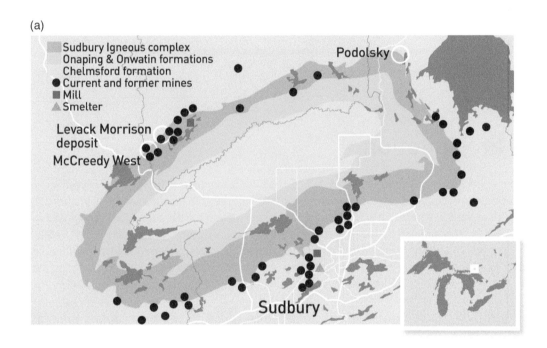

(b)

FIGURE 13.18 (a) Geologic map of the Sudbury, Ontario area and the location of the mines, refineries and smelting operations. Inset shows the location of Sudbury in central North America. (b) Photo of the mining and smelting operations of Sudbury where sulfur fallout has killed virtually all of the vegetation. Source: British Library Board / Wikimedia Commons / Public Domain.

around Sudbury. In addition, by 1984, mine tailings and slag waste from smelting covered more than 3000 acres (13 km²) of the Sudbury area, with waste being added at the rate of eight million tons (7.3 million mt) per year.

Acid fallout and acid precipitation was disastrous in the Sudbury area. By 1967, virtually all vegetation within 40 square miles (103.6 km²) of Sudbury was dead (Figure 13.18b). Soil was washed away due to the lack of vegetation to hold it. Soil near the smelters was toxic from copper and nickel fallout. The landscape was so bleak and barren that NASA trained astronauts for moon landings there during the 1970s. Outside of the dead area but within 140 square miles (362.6 km²), only sparse shrubs were able to grow and damaged vegetation covered a 1700 square mile (4403 km²) area outside of that. The soil in the worst area had an acidic pH of 2.0–4.5 and was devoid of essential plant nutrients. Soil near the smelters had copper and nickel concentrations of 1000s of parts per million, or hundreds of times the regulatory limits.

There have been several attempts to address this pollution. In 1947, residents of nearby Copper Cliffs planted vegetation on a mine tailings pile to reduce mine dust. During the 1950s, trees were planted in imported soil. In 1969, the Sudbury Environmental Enhancement Programme (SEEP) between Ontario Department of Lands and Forests and Laurentia University investigated restoring the area. They aimed to improve soil fertility by liming and fertilizing soil, and planting thousands of trees. By 1986, nearly 10 square miles (25.9 km²) of land was actively being restored. By 1995, more than 230 000 trees had been planted and a number of insects, birds, and mammals had returned. These efforts continue today on about 30% of the area. It will take hundreds of years of effort for the local ecology to be self-sustaining.

Reference

Peel, M. C., Finlayson, B. L., & McMahon, T. A. (2007). *Updated world map of the Köppen-Geiger climate classification. Hydrology and Earth System Sciences, 11(5), 1633–1644.* Doi:10.5194/hess-11-1633-2007. Public Domain.

Mining and Earth Resources

Polluted Earth: The Science of the Earth's Environment, First Edition. Alexander Gates.
© 2023 John Wiley & Sons, Inc. Published 2023 by John Wiley & Sons, Inc.
Companion website: www.wiley.com/go/gates/pollutedearth

Words you should know:

Acid mine drainage – Water that drains from a mine or filters through tailings and takes up sulfur to make sulfuric acid.

Blowout – An uncontrolled to explosive release of oil and/or gas from a well.

Crude oil – Petroleum directly from the Earth and in its native state and used as fuel and chemicals.

Gangue – The non-economic and non-target minerals occurring with ore in a mineral deposit.

Mining – Removal of ore from the earth for economic use.

Natural gas – Flammable gaseous hydrocarbon released from the earth and used as fuel. Primarily methane.

Ore – Target minerals or rock in a mining prospect.

Pipelines – Large and extensive pipes that transmit oil and/or gas fuel.

Refineries – Industrial plants that process crude oil into usable petroleum and related products.

Smelting – High-temperature chemical reaction to process ore into usable metal.

Strip mining – Highly destructive surface mining that strips away overlying material to retrieve ore.

14.1   Mining of Minerals and Rocks

Mines, petroleum wells, and processing plants are small and localized pollution sources but they are among the most damaging activities to the environment. They destroy ecosystems and make productive areas devoid of life. Recovery of these areas to a natural state takes centuries by natural processes. Yet they are probably the most profitable economic ventures. In 2019, the nonfuel mineral production in the United States was $83.7 billion alone. Over a lifetime, every American uses 3.5 million pounds (1.6 million kg) of minerals, metals, and fuels as in other developed countries. The waste from mining operations produces contaminated and acid waters, waste rock and tailings, leaks and spills of chemicals stored at mine sites, and damage to vegetation and wildlife from air pollution fallout from mining and smelting.

Natural chemical weathering of rock surfaces degrades the less stable rocks, minerals, and soils, leaving relatively stable rocks and minerals exposed at the surface. The more stable minerals such as quartz, feldspar, and clays are less dangerous to the environment and human health. They are less soluble in water and global life has adapted to them. Most dangerous chemicals are removed by weathering. Mining, on the other hand, excavates underground rock that is not chemically weathered and dumps it on the surface where it is bioavailable. Highly reactive minerals quickly release hazardous chemicals to the environment. The speed of dissolution of the minerals is enhanced by crushing them into small pieces during mining and processing which produces enormous amounts of reactive surfaces.

Although some mined minerals are relatively safe, most are dangerous. The reason for mining ore minerals is commonly for their abundance of heavy metals, radioactive elements, asbestos, and sulfur among other dangerous chemicals. Even salt, lime, and alkalis can be damaging to the environment in high concentrations.

Mined deposits require high concentrations of target minerals relative to the matrix of other minerals to be economic. The target mineral is ore and is metallic, nonmetallic, precious, industrial, or strategic. The non-economic matrix minerals to the ore, or gangue, are separated from the ore and discarded during processing. The whole rock is assayed to determine if mining is economic. Assaying is a chemical analysis that determines the percentage of ore in the rock, which determines its grade. Mined rock that is not high enough grade is waste rock and is dumped around the mines as spoils or tailings.

There are several types of economic mineral deposits depending on the target mineral or material. Table 14.1 shows the metal of interest, the ore minerals it is found in, and the general geologic occurrences. Metals such as iron, aluminum, and copper are abundant whereas other metals like chromium, mercury, and tin are relatively scarce. Other metals such as platinum, palladium, and rhodium among others are strategic, though this category changes with need.

Table 14.2 shows non-metal ores, the uses of these minerals, and the general geological occurrences. Non-metal minerals are used primarily for chemical production such as halite (rock salt), sulfur, and spodumene, for agriculture such as calcite (lime) and apatite (phosphate), and for building materials such as gypsum (sheetrock), clay, sand and gravel, and crushed rock. Less commonly but still important, they are used for ceramics (clay), abrasives (diamond, garnet), and for jewelry such as diamonds, corundum (ruby and sapphire), and beryl (emeralds) among others.

14.1.1   Geological Occurrences

Igneous rocks and processes form most metallic and precious mineral deposits and can form nonmetallic, strategic, and industrial mineral deposits. Igneous formation includes pegmatites, crystal settling, disseminated, hydrothermal, and volcanogenic. Pegmatites are igneous rocks with very large minerals that form underground from magma. Pegmatite magma forms late in igneous

TABLE 14.1 Metals and mining.

Metal	Ore minerals	Main occurrence
Aluminum	Bauxite	Deep weathering
Chromium	Chromite	Magma settling
Copper	Chalcopyrite, bornite, chalcocite	Hydrothermal deposits;
Gold	Native gold	Hydrothermal deposits, placer
Iron	Hematite, magnetite, limonite	Banded formations, sedimentary cement, hydrothermal
Lead	Galena	Hydrothermal deposits
Magnesium	Magnesite, dolomite	Hydrothermal deposits, sedimentary
Mercury	Cinnabar	Hydrothermal deposits
Molybdenum	Molybdenite	Hydrothermal deposits
Nickel	Pentlandite	Magma settling
Platinum	Native platinum	Magma settling, placer
Silver	Native silver, argentite	Hydrothermal deposits, placer
Tin	Cassiterite	Hydrothermal deposits, placer
Titanium	Ilmenite, rutile	Magma settling, placer
Tungsten	Wolframite, scheelite	Pegmatite, metamorphism, placer
Uranium	Uraninite	Pegmatite, sedimentary cement
Zinc	Sphalerite	Hydrothermal deposits

TABLE 14.2 Uses and occurrence of nonmetallic minerals.

Ore minerals	Uses	Occurrence
Apatite	Phosphorus fertilizer	Sedimentary
Asbestos (chrysotile)	Fire-proof fibers, insulation	Metamorphic, hydrothermal
Calcite	Aggregate, cement, flux, soil conditioning, chemicals	Sedimentary deposits
Clay minerals	Ceramics, china, bricks, paper	Deep weathering
Corundum	Gemstones, abrasives	Metamorphic
Diamond	Gemstones, abrasives	Kimberlites, placer
Fluorite	Flux, glass, chemicals	Hydrothermal deposits
Garnet	Abrasive, gemstones	Metamorphic
Graphite	Pencil lead, lubricant,	Metamorphic
Gypsum	Plaster, sheetrock	Evaporite deposits
Halite	salt, chemicals	Evaporite deposits
Quartz	Glass, abrasive	Hydrothermal, sedimentary
Spodumene	Lithium-ion batteries	Pegmatite
Sulfur	Chemicals	Sedimentary, hydrothermal deposits
Talc	Paints, cosmetics	Metamorphic deposits

processes and contain high amounts of water and elements that are uncommon in igneous rocks. The water reduces nucleation of minerals, allowing the few that form to grow large. A pegmatite from the Midwestern United States has crystals to 40 ft (12 m) long. The abundance of uncommon or incompatible elements produces uncommon minerals like boron-rich tourmaline and lithium-rich spodumene among others. Pegmatites may also contain beryl (including emeralds), uranium minerals and garnet, among others.

The intrusion of mafic or ultramafic magma into the crust produces a pluton that may form a layered mafic intrusion. These plutons cool slowly, allowing dense minerals to settle through the magma to the bottom of the chamber through gravity and accumulate as an ore deposit. These deposits include platinum, chromite, nickel, and magnetite (iron).

Disseminated ores occur in small cracks and pore spaces above plutons. Also called porphyry deposits, these ores occur in low concentrations in large volumes of rock. Disseminated ores can produce large deposits but contain so much gangue material that the cost of extraction is high. Diamonds occur as disseminated ore but they form 90 to 120 miles (150–200 km) deep in the mantle under extreme pressures (Figure 14.1). They form in gas charged fluid-liquid-rock mixtures that intrude explosively through the crust at speeds up to Mach 2 (twice the speed of sound). They form funnel-shaped bodies called kimberlites at or near the surface in a mass of crushed rock. This is the only way for diamonds to reach the surface because if brought up slower, they convert to graphite.

Hydrothermal deposits are made by hot aqueous fluids carrying ore chemicals into an area that is chemically favorable. As magma crystallizes, it releases hot corrosive fluid that dissolves and transports metals and other ions. If these brines encounter favorable chemical, temperature, and/or pressure conditions, they precipitate the dissolved chemicals as minerals into cracks, faults, and pores. They can also replace rocks by dissolving them and precipitating the chemicals in their place. The ore minerals precipitate in cracks as veins called lodes.

Undersea volcanism produces chimneys that spew gas, fluid, and particles through pipes called black smokers. These emissions cause the surrounding ocean water to become saturated with metals and other ions which then precipitate onto the seafloor. The result is interlayered volcanic rocks, deep sea sediments, and base metal and/or massive sulfide ore deposits. The ores include metals such as copper, zinc, iron, nickel, and magnesium among others.

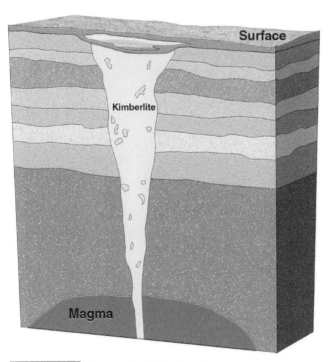

FIGURE 14.1 Diagram of a kimberlite diatreme.

Sediments are formed by weathering of existing rocks into particles or dissolved compounds that are transported, deposited, and lithified into rocks. The precipitated dissolved chemicals form rocks called precipitants. Precipitation is from supersaturation of the dissolved chemicals by evaporation of water from the solution or biological processes. Some precipitation occurs in the deep ocean but most occurs in small ocean basins with limited circulation. Most precipitants are rock salt, carbonates, phosphates, and agates, primarily in tropical areas. Manganese nodules are deposited in the deep ocean. Phosphates and most carbonates are biologically enhanced shallow marine deposits.

Evaporites form like precipitants by supersaturation of water through evaporation. Evaporites form in terrestrial or restricted shallow marine conditions. On land, evaporites form in lakes and shallow seas not connected to the ocean. The water supersaturates with ions through evaporation and precipitates evaporite minerals. Deposits of evaporite minerals include gypsum, carbonates, borax, and alkali salts. Coal is an energy source but it is mined as a solid precipitant sedimentary rock (Figure 14.2). Coal forms where dead vegetation collected in swamps and other wetlands. It was slowly buried, compressed, and heated, turning it into peat and then into coal. Coal is classified in ranks by age and burial as lignite, bituminous coal, and anthracite.

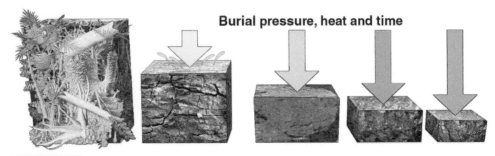

Burial pressure, heat and time

FIGURE 14.2 Diagram showing the development and ranks of coal. Light green = lignite, dark green = bituminous coal, blue = anthracite.

Surface exposures of ores are eroded and carried as particles in flowing water. Particles of metal ores are dense and are consequently deposited in sand and gravel bars in higher concentrations where stream flow slows. This concentrates ore particles into these areas called placer deposits. Most mined placer deposits are gold, platinum, and silver because they are dense. There are also placer deposits of diamonds but they are not dense so do not concentrate. The other sediment particles can concentrate into sand and gravel deposits which are economic. Sand and gravel are concentrated in streams and glacial outwash deposits, as the result of water flow. Sand and gravel are not rare but they are used in construction in poured concrete and asphalt in massive amounts in all cities. The problem is that transporting it great distances from the sources to the cities is expensive. This makes deposits closer to cities worth a lot of money.

Clay is a high-volume industrial mineral, used in bricks, ceramics, cat litter, and for the slick coating on paper among others. Clays are primarily from lagoons and bays in marine settings, lake deposits, glacial lakes, and floodplain deposits around rivers. Large, pure clay deposits in Georgia are called "white gold" because they are worth so much money.

Weathering processes remove minerals by dissolving and transporting them. This can create economic deposits through weathering. Deep weathering of soils in tropical climates with high precipitation removes unstable and semi-stable minerals. Warm temperatures and acidic water enhance chemical alteration and dissolution. Immobile elements and compounds are left behind. Aluminum is one of the more immobile elements under these conditions, and it accumulates in laterite soils that form bauxite, which is mined for aluminum.

14.2 | Types of Mining

The two types of mining operations are surface and underground with several methods of each. The type of mining is based on several factors such as the type of ore and gangue, shape and depth of the ore body, population around the area, grade of ore, local mining regulations, technology available, and groundwater in the area. Most mining is underground unless the ore is shallow and economic and all limiting factors are minimized.

Underground mining is used for ore bodies at depth. Vertical and subvertical ore bodies are mined underground though the top of the body. The tunnels into ore bodies are mostly ramps, shafts, or adits (Figure 14.3). Ramps are sloped spiral tunnels that circle the deposit. The entrance may be lined with a galvanized steel culvert. Shafts are vertical tunnels adjacent to the body that use elevators. Adits are horizontal tunnels into a hill used for flatter ore bodies. There are several methods of mining the ore through these tunnels.

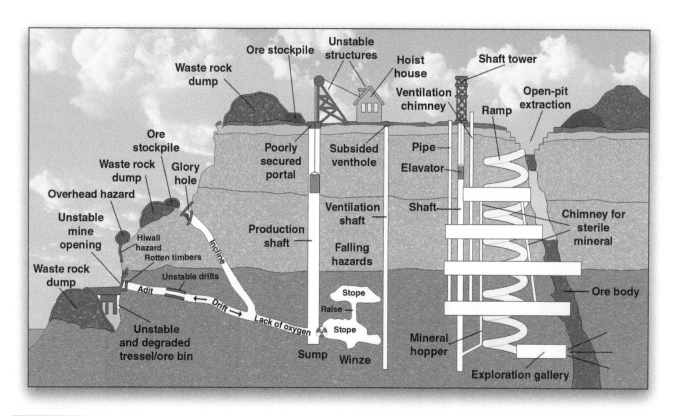

FIGURE 14.3   Diagram showing the access and mining structures in underground ore bodies.

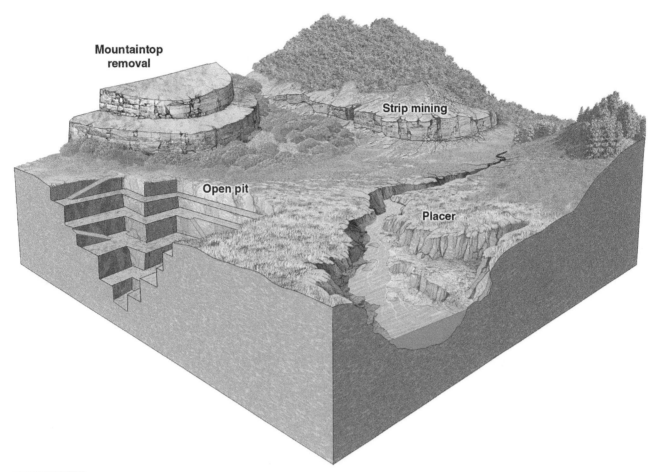

Mountaintop removal

Strip mining

Open pit

Placer

FIGURE 14.4 Diagram showing the different types of surface mining types.

Hard rock mining is underground mining into solid rock where tunnels must be drilled and blasted with dynamite. Hard rock mines contain metals like gold, copper, and lead or gems such as diamonds, emeralds, and rubies. Drift mines are used if ore is on the side of a hill and accessible using horizontal tunnels or drifts. In borehole mines, a shaft is drilled into the ore body and a tube-like tool is lowered into it. Water is forced down the tube and the stream breaks the rock, forming a slurry that is pumped up to the surface.

There are several types of surface mining and strip mining is the most common (Figure 14.4). In strip mining, a shallow ore body is exposed by removing overlying soil and rock. The ore is removed from the surface using huge machines. Strip mines include area stripping and contour stripping. Area stripping is used in flat areas. Strips are removed and the overlying soil fills previous excavations. In contour stripping, in hilly terrains, soil is removed from above the ore body if it follows the land contour and leaves terraces on hillsides. Open-pit mining removes ore from an open pit in benches. Mountaintop mining reduces mountains to flat surfaces.

Gold mining mainly uses chemical methods. In placer deposits, most of the gold occurs with gangue minerals. Mercury is mixed with the sediments removing the gold to an amalgam that can be separated. This method was used extensively in nineteenth-century California, where 0.75 pounds (0.34 kg) of mercury was released to the environment for every 1000 pounds (455 kg) of ore processed. This caused mercury pollution of 10–100 times background and is still causing problems. Cyanide leaching replaced the mercury amalgam method in which ore is soaked in cyanide, dissolving the gold into solution. Zinc shavings are added to the solution, which bonds to the gold, and the cyanide solution is drained away. The shavings are heated vaporizing the zinc and leaving the gold as residue. Cyanide is deadly to both humans and wildlife if released.

CASE STUDY 14.1 Aurul Gold Mining Disaster

Widespread Cyanide Poisoning

Although the gold mining industry does not use highly polluting mercury to separate gold from gangue, the replacement process using cyanide is even more dangerous. Cyanide is lethal to humans and most life

in even small doses. The cyanide separation process was developed in the 1900s and was modified to cyanide heap leaching in the 1960s. Although the process removes more gold quickly, the impact of gold mining on the environment has skyrocketed. In the process, a cyanide solution is poured over crushed gold ore. The cyanide solution

(a)

FIGURE 14.5 Map showing the rivers impacted by the by the Aurul gold mine spill including timing of the spill and concentrations of cyanide. Inset shows position in Europe. *Source:* Adapted from Environmental Justice Atlas (2015), Pollution from hog farming (CAFOs), USA. (a) Photo of volunteers clearing dead fish from the Taiza River. *Source:* Délmagyarország / Karnok Csaba / Public Domain.

dissolves the gold and it is carried to solution ponds. This process requires large quantities of cyanide. The used cyanide-rich solution is reused, stored in a dam, or discharged untreated into rivers. Accidents involving the release of the cyanide solution can be devastating and the practice is banned in most European Union countries.

The Aurul gold mine and precious metal recovery plant is located in Baia Mare, Romania (Figure 14.5). The mine is owned by Remin, a Romanian group, and the Australia-based Esmeralda mining company. The mine had environmental accidents, especially after a tailings lagoon dam was built in 1998. In 1999, former employees of the mine reported that serious mistakes were made in constructing the dam and Romanian authorities warned the mining group about them. The walls were constructed of sand-rich soil rather than rock, which make it unstable. In fall 1999, cyanide-rich solution was released from the mine's pipe system. As a result, five cows died in a nearby village after drinking cyanide-contaminated water. In December 1999, the dam was reported leaking and Baia Mare paid to have it covered up.

Heavy rain and rapidly melting snow in January 2000 caused the water level in the waste lagoon to rise quicker than the Aurul dam was designed for. As a result, on the night of 30 January 2000, the dam burst, producing an 82 ft (25 m) hole and releasing 3 500 000 cubic feet (100 000 m³) of cyanide-contaminated water that flooded nearby farmland and the Lapus and Someş Rivers with cyanide concentrations up to 700 times the maximum safe level (0.1 mg/l). The US Environmental Protection Agency limits cyanide concentrations to less than 0.07 mg/l in mine wastewater. It is estimated that the waste contained about 110 tons (100 mt) of cyanide.

The Lapus and Someş Rivers are tributaries to the Tisza River, which is the second-largest in Hungary. As the spill spread, the upper part of the Tisza had cyanide concentrations 100 times the limit for drinking water. By mid-February 2000, the spill entered the Romanian part of the Danube River. The large volume of Danube River water diluted the cyanide but in some areas it was still 20–50 times the maximum allowed. The Danube River flows through Serbia, Bulgaria, and Romania and empties into the Black Sea. The spill badly contaminated 250 miles (402 km) of rivers in Hungary and Yugoslavia and spread into Serbia. As a precaution, Ukraine forbade use of water from the Danube. Even tributaries, like the Sasar River, had cyanide concentrations nearly 88 times allowable. The cyanide spill was still evident at the Danube delta in the Black Sea, four weeks after the spill and 1243 miles (2000 km) from the source.

Fortunately, news of the spill spread quickly, which avoided a certain public health disaster. In Hungary, the spill polluted rivers as well as a reservoir that supplied drinking water to the city of Szolnok. This reservoir is just 50 miles (80 km) southeast of Budapest. Residents were warned not to drink tap or surface water. Nine days after the spill, cyanide was still 2.8 mg/l or 28 times the safe limit. It is estimated that the spill contaminated the drinking supplies for more than 2.5 million Hungarians. The interaction between surface water and groundwater

contaminated private wells, with cyanide levels up to 50–80 times the safe limit near the river.

On the other hand, the news was not so good for the ecosystems of the impacted area. The ecological impact was greatest near the mine and the Szamos and Someş Rivers but spread to Tisza River and the Danube. It was estimated that in the area between the source of the spill and Szolnok on the Tisza River, 80–100% of the fish were killed (Figure 14.5a). Virtually all aquatic organisms were killed in this area and farther south, in Serbia, about 80% of the aquatic life was killed. It is estimated that between 200 and 1240 tons (181–1125 mt) of fish were killed in these rivers, many of which washed downstream. Some 62 species of fish including 20 protected species were impacted. These fish contained 2.6 mg of cyanide per kilogram of weight. Many of the contaminated fish, such as carp, catfish, and pike, are eaten by residents. Other wildlife was also affected, including mute swans, black cormorants, foxes, otters, ospreys, and other carnivores through direct contact and biomagnification. It was reported that on 19 February Lake Tisza in the Hortobágy National Park, a recently approved World Heritage site, was impacted by the spill as well as areas protected under the Ramsar Convention of UNESCO.

Residents of Szolnok were warned not to drink their tap water because of high cyanide levels in water supply reservoirs. The drinking of Danube River water was also banned. Fishing in the impacted rivers was banned both by Romania and Hungary in the impacted rivers. The local fishing industry was shut down. Restaurants in the affected area removed fish from their menus. This contaminated water also damaged cattle and horse production.

In Hungary, volunteers removed dead fish from the rivers. Fishermen on Lake Tisza pitchforked dead fish, including three- and four-foot (0.9–1.2 m) long catfish, from the contaminated water. It is estimated that up to 45 tons (41 mt) of fish was removed from the Tisza River in a single two-day period. Even with all of this effort, it was reported that significant numbers of fish were still floating on the rivers.

If this accident was not bad enough, another spill occurred on 10 March 2000, from a different mine in the area. More than 22 000 tons (20 000 mt) of tailings loaded with heavy metals were spilled in the Tisza River tributaries from a dam that failed following a heavy rain.

This disaster shows that an industrial accident can cross international borders. In this case, the pollution was in transboundary rivers and groundwater. The UN established the Economic Commission for Europe (UN/ECE) on the Transboundary Effects of Industrial Convention for the Protection and Use of Transboundary Watercourses and International Lakes to deal with such situations. The agreement compels countries to notify their neighbors immediately in case of an accident. This gives the neighbors time to respond. Unfortunately, most of the countries with the worst effects of this spill did not sign up to the international conventions. Of the countries impacted by the disaster, only Hungary ratified the convention.

14.3 Pollution from Mines

There are many examples of heavy metal, non-metal, asbestos, and radioactive pollution from mining already described and to be described. There is also pollution from processing such as that from gold mining. However, acid mine drainage (AMD) is among the most environmentally damaging products of mining by virtue of its effects and broad scale. The waters draining from a mine or percolating through the tailings can react with sulfur, producing sulfuric acid in the surface water, soil, and groundwater. This hazard is prominent in coal fields and sulfide deposits. Plants and animals do not survive in the acidic conditions and the area is ecologically damaged or dead. The acidity can also dissolve and remove heavy metals like lead, chromium, and arsenic from the tailings, creating an added threat.

Many mineral deposits contain sulfides. These sulfides can be economically valuable if the metal in the ore has industrial applications. Sulfides with zinc, copper, and nickel among others have commercial value. In some cases, sulfur itself is valuable for use in sulfuric acid, matches, explosives, or chemical production. In other cases, sulfides occur as gangue or as accessory minerals and are waste components of an ore deposit. They are removed before ore processing begins and left in waste rock piles around the mine. The sulfur undergoes oxidation reactions to produce sulfurous and sulfuric acid and is flushed out in surface water runoff. Sulfate mineral deposits can produce AMD, but not as bad as that from sulfides.

Acidic runoff is harmful to aquatic life. Freshwater lakes, streams, and ponds typically have a pH range of 5.6–8 depending on the surrounding rock and soil. Acid water has harmful ecological effects if the pH is below 5.5–6. Snails and clams are absent, and populations of many fish decline at pH between 5.5 and 5. Bottom-dwelling bacteria die, allowing leaf litter and detritus to accumulate. This reduces essential nutrients as well as the availability of carbon, nitrogen, and phosphorous for use by other organisms. Fungi begin to replace these bacteria on the substrate. Normal plankton disappear and undesirable plankton and moss can invade the ecosystem. Metals such as aluminum and lead, toxic to aquatic life but normally fixed in sediments, are released into the acid water. For pH below 5, fish populations may disappear. The water bottoms are covered with undecayed vegetation, and mosses spread to near-shore areas. Most fish eggs do not hatch and many insects are unable to survive. At pH below 4.5, the water is devoid of fish, most frogs, and insects. At pH between 3 and 4, fish can survive for no more than a few hours, but some unwanted plants and invertebrates survive. In AMD from coalfields, pH of 3–4 is not uncommon and in sulfide mines, it can be less than 2.

CASE STUDY 14.2 Aberfan Coal Waste Disaster

Extreme Grief from Mining Spoils Failure

One of the most horrifying disasters involving underground mining was at Aberfan in Wales, UK (Figure 14.6). This is a large coal mining area so it involves both energy and underground mining. Wales was a major coal mining district in the Industrial Revolution, with 271 000 workers in the coal mines at its peak in 1920. Aberfan is located on the western valley slope of the Taff Valley, about four miles (6 km) south of Merthyr Tydfil. The first coal mining venture began on 23 August 1869 when Aberfan was barely inhabited with only a few houses. Demand for coal was so strong and growing that by 1966 the population of the town had increased to 5000–8000, primarily employed in the coal industry. The British coal industry was nationalized in 1947 and controlled by the National Coal Board (NCB). The coal industry was regulated by inspectors who had worked in the coal industry and were former employees of the NCB.

Coal mine waste was first dumped on the valley's lower slopes but during the 1910s the first waste piles or tips were begun on the slopes above the village. By 1966, there were seven waste piles containing more than 2.6 million cubic yards (2.0 million m³) of tailings. Tips 4 and 5 were located at the top of the slope. The other five piles were lower on the slope. All tips were located directly above the village. In 1966, only tip 7 was being used. This waste pile was initiated in 1958 and was 111 ft (34 m) high and contained 297 000 cubic yards (227 000 m³) of debris by 1966. This waste included 30 000 cubic yards (23 000 m³) of tailings from chemical extraction of coal, including fine coal dust and ash which became quicksand-like when wet.

Piles 4, 5, and 7 were dumped on top of streams or springs. The large pile/tip 4 was used between 1933 and 1945 and was on soggy ground between two streams. An oversight engineer deemed that despite the water, it was unlikely to fail. However, there was some instability in the early 1940s and a drainage channel was added in 1944. It was not enough because in November 1944 part of the pile slid 1600 ft (490 m) down the mountain, stopping just 500 ft (150 m) from the village. In May 1963, pile 7 shifted and then slid substantially in November. The NCB declared the movement to be a "tailings run" which did not affect the stability and allowed dumping to continue on it.

Aberfan receives an average of 60 in. (150 cm) of precipitation per year. In 1960, it received a recent record of 70.5 in. (179 cm). The heavier precipitation resulted in 11 floods between 1952 and 1965 in and near Aberfan. The flood waters were black and greasy from coal dust. Officials contacted the NCB between July 1963 and March 1964 about coal slurry concerns. In 1965, meetings were held at which the NCB agreed to unclog drainage pipes and ditches to relieve the flooding, but no action was taken.

It rained 6.5 in. (17 cm) during the first three weeks of October 1966 in Aberfan. On 20 October at night, tip/pile 7 subsided 9–10 ft (2.7–3.0 m) and the rails to transport the waste fell into a sinkhole. At 7:30 a.m. on 21 October workers on the heaps found waste slippage. One worker reported the slip and the supervisor decided that no work on the pile would be done that day. At 9:15 a.m., about 140 000 cubic yards (110 000 m³) of water-saturated waste from tip 7 appearing like black quicksand, slid down the pile and flowed down the hill at speeds of 21 miles per hour (34 kph) in waves 20–30 ft (6.1–9.1 m) high (Figure 14.6a). Some reports claim that the wave was 40 ft (12 m) high and traveled at 80 miles per hour (129 kph).

The mass flowed 2100 ft (640 m) down the slope destroying two houses and killing the occupants. About 50 000 cubic yards (38 000 m³) of the mass continued into the village destroying two water mains which further saturated it. As it flowed into town, the mass followed the worst possible track, directly into the Pantglas Junior School while it was in session. The flow destroyed the school engulfing it and filling and crushing classrooms with thick, black sludge and rubble. It is reported to have sounded like a jet taking off as it flowed. It crushed and drowned 109 young children of the 240 students in the school as well as five teachers. It also damaged the adjacent secondary school as well as 18 surrounding houses. The mud flooded other houses, leading to extensive evacuations. The slide finally halted and solidified, leaving a huge mound up to 30 ft (9.1 m) high that blocked the area.

When the waste slide stopped, horrified residents rushed to Pantglas School and began frantically digging through the rubble. Desperate parents clawed through the mud with their bare hands in search of their children. At 9:25 a.m., the disaster was reported to the Merthyr

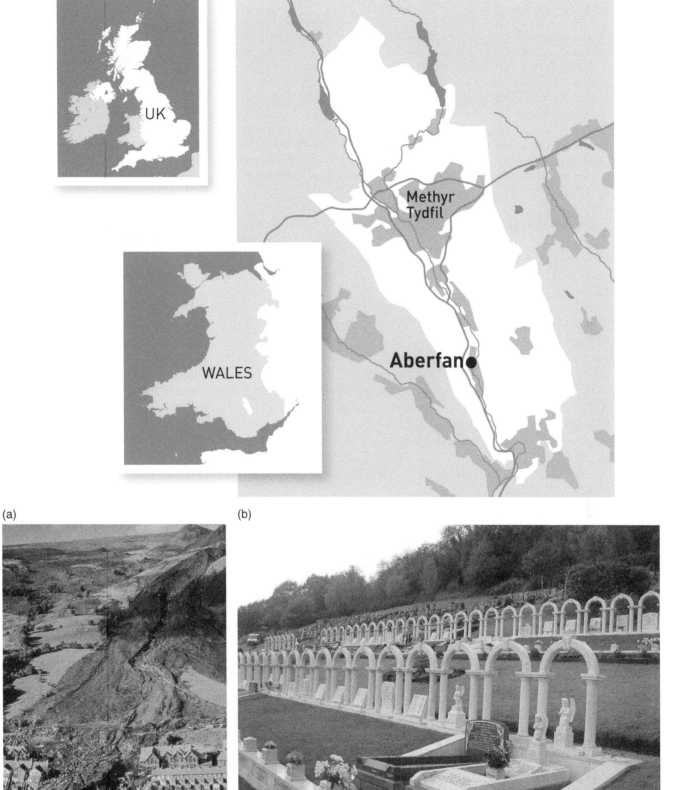

(a)

(b)

FIGURE 14.6 Map of the Aberfan, Wales area. Inset shows the position of the map in a map of Wales and the United Kingdom. (a) Photo showing the coal debris flow down the hill and into Aberfan. *Source:* British Geological Survey. (b) Photo of headstones of the mass grave of the children killed in the Aberfan disaster. *Source:* Llywelyn2000 / Wikimedia Commons / CC BY 3.0.

Tydfil police and the fire brigade. Then local hospitals and ambulance service were informed. Coals miners raced into Aberfan, arriving within 20 minutes of the disaster. They participated in the digging, making sure not to cause further shifting of the unstable mass.

As the digging progressed, it became clear that some staff died trying to protect children. One woman shielded five children using her body and all of them survived. She did not. The deputy headmaster shielded himself and five children from the slurry using a blackboard but he and all 34 students in his class were killed.

The first school casualties arrived at the Merthyr Tydfil hospital at 9:50 a.m. The rest of the rescued survivors arrived before 11:00 a.m. In all, 22 children and 5 adults were in this hospital and 9 other survivors went to another hospital. After 11:00 a.m., no more survivors were found. In all, 144 people died including 116 children, mostly 7–10 years old (Figure 14.6b). Five of the adult victims were teachers at the school. There were also 29 children injured.

The Aberfan disaster was among Britain's worst coal mining accidents but it was, by far, the most heartbreaking. Money was donated from around the world to help the town. However, a lot of this money did not help the bereaved villagers. Instead, much of the money was allocated to remove the remaining waste piles because the NCB refused to cover the costs. There were even reports of fund managers considering distributing compensation based on how close parents had been to their deceased children. Queen Elizabeth also neglected to visit the disaster, an oversight she would regret for years. This incident was emphasized in the 2016–2023 miniseries *The Crown*. In the end, the NCB took responsibility, the impacted villagers received compensation, and Queen Elizabeth paid several visits to Aberfan.

14.4   Smelting and Refining

Most ores require chemical processing for industrial use. The majority of metals are contained in compounds and require chemical separation from other the elements. Only native metals such as gold, silver, platinum, and some copper and iron do not require chemical processing.

There are several steps in the smelting and refining process and they vary depending upon the ore. Smelting is the first chemical process and it is done in a furnace. Ore, fuel, and flux are required for smelting (Figure 14.7). Blast furnaces for smelting are generally operated near the fuel source because so much is required for processing. In most cases, the fuel is coal or coke (processed coal), which is why Pittsburgh, Pennsylvania in the Appalachian coal belt is associated with steel production. Before coal, charcoal was used. The fuel is burned in a blast furnace that is stoked and pumped with air to make it as hot as possible. The ore is then mixed with a chemical flux and placed in the smelter. This drives a chemical reaction that removes unwanted elements and compounds from the ore.

For example, iron is mostly made from magnetite or hematite ore, both of which are iron oxides. Smelting these ores removes the oxygen to make usable iron. Limestone is the flux for this process which is widely available. The limestone is crushed and baked or calcined to remove water and some oxygen before being added to the ore. Once in the furnace, the reaction is:

$$\text{magnetite} + \text{limestone} + \text{energy}(\text{coal}) = \text{iron} + \text{slag} + CO_2$$

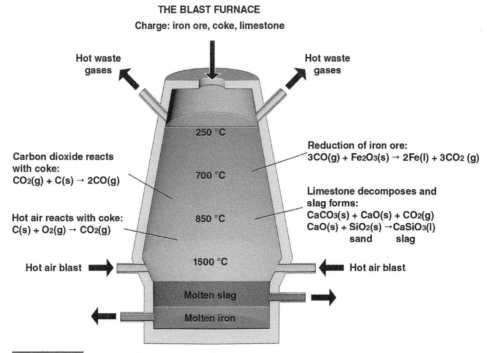

THE BLAST FURNACE
Charge: iron ore, coke, limestone

Hot waste gases

Hot waste gases

250 °C

Carbon dioxide reacts with coke:
$CO_2(g) + C(s) \rightarrow 2CO(g)$

700 °C

Reduction of iron ore:
$3CO(g) + Fe_2O_3(s) \rightarrow 2Fe(l) + 3CO_2(g)$

Limestone decomposes and slag forms:
$CaCO_3(s) + CaO(s) + CO_2(g)$
$CaO(s) + SiO_2(s) \rightarrow CaSiO_3(l)$
      sand   slag

Hot air reacts with coke:
$C(s) + O_2(g) \rightarrow CO_2(g)$

850 °C

Hot air blast ➤

1500 °C

◄ Hot air blast

Molten slag ➤

◄ Molten iron

FIGURE 14.7   Diagram of a blast furnace showing inputs, temperatures, and outputs.

The CO_2 is released to the atmosphere, the dense molten iron sinks to the bottom of the furnace and flows into casting molds. The slag is a compound of leftover elements and is removed and disposed of. This smelter iron has a lot of impurities and must be refined to remove them. This was historically done in a steel mill or forge, where the impurities are driven out by hammering the reheated smelter iron. Today, high-quality steel is produced in a single process.

Mining the iron ore, limestone, and coal is severely impacting to the environment, as is the massive release of CO_2. Additionally, if the iron ore has impurities like chromium, they can be released to the environment. The slag can also contain pollutants that if not disposed of properly, can be damaging. The production of iron, however, is much less damaging than some other metals. Among the most damaging ores are sulfide minerals. Most copper, lead, and zinc are from sulfide ore. In this case, the sulfur is removed from the target metals in smelters. Until recently, sulfur emissions were released to the atmosphere from smelter smokestacks and settled to the ground around the smelter. The sulfur fallout combined with water to make sulfuric acid causing the soil and surface water to be so acidic that it killed all life around the plant. The emissions also contributed to acid precipitation. Today, filters are required to prevent the release of sulfur.

14.5   Petroleum Production and Processing

Petroleum, or oil, and gas extraction, transport, processing, storage, and distribution are probably the most polluting human activities. On the other hand, they are essential to society not only for powering transportation and producing electricity but also for the multitude of essential chemicals that are produced from it. The petroleum industry produces most methane, a powerful greenhouse gas that causes global warming 25 times more than carbon dioxide. It is also the largest source of volatile organic compound (VOC) emissions which help form ground-level ozone or smog. Petroleum-produced VOCs also include air toxins such as benzene, toluene, ethylbenzene, xylene (BTEX), and n-hexane. These air pollutants are either known or suspected of causing cancer and other serious health effects. Some VOCs are even toxic if inhaled.

All crude oil contains these VOCs among other pollutants. At oil spills, VOCs threaten the health of nearby residents and responders to the spill but also all air-breathing mammals, birds, amphibians, and reptiles both on land and in the water. They will also kill most insects and microbes on contact. They are the main reason why spills are so deadly to the natural environment. However, VOCs are primarily a concern as an air pollutant right after the oil is spilled because they quickly evaporate, especially in water. This is because they float. On land, however, the VOCs can be absorbed into the soil and cause damage for much longer periods.

There are other toxic and dangerous substances in crude oil such as polycyclic aromatic hydrocarbons (PAHs). In contrast to VOCs on the surface, PAHs can persist in the environment for many years, potentially continue to cause damage long after the oil is spilled. Oil can also contain creosote which is carcinogenic among other chemicals.

The physical aspects of crude oil can also cause damage the environment. Birds and mammals will die if oil coats fur and feathers so that they no longer insulate. Small organisms can be smothered by oil as can most plant life. Even small amounts of oil can damage the development of fish eggs and embryos in aquatic and marine environments. Natural surface seeps of crude oil can form tar pits that are so sticky that they trap and kill any wildlife that comes into contact with them.

Huge amounts of oil and gas are produced on a continuous basis. Global production of oil peaked at 95 million barrels (15.1 million m³) per day in 2019 but then dropped to 88.4 million barrels (14.1 million m³) per day in 2020 because of the COVID-19 pandemic. This amounted to a 6.1% drop in production. The top oil-producing country surprisingly was the United States because prior to 2005 they imported more than 40% of their oil. Advances in drilling technology helped them produce much more. The other two top producing countries were Russia and Saudi Arabia, followed by Canada, Iraq, China, United Arab Emirates, Brazil, Iran, and Kuwait.

Total natural gas production was 136 trillion cubic feet (3.85 trillion m³) in 2020 on a global scale. This also decreased from more than 141 trillion cubic (4 trillion m³) in 2019 but the decrease was only 2.5%. The top natural gas producer was the United States thanks to fracking, followed closely by Russia. The next producing countries in decreasing order were Iran, Canada, Algeria, Qatar, Norway, China, and the United Arab Emirates.

About 80% of the world's oil reserves are in the Middle East meaning that it will continue to be the major producer into the future. The United States is the biggest oil and gas producer and has several "plays" including from larger to smaller, the Permian Basin in West Texas and New Mexico, the recently opened Williston Basin and Bakken unit in North Dakota, Montana, and South Dakota, Gulf of Mexico in coastal and offshore Texas and Louisiana, the Anadarko Basin of Oklahoma, the Appalachian Basin in Pennsylvania, West Virginia and parts of the states surrounding them like New York, Kentucky, and Ohio, the Denver-Julesburg Basin of Colorado, Kansas, Nebraska, and Wyoming, and the Fort Worth/Barnett Basin, Texas.

Crude oil is a complex combination of many chemicals that vary in type and quantity depending on the rocks in which it was formed. Typically, this is determined by the area where it comes from. Crude oil is processed into gasoline, diesel fuel and heating oil, jet fuel, petrochemicals, waxes, lubricating oils, and asphalt among other (Figure 14.8). Basically, crude

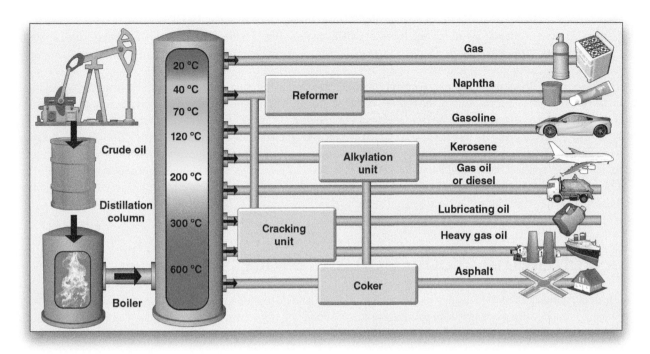

FIGURE 14.8 Diagram of a distillation tower in an oil refinery.

oil consists primarily of the long-chained organic compounds groups of paraffins, naphthenes, and aromatics in various proportions but there are other common compounds as well. Paraffins are the most abundant hydrocarbons in most crude oils and some liquid paraffins are the major components of gasoline. Naphthenes form heavy asphalt-type residues. Aromatics form the smallest part of most crude oils. The most common aromatic is benzene, which is heavily used in the petrochemical industry.

This mixture of components in crude oil are separated into the usable products in an oil refinery through two main procedures. The first step is through distillation. The crude oil is heated in the base of a distillation tower until it evaporates. The gases rise up in the tower and condense at different levels or heights based on their density and condensation temperature. The lighter liquids, like jet fuel, condense in the higher parts of the tower and the heavy material, like tar, are at the bottom of the tower. The amounts of each product depend on the type of crude oil and usually do not match market demand. To address this, the refinery engineers these long-chained organic molecules into the desired proportions at the time. In most cases, the long chain molecules are broken into shorter chain molecules, using catalysts in a process called "cat cracking." In other cases, molecules can be combined to build needed compounds. As part of the refining process, natural gas is produced, which has historically been flared or burned off. Elemental sulfur is also produced. Products can also be combined to build other compounds.

Crude oils are classified on a number of systems from location types to more universal systems. The widely used American Petroleum Institute (API) gravity scale assigns pure water API gravity of 10°. Liquids lighter than water have API gravities higher than 10. Crude oil is classified as heavy, medium, and light based on API gravity. Heavy crude has API gravity of 10–20°, medium has 20–25° API gravity, and the API gravity of light oil is above 25°. In addition, crude oil can be categorized as "sweet" or "sour" if the sulfur level is low or high, respectively.

14.5.1 Exploration and Production

The process of obtaining oil and natural gas begins with exploration. Geologists search for a deep underground "trap," an oil or gas accumulation that can be extracted. Although oil and gas can reach the surface, at this point, most prospects are between 10 000 and 30 000 ft (3000–9700 m) deep. At this depth, traps must be identified mainly using seismic reflection surveying, basically, an underground sonogram. Once a subsurface trap is identified, exploratory wells are drilled to determine if economic quantities of oil or gas is present. If it is, development or production wells are drilled to extract it. These wells are drilled using a drilling rig. On land, these can be drilled individually or from a drilling center.

Offshore oil wells are more involved. They begin with a drillship with drilling equipment mounted on the deck that can drill through an opening in the hull. Drillships operate in waters 2000 ft (610 m) deep or more. Next, a jack-up rig is used.

This floating barge is equipped with extendable legs and is towed to the drill site. The legs raise or "jack-up" the barge off the water surface, making it a stable drilling platform. This type of rig is used in relatively shallow water less than 500 ft (152 m) deep. A semi-submersible rig is also on a barge, but the barge contains large pontoons instead of legs. The pontoons are flooded to submerge the hull of the barge just below the waterline and it is anchored to the seafloor with cables. They are used in waters 500–2000 ft (152–610 m) deep.

Once the location of an offshore oil or natural gas deposit is confirmed, a permanent drilling platform is constructed. Platforms are large, steel structures with a deck just above the water surface. The deck is attached to large-diameter pipes extending to the sea floor to hold it in place. Several wells can be drilled into the accumulation to produce it. The oil or gas are then commonly transported to the shore using a subsea pipeline. Many of these pipelines are hundreds of miles long and connect platforms to a refinery or distribution center.

The oil and gas in the reservoir or trap is under high pressure as a result of burial beneath thousands of feet of sediments. Careful precautions must be taken during drilling to prevent oil and gas from rushing to the surface in an uncontrolled release, or blowout. These commonly produce intense fires because the oil and gas ignites. The fires are dangerous and difficult to extinguish. Blowouts in offshore wells commonly result in profound ecological damage as the oil contacts marine life and their fragile habitats. These blowouts are avoided using drilling mud and mechanical valves called blowout preventers.

During drilling, the rock grindings are removed using drilling mud, a mixture of clay, water, and sometimes barium to add density. The mud is circulated up the well to carry the ground rock chips to the surface. Drilling mud also is used to keep the well pressurized to prevent oil and gas from entering the well until drilling is finished. Unexpected high pressures in the well can still cause a blowout. Release of the mud into the ocean can also cause problems. Oil-based mud can be toxic and its release into the ocean is prohibited. Spills of this mud are therefore serious. Water-based muds and synthetic-based muds are much safer.

14.5.2 Transportation

The largest and most devastating oil pollution events involve spills of petroleum from tanker accidents, from broken pipelines on land or barges or ships on major inland waterways, in addition to blowouts of wells and drill platforms at sea during acquisition. Tankers are huge ships that carry enormous amount of crude oil that leave no room for error. Many get caught in storms, strike unanticipated rocks and reefs, or strike ships and structures in harbors. They typically spill the oil at the shoreline where it does the most damage. The fresh oil still contains its most toxic VOCs when it is first spilled in the most fragile of coastal marine and shoreline terrestrial habitats with high biodiversity. Although they can spill much more oil than tanker accidents, blowouts are typically far from the shoreline. By the time the oil slick reaches the shore, it has lost the vast majority of its toxic components.

A spill on land can include severe blowouts, but the most common involve a pipeline rupture. The causes of pipeline ruptures include faulty pipe seam welds, faulty pumping equipment, earthquakes, sabotage, deliberate spills, and even hunters using above-ground pipelines for target practice. The amount of oil spilled from pipelines is not well quantified in many parts of the world. In 2016, the United States had more than 73 000 miles (117 500 km) of crude oil transporting pipelines. These pipelines transport more crude oil than tankers or trains. These pipelines frequently experience spills totaling nearly 9 million gallons (34 million l) between 2010 and 2016. The highest amount of crude oil spillage from pipelines in a single year was more than 60 000 barrels (9.5 million l) in 2016. In 2020, it decreased to 43 157 barrels (6.9 million l) accidentally spilled. Sensors are installed in pipelines to prevent leaks but the amounts remain unacceptable and extremely damaging to the environment.

14.5.3 Refining

Refining crude oil into usable products also produces releases, spills, and leaks. Refineries have been a major source of hazardous and toxic air pollutants though newer procedures have reduced them significantly. VOCs are directly dangerous, ubiquitous in air, and primary reactants in the formation of ground-level ozone, which is perhaps the most dangerous broad-scale air pollutant both to human health and the environment. VOCs include BTEX compounds (benzene, toluene, ethylbenzene, and xylene) among others and are the reason for the odor of gasoline, paint thinner, oil paint and stain, adhesive, and industrial solvents among others. Refineries are also a major source of other air pollutants such as particulate matter (PM), nitrogen oxides (NO_x), carbon dioxide (CO_2), carbon monoxide (CO), sulfur dioxide (SO_2), and hydrogen sulfide (H_2S). Refineries also release the potent greenhouse gas, methane, though most is flared off. Some of these chemicals are threats to public health being known or suspected cancer-causing agents, and responsible for developmental and reproductive problems.

Accidental releases at refineries are common, severely polluting the soil and groundwater beneath and around them. They are among the worst polluted sites as a result, requiring near constant remediation and monitoring. VOCs can be released into the soil before they evaporate and percolate all the way to the water table, killing most microbes in the process. The soil and groundwater are contaminated in the process. There are many other dangerous chemicals in a refinery that follow the same path. In addition, refinery products are sent to factories in bulk for industrial use or chemical manufacturing. Many refineries are on the coast or large rivers so the products may be transported by ship. The transfer process from refinery to ship is fraught with accidental spills damaging the surface waters and marine or aquatic ecology around the refinery as well. Transfer to tanker cars on trains or to tanker trucks is also messy, further adding to the soil and groundwater pollution. These issues typically make refineries devoid of natural life.

CASE STUDY 14.3 2005 BP Refinery Explosion

Deadly Industrial Accident

The BP refinery in Texas City, Texas was the site of recurring safety violations but nothing like the accident that occurred on 23 March 2005. The refinery was built in 1934 and was operated by Amoco until 1999 when the company became part of BP. By 2005, the refinery was processing about 437 000 barrels (69 500 m³) per day. It was the second-largest oil refinery in Texas, and the third-largest refinery in the United States. It produced 3% of the United States gasoline supply.

In January 2005, numerous safety issues were found at the refinery including broken alarms, thinned pipes, broken concrete and bolts, and dangerous fumes. In response, several overhauls and repairs were made. Work was initiated on the raffinate splitter, the adjacent Ultracracker Unit (UCU), and at the Aromatics Recovery Unit (ARU) on 21 February 2005. The isomerization (ISOM) unit converts low octane hydrocarbons into higher octane varieties to be blended into unleaded gasoline. One component of ISOM is the raffinate splitter, a 170-ft (50-m) tower to separate lighter hydrocarbon components and pump them into the light raffinate storage tank. Heavier components were lower in the splitter and pumped to a heavy raffinate storage unit. It had a capacity of 45 000 barrels (7200 m³) per day.

Other serious safety issues had not been repaired, including a broken pressure-control valve, a broken alarm in the splitter tower, and a defective fluid level sight glass. Also, the splitter-tower level transmitter was not calibrated.

Nonetheless, on 23 March 2005, at 02:05 a.m., an ISOM unit at the Texas City refinery was restarted after being offline for maintenance. The BP procedures were followed. However, the raffinate level was not calibrated for the specific gravity of the fluid. As a result, the fluid level indicated 9 ft (2.7 m) when it was really 13.3 ft (4 m) and the sounding high-level alarm was ignored. The secondary high-level alarm was broken. The restart was stopped for the day at that point, but the information about the ordered delay was not passed on to the day shift and the start-up procedure was resumed at 9:30 a.m. By 10 a.m. raffinate was being fed into the tower where the level was already too high and it could not be checked.

The furnace was heating the raffinate and it was turned up at 11:16 a.m. The heating procedure was not followed and a defective level transmitter indicated a safe level even though it was too high. Just before noon, with heat increasing in the tower, pressure started building in the tower and pipework. The operations crew simply released the pressure. At 12:42 p.m., the furnaces were turned down and heavy raffinate was drained from the splitter tower. However, there was far more raffinate in the system than believed and it was still expanding. At 1:13 p.m., the three pressure relief valves burst open, allowing 52 000 gal (196 000 l) of heated raffinate into the collection header in six minutes. This collection system was installed in the 1950s and it was antiquated and unsafe. It had never been tested to safely contain liquids. Liquids flowed into ISOM drains and then shot straight up into the air in what was described as a geyser at least 30 ft (9 m) high. The petroleum rained around the ISOM unit, releasing clouds of flammable vapors that spread quickly throughout the area.

A truck was idling 25 ft (8 m) from the blowout. The vapor cloud reached the truck and caused the engine to race. Nearby workers tried to shut off the truck, without success as the cloud spread across the ISOM area. At 1:20 p.m., the truck backfired, igniting the vapor cloud. This triggered a massive explosion that was heard for several miles (Figure 14.9). The blast struck the contractor trailers, completely destroying them, and flying debris instantly killed 15 people and severely injuring 180 others. The blast blew out windows and damaged homes up to three quarters of a mile (1.2 km) away. More than 200 000 square feet (19 000 m²) of the refinery was badly burned by the fire that followed the explosion, damaging equipment worth millions of dollars. The ISOM unit was badly damaged by the explosion and fire, and it was shut down for over two years.

The explosion spewed black tar all around the plant and black smoke into the air, all loaded with PAHs and other cancer-causing compounds (Figure 14.9a). This toxic fallout caused extensive environmental damage in the area and officials issued a shelter-in-place order. It required 43 000 residents to remain indoors for several days. The accident is considered among the worst industrial disasters in recent history.

The US Chemical Safety and Hazard Investigation Board identified numerous technical and procedural problems at the refinery. On 4 February 2008, BP offered to plead guilty to a federal environmental crime and pay a $50 million fine. Victims and their families objected to the

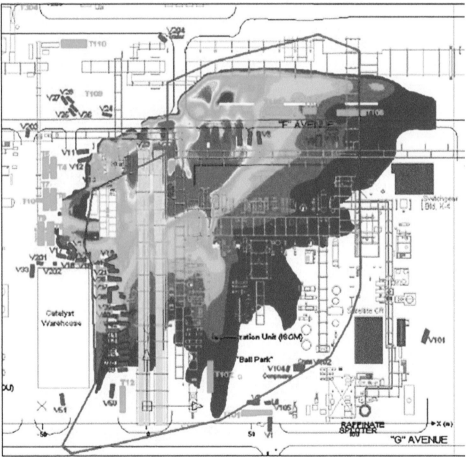

Map of the Texas City BP refinery showing the contoured toxic cloud spreading from the leak (Source: John Nordin / Aristatek, Inc.). Inset map of Texas showing the location of Texas City. (a) Aerial photo of the BP Texas City refinery with fire fighters extinguishing the fire from the 2005 explosion. *Source:* Courtesy of US Chemical Safety and Hazard Investigation Board.

(a)

FIGURE 14.9 (Continued)

proposed fine for being so small. On 30 October 2009, the Occupational Safety and Health Administration (OSHA) imposed a record $87 million fine on BP for failing to address safety hazards uncovered in the 2005 accident. OSHA also cited over 700 safety violations. BP claimed to have paid in excess of $1.6 billion in compensation to the victims.

Eva Rowe lost both of her parents in the explosion and attracted nationwide attention because she said that she would not accept a settlement from BP and would bring them to court. The television magazine "60 Minutes" host, Ed Bradley, published her case.

Military and Pollution

CHAPTER OUTLINE

Polluted Earth: The Science of the Earth's Environment, First Edition. Alexander Gates.
© 2023 John Wiley & Sons, Inc. Published 2023 by John Wiley & Sons, Inc.
Companion website: www.wiley.com/go/gates/pollutedearth

Words you should know:

Agent Orange – A defoliant that was sprayed from the air to increase visibility and speed of passage in the jungles of Vietnam. The active ingredient is dioxin.

Atomic weapon – A highly destructive weapon in which nuclear fission provides the energy for the explosion.

Chemical weapons – Any weapon in which chemical reactions and interactions disable, injure, or kill soldiers. Toxic gas is the most common.

Conventional weapons – Gunpowder, TNT, and other non-nuclear or chemical-based weapons.

Defoliant – A chemical herbicide designed to remove vegetation for warfare.

Military base – Military facility to train soldiers and to maintain a strong and ready force for wars.

Nerve gas – Organophosphorus cholinesterase inhibitors that attack the nervous system and kill quickly.

Nuclear weapons – Weapons in which nuclear fission or fusion provides the energy for the explosion.

Underground storage tanks (USTs) – Large tanks used to store fuel or other chemicals underground.

15.1 War and Pollution

War is meant to kill people with basically no concern for the environment or public health. With the number of human casualties, the focus is understandable. During D-Day in World War II, tons of lead bullets were shot into Normandy beaches, bombs and flares emitted sulfur and phosphorous compounds into the water and air, landing craft and ships spilled diesel fuel into the ocean, and explosions released polycyclic aromatic hydrocarbons (PAHs) and other chemicals into the air and water. However, 5500 young American men lost their lives in the battle, making the environmental aspects far less important. The marine life even suffered more from the exploding bombs than it did from all of the pollutants. Some of the worst environmental disasters, however, are related to military action or its preparation.

Early conflicts were not friendly to the natural environment. Vegetation was trampled and killed during battles, which exposed bare soil. This enhanced erosion of the area during rains silting streams and ponds. In dry weather, wind blew soil as particulate air pollution. Settlements were commonly burned which produced air pollution containing PAHs, particulate, methyl ethyl ketone (MEK), and other pollutants depending on the fuel. Fire was also used as a weapon, burning wood, straw, and other vegetation. Burning oil and other flammable substances were also used as weapons. Even chemical warfare was used by the Greeks in 431 BCE who mixed sulfur with pitch resin to produce suffocating fumes during the Trojan War. This burning increased particulate and other air pollutants. The burned oil added to soil pollution and water pollution through runoff. The partially burned oils produced PAHs, creosote, and other organic pollutants such as benzene and xylene depending on the oil.

Even preparation for war damaged the environment. In the Bronze and Iron Ages, smelting of ore to produce the metals for weapons, shields, and armor produced extensive pollution in the mining and processing areas. Waste rock from mines polluted soil and the weathering of fresh rock and ore surfaces increased heavy metals and other compounds in the soil and runoff. Heavy metals included lead, chromium, cadmium, mercury, arsenic, and zinc depending on the ore. Smelting increased air, soil, and water pollution with these heavy metals in addition to sulfur, phosphorous, and other compounds. Fire for processing added particulate and organic pollutants to the air as well.

Chemical weapons were developed to increase effectiveness of war. The European Dark Ages had only small advances but in China gunpowder was developed, which changed war forever. Mining and processing to obtain, refine, and produce sulfur, saltpeter, and charcoal and to process them into gunpowder locally increased pollution. Gunpowder produces smoke or particulate that contains sulfur compounds that pollute the air. After battles, this smoke hung over the battlefield. However, considering the number of deaths from the projectiles shot using gunpowder, concern for pollution was non-existent. There was also no concern for lead in the ground from bullets.

In the nineteenth century, great advances in chemistry were quickly used on the battlefield. The first chemicals were medicinal, including drugs and anesthesia. However, the new chemicals were also quickly used as weapons.

15.1.1 World War I Pollutants

Chemical weapons were widely used and produced horrifying results in World War I (Figure 15.1). Approximately 88 500 people were killed by exposure to chemical weapons and 1 240 850 were injured. The weapons were called gas but were primarily vaporized liquids. Although they were feared and caused the hideous deaths, they were not that effective. They caused about 4–5% of the casualties and were ineffective in windy and rainy conditions and later protective clothing made them marginally effective. The first chemical weapon was tear gas launched in shells to temporarily incapacitate the enemy, making them easier targets for infantry. Ethyl bromoacetate was used as tear gas by the French on 19 August 1914 and then by the Germans on 19 October 1914. Neither of these efforts was effective, nor was the Germans' use of xylyl bromide (t-stoff) on 19 January 1915. It was easy to avoid the effects by wearing goggles.

Chemical weapons were used to kill enemy troops on 22 April 1915 when the Germans used chlorine gas against French troops at Ypres, France. The gas was released from canisters forming a cloud that drifted over enemy positions, killing 1000 soldiers and injuring 4000 others. Chlorine forms hydrochloric acid when moistened, which strips off the membranes in the eyes, nose, throat, and lungs. At high exposure, it causes pulmonary edema and death by asphyxiation. Initial releases were especially effective because soldiers tried to outrun the cloud and so spent more time inhaling the chlorine as the cloud overtook them. They would then inhale the chlorine more deeply from panting. The practice of canister release also proved to be a problem at Loos, France, when the chlorine gas cloud was blown back over the British troops who released it. In response, the gas was loaded into shells and shot into enemy positions. Chlorine gas was found to be filtered by simple masks so was not used alone for long.

In 1915, the Germans used the much deadlier phosgene gas. Phosgene changes from a liquid to a gas at 47 °F (8 °C), which is four times as dense as air. It was commonly mixed with chlorine to help spread it better. About 80% of chemical warfare deaths were the result of phosgene. The high density allowed it to travel along the ground and fill the trenches, which became the battle tactic of the war. Phosgene can only be smelled if concentrations are high enough to cause irreparable damage. It causes burning of the eyes and nose, blurred vision, lesions, nausea, and vomiting, difficulty in breathing, pulmonary edema, and death from heart failure. It is even deadly at concentrations below detectability. Soldiers were unknowingly exposed and did not show symptoms for as much as two days. At that point, pulmonary collapse occurred and was ultimately fatal.

The most notorious chemical weapon of World War I was mustard gas. Mustard gas is bis(2-chloroethyl) sulfide and first used by the Germans in September 1917. At high doses, mustard gas strips flesh to the bone. Symptoms appeared 2–24 hours after exposure, beginning with a headache and fever followed by the skin and mucous membranes blistering before being stripped away.

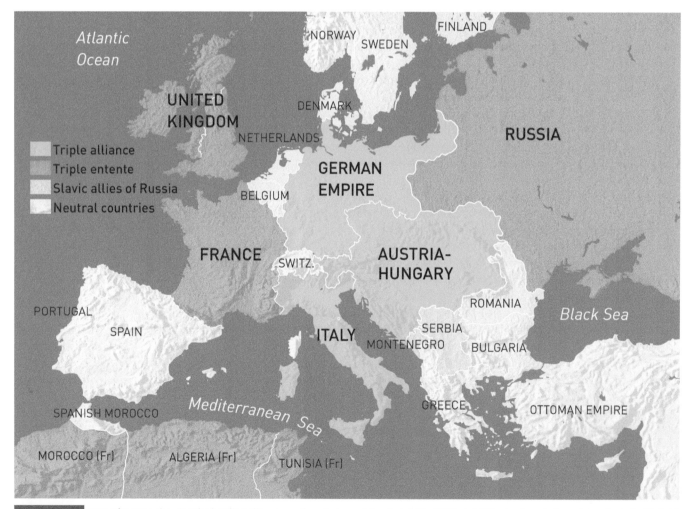

FIGURE 15.1 Map of Europe showing the battling alliances and neutral countries during World War I. (a) Photo of gas being released on a battlefield during World War I. *Source:* Unknown author / Wikimedia commons / Public Domain. (b) Photo of soldiers being exposed to chemical weapons on a battlefield during World War I. *Source:* Courtesy of US National Archives and Records Administration (NARA) / Wikimedia Commons / Public Domain.

(a)

(b)

FIGURE 15.1   (Continued)

Pneumonia developed but death took four to five weeks. During that time, the soldier was in such searing pain that they had to be strapped to the bed. If anything touched their skin, even clothing or bandages, it put them in sheer agony. It was a horrible way to die. Fortunately, by the end of the war, soldiers were equipped with relatively sophisticated gas masks and protective clothing, so gas attacks were uncommon. Mustard gas was classified as a known human carcinogen by the International Association for Research on Cancer (IARC), causing increased risk of lung and other respiratory cancers. More than 28 000 American soldiers were exposed to non-lethal doses of mustard gas. The United States developed an even deadlier gas called Lewisite (β-Chlorovinyldichloroarsine) but the war ended before it could be deployed.

The horror of chemical weapons led them to being banned by the Geneva Convention of 1925. The United States still produced 40 000 pounds (18 000 kg) of chemical weapons by the end of World War I and continued to produce them until 1968. By then more than 34 million pounds (15 400 mt) had been stockpiled. In 1997, the United States signed the Chemical Weapons Convention treaty requiring the destruction of all chemical weapons by 2007. The treaty included 183 countries.

World War I is considered the most damaging to the environment because of changes to the landscape from trench warfare. Digging trenches caused destruction of grassland and churning of soil. Forest logging to build the network of trenches resulted in excess erosion and natural soil structures were severely altered.

The artillery bombardment destroyed numerous villages, rendering them uninhabitable and the land unusable. These areas were littered with human and animal remains, unexploded ordnance, and the soil was polluted with toxic chemicals. The French rebuilt some villages but others were known as "villages that died for France" and were not rebuilt. Environmental pollution also resulted from damaged or corroded munitions. Bullets and shrapnel were mostly made of lead and percussion caps used mercury or lead as detonators. The British alone used billions of cartridges containing heavy metal pollutants.

More than 125 million tons (113 million mt) of toxic gases were used in World War I. Organoarsenic and other toxic compounds have contaminated the soil and surface water in many areas. Recently, the warfare chemical 1-oxa-4,5-dithiepane was measured at levels up to 250 µg/l in the groundwater in the German city of Munster, exceeding safe levels more than 100 years later. This compound is classified as toxic to aquatic organisms. The war-generated toxic compound 1,2,5-trithiepane was also found in groundwater at excessive levels and arsenic-bearing weapon compounds have been found in soil at high enough levels that they required remediation.

Mustard gas is relatively persistent in the natural environment. It is extremely toxic to all animals but its environmental impact is limited by its low solubility. Fish are the most sensitive species to mustard gas. Large quantities of it persist underwater for long periods of time and retain toxicity. Mustard gas was disposed of in the oceans. Once spilled into seawater, it sunk and remained on the seafloor, where it slowly dissolved. Because of the low solubility, mustard gas and intermediate decay products are persistent, and remain underwater for some time.

Unfortunately, disposal of munitions into the ocean was regarded as safe and environmentally benign for many years. It was deduced that any leak of toxic chemicals would be diluted before they could cause any damage. With this assumption in mind, the British dumped more than 120 000 tons (109 000 mt) of chemical weapons off their coast in the mid-1950s, most produced during World War I. In total, over 300 000 tons (272 000 mt) of chemical weapons were dumped into the ocean worldwide after World War II. Research has since shown that bioaccumulation of arsenic in seafood is likely the result of ocean-dumped chemical munitions. Chemical weapons have wound up in fishermen's nets and washed up on beaches worldwide, resulting in numerous deaths and injuries. It was not until 1972 that ocean dumping was banned, but it was not enforced until 1975.

In 2005, the amount of phosgene in the atmosphere was found to be significantly higher than anticipated. Phosgene contributes to the destruction of the stratospheric ozone layer.

15.1.2 World War II Pollutants

Chemical weapons were not used in World War II but there were newly developed pollutants that were significant (Figure 15.2). With the removal of horses and other animals from the war and the larger weapons, there was a greatly increased use of petroleum both for transportation and weaponry. Warfare involved extensive use of ships, tanks, airplanes, and trucks and jeeps, all of which

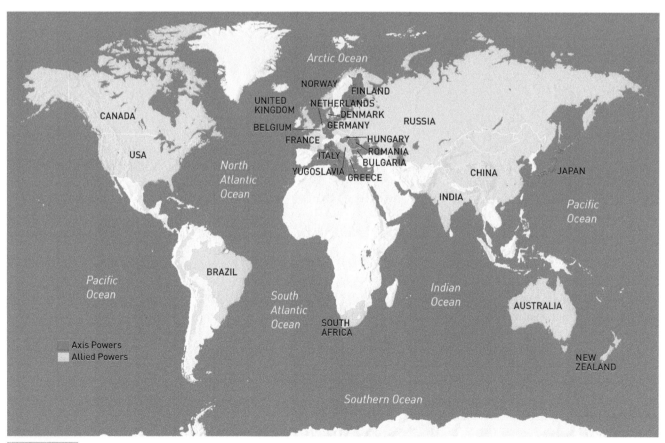

FIGURE 15.2 Map showing the allies, axis, and neutral countries during World War II.

required fuel. Transportation mishaps and destruction of vehicles during the war resulted in huge amounts of toxic fuel being released to the environment. Fuel was also used in flame throwers and pumped into the cave systems in the volcanic islands of Japan (Guadalcanal, Iwo Jima, etc.) and ignited. Gasoline was used in flame throwers but later a compound of naphthalene and palmitic acid called napalm was used. Bombs were sophisticated and abundantly used during World War II. Incendiary bombs used phosphorous and napalm among other dangerous substances. Napalm produces enough carbon monoxide to cause people to pass out near the explosion. Pesticides such as DDT were heavily used to rid areas of disease-carrying mosquitoes in swamps in numerous campaigns. When chemical factories were destroyed during bombing, large quantities of dangerous chemicals were released to the environment.

Zyklon-B or cyanide was not used in battle but it fueled the gas chambers in the Nazi concentration camps. The Nazis crowded people into small tightly sealed rooms and released Zyklon-B which vaporized causing quick death by asphyxiation. Millions of civilians were executed and burned in crematoriums. This atrocity, including the Holocaust as its major component, is one of the darkest chapters in history.

Nerve gas is a chemical weapon developed but not used during World War II. The chemical weapons used in World War I allowed soldiers to fight for a day or two before symptoms disabled them. This time lapse could allow soldiers to win a battle or be replaced. Nerve gas incapacitated troops immediately. In anticipation of deploying it, much nerve gas was produced and stockpiled. The first nerve agent was discovered by accident on 23 December 1936 in a pesticide laboratory in Germany. This agent, tabun, was very potent on insects but an accidental spill proved that it was also lethal for humans. A sample sent to the Nazi high command in May 1937 led the German military to relocate the research team to a secret research lab for the duration of the war. The research team developed additional nerve agents including sarin in 1938, soman in 1944, and cyclosarin in 1949.

Nerve agents are organophosphorus cholinesterase inhibitors that attack the nervous system. They inhibit acetylcholinesterase, which results in excess acetylcholine and does not permit muscles to relax. Symptoms of exposure to gas appear within seconds to minutes. Symptoms initially include runny nose, tearing, constricted eye pupils, and eye pain, followed by blurred vision, coughing, excess salivation, tightness of the chest, and rapid breathing. The victim becomes confused, drowsy, and weak, with muscle spasms, headache, nausea, vomiting, and irregular heart rate and blood pressure. Finally, the victim passes out and has convulsions and paralysis followed by death by respiratory failure. This reaction takes about four minutes.

A second group of nerve agents were also discovered by mistake in Great Britain. The first compound was Amiton, released as a pesticide in 1954, but it was too toxic and removed from the market. The military assumed control of research on Amiton and renamed it VG. The rest of the V-series include VE, VM, and the infamous VX. The V-series nerve agents are 10 times as powerful as the German agents and far more persistent. The United States traded information on developing thermonuclear weapons with the British for VX technology in 1958. By 1961, the United States began research and production on its own chemical weapons.

The only use of chemical weapons was the Iran-Iraq War of 1981–1988. Sarin was used in a terrorist attack in a subway in Tokyo, Japan in 1995. Until 1972, the United States military disposed of chemical weapons by ocean dumping. They dumped 32 000 tons (29 090 mt) of mustard and nerve gases in 26 poorly located sites off the east and west coasts. If the dumped steel drums leak, the nerve agent could persist for up to six weeks in the ocean, killing every marine organism. Part of the drastic decline in marine life on the coastal United States may be the result of leakage.

15.1.3   Vietnam War Pollution

The Vietnam War primarily used conventional weapons, meaning most environmental effects were similar to previous wars (Figure 15.3). The two main differences were napalm-B and Agent Orange. Napalm-B or "super napalm" was made of benzene and polystyrene. Napalm-B produces an abundance of partially burned hydrocarbons like PAHs and carbon monoxide. Agent Orange is a herbicide that was sprayed from the air to increase visibility and speed of passage in the jungle. It was applied from 1961 to 1971 before it was banned (Figure 15.3a). By then Agent Orange had contaminated about 50% of Vietnam's forests and farmland, most of which is struggling to return to productive use. Agent Orange is part of a lawsuit by Vietnam veterans because the dioxin it contains causes several types of cancer among other health problems (Figure 15.3b).

15.1.4   Nuclear War

Nuclear radiation is the deadliest military threat to the environment. Radiation was discovered in 1895 and by the late 1930s it was being tested as a weapon. It was used as a weapon by the United States against Japan to end World War II. This radiation was produced by atomic bombs developed in the United States by a team of research scientists in the Manhattan Project. The Manhattan Project was a concerted scientific effort to develop a powerful weapon. The project was named for the US Army Corps of Engineers district in New York City and involved British and American scientists without the Soviet Union. American physicist

Robert Oppenheimer was the leader and he enlisted the participation of prominent scientists fleeing Europe. Some of his extensive laboratories developed into small cities, including Los Alamos, New Mexico, Oak Ridge, Tennessee, and Hanford, Washington, each of which have significant pollution issues.

The first nuclear weapon test was in New Mexico on 16 July 1945. It was a plutonium implosion device that produced an explosion with an equivalent power of 19 kt (17 300 mt) of TNT. Because of this success, President Truman ordered the bombing of Japanese cities to avoid a lengthy invasion which would cost millions of lives. On 6 August 1945, the city of Hiroshima was destroyed using a uranium gun assembly bomb but Japan still refused to surrender (Figure 15.4a). As a result, on 9 August 1945, the city of Nagasaki was destroyed using a plutonium implosion assembly bomb (Figure 15.4b). The blasts killed over 100 000 people each and nuclear fallout killed tens of thousands from radiation sickness over the next few years. This remnant radiation posed a significant environmental threat for many years.

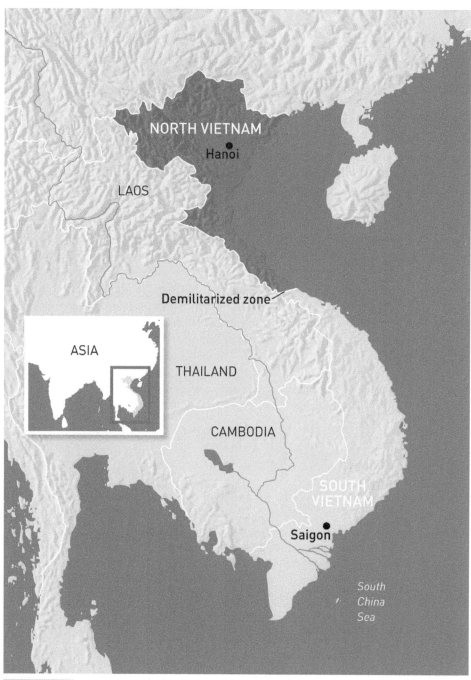

FIGURE 15.3 Map of North and South Vietnam before and during the Vietnam War. Inset shows location in Southeast Asia. (a) Top photo shows an area of Vietnam prior to the application of Agent Orange defoliant and the bottom photo show the effects on the same area. *Source:* Wikimedia. (b) Photo of aircraft spraying Agent Orange on the jungles of Vietnam. *Source:* USAF / Wikimedia Commons / Public Domain.

(a)

(b)

FIGURE 15.3   (Continued)

On 29 August 1949, the Soviet Union tested an atomic bomb. The fear of this enemy advance spurred a new research effort. As early as 1942, it was recognized that nuclear fission could generate enough energy to initiate nuclear fusion which could produce a more powerful weapon, but the Manhattan Project group did not favor pursuing it. Edward Teller, however, continued to research it despite the wishes of his superiors. Even after the Soviet test, Robert Oppenheimer still did not favor development of a fusion bomb. He was overruled by President Truman, who announced on 31 January 1950 that the United States would develop a fusion bomb.

The program, led by Teller, detonated the first hydrogen bomb on 1 November 1952 in the Marshall Islands (Figure 15.4c). The device yielded an explosion equivalent to 10.4 mt (9.5 million mt) of TNT. It was 450 times as powerful as the Nagasaki bomb and obliterated an island, producing a crater 6240 ft (1.9 km) wide and 164 ft (50 m) deep. The products of fusion are not dangerous but a primary fission reaction initiates the fusion reaction. When the fusion reaction begins, it pulverizes the radioactive fission material into the explosion cloud. The environmental danger of the radioactive cloud was realized on 28 February 1954 when the United States detonated a thermonuclear weapon at Bikini Atoll, Marshall Islands. The explosion was equivalent to 15 mt (13.6 million mt) of TNT and produced a radioactive cloud of 7000 square miles (18 130 km²). A mist of nuclear fallout covered several islands that were evacuated and are still uninhabitable.

In the late 1950s and early 1960s, numerous thermonuclear tests yielded extensive nuclear fallout that was distributed worldwide by weather systems. This fallout settled on soil, surface water, and vegetation and was consumed by humans and animals throughout this period. Early nuclear tests were either underwater or above ground. In 1963, 116 countries signed the Limited Test Ban Treaty eliminating the testing of nuclear weapons underwater, in the atmosphere, and in outer space. It did not limit underground tests, which were still conducted frequently. In 1968, 133 countries signed the Nuclear Non-Proliferation Treaty, which was made permanent in 1995. The limitation and reduction of nuclear weapons began in the 1970s with the Strategic Arms Limitation Treaties (SALT). These were started because there was more killing power stockpiled in nuclear weapons than there were people.

(a)

FIGURE 15.4 (a) Photo of the damage done to the city of Hiroshima, Japan from the first atomic bomb. *Source:* Courtesy of the US National Archives. https://museum.archives.gov/featured-document-display-atomic-bombing-hiroshima-and-nagasaki. (b) Photo of the mushroom cloud from the detonation of the Nagasaki, Japan atomic bomb. *Source:* Courtesy of the US National Archives. https://catalog.archives.gov/OpaAPI/media/535795/content/arcmedia/media/images/28/14/28-1378a.gif. (c) Photo of the mushroom cloud from the detonation of the thermonuclear bomb at Bikini Atoll. *Source:* Courtesy of the US Library of Congress. LOC https://en.wikipedia.org/wiki/Operation_Crossroads#/media/File:Operation_Crossroads_Baker_(wide).jpg.

(b)

FIGURE 15.4   (Continued)

(c)

FIGURE 15.4 (Continued)

CASE STUDY 15.1 1991 Gulf War Oil Spills

Pollution Damage in Excess of War

In most wars and battles, the loss of human life is so great that it dwarfs the environmental effects in the eyes of the public, media, and government to the point that it is ignored. The best example of environmental impacts exceeding those of the war is the 1990–1991 Gulf War in Kuwait and Iraq. The images of the disaster were so striking that they appeared in several popular movies. Most impressive of these is *Jarhead*, which showed the fires and fallout.

The Iraqi army invaded Kuwait on 2 August 1990, ostensibly to reunite it with Iraq, reversing their separation in 1913 by the British. Within a few days, the small Kuwaiti army had been destroyed, its government dissolved, and its monarchy in exile (Figure 15.5). Working through the United Nations, it took the United States about seven months to build a coalition among Iraq's Arab neighbors, Russia, China, and Western European countries to expel Iraq from Kuwait. The air war began on 17 January 1991 and lasted about 40 days. On 24 February the US and coalition forces began a land offensive that ended three days later with Kuwait liberated and the Iraqi army in retreat.

Shortly after the air war against Iraq began, the Iraqi army sank five oil tankers at the port of Mina Ahmadi, Kuwait's main oil shipping facility. Then they blew up an offshore oil loading terminal, another major oil transfer station and tanker embarkation point. Together these actions created a huge oil slick in the Persian Gulf that was about twice the size of the world's previous largest slick. As the air attacks continued into February, the Iraqi Army damaged oil distribution centers in Abu Halifa and Shuaiba, destroying the oil well safety systems, allowing oil to gush onto the surface and form enormous oil lakes. There was no strategic or military advantage from polluting Kuwait's

coastline, soil, or atmosphere. It did not stop or even delay the air war and land invasion.

The oil slick had a volume of about 11 000 million barrels of oil (1.6 billion l) from the coastal terminals and land-based oil wells. It polluted 300 miles (483 km) of beach along the northern coast of Saudi Arabia. The oil slick destroyed fragile salt marshes, mangroves, coral islands, and other important wildlife habitats and breeding grounds. At least 30 000 sea birds were oiled and killed, as were thousands of shore birds. The fishing industry was shut down indefinitely. Oil was visible across the entire field of view in this part of the Persian Gulf. Last minute action by the Saudi Arabian government saved its largest desalinization plant from having the intakes flooded with oil. This plant supplied 40% of the country's fresh water.

Kuwait had about 1100 land-based, operating oil wells. As the Iraqi army retreated, it set fire to 850 of them (Figure 15.5a). The well heads were blown up with dynamite and the gushing oil was set on fire by the explosion. In some wells, the oil flooded and saturated the sand around well head and caught fire, forming blazing lakes of fire. Several hundred million gallons of oil was catastrophically released. At the peak of this disaster, the Kuwait economy lost $2500 per second. Nearly one third of Kuwait's land surface wound up covered by a layer of oil, ranging from a thick, asphalt layer adjacent to the well fires to a thin tar crust downwind of them.

The oil well fires emitted hundreds of tons of particulate, sulfur dioxide, and PAHs into the atmosphere. Fallout from this air pollution was deposited across Kuwait, southern Iran, parts of Pakistan, and northern India. Snow was black in parts of the Himalayas and oil odor was detected more than 1000 miles (1600 km) away. Coalition and Iraqi soldiers as well as Kuwaiti residents suffered from fatigue,

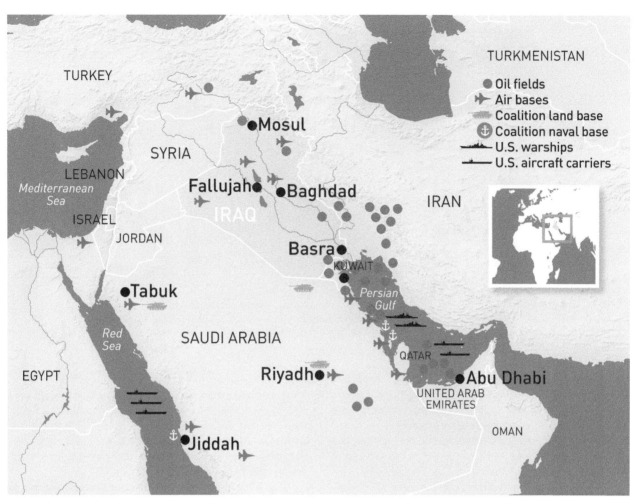

(a)

FIGURE 15.5 Map showing the location and actions of the 1991 Persian Gulf War. (a) Photo of the oil well fires during the 1991 Persian Gulf War. *Source:* Tech. Sgt. David McLeod / Wikimedia Commons / U.S. Department of Defense / Public Domain. (b) Photo of workers attempting to contain a gushing oil well in Kuwait. *Source:* United States Marine / Wikimedia Commons/ Public Domain.

(b)

FIGURE 15.5 (Continued)

headaches, joint pain, indigestion, insomnia, respiratory disorders, memory problems, and fibromyalgia, collectively known as Gulf War Syndrome. Some even had higher incidence of brain damage and certain types of cancer likely the result of exposure to smoke and chemicals from oil field fires. Some 250 000 US and 33 000 UK Gulf War veterans are reported afflicted and about 30% of all exposed people may have developed the syndrome.

There were 788 damaged oil wells when the war ended, 613 of which were burning and 76 were gushing oil. Flames shot hundreds of feet into the air emitting dense, black smoke that reached heights of 0.63–2.5 miles (1–4 km) and extended hundreds of miles. The desert was covered with a black crust and ponds of oil around some of the damaged wells, posing a threat of fires and groundwater pollution. Oil fell like rain around the wells and there were unexploded land mines, bombs, and shells all around. There were also shortages of water, electricity, and equipment.

The internationally famous Red Adair, Boots and Coots, Safety Boss, and Wild Well Control, Inc. from the United States were the first companies to attempt to control the wells (Figure 15.5b). They began activities on 16 March 1991, but after several months, the Kuwaiti Government enlisted more help. By the end of August, there were firefighting teams from France, Iran, Hungary, China, USSR, and UK as well. By September there were 27 firefighting companies involved. In May, the effort was capping three wells per day but by October they were up to eight per day. By 10 October 566 of the damaged wells had been capped. By 9 November 1991, all burning wells were extinguished and capped, thereby ending the operation.

The firefighters quickly developed innovative techniques to deal with the disaster. Seawater from the Persian Gulf was pumped onto the fires through an oil pipeline at 4000 gal (15 200 l) per minute. Another innovative method was attaching a metal casing 30–40 ft (9.2–12.3 m) to the well head to raise the height of the flame. Liquid nitrogen and water were pumped into the casing, reducing the oxygen supply to the fire. Still another innovation was to drill relief wells into the underground part of the burning wells. The relief wells diverted some of the flowing oil and reduced the pressure at the main well, allowing it to be capped. Dynamite was also used to extinguish the fires and relieve wellhead pressure. They even mounted Mig-21 turbine engines on old Soviet T-62 tanks to blast high-pressure water and air into the burning well.

15.2 Pollution at Military Facilities

National defense is the primary goal of most countries, as it should be. It is such a high priority that other important goals like environmental health are relegated to distant secondary concerns when there is a decision to be made. A large portion of national budgets is devoted to planning and development of military actions. Nowhere is this more the case than in the United States, which budgeted more than $760 billion toward military spending in 2022. This is more money than the next 11 countries combined. Such focus on national defense preparation at the expense of almost all else has resulted in the most profoundly polluted sites in the United States around military facilities. This pollution has not only endangered the environment but also the health of military personnel and even civilians, some quite severely. In fact, about 145 military facilities are listed as current Environmental Protection Agency (EPA) Superfund sites, though with an expanded definition, some claim that about 900 of the approximately 1350 Superfund sites ever listed are related to military facilities and activities. A new major pollutant and resulting public health concern is per- and polyfluoroalkyl substances (PFAS), and these have already been identified at nearly 700 military facilities, some with astounding concentrations.

15.2.1   Military Bases

There are many types of military-related facilities. The most polluted types are camps and training facilities; research and development and production facilities; and testing or military proving ground facilities. Other facilities can also be quite polluted, but not to the degree of these three. Military academies, such as West Point and Annapolis, are impacted by contamination but at a lower level. Other facilities such as military and Veteran's Administration (VA) hospitals, supply depots, investigative units (like on television shows *NCIS* or *JAG*), administrative and recruitment stations are no more polluted than their civilian counterparts. Pollution issues at military contractor facilities vary from none to extensive.

The location and development of military facilities is based on factors such as availability of a reliable fresh water supply, transportation access, infrastructure, power availability, and location relative to population centers. Considering the reason for military facilities, choosing a site for a military base or weapons complex, effect on the local environment is not given much consideration. A second factor that makes military facilities especially polluting is that the development, manufacture, storage, and maintenance of weapons involves the use of dangerous and hazardous chemicals. The weapons themselves are designed to kill people, so public health is not a major concern.

A major source of pollution on military bases is fuel usage and storage. The US military used 4.13 billion gal (15.6 billion l) of fuel annually or an average of 11.3 million gal (42.8 million l) of fuel per day in 2017. The consumption decreases steadily every year and is at the lowest level since the 1970s. The consumption by branch is 53% by the Air Force, 32% by the Navy and Marines, and 14% by the Army. At installations, however, consumption is 34% by the Army, 30% by the Air Force, 28% by the Navy and Marines, and 8% by other military uses. All bases use fuel for similar purposes and have the same environmental issues.

The use, transfer, and storage of gasoline, jet fuel, marine diesel, and other liquid fuels results in significant pollution. Liquid fuels are kept in surface and underground storage tanks (USTs) that leak, are punctured, or are otherwise damaged on an occasional basis and contaminate the soil and groundwater. At military airports, fuel is commonly distributed using a buried, high-pressure pipe system from central tanks. Aircraft can be re-fueled much quicker and supply tankers re-loaded without using slow and limited tanker trucks. However, even a pinhole leak in the lines can leak enormous amounts of gasoline or jet fuel.

The U.S. Navy's base on Diego Garcia, a small island reef, about 6700 acres (2711 ha), in the Indian Ocean some 1000 miles (1609 km) south of India, exemplifies the impact of bases. In the early 1970s, the Navy built a small communications facility on the rural island as the Persian Gulf became more critical to the US. As tensions built in the region, military facilities on Diego Garcia rapidly expanded. By the 1991 Gulf War, the island had an advanced airfield and housed 1700 military personnel and 1500 civilian contractors. An underground pipe and hydrant refueling system was installed to support airfield operations.

An 18-in. (45.7-cm) buried pipeline at the airfield connecting the fuel system to the supply tanks cracked and leaked more than 160 000 gal (605 666 l) of jet fuel underground before the leak was discovered and repaired. The spilled fuel was underneath a concrete airport tarmac and it spread to about 9 acres (3.6 ha). The problem was the spill had to be addressed without interfering with airfield operations. Six recovery wells were drilled in non-critical areas and vacuum pumps removed fuel from the water table. The operation removed 2000 gal (7571 l) of fuel in the first month so it was expanded to a 50 well system. More than 100 000 gal (378 541 l) of fuel was recovered using this system.

Another major source of contamination is military equipment operation and maintenance. User or direct support operations and maintenance include cleaning, lubrication, and minor adjustments performed by equipment operators. Direct support maintenance is also carried out by specialized teams, if needed. Intermediate or general support is performed by unit personnel and includes inspection, troubleshooting, replacement of parts and assemblies, and repair of equipment. This maintenance is performed in specialized repair centers. Full maintenance is carried out by highly specialized personnel in specialized facilities and includes the rebuilding and renovation of major assemblies.

Pollution from maintenance activities can be significant, especially if it involves ships, aircraft, trucks, other vehicles, or weapons. Paint application and removal, engine and hydraulic system repair and refurbishing, and metal working require the use of dangerous chemicals and generate hazardous wastes such as solvents, used oils, acids, and plating residues. Repairs and maintenance in the field can result in local environmental impacts but the potential for widespread environmental damage is the greatest at large maintenance centers.

CASE STUDY 15.2 Marine Base Camp Lejeune, North Carolina

Pollution Health Risks to Servicepersons

When people join the US Marine Corps (USMC), they commit to serving their country to the point of making the ultimate sacrifice, if need be. They do not, however, anticipate that this sacrifice will be sickness or even death as the result of exposure to pollution at a US military base. This was and is the case, however, at Camp Lejeune, North Carolina, the largest Marine base east of the Mississippi River and the second largest in the United States (Figure 15.6).

Construction of the base began in 1941 on a 11 000-acre (45-km²) tract of land in Jacksonville, along the North Carolina coast. The base was named Camp Lejeune in 1942 and has since grown into a 246-square mile (640-km²) installation with 14 miles (22.5 km) of beachfront where amphibious landings can be practiced. About 150 000 active duty and civilian employees, retirees, and their families live on the base. The base

includes military commands, training centers, motor pools and waste disposal facilities, as well as schools, day care centers, nine family housing areas, libraries, gyms, a shopping center, and a hospital. Until 1985, it also had eight potable water systems with 100 wells.

Camp Lejeune is also home to perhaps the worst water contamination at any military base as well as a poor handling of the problem once it was discovered. Water contamination occurred at the base from 1953 to 1987, during which time as many as one million people were exposed through drinking and bathing. As early as 1980, the base began testing drinking water for trihalomethanes following new regulations from the EPA. In addition, a US Army Environmental Hygiene Agency laboratory found excessive levels of volatile organic compounds (VOCs) in the water. They reported these results to USMC leadership in March 1981 and a period of water testing began that continued through the 1980s.

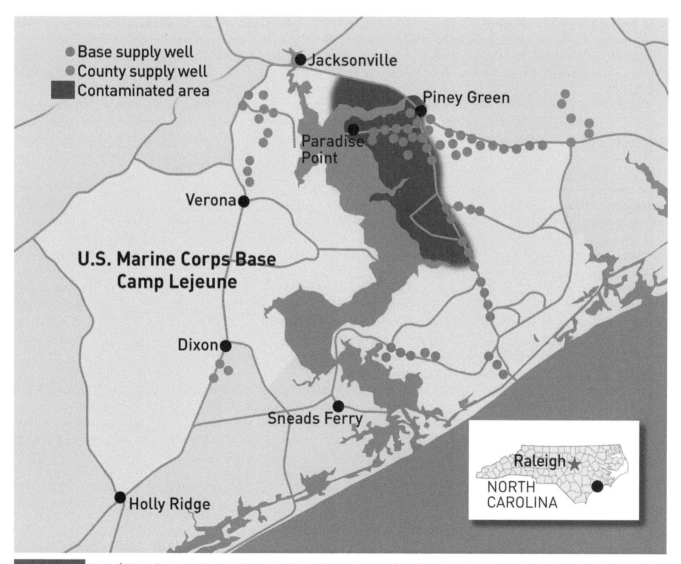

FIGURE 15.6 Map of US Marine Corps Camp Lejeune, North Carolina, water supply wells and polluted areas. Inset shows location in North Carolina. *Source:* Finlay McWalter / NASA WorldWind / Wikimedia Commons / CC BY-SA 3.0.

Testing of water at water treatment and distribution plants yielded high levels of VOCs. The Hadnot Point plant began operating in 1942 and served barracks and four family housing units. Drinking water testing found 1400 ppb of TCE (trichloroethylene) in May 1982 when the limit was 5 ppb. Other contaminants found in the testing included PCE (perchloroethylene or tetrachloroethylene), DCE (trans-1,2-dichloroethylene), vinyl chloride, and benzene. The most contaminated wells were shut down in 1985. The Tarawa Terrace plant began operating in 1952 and served family housing and a trailer park. PCE level detected in drinking water was 215 ppb in February 1985 when the limit for PCE in drinking water was 5 ppb. The contaminated wells were shut down in March 1987.

The source of the PCE contamination was the ABC One-Hour Cleaners, a busy, off-base dry-cleaning company. It operated from 1964 to 2005 using two to three 55-gal (208 l) drums of PCE per month and was located just 900 ft (274 m) from a base water well. TCE was widely used by the military and industry as an engine degreaser and solvent and it degrades to trans-1,2-DCE and vinyl chloride in groundwater. The other contaminant sources were leaking USTs with gasoline and other chemicals, waste disposal sites, and industrial spills. More than 70 other drinking-water contaminants were identified, including 1,1-DCE, methylene chloride, benzene, and toluene.

The study of contaminated drinking water determined that at Camp Lejeune's main base, barracks, and family and temporary housing, people drank and bathed in water with pollutants at concentrations from 240 to 3400 times the levels permitted by EPA for 30 days or more. TCE was found in concentrations up to 180 000 ppb (EPA limit of 5 ppb) and PCE was as high as 215 ppb (EPA limit of 5 ppb). The Agency for Toxic Substances and Disease Registry (ATSDR) modeled PCE concentrations at Tarawa Terrace and determined that they exceeded the current EPA maximum in drinking water for 346 months between November 1957 and February 1987. It also determined the Hadnot Point plant used contaminated drinking water intermittently during spring and summer months when demand was high from 1972 to 1985.

In 1988, a massive fuel leak from USTs was discovered. They had been leaking for a long time, contributing to the contamination of the groundwater. At the time, the tanks were leaking 1500 gal (5678 l) per month, which deposited a 1.1 million gal (4.16 million l) layer of gasoline on top of the Camp Lejeune water table. The benzene found in water likely resulted from the fuel leak from USTs.

In 1989, the EPA declared 236 square miles (611 km²) of coastal North Carolina at Camp Lejeune as a Superfund site. They listed groundwater, surface water, sediment, soil contaminated by base operations, and waste handling as the reasons.

Although contamination of drinking water was reported to USMC leadership in 1981, no actions were taken to address it for years. Many contaminated wells were shut down in the mid-1980s but placed back in use despite it being in violation of the law. Other agencies contributed to the negligence as well. ATSDR released a report in 1997 finding no relationship between the water contamination and health effects on the base. The National Research Council then assembled a multidisciplinary committee of scientists to review the associations between adverse health effects and exposure to contaminated water at Camp Lejeune. They did not find sufficient evidence to justify the water contamination as the source for any of the health effects. However, on 28 April 2009, ATSDR changed their position, admitting that the water had been contaminated with benzene, a known carcinogen, and withdrew the 1997 report.

This began a huge media and legal response. Victims maintain that USMC leaders concealed the water contamination and did not act to resolve it or notify current and former base residents that their health could be at risk. More than 850 former residents filed claims for nearly $4 billion. In 2014, ATSDR completed a study that found an enhanced incidence of cervix, kidney, esophageal, and liver cancers, Hodgkin's lymphoma, and multiple myeloma. An ATSDR study in 2015 also linked 64 cases of male breast cancer to water contamination. They further found an increase in serious birth defects such as spina bifida as the result of exposure of mothers to water contaminants. A book entitled *A Trust Betrayed: The Untold Story of Camp Lejeune* by Magner described the contamination disaster.

In response to these developments, the Veteran's Administration (VA) began special treatment for those exposed to water contaminants at Camp Lejeune from 1 August 1953 through 31 December 1987 who developed adult leukemia, aplastic anemia, bladder, kidney and liver cancer, multiple myeloma, non-Hodgkin's lymphoma, or Parkinson's disease. They later added qualifying health conditions of esophageal lung and breast cancer, renal toxicity, female infertility, scleroderma, leukemia, myelodysplastic syndromes, hepatic steatosis, miscarriage, and neurobehavioral effects. At last count, 71 397 Camp Lejeune veterans had enrolled in VA health care for the Camp Lejeune Veterans Program. Five percent of these veterans were treated for one or more of the 15 Camp Lejeune medical conditions, most commonly bladder and kidney cancer.

If these problems were not enough, an assessment was initiated in 2019 to identify potential sources of a new contaminant of concern, per- and polyfluoroalkyl substances (PFAS). PFAS was used extensively in fire extinguishing foam and is extremely persistent, being dubbed "a forever pollutant" as a result. PFAS exposure increases incidence of many diseases including leukemia, lymphoma, and kidney, prostate, and bladder cancers.

The 2019 assessment identified 59 sites for investigation and the resulting 2021 report recommended 51 sites undergo further investigation or remediation. The safe limit for PFAS is 70 ppt (parts per trillion) but was found at Camp Lejeune at 35 100 ppt near the Firefighting Training Pit. A 47-acre (19 ha) area where a fire was extinguished with PFAS, and 100 acres (40.5 ha) where a fly ash dump was doused with PFAS were recommended for further investigation. These may yield additional health problems.

15.2.2 Testing and Proving Grounds

The cleanup of environmental contamination associated with military munitions testing and development is a great concern for the Department of Defense. These facilities come in several categories. Conventional munitions that are not biological or chemical agents or nuclear/radioactive. Firing of small arms contaminates the soil and groundwater with lead and propellants from projectiles, which can leach heavy metals and organic chemicals upon launch and impact. Finally, onsite disposal of unexploded ordnance (UXO) by burial, open burning, or open detonation can result in an explosive hazard and ecological damage.

The Jefferson Proving Ground (JPG), Indiana is a 55 000-acre (22 258-ha) army facility. From 1940 to 1995, JPG tested ammunition and weaponry. JPG was closed and decommissioned in 1996 and environmental cleanup began. The base had a 51 000-acre (20 639-ha) northern firing range area and a 4000-acre (1619-ha) southern cantonment area. The southern area was used for administrative support, ammunition assembly, and testing. The soil and groundwater of this area was impacted by polychlorinated biphenyls (PCB) leaks, fuel leaks from USTs, cleaning solvents, and heavy metals, primarily used as a propellant. The waste was dumped into pits, and open-burned in designated areas or buried in onsite landfills.

The northern area is undeveloped and wooded. More than 25 million projectiles and explosive charges were launched into this area and it includes clearings that were targeted during munitions tests. It is estimated that 1.5 million fragments of expended ordinance, including 150 000 pounds (68 039 kg) of depleted uranium, were present in the northern firing range. Only dense, depleted uranium was used in projectiles and other weapons designed to penetrate armor.

Test facilities for biological and chemical weapons are typically very tightly controlled. Most testing is only done in airtight facilities under tight security depending on the agent. Low-level chemicals such as tear gas can be tested in open spaces but many of the components of the dangerous chemical weapons are not. However, there is some open-air testing of chemical weapons and even accidents. The infamous Dugway sheep incident began on 13 March 1968 with the aerial spraying of nerve agent about 27 miles (43 km) west of Skull Valley, Utah though several other activities may have contributed, including disposal burning of nerve agent and the testing of artillery containing nerve agent. The result is that 6249 sheep died in Skull Valley and VX was identified as the cause, though it was initially denied by the military. Nuclear testing is even more restricted.

CASE STUDY 15.3 Nevada Nuclear Test Site

Nuclear Testing and Fallout

The Nevada Test Site is a 1360-square-mile (3500-km^2) area in which the United States conducted its nuclear weapons arsenal testing during development. It is part of a larger 5500-square-miles (14 245-km^2) tract of the Mojave Desert controlled by the US military (Figure 15.7). It also serves as a gunnery range and an outdoor laboratory and experimental center for testing of hazardous spills to ecological habitat protection techniques. The town of Mercury, Nevada lies along the south-central border of the site, serves as the base camp for the area, and contains many of the needed facilities. There are more than 1100 buildings in Mercury, including housing for personnel.

The United States conducted 925 nuclear tests at the Nevada Test Site from 1951 until 1992. Of these, 825 were below ground and 100 were above-ground or atmospheric tests (Figure 15.7a). The yield of these devices ranged from 500 000 tons (454 000 mt) to 1 million tons (0.9 million mt) of TNT. Atmospheric tests involved nuclear explosive devices that were detonated suspended from balloons, dropped from aircraft, or in a rocket. Underground tests were in a drilled vertical shaft sealed with concrete. In underground testing, the explosion and fireball were better contained and the release of radioactivity to the atmosphere was minimized (Figure 15.7b). Troops were stationed a few miles from these tests to allow the army to evaluate military preparedness for nuclear warfare and radioactive fallout. These tests were just 65 miles (104.6 km) from Las Vegas.

The atmospheric tests between 1952 and 1957 resulted in elevated levels of radioactive iodine 131 across most of the United States. Health effects from iodine and other radioactive elements were concentrated in population centers downwind of the test (Figure 15.7c). In 1990, the US Congress established a fund to compensate residents afflicted with illnesses potentially caused by fallout from testing. By 2006, 11 000 claims had been approved and more than $525 million in compensation had been awarded.

Of the 825 underground nuclear tests, approximately 275 were near or below the water table. This resulted in contamination of groundwater in some areas. Approximately 57 000 000 curies (Ci) of radioactive tritium were deposited below ground from the tests. Tritium is a radioactive isotope of hydrogen. It emits low-energy beta radiation that cannot penetrate skin and is harmful only if ingested or inhaled. It reacts to form tritiated water, which is highly mobile in the environment. Fortunately, tritium has a relatively short half-life of 12.3 years. This means that it loses radioactivity in 123 years (10 half-lives). Tritiated groundwater is found at six sites along the northern and eastern sides of the test area. Groundwater flows to the south and southwest in this area and the water table is at least 500 ft (152.4 m) below surface. The Test Site monitors the occurrence and movement of this contamination to ensure it will not impact down-gradient external groundwater users. The tritium should decay to safe levels by 2115 for the entire area.

At least 7 million cubic feet (203 914 m^3) of contaminated soil contains radioactive americium 241, cesium 137, plutonium 238, 239, and 240, and strontium 90 with greater than 1000 picocuries per gram (pCi/g) at 87 sites. These radioactive elements can cause cancer and other health effects if inhaled or ingested. The action level for soil is 1000 pCi/g, above which remediation or control measures for protection of public health are required. Soil cleanup levels at contaminated sites accessible to the public or with potential for movement of contaminants offsite, have action levels of 3–50 pCi/g.

Although nuclear weapons testing at the Nevada Test Site ended in 1992, airborne radioactive contamination still occurs through the resuspension of contaminated dust from areas where devices were detonated. Air quality within the site is monitored and concentrations of airborne radiation ranged from 48 to 2000 Ci for tritium, 0.24 to 0.40 Ci for plutonium, and 0.39 to 0.049 Ci for americium since 1992. In addition to soil, air, and groundwater, there are more than 1600 sites containing discarded and contaminated equipment. This remedial activity produces low-level radioactive waste. In addition, research and testing of nuclear furnaces, rockets, and airplanes have generated high volumes of low-level radioactive waste. These and similar materials have been stabilized and permanently disposed of on the site.

The operation of the site was not without accidents. On 18 December 1970, the 10 kt "Baneberry" subsurface test took place under Yucca Flat. The weapon detonated at the bottom of a sealed 900 ft (274 m) deep shaft, but the surface soil cracked causing a fissure and releasing hot gases and radioactive dust for several hours. Six percent of the test's radioactive products were vented, releasing 6.7 MCi of radioactive material. These radioactive materials included 80 kCi of Iodine-131, much of which rained as fallout on workers. The lighter particles were carried to high altitudes by jet stream layers and deposited as radionuclide-laden snow in Lassen and Sierra counties, California, and in northern Nevada, southern Idaho, and eastern Oregon and Washington.

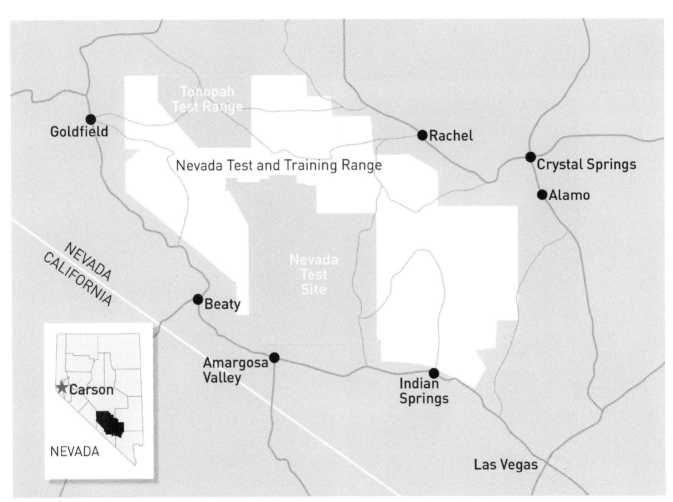

FIGURE 15.7   Map of the Nevada Test Site in the southern Nevada military complex. Inset map shows location in Nevada and surrounding states *Source:* Finlay McWalter / NASA WorldWind / Wikimedia Commons / CC BY-SA 3.0; National Cancer Benefits Center. (a) Satellite image of the Nevada Test Site showing craters from surface nuclear tests. *Source:* Image courtesy of Google maps. https://www.google.com/maps/place/Las+Vegas,+NV/@37.03853,-116.024809,3253m/data=!3m1!1e3!4m5!3m4!1s0x80beb782a4f57dd1:0x3accd5e6d5b379a3!8m2!3d36.1699412!4d-115.1398296?hl=en. (b) Photo of a nuclear explosion at the Nevada Test Site. *Source:* National Nuclear Security Administration. (c) Map of the United States showing the fallout released by nuclear detonations at the Nevada Test Site.

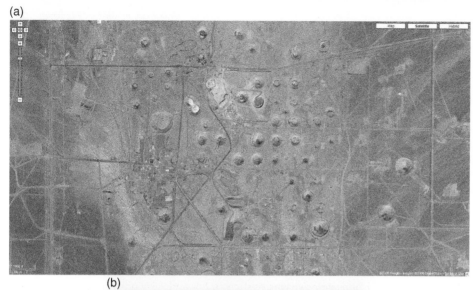

(a)

(b)

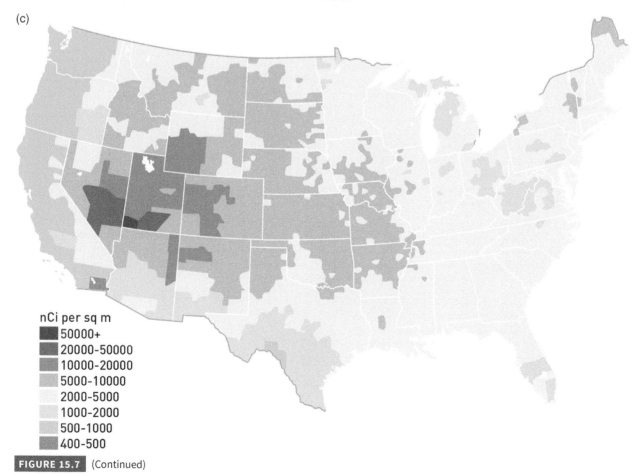

(c)

nCi per sq m
- 50000+
- 20000-50000
- 10000-20000
- 5000-10000
- 2000-5000
- 1000-2000
- 500-1000
- 400-500

FIGURE 15.7 (Continued)

15.2.3   Research, Development, and Production Facilities

The other major source of pollution from military operations is in research and development facilities and in production facilities. Conventional weapon development and production plants have similar pollution issues as bases and testing facilities in most cases. These generally include lead for projectiles, phosphorous for tracers, explosives, propellants, metal/gun oil, fuel, and solvents for cleaning the weapons and equipment and other related chemicals.

Munition production facilities at the Badger Army Ammunition Plant, Wisconsin intermittently manufactured single- and double-base propellant for cannon, rocket, and small arms ammunition for more than 30 years. Double-base propellants consist of nitro-glycerine or a similar compound (ethylene glycol dinitrate) with two active ingredients, oxygen and fuel, in the same compound. Single-base propellants are smokeless powders of either gun cotton or nitro-glycerine.

At the time the Badger plant opened, it was the largest propellant facility in the world. It employed about 20 000 people and included housing, schools, and recreation facilities for workers. It operated during World War II, Korea, and Vietnam but was declared surplus property in 1997 and it was decommissioned.

There was extensive soil and groundwater contamination around the facility. EPA scientists identified 312 areas that had been contaminated. Carbon tetrachloride and TCE solvents, lead, and explosive wastes such as 2,4- and 2,6-dinitrotoluene, and nitrates contaminated soil and groundwater around the plant and especially in the propellant and deterrent burning grounds. In these areas, excess or off-spec gunpowder and their ingredients were ignited to dispose of them. The combustion was commonly incomplete and residues were covered with fill. Rain and snowmelt percolated through the residues and formed leachate which contaminated soil and groundwater.

A plume of contaminated groundwater 3 miles (4.8 km) long extended offsite from the Propellant and Deterrent Burning Areas and into private wells. To address this, a vapor extraction system was installed to remove VOCs. A pump and treat system was also installed. The cost of the investigation and remediation was about $250 million.

Chemical and nuclear weapon research facilities could generate much more dangerous contaminants. The United States made a major investment in developing the most advanced nuclear weapons during World War II and through the Cold War. These began during the Manhattan Project in 1942. Initially, there were three research labs, Los Alamos, New Mexico, the main research and design branch of the Manhattan Project, Oak Ridge, Tennessee, a 60 000-acre (24 281-ha) facility to produce enriched uranium, and Hanford, Washington, a 700-square-mile (1813-km^2) facility to produce plutonium for 60 000 atomic weapons. As the arms race ballooned during the Cold War, a second plutonium facility was opened at the Savannah River Site, South Carolina in the 1950s. Later, many other supporting labs were developed around the United States.

These facilities were not without accidents and contamination, and many were very serious. A fire at the Rocky Flats Plant, Golden, Colorado, on 11 September 1957 resulted in plutonium contamination near Rocky Flats at 400–1500 times higher than normal. It was the highest ever recorded near an urban area, including Nagasaki, Japan. On 3 January 1961, at the Idaho National Lab, a critical reaction in an experimental nuclear reactor caused core materials to explode killing three people, but the radioactive material did not leak.

Hanford is the most contaminated nuclear site in the United States and has been named the most contaminated site in the Western Hemisphere. This contamination has been taking place since the plant first went into operation and continues today. It is estimated that 685 000 Ci of radioactive iodine-131 was released into the air and the Columbia River from 1944 to 1947. In 1949, the plant intentionally released 8000 Ci of iodine-131 over two days. Releases now are all leaks from storage containers. During a 2002 survey, tritium concentrations of 58 000 pCi/l were found in riverbank springs at Hanford Townsite. There are still 53 million gal (200 000 m^3) of high-level radioactive waste sludge in 177 storage tanks in addition to 25 million cubic feet (710 000 m^3) of solid radioactive waste. More than 14 inadequate single-shell tanks have been leaking about 640 gal (2400 l) per year since 2010.

CHAPTER **16**

Agricultural Pollution

Words you should know:

Dead zones – Areas in the ocean that are devoid of marine life because of fertilizer-induced hypoxic conditions.

Eutrophication – The biochemical process by which surface water becomes hypoxic as the result of overfertilization.

Fertilizer – Organic material used to enhance the growth and productivity of plants.

Herbicide – A chemical or biologic substance that removes unwanted plant species.

Hypoxia – A condition of low levels of dissolved oxygen in surface water.

Pesticide – A chemical or biologic substance that removes unwanted species.

16.1 | Introduction

In general, people avoid considering the idea that perhaps the most important activity for human survival is among the worst sources of pollution. Agriculture disrupts the natural ecosystem of an area by removal of native plants and addition of water, fertilizers, pesticides, and non-native plants. The process typically involves poorly conceived and implemented soil treatment causing catastrophic environmental damage and is a major source of pollution.The introduction of non-native and potentially damaging plants and animals at the expense of native species as well as dealing with their waste is equally detrimental. The major problem is that the human population is more than seven billion and consuming resources at such a breakneck pace that society has adopted non-sustainable practices in agriculture. In attempt to get even more from agriculture, humans have attempted the potentially risky practice of genetically engineering food (GMO) to increase the crop yields. Despite the fears of unanticipated side effects from GMOs, to serve the ever-increasing world population, there may be no choice.

In addition, habitat destruction both for agricultural purposes and for the development of human communities fragments land tracts that are required to maintain ecosystems. This adds to the demise of plant, animal, and insect species that is rampant at a global level. Indeed, we are witnessing one of the worst extinction events in the Earth's history and agriculture is a major cause. Flying over the central United States in an airplane, window seat passengers can see about 500 continuous miles (800 km) of cultivated land with essentially no native stands. There is no way that this has not caused irreparable damage.

16.2 | Agricultural Threats to the Environment

The US Environmental Protection Agency (EPA) regards agricultural pollution as the primary threat to river and lake water quality, the third greatest cause of estuary damage, and among the primary sources of groundwater contamination and wetland degradation. There are several types of agricultural pollution and their effects that are specific to crops and types specific to livestock. Ultimately, crops are the greatest source of pollution in agriculture because livestock is also fed on them. This pollution from crops is primarily from pesticides, including herbicides, and chemical fertilizers. Lesser impactful sources include energy consumption in farming and in transportation of supplies and produce, plant waste generation and burning of waste and for clearing. There are several farming practices that cause these primary problems either now or in the past including deforestation, soil preparation, and water use in irrigation and crop processing.

16.2.1 Deforestation

Trees block sunlight that crops require and they commonly shed leaves and chemicals that reduce the underlying vegetation so they can flourish. These characteristics make all trees undesirable for ground crops and virtually all native trees undesirable for all agriculture. For this reason, trees are summarily removed as the first step in an agricultural program. Most of the logging debris produced in this removal is burned. In some cases, the vegetation in an area to be farmed is simply set ablaze and all of it is incinerated. This burning of the trees and ground vegetation produces smoke composed primarily of particulate but also including chemical pollutants such as creosote, carbon dioxide, benzene and polycyclic aromatic hydrocarbons (PAHs). If trees are removed, the deep roots no longer hold the soil in place and it is eroded, only to silt up streams and lakes.

A current concern about deforestation is that it is accelerating climate change. Trees convert more carbon dioxide to oxygen through photosynthesis than most ground vegetation. Removal of trees to grow smaller and more widely spaced crops significantly reduces photosynthetic oxygen production. Burning the deforestation waste increases the carbon dioxide in the atmosphere. This is currently a very serious problem in tropical rainforests and especially those in Central and South America. Whole forests are being bulldozed, burned, and replaced with agricultural lands at a horrific pace. It has been estimated that just 36% of the Earth's 5.4 million square miles (14.6 million km²) of tropical rainforest is intact, 34% is destroyed, and the remaining 30% is degraded. At the current rate of destruction, only 10% will remain by 2030. What makes this even more impactful to the carbon dioxide problem is that these trees and vegetation grow and photosynthesize throughout the year in contrast to higher-latitude trees, which only grow in warm seasons. It is estimated that rainforests traditionally produce 20% of the oxygen from photosynthesis on Earth.

16.2.2 Soil Preparation

Soil preparation begins with plowing soil to soften it and to cover and kill the native vegetation. The softened soil facilitates planting of seeds and penetration of crop roots into the ground to allow them to grow faster and larger. The danger is that the native vegetation evolved to best protect the soil. It holds moisture to keep the soil from drying and fixes it in place to keep it from eroding. The soil preparation for the crops leaves exposed, bare soil at the mercy of the elements. Fortunately, normally the elements are generally predictable and farmers can plant and harvest their crops without problem but not always.

In the south-central United States, there was an unusually hot and dry period from 1926 to 1934. Exposed soil dried out on the fields and no cover vegetation grew during the off seasons, turning the most fertile of soil into a dry powder. When the winds came, they removed the powdery soil into great clouds and storms of dust or particulate that wreaked havoc across the United States. This event was named "Dust Bowl" as a result. It permanently removed fertile soil that had taken thousands of years to develop. The hardest hit areas to this day have a red sandy soil on the surface that can barely support vegetation much less crops.

When the Dust Bowl ended, new farming practices were introduced to prevent such events in the future. The same problem has happened and is happening in many other places around the world but in most cases it is more subtle and as a result is not acted upon as quickly or effectively as it should be.

CASE STUDY 16.1 1930s Central United States Dust Bowl

Damage from Poor Soil Practices

When the central plains of the United States were settled, they were covered with tall, deep-rooted, natural grasses (Figure 16.1). These grasses fixed the soil and could survive in the dry, cold, and windy conditions typical for this region rather than the unusual wet, warm weather that the region experienced in the mid-1800s. Farmers immediately removed these grasses and replaced them with crops like corn, soybeans, and alfalfa, which were not conducive to the normal Midwest weather. They also made the soil less cohesive and fixed to the surface. By the early 1930s, a little drought and wind would wreak havoc.

For nearly a century, the Midwestern United States had experienced above-average rainfall and moderate temperatures. Combined with demand for agricultural production that accompanied World War I, the demands on farmers to increase production and to cultivate less desirable land increased dramatically. The less desirable areas were along stream banks, on steep ridges, and areas covered with trees. The intense level of agriculture they instituted included several dangerous farming techniques such as no rotation of crops, plowing entire fields after each harvest, and allowing animals to graze on crop residues.

Periods of low rainfall in the central United States occurred in 1890 and again in 1910 but they were nothing compared to the dramatic drop in precipitation between 1926 and 1934. Areas that typically received 30–40 in. (76.2–101.6 cm) of rainfall per year received as little as 15–20 in. (38–50.8 cm) annually. The reasons for this drought appear to be that changes in the jet stream and ocean currents may have altered atmospheric circulation and moisture patterns. Precipitation levels in the area did not reach pre-drought levels until the early 1940s.

As the agricultural fields dried out in the hot sun, brisk winds stripped up the fertile topsoil and carried it in massive, blowing clouds of dust (Figure 16.1a). With no trees or grass to bind the soil or obstructions to slow the wind, the blowing topsoil was blown everywhere like black snow (Figure 16.1b). In some places, it buried farms in layers of 3 ft (1 m) or more and turned previously productive fields into barren wastelands (Figure 16.1c). In 1932, 14 dust storms were recorded in the central plains, and by 1933 there were 38 storms. By 1934, the peak of the drought, more than 100 million acres (40.5 million ha) of prime farmland had lost its topsoil to wind erosion. At this point, the effects of the Dust Bowl encompassed an area approximately 399 miles (644 km) long and 298 miles (483 km) wide in parts of Colorado, New Mexico, Kansas, Texas, and Oklahoma.

The worst dust storm took place on 14 April 1935 which was named Black Sunday. With wind speeds of 60 miles per hour (96.6 kph), dust was lifted so high that it blocked out the sun, making the day as dark as night. Traffic came to a standstill on the roads. Regardless of how well sealed, a layer of dust that could be measured in inches seeped into every home in the area, causing severe health effects. The Black Sunday storm blew eastward, depositing a thick layer of black dust over Chicago equal to 4 pounds (1.8 kg) for every person living there. It was blown all the way into New England, New York City, and even Washington, D.C., where legislators experienced just a hint of what almost one third of the United States was suffering through. It was the result of this storm that an Associated Press reporter coined the term "Dust Bowl" for the first time to describe this disaster.

Human costs associated with the Dust Bowl were enormous. Much of the United States was suffering through the effects of the Stock Market Crash of 1929 and onset of the Great Depression, but, until then, the central plains had been relatively insulated from this economic disaster. Farmers grew their own food and maintained their lifestyle. The Dust Bowl destroyed their livelihood, and they could no longer make mortgage and tax payments on their farms. Many people were forced to abandon their homes and seek a new life somewhere else. The largest mass migration in American history resulted, with more than 2.5 million people leaving the central plains states and settling elsewhere by 1940. About 15% of the people in Oklahoma left the

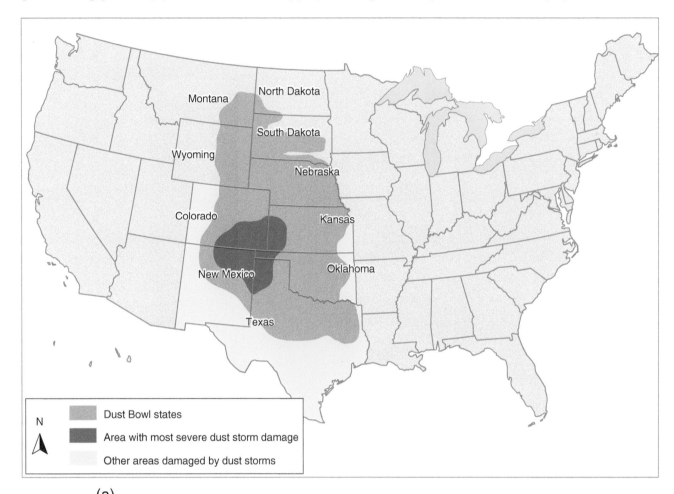

(a)

FIGURE 16.1 Map of the United States showing the distribution of Dust Bowl impact. (a) Photo of a dust storm approaching a small town during the Dust Bowl. Source: Courtesy of the US Public Health Service. (b) Photo of a family walking across a farm covered in windblown dust. *Source:* Courtesy of the US Department of Agriculture. (c) Photo of sand dunes engulfing a farm homestead. *Source:* Courtesy of US Library of Congress.

(b)

(c)

FIGURE 16.1 (Continued)

state, many moving westward and settling in California. Other parts of the country were overwhelmed with the influx of Dust Bowl refugees, collectively called "Okies," even though they were from several states. The Great Depression was in full swing by this time and these transient newcomers stressed local relief services and competed with local residents for jobs, often willing to work for greatly reduced wages. The novel "Grapes of Wrath" by John Steinbeck describes the personal struggles these people endured.

Government response to the disaster was slow. Eventually political leaders understood that a large percentage of the productive farmland in the United States was becoming permanently unproductive and finally took action. Congress formed the Soil Conservation Service (later changed to the Natural Resource Conservation Service) in 1935 that investigated and implemented cultivation techniques to reduce topsoil erosion by wind and water. As New Deal legislation started to be enacted, the Works Progress Administration (WPA) and other newly formed government agencies and projects began to provide emergency supplies, funds, livestock feed, and transportation to maintain

the livelihoods of farmers and ranchers who had not fled. Health care facilities were also established and medical supplies provided to impacted communities to meet emergency medical needs. The government established reliable markets for farm goods, imposed higher tariffs on imported foodstuffs, and established loan funds for farm market maintenance and business rehabilitation. By 1934, Congressional appropriations for drought relief exceeded $500 million.

Later, additional soil conservation measures and policies were implemented. Water supply and irrigation systems were improved and a federal crop insurance program was established. The average size of farms increased and low-yield, ecologically sensitive and buffer lands were removed from production. These and other soil conservation programs helped to mitigate the impact of future droughts and establish a pattern of soil conservation measures. However, the topsoil has not been replenished to much of the central plains and the soil is red, acidic, and nowhere near as productive as it was prior to the Dust Bowl. Natural processes will take thousands of years to replace what was lost in less than a decade.

16.2.3 Overuse of Water

Crop plants require water to grow. Natural precipitation (rain and snow) is the ideal way to provide this water but it is not reliable enough to insure an adequate supply. As a result, farmers water their fields. Planting fields near rivers and other bodies of water provides a solution unless there is a flood and an entire year's crop may be lost. The land around surface water is also quite limited further reducing its usefulness. In response, farmers developed irrigation systems to move water to the crops in a much more controlled environment. If the crop land is near a river or other surface water body, a combination of surface water and well water is used to water the crops. If the crop land is distant from a surface water source, irrigation is primarily from well water in most cases.

The main pollution problems from agricultural water usage are drawdown of the groundwater table and possible desertification and resulting degradation of the water source. An excellent example of desertification resulting from overuse of surface water for irrigation is the Aral Sea. Agriculture in the region requires extensive irrigation because the area is so dry. As a result, the Aral Sea became one of the worst agricultural disasters in history.

CASE STUDY 16.2 Aral Sea Disaster, Kazakhstan, and Uzbekistan

Overuse of Water for Irrigation

The Aral Sea environmental disaster is considered to be one of the Earth's worst. The Aral Sea is a saline inland lake that was called a sea because of its high salt content. It lies between Kazakhstan to the north and Uzbekistan to the south in Central Asia. Previously, the Aral Sea was the fourth largest lake in the world having a surface area of 26 300 square miles (68 000 km²). However, as a result of poor agricultural practices and overuse it began shrinking in the 1960s and by 1997, it was only 10% of its original size in four small bodies (Figure 16.2).

The Aral Sea has traditionally supported a thriving human and animal community in an otherwise arid region. The lake is fed by several sources, including the region's two major rivers, the Syr Darya and the Amu Darya, which are fed by snowmelt and precipitation in distant mountains. They flow northwest across the Kyzylkum Desert and pool at the Aral basin. In 1900, in excess of 7.4 million acres (3 million ha) were irrigated for agriculture in the Aral Sea basin. Construction of extensive irrigation canals started in the 1930s but accelerated

in the 1960s. Most canals were not well constructed and they leaked and experienced excessive evaporation. The Qaraqum Canal was the largest in Central Asia but is estimated to have lost 75% of its water as a result of this poor construction. By 1960, this diversion of water from the Aral Sea was between 4.8 and 14.4 cubic miles (20 and 60 km³) each year. However, up to 1960, the removal was still balanced by input and it had not yet begun to shrink.

The balance changed in 1960 when irrigated land was increased to 12.4 million acres (5 million ha) and it continued to grow. By 1980, the irrigated land in the Aral Sea basin was nearly 16.1 million acres (6.5 million ha). Further, the irrigation canals to distribute water wasted even more. It was estimated that only 12% of the main irrigation canals length, only 28% of irrigation channels between farms, and 21% of within farm channels had anti-infiltration liners. Water removal from the Amu Dar'ya and Syr Dar'ya Rivers had reached 31.7 cubic miles (132 km³). By 1987, the irrigated area had increased to about 18.8 million acres (7.6 million ha).

This massive irrigation made the surrounding desert bloom but it devastated the Aral Sea. From 1961 to 1970, the Aral's lake level

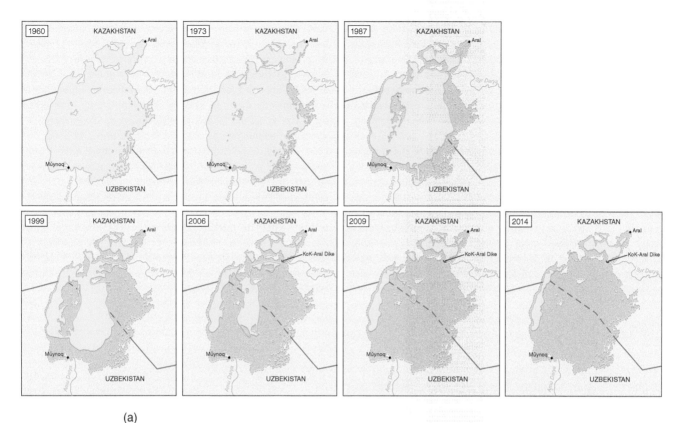

(a)

FIGURE 16.2 Map of the Aral Sea area showing its shrinkage over 54 years. (a) Photograph of ships previously lined up at moorings Aral Sea but now that the water is gone, lying in the desert. *Source:* Sebastian Kluger / Wikimedia Commons / CC BY-SA 3.0.

declined an average of 7.9 in. (20 cm) per year. In the 1970s, the rate radically increased to 20–24 in. (50–60 cm) per year and in the 1980s it was 31–35 in. (80–90 cm) per year. The reason is that the amount of water diverted for irrigation from the two rivers doubled between 1960 and 2000. Prior to the irrigation, the lake level was 174 ft (53 m) above sea level. By 2010, the main Aral body was 88.5 ft (27 m) above sea level, a reduction in lake level by a disastrous 85.5 ft (26.1 m) (Figure 16.2a).

The Aral Sea size reduced in proportion to the reduction in lake level. From 1960 to 1998, as the volume reduced by 80%, the surface area shrank by 60%. In 1960, the volume was 260 cubic miles (1100 km³) but by 1998 it was 50.4 cubic miles (210 km³). The surface area was 26 300 square miles (68 000 km²) in 1960 and 11 076 square miles (28 687 km²) in 1998 and it split into two bodies in 1987. In 2003, the south body of the Aral Sea further divided into eastern and western lakes and it later split into four bodies.

To attempt a remediation of the northern body of the Aral Sea, the Dike Kokaral dam was constructed in 2005. By 2008, the lake level in this body had risen by 39 ft (12 m), producing a maximum depth of 138 ft (42 m). The rest of the bodies, however, have continued to recede. By 2009, the southeastern lake disappeared and the southwestern lake shrank to a thin strip at the western edge of the former southern lake. In August 2014, the entire eastern body of the Aral Sea had completely dried up and it is now called the Aral-kum Desert.

With the high level of evaporation and diversion of fresh water in the rivers to irrigation, the salinity level increased from approximately 1.3 oz/gal (10 g/l) in 1960 to often more than 13.3 oz/gal (100 g/l) in 1987 in the remaining southern Aral basin. For comparison, the salinity of seawater is about 4.7 oz/gal (35 g/l). By 1990, the salinity in the remaining lake increased to 50.2 oz/gal (376 g/l).

The increasing salinity of the lake water coupled with runoff from the agricultural fields that were heavily covered with fertilizer and pesticide caused severe pollution. As a result of evaporation, concentrations of these pollutants have risen drastically in remaining water and dry beds of salt and sediment. Prior to 1960, the Aral Sea supported about 24 species of fish, over 200 species of free-living macroinvertebrates, and 180 land animal species. With shrinking of the sea and increase in salinity and toxicity, none of the fish species survived. Less than 30 macroinvertebrate species have survived and only a few dozen of the land animals remain.

As the waters evaporated and salinity increased, the fisheries and communities that depended on them collapsed. The increasingly saline water became polluted with fertilizer and pesticides ultimately drying to contaminated dust. The dust in the exposed lakebed was blown around and became a public health hazard. The dust settled onto fields, degrading the soil. A 2001 UN report estimated that 46% of the area's irrigated lands have been damaged by salinity, an increase from 38% in 1982 and 42% in 1995. To mitigate the salinity of the croplands, soils are flushed at least four times. This process also eliminates minerals and salts that are needed for productivity. To compensate for the loss of nutrients in soil, farmers use excessive amounts of fertilizers and pesticides. Pesticide use is more than 20 times the national average and some crops exceed allowable limits in nitrate and pesticides by two to four times. The quality of local groundwater has also suffered from the increased salt deposition. Salt concentrations reaching 0.8 oz/gal (6 g/l) have been recorded

in drinking-water wells. This is six times higher than the maximum acceptable concentration established by the World Health Organization.

Strong northeasterly winds in the Aral Sea area pick up the contaminated sand, salt, and dust from the dried lakebeds, forming extensive dust storms. The salt content of the dust is about 30–40% in the summer, and as high as 90% in the winter months. The dust storms are huge, often as large as 93.2–186.4 miles (150–300 km) wide. This dust is distributed well beyond the Aral Sea region and has been found 310.7–621.4 miles (500–1000 km) away from the original source. It is estimated that the salt removed from the dried seabed was about 47.4 million tons (43 million mt) between 1960 and 1984. In 2002, it was found that the windstorms carried 200 000 tons (181 437 mt) of salt and contaminated sand and dust daily throughout the Aral Sea region and as far as Russia's Arctic to the north.

Between the wind-borne toxic dust from the dried lakebed and the damage to groundwater, the Aral Sea area has been enduring a public health crisis. People living in the area ingest pollutants primarily through drinking water and inhaling the toxic dust. However, because of absorption of contaminants by plants and plant consumption by livestock, the toxins are also in the local food. As a result, residents are suffering from a number of health problems. In the late 1990s, infant mortality in the Aral Sea region was 60–110/1000, compared with 48/1000 in Uzbekistan, and 24/1000 in Russia. By 2009, it was still 75/1000 in the region and maternity death was an elevated 12/1000. Residents suffer from anemia, brucellosis, bronchial asthma, and typhoid at eight times the national averages. They also have elevated occurrences of liver, kidney, and eye problems, digestive disorders, infectious diseases, and tuberculosis. During the 1980s, the occurrence of liver cancer doubled, and the incidence of esophageal, lung, and stomach cancer spiked. It has been estimated that the Aral Sea disaster displaced more than 100 000 people and impacted the health of more than five million people.

All of these health issues occurred while more people came to the region. Between 1950 and 1988, the population of the Aral Sea area increased from 13.8 to 33.2 million people. This at the same time that the region's fishing industry was being devastated, which brought these people unemployment and economic hardship making it a humanitarian disaster as well.

Overuse of water can also affect river systems as well as lakes and groundwater. Much of the water in a river is typically used as drinking water for settlements along its banks, however, the agricultural draw can be significant. The mighty Colorado River is known for its raging class 5 whitewater. However, by the time it reaches the Gulf of California in Mexico, it is a trickle, if anything. It is primarily drained for drinking water in Las Vegas and Los Angeles among other communities. The United States was forced to build a desalination plant for Mexico as a result of this incredible use. The primary effect on the river along its course is the decrease in discharge because of the reduced volume. However, a secondary effect is a degradation of the water quality. The lowered elevation of the water in the river forces more base flow to be drawn from the water table into the river, making it more effluent. Polluted groundwater can then be drawn from greater distances to the river. More polluted, unfiltered precipitation runoff becomes a larger volume of the river water. The shallower water level in the river results in more evaporation, thereby increasing the salinity in the water. Lakes and ponds suffer the same issues.

However, the most pressing danger, by far, is the overuse of groundwater and its effect on major aquifers. Some aquifers, such as the High Plains or Ogallala Aquifer in the Midwestern United States, are primarily used for irrigation of crop lands. The Ogallala is the largest aquifer in the United States. It is used 94% for irrigation and the heavy use has dropped the water table in the aquifer by 100 ft (30 m) on average and as much as 175 ft (54 m) locally. It has reduced the height and dynamics of streams and rivers by reducing base flow and many have dried out. It also draws more water from saline sources as well as those from the surface and as a result, the quality of the remaining water has suffered. The Ogallala receives much recharge from the surface but part was charged during the last ice age, 13 000 years ago, which is irreplaceable. It will not be long before these aquifers will not be able to irrigate the farmlands, resulting in a true crisis.

CASE STUDY 16.3 Ogallala Aquifer, Central United States

Overuse of Groundwater

Perhaps the major aquifer in the United States, the High Plains Aquifer, is primarily comprised of the Ogallala Aquifer. The aquifer produces 30% of all water for agricultural irrigation in the United States. It serves parts of eight states including Colorado, Kansas, Nebraska, New Mexico, Oklahoma, South Dakota, Texas, and Wyoming (Figure 16.3). The total area of coverage is 174 000 square miles (450 000 km²), which contains 20% of the nation's crop land in among the most important agricultural areas. Water withdrawn from the High Plains Aquifer is 94% for irrigation, helping to produce 20% of the corn, wheat, cotton, and cattle in the United States. Not all of the water in the aquifer, however, is appropriate for use. Parts of the aquifer contains water that is so saturated with natural contaminants that it is of questionable safety. This component is increasing. The crisis is that the huge removal of water from the Ogallala Aquifer is reducing water levels at a breakneck pace, creating a potentially catastrophic situation that is being investigated by the US Congress.

After the Rocky Mountains were built by plate tectonic activity, large rivers began draining the eastern slopes of the Rocky Mountains.

These rivers flowed eastward, carving deep valleys and channels into the preexisting rocks and sediment and depositing what would be the Ogallala. The earliest deposits are gravels and coarse sands in the base of the river valleys, but aprons of sediment covered the entire area, and braided streams formed a large depositional plain across the western edge of the Great Plains. Sediments covered the entire area, reached a maximum thickness of 900 ft (274 m), and formed the aquifer.

The Ogallala Aquifer contains an immense amount of groundwater. The water-saturated thickness of the aquifer is up to 525 ft (160 m). The thicker water-bearing section is in the northern plains and in the southern plains, it ranges from 50 ft (15 m) to 200 ft (61 m). Depth to the water table ranges from about 400 ft (123 m) in the northern plains to about 100–200 ft (30–61 m) in the southern plains.

The quality of the water varies by location. Dissolved solids from natural sources range from about 250 mg/l to greater than 1000 mg/l of water with the vast majority below 500. Crops can tolerate water below 500 mg/l, but above that, it will affect their growth. In the north, the main dissolved solid is sulfate from gypsum but to the south, it is mainly calcium-magnesium bicarbonate.

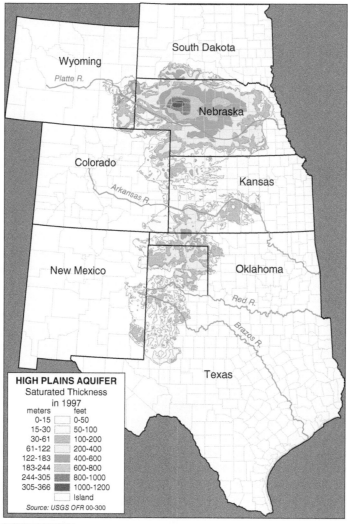

FIGURE 16.3   Map of the High Plains Aquifer in the central United States showing the thickness of the water-bearing zone. *Source:* Courtesy of US Geological Survey.

Large-scale withdrawal of water from Ogallala began after the great drought and Dust Bowl of the 1930s. By 1949, 10% of the land covering 2.1 million acres (850 000 ha) of the aquifer area was irrigated. By 1980, the irrigated area was 13.7 million acres (5.54 million ha), increasing to 13.9 million acres (5.63 million ha) by 1990. Removal of water from the aquifer has exceeded recharge since the 1960s. The elevation of the water table in the Ogallala Aquifer has decreased an average of 100 ft (30.8 m) but as much as 175 ft (54 m) in some areas. The rate of subsidence of the water table ranges from 1.4 ft (0.4 m) to 1.7 ft (0.5 m) per year. It is estimated that there has been a 60% subsidence of the water table in the Ogallala to date and it will reach 70% by 2060 at the current removal rate. It would take 6000 years to recharge the loss by natural processes.

The rapid drop in water table changed the Arkansas River from effluent to influent. The Ogallala and related aquifers previously fed water to the river from springs. However, between 1995 and 2000, the Arkansas River significantly recharged the aquifer and dried up on occasion. The Canadian River and South Platte River are currently still effluent but receive less groundwater every year. At the rate of subsidence of the water table in the aquifer, they will also become influent rivers in the near future. Even canals, ponds, and irrigated land are now recharging the aquifer in most areas.

Evaporation of water above the Ogallala Aquifer is concentrating ions and contaminants. Excessive use of fertilizers and pesticides to improve crop output accompany the heavy water usage. The aquifer recharge zone is widespread over a large area allowing these chemicals to permeate the aquifer and further contaminate the water. With the increasing degradation of surface water quality and increase in infiltration, the quality of the water in the aquifer will continue to decline. As a result, the US Congress is considering bills to regulate water use from the Ogallala.

16.3 Agricultural Chemicals

Agricultural chemicals are manufactured compounds or cultivated organisms designed to destroy or repel agricultural pests (plant, animal, or insect), inhibit the consumption of crops by pests, or to prevent pests from attacking vegetation or related products. These chemicals destroy plants or modify their physiology, modify the effect of other agricultural chemicals, or attract pests to destroy them. Agricultural chemical products are pesticides including herbicides, insecticides, and fungicides. Some government agencies classify plant growth regulators and veterinary medicines as agricultural chemicals and they require special registration.

Agricultural chemicals are among the main reasons that farmers can produce more crops from a given amount of land. China, for example, produces food for 21% of the world's population on only 9% of the world's cultivated land. This is because 75% of crop nutrients are from chemical fertilizers and managed by agricultural chemicals. Agricultural chemicals, however, can have severe ecological and environmental impacts.

16.3.1 Pesticides

Pesticides have been among the greatest advancements in human survival and yet may cause catastrophic problems as pollutants. Pesticides have controlled or eliminated most of the insects and other pests that reduce crop yield. American farmers have lavishly applied pesticides to increase crop yield and turned the Midwest into the breadbasket of the world. This has allowed enough food to be produced to facilitate the burgeoning human population.

The overuse of pesticides, however, has very negative aspects. Rachel Carson is the pioneer of the American environmental movement and her main focus was the damage caused by pesticides. Her 1962 book *Silent Spring* identified the most dangerous of pesticides at the time and focused on DDT. Her leadership in the environmental movement played a role in the formation of the US EPA and the banning of DDT and many other agricultural pesticides such as aldrin/dieldrin, carbofuran, chlordane, DBCP, diazinon, endosulfan, endrin, HCB, HCH or lindane, heptachlor, methyl parathion, PCP, and toxaphene among others.

The main system of classification for pesticides is based upon the pest to be controlled. The primary pesticide type is insecticide which primarily targets insects. Insecticides are used in most agricultural applications as preventative measures and in individual applications around homes. They are applied to seeds, soil, or directly to plants by spraying or crop dusting. Large-scale preventative applications mostly involve spraying and are done to control destructive or potentially disease-bearing insects like mosquitoes. Household insecticides are used to control insects that are threats to health such as wasps and cockroaches or threats to property such as termites and carpenter ants. Use of household insecticides must be carefully monitored because it carries the threat of accidental poisoning.

Herbicides and including algicides and fungicides, can be considered separate from pesticides but are briefly included here. Herbicides control the growth of unwanted vegetation. They are used in agricultural applications, ornamental plants, and household applications. In agricultural fields, they control plants that compete with the crops. The most common use of herbicide is at private residences. This herbicide is spread throughout suburban areas to remove and prevent weeds that compete with grass. Herbicides are also commonly sprayed on undergrowth along roadsides, on plants that threaten health such as poison ivy and in lakes and ponds to control weeds.

Chemical classifications of pesticides divide them into inorganic, organic, and biological agents. Inorganic pesticides contain heavy metals like lead, mercury, copper, or arsenic. They can be highly toxic and are essentially indestructible. They are used as seed coatings but are more commonly used on wood and in nonedible applications. Biological pesticides are living organisms or toxins from them that are used in place of conventional pesticides. They include both insects and bacteria that control unwanted species and are especially used in organic gardening

Organic pesticides include four main types:

1. Organophosphorus including parathion, malathion, and dichlorvos.
2. Carbamate including carbaryl (a.k.a. Sevin), aldicarb, and carbofuran.
3. Organochlorine including chlordane, Toxaphene, heptachlor, DDT, methoxychlor, Endrin, aldrin, dieldrin, Lindane.
4. Botanical including nicotine (from tobacco), pyrethrum (from chrysanthemums), and rotenone (root extract).

Each of these has different properties and they are designed to control specific insects under specific conditions. Some of the properties are toxicity, selectivity, chemical stability, persistence in the environment, and mobility. Organophosphorus pesticides are usually extremely toxic, ranging from 10 to 100 times more effective than the others but generally degrade quickly after application. They target the nervous system and cause rapid death in insects, birds, mammals, fish, and humans. They are actually the precursor to the chemical weapon of nerve gas. Carbamate toxicity ranges from moderate to very high but are relatively mobile and only moderately persistent. Other pesticides are similar to organophosphates in terms of action and toxicity, but they tend to be more persistent and less potent. Organochlorine pesticides are highly persistent in the environment and very toxic. In some cases, they can remain active for several decades. A number of them are banned because they are a threat to human health or the environment. Botanicals are natural organic pesticides and commonly used by organic gardeners because they are safe or minimally damaging to the environment.

Fumigants are also organic pesticides but the chemical type and include carbon tetrachloride, carbon disulfide, and methylene bromide among others. They range from moderately to highly toxic and are very dangerous to work with. Many are restricted or banned.

DDT is in this group and was regarded as one of the greatest discoveries in modern civilization, earning the discoverer of its properties the Nobel Prize. On the other hand, Rachel Carson, the pioneer of the modern environmental movement, regarded DDT as the most dangerous pesticide to the environment. Through her influence, the US EPA was created and among its first efforts was to ban DDT.

Pesticides are responsible for much of the success of the human race in the 19th and 20th centuries. DDT banished typhus and malaria from Europe and North America. It controlled the insects carrying these and other diseases worldwide, literally saving hundreds of millions of human lives. Pesticides also control insects, plants, and other threats that can consume and destroy human food supplies. The success in feeding the human population over the past two centuries is largely the result of pesticide effectiveness. Agricultural effectiveness has increased worldwide through pesticides by orders of magnitude.

Enormous quantities of pesticides are manufactured and applied to crops in order to achieve these results. Some classifications include the chemicals used in water purification as pesticides. These would account for half of all pesticides. Only pesticides applied to food, seeds, and living plants and their surroundings are considered here. This constitutes about 449 500 tons (407 779 mt) per year in the United States. Global consumption of conventional pesticides is about 4.5 million tons (4.1 million mt), the United States portion is about 10%.

Pesticides are very harsh on the environment and the damage depends on toxicity, persistence, and mobility. Upon application, pesticides evaporate into the air or enter the soil/water system. They may break down by photodegradation or settle to the ground as fallout or wash out in precipitation. Pesticides can travel long distances to remote areas. Persistence is the resistance of the pesticide to chemical breakdown by reaction the with air, soil, or water to produce less toxic by-products. Biological removal is typically accomplished through microbes that dissociate the pesticide. Persistent pesticides are commonly toxic to microbes. Low persistence means half of the pesticide will degrade (half-life) in 30 days or less, moderate persistence is 30–100 days half-life and high persistence has a half-life greater than 100 days. More persistent pesticides can be spread farther than lower persistence ones.

Mobility of pesticides also determines impact on the environment. Pesticides can adhere to soil particles in a process called adsorption which determines its mobility. Pesticides are most commonly adsorbed by attaching to organic particles or clays where they are immobile. Pesticides that are not adsorbed are mobile and can leach into the groundwater system where they can travel long distances and impact human health and the environment. In sandy soils, many pesticides are mobile. Surface runoff can erode soil with otherwise immobile pesticides and add it to surface waters and eventually, other sediments.

The main problems with pesticides are (i) damage to beneficial species, (ii) imported food from areas with less stringent regulations, and (iii) evolving insect and bacterial immunities to pesticides. The global demand for food is so great that the yield of

Agricultural Chemicals **305**

farmland is pressed to its limit. This demand is so high that the loss of crops to pests is unacceptable. In response, farmers tend to overuse pesticides. Further, American homeowners are so intent on well landscaped and insect-free communities that they apply excess pesticides also degrading the environment.

Pesticides kill many insects and other small organisms in addition to the target insects. As target populations are reduced, their predator populations are similarly reduced because they have less food. This eventually leads to food shortages for humans. This damage is even more dangerous to beneficial species like bees. The pollination of flowering plants, trees, and shrubs is necessary for the growth of fruits and nuts and mostly accomplished by bees. Bee populations have been steadily declining for most of the past century. North American bees went into crisis starting in 2004, when colony collapse disorder (CCD) began causing the destruction of entire bee colonies. In 2007, one third of the North American bee population perished in one year including 50% of the Canadian bee colonies. CCD has been blamed on parasites and disease, as well as the pesticide imidacloprid which was banned in most of the European Union as a result. It is possible that even if CCD results from disease or parasite, declining health of bees from years of pesticide exposure could have weakened them.

The reason that many pesticides have been banned was not because of their effect on pests but instead because of the damage they did to other species. One reason Rachel Carson opposed DDT was a large bird kill in Massachusetts as a result of the overuse of it. The bald eagle almost went extinct because DDT in its system caused its eggs to have such a thin shells that they broke before hatching. Frog and amphibian populations also decreased from pesticides and destruction of habitat. Some 122 species of amphibians have gone extinct since 1980 and 32% of the remaining species are on the brink of extinction. In Central America, two thirds of the frogs are now extinct. Recent studies estimate that flying insects have decreased by 75% in some areas and at least 40% of all insects are lost with and annual reduction of 1% per year. Bird populations are no better, with numbers having decreased by 2.9 billion across North America since 1970.

Pesticides get into the drinking-water supply by washing from fields during precipitation. This surface runoff enters lakes and rivers which may be used for human consumption. Some pesticides are systemic, being taken through the roots and distributed throughout the plant. This pesticide cannot be removed by washing. Pesticides are also blown from the surface into the atmosphere around the world. Pesticide residues are in glacial ice in Greenland. This continued exposure to persistent pesticides increases their level in humans. By the 1970s, levels of pesticides in human mother's milk was so high that nursing was discouraged. The long-term health effects of this persistent human exposure to pesticides is unknown.

16.3.2 Fertilizers

Fertilizers are nutrients used for growing healthy plants. There are two basic types, macronutrients which are needed in large quantities and micronutrients which are needed in very small amounts. Macronutrients for plants are nitrogen (N), phosphorus (P), and potassium (K). Nutrients required in small amounts include boron (B), copper (Cu), iron (Fe), chloride (Cl), manganese (Mn), molybdenum (Mo), and zinc (Zn). Fertilizers deliver both macro and micro nutrients to the soil in a form that is readily used by plants. All nutrients are present in nature, but not in sufficient quantities to meet the needs of the world's population. Global manufactured fertilizer consumption is 210.3 million tons (190.8 million mt) annually as of 2019. The good news is that it did not increase appreciably between 2015 and 2019.

Natural fertilizers have been used historically but are limited in effectiveness. Chemical fertilizers have been developed and tested and are very effective at growing crops with the highest possible yield in the shortest time. This allows farmers to produce more food to support a larger human population. As the population increases, there is more demand on farmers to produce food. Consequently, farmers are encouraged to use fertilizers to excess. Some of this excess is washed from the soil by runoff after a rainstorm and into a surface water system. As a result, essentially every pond and small lake in the United States is impacted by excess nutrients. These excess nutrients cause eutrophication in the water body. This condition involves the added fertilizer, producing an algal and plant bloom in the surface water. As the growth peaks and the plants die off, an oxygen-consuming bacterial bloom ensues to consume the dead vegetation. This reduces the oxygen content of the surface water. The surface body becomes hypoxic which kills much of the native aquatic life.

Minimal efforts were made to control eutrophication because the impacted water does not provide much food. However, an area of eutrophication developed in the Black Sea, USSR in the 1960s from runoff from overfertilized agricultural land. The area was designated as a "dead zone" because the fish and invertebrate life either died or fled the area, making it devoid of life. Dead zones appeared around the world through the 1980s and 1990s and are continuing to appear today. They mainly occur where rivers empty into seas or oceans with restricted circulation. Fertilizer is washed into the rivers and they carry it to the sea where it accumulates at the mouth. For example, the Mississippi River drains from the Midwestern United States creating the Gulf of Mexico dead zone at the mouth of the river. There are currently more than 400 dead zones worldwide and most from overuse of agricultural fertilizer. Many appear in the most productive of marine areas and are reducing our marine food supply which could lead to global famine.

CASE STUDY 16.4 Gulf of Mexico Dead Zone

Eutrophication of Oceans by Fertilizer

At the mouth of the Mississippi River in the Gulf of Mexico, there is an area of ocean that suffers from hypoxia. The zone appears in February and expands through the summer when it peaks before dissipating in the fall as large storms develop (Figure 16.4). The average size of the zone has been 5380 square miles (13 934 km²) from 2016 to 2021, but in 2017 it grew to a record 8776 square miles (22 730 km²). In 2021, it was approximately 6334 square miles (16 405 km²). This is a dead zone because the oxygen levels are so low (<2 mg/l) that immobile and slow-moving marine life such as clams, crabs, snails, worms, and others slowly suffocate. The swimming animals flee from the area. There are few animals living in the zone of hypoxia.

The Gulf of Mexico dead zone is produced by the influx of nitrogen as nitrate from fertilizer-laden runoff in the Mississippi River water (Figure 16.4a). The river carries 140 cubic miles (580 km³) of water from 31 states and draining more than 41% of the continental United States. At least half of the nation's agricultural lands drain

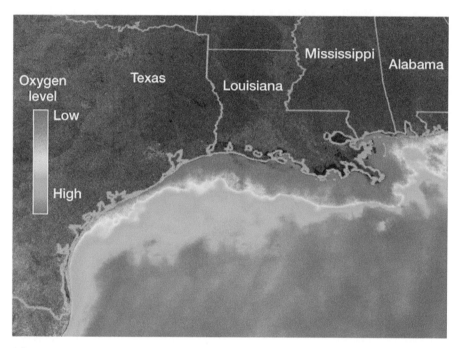

(a)

FIGURE 16.4 Map of the Dead Zone in the Gulf of Mexico. *Source:* Courtesy of NASA. (a) Photo of the Mississippi River (left) meeting the Gulf of Mexico water (right). *Source:* Courtesy of NOAA. (b) Diagram showing the difference in a receiving salt water body between input of fertilizer-rich river water (left) and clean river water (right).

(b)

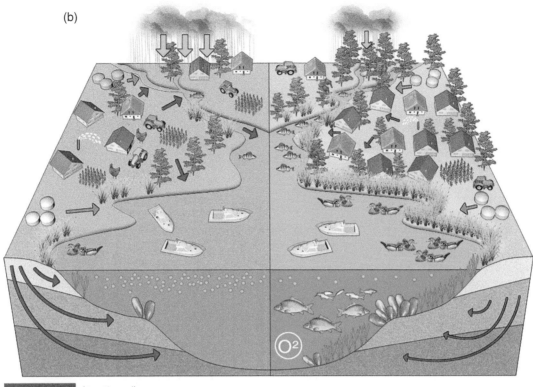

FIGURE 16.4   (Continued)

into the river either directly or through tributaries. The amount of fertilizer used on these lands has ballooned since the 1950s. The amount of nitrate from fertilizer carried to the Gulf of Mexico by the river tripled between 1960 and 1997 and phosphate fertilizer doubled. More than 1.75 million tons (1.6 million mt) of nitrate was delivered by the river per year by the late 1990s, 30% from agricultural fertilizers, 30% from natural soil decomposition, and 40% from other sources such as sewage treatment plants, animal waste, and air pollution fallout.

When nitrates and phosphates enter the Gulf of Mexico, they are nutrients for algae which grows rapidly in a "bloom" (Figure 16.4b). Organisms that feed on the algae also grow rapidly and produce waste in large quantities. There is rapid growth of bacteria that feed on the waste from both algae and consuming organisms, thereby consuming the dissolved oxygen in the water and producing hypoxia. The hypoxic zone expands from deeper levels to the near surface. As a result, this part of the Gulf of Mexico undergoes eutrophication, which kills or drives off most marine life. The hypoxic zone grows

larger as more nutrients are added as long as there are no storms that disrupt the system.

The dead zone in the Gulf of Mexico was first recognized in the 1970s but the extent was not accurately mapped until the 1990s. The zone shifts from year to year and changes size and shape. It has been accurately mapped for the past 21 years. The shape, size, and location is controlled by the amount of fertilizer used in Midwestern agriculture areas, the amount of precipitation and runoff there relative to other areas in the Mississippi drainage basin, water temperature in the Gulf of Mexico and the number of weather events in the gulf. The situation that results in the largest dead zone is heavy fertilizer usage and precipitation in agricultural areas along the river and drought everywhere else and calm, warm water in the Gulf of Mexico.

The dead zone devastates the gulf fishing industry and results in unexpected consequences such as more shark attacks along the coast because of the lack of food. There is now a federal effort to restrict the size of the dead zone but it is not going well as evidenced by the record size in 2017.

16.3.3 Livestock

In order to meet the needs of the human population, the raising of livestock has also increased dramatically. Many of the crops are grown for livestock meaning some of that pollution is partly attributable to livestock. Direct pollution from livestock is largely from animal waste both from raising and processing them. If waste comes into contact with food there can be outbreaks of *E. coli* and *salmonella* among other bacteria on a national level, causing sickness and sometimes death.

The main environmental problem is from disposal of waste. In small farms, runoff from waste flows into streams and rivers, causing eutrophication. In large livestock operations, waste is dumped into lagoons. The breakdown of waste is a significant source of methane or natural gas. Methane is 23 times more potent a greenhouse gas than carbon dioxide. Flatulence from cattle also contributes to methane issues. In the United States alone, cattle emit about 6 million tons (5.5 million mt) of methane per year, accounting for 20% of total methane emissions.

Catastrophic events can be devastating to livestock operations at maximum capacity. The problem is that domestic animals lose their native instincts when they are farmed. In catastrophic events, animals need those instincts to survive. The eruption of the Laki volcano in Iceland, 1783 emitted fluorine gas that settled on and poisoned plants. Livestock ate the poisoned grass and hundreds of thousands of them died. Animal carcasses must be disposed of by burning or burying them as quickly as the human bodies or they can pollute water supplies and generate disease, causing outbreaks of diseases like cholera and dysentery.

CASE STUDY 16.5 Hurricane Floyd and Hog Waste in Eastern North Carolina

Extreme Case of Livestock Pollution

North Carolina developed a large hog farming industry. It expanded from about 2 million hogs in 1990 to more than 10 million hogs by 1997 (Figure 16.5). This made North Carolina the second-largest hog

producer in the United States. This rapid growth was largely accomplished by building large, automated farms that contained hundreds to thousands of hogs. The hog farmers had a lot of political power so expanded farms and facilities with minimal restriction. Many of the factory-type farms were situated on floodplains of rivers and in

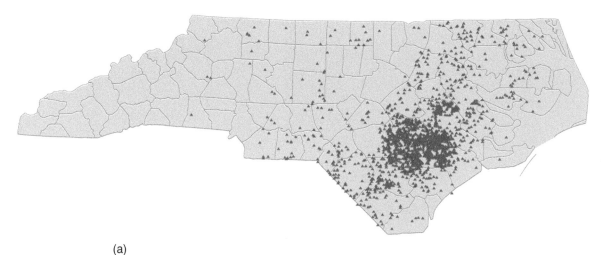

(a)

FIGURE 16.5 Map of North Carolina showing the distribution of large hog farming operations. (a) Photo of a waste lagoon for a large hog farm. *Source:* Bob Nichols, USDA Natural Resources Conservation Service / Wikimedia Commons / Public Domain.

environmentally sensitive reclaimed wetlands that were prone to flooding. Each hog produces about 2 tons (1.8 mt) of waste per year and the total output for the area was about 20 million tons (18 million mt) of waste per year. Disposal of this waste was a serious issue because the amount was growing so quickly and it posed a risk to public health. As a solution, farmers dug huge unlined lagoons and dumped the liquid and solid waste into them. The lagoons were huge, on average 300 ft (92 m) long by 150 ft (46 m) wide, and they were filled to about 10–15 ft (3–5 m) in depth (Figure 16.5a).

Environmentalists protested the practice, claiming that it was a disaster waiting to happen. In this part of North Carolina, heavy rains occur regularly, including an occasional hurricane. There were several collapses of waste lagoons that released tons of hog waste into rivers which flowed through towns causing health hazards and unimaginable odor. The problems were widely publicized, even appearing on the popular television show, "60 Minutes." The growing public pressure finally compelled the government to act. In 1997, a law was enacted that prohibited construction of new waste lagoons on flood plains. In 1999, the governor of North Carolina devised a 10-year plan to introduce new technologies for waste treatment and to replace the now 4000 waste lagoons.

Unfortunately, nature did not wait for the legislation. In late August 1999, Hurricane Dennis struck North Carolina, completely saturating the ground, and filling lakes, rivers, and reservoirs to capacity. A mere two weeks later, a much larger and more powerful Hurricane Floyd struck the same area. Precipitation exceeded 20 in. (50 cm) in eastern North Carolina, 50 people were killed, 48 000 people were forced into shelters, and the storm destroyed 2.3 million acres (931 000 ha) of crops. The environmental disaster part of the event involved the hog farming industry. More than 38 waste lagoons were completely washed away and 250 hog operations were flooded or had their lagoons overflow. More than 250 million gallons (946 million l) of untreated hog waste was washed into rivers, creeks, and wetlands by the storm. In addition, more than 30 000 hogs were killed as were two million chickens and 735 000 turkeys.

The dead birds were piled up and left to rot. North Carolina sent mobile incinerators to destroy the decaying hog carcasses but there were too many for the incinerators to handle. Consequently, farmers were instructed to bury as many of the carcasses as possible. They dug pits that were supposed to be placed on dry ground and be at least 3.3 ft (1 m) deep. The problem was that there was not much dry ground available at the time and there was no oversight on the operations to make sure they followed directions. The flooded waste coupled with the rotting carcasses, many of which floated in rivers through towns, thoroughly contaminated both the surface and groundwater in the area. Water was inundated with bacteria and disease and it remained that way for months. Residents were forced to use bottled water due to the threat of disease or even death. The stench that accompanied this disaster has been described as unbearable, causing residents to get sick just by being outside.

As a result of this disaster, in July 2007, North Carolina became the first state to ban construction or expansion of waste lagoons and sprayfields on hog farms. This ban is part of the Swine Farm Environmental Performance Standards Act.

CHAPTER **17**

Nuclear Energy and Dangers

CHAPTER OUTLINE

Words you should know:

Decay series – The sequence of parent to daughter elements/isotopes from a radioactive element to a stable element.

Fuel rods – Rods of uranium fuel pellets used in a nuclear power plant.

High-level nuclear waste – Radioactive wastes with radioactivity ≥10 Ci/kg, mainly discarded by the military and some nuclear power plants.

Low-level nuclear waste – Radioactive wastes with radioactivity ≤0.01 Ci/kg, largely from medical treatments, radiation equipment, materials exposed to radiation, and handling of radioactive materials.

Nuclear device – Basically a nuclear bomb but not necessarily in a bomb shape.

Nuclear powerplant – An electrical generating power station that uses nuclear power.

Nuclear reactor – A facility that houses a controlled nuclear reaction in a core.

Radiation – Electromagnetic fields and particle emissions from elements, devices, and outer space that can be damaging.

Radioactivity – The release of radiation during decay of radioactive isotopes.

Radioactive isotopes – Isotopes are elements having different atomic masses; some are radioactive and decay to daughter isotopes.

17.1 Radiation Types and Sources

People are exposed to numerous forms of radiation on a daily basis. The major sources of radiation are cosmic or extraterrestrial, earth materials, and equipment and devices. Cosmic radiation also includes solar radiation and is a mixture of several types of radiation with varying intensity and health implications. Earth materials produce radiation from the decay of radioactive isotopes in minerals and other materials present everywhere on Earth. Equipment produces several types of dangerous and benign radiation. They can be directed such as X-ray machines and CT scanners or they can be electric and magnetic fields (EMFs) emanating from equipment or power lines.

The amount of radiation that a person receives is the radiation dose or dosage. The absorbed dose is the amount of radiation energy per unit mass of a target, measured in Roentgen Absorbed Dose or rads. Grays are equal to 100 rads. Equivalent dose is the absorbed dose adjusted for biological effects produced by radiation and is measured in Roentgen Equivalent Man or rems. The committed dose measures continued exposure over decades to a lifetime. Daily doses of radiation to people are in millirems or mrems. Human X-ray exposure is also commonly measured in milliSieverts (mSv), where 0–3 is low, 3–20 is moderate, and above 20 is high.

The most common radiation is electromagnetic (EM) and is emitted by the Sun, stars, and many electrical devices. The EM spectrum varies by wavelength and frequency of the waves in the radiation (Figure 17.1). Non-ionizing radiation has long wavelength, low frequency, and low energy such as AM radio, shortwave radio, FM radio, and television with increasing frequency and shorter wavelengths. The next group includes microwaves, radar, infrared light (IR) and visible light. Shorter wavelength, higher frequency and higher energy EM radiation is ionizing radiation including ultraviolet radiation (UVR), X-rays, and gamma radiation in increasing order. Ionizing radiation is dangerous to humans.

EM radiation is also included as particle emission radiation. Radioactive elements decay by converting from a parent to a daughter isotope. This includes emission of particles with atomic number, mass, and charge. The atom converts from one element to another in a radioactive decay series through this emission. The largest emitted particle is an alpha particle, with the mass and charge of a helium atom. The decay is energetic, but particles travel only about one half inch (1 cm) because of mass. They can only be damaging to humans in cuts and internal organs. A beta particle or betatron is equivalent to an electron emitted during beta decay from the nucleus that converts a neutron into a proton. Betatrons are only hazardous to human health if ingested. Gamma and X-radiation do not yield particles and are hazardous to health.

EMF are lines of force surrounding electrical devices and electrical distribution equipment. EMF includes extremely low frequency (ELF) fields with frequencies up to 300 Hz, intermediate frequency (IF) with frequencies between 300 Hz and 10 MHz, and radiofrequency (RF) between 10 MHz and 300 GHz. ELF is emitted from electric appliances and power lines, IF is emitted by security systems and computer screens, and RF is from cell phones, microwave ovens, television, and radio.

Extraterrestrial radiation is both solar and cosmic (Table 17.1). Cosmic radiation constantly bombards the Earth providing 8–13% of the background radiation. Cosmic rays are charged particles that impact the Earth's atmosphere. Cosmic rays originate from neutron stars, supernovas, and black holes outside of the solar system. As cosmic rays strike the Earth's atmosphere, they interact with oxygen and nitrogen molecules. These interactions change nitrogen-14 into carbon-14 which is used for isotopic dating. The cosmic rays are deflected by the magnetic field and form the aurora borealis. They are stronger at higher elevations. Airplane travel

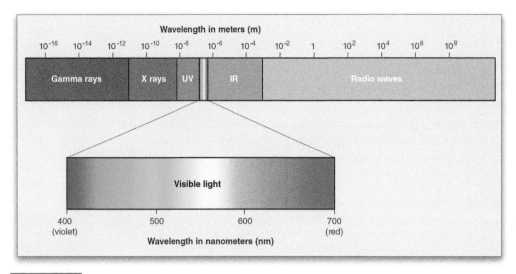

FIGURE 17.1 Diagram showing the range of wavelengths for electromagnetic radiation and the designations for the subranges. UV = ultraviolet, IR = infrared.

TABLE 17.1 Sources and amounts of radiation exposure.

Source	Radiation
Five-hour jet plane ride	3 mrem
Building materials	4 mrem
Chest X-ray	8 mrem
Cosmic rays	30 mrem
Soil (naturally occurring isotopes)	35 mrem
Mammogram	138 mrem
Radon gas (from natural sources)	200 mrem
CT scan	2500 mrem
Smoking (pack/day)	5300 mrem (to lung)
Cancer treatment	5 000 000 mrem (to tumor)

Radiation sickness in humans occurs at short-term exposures of 100 000 mrem or more.

increases the dosage from the cosmic radiation. Cross-country flights result in a radiation exposure dose of 2–5 mrems. People who fly a lot or fly near polar regions are exposed to enough radiation to impact their health.

There are radioactive elements in rock and soil but generally in small amounts. The most common radioactive elements are uranium, radium, thorium, potassium, strontium, and radon. The elements of a radioactive decay series occur together. Parent uranium decays in many steps to radium and radon. All radioactive isotopes emit gamma radiation but they may also emit alpha or beta particles. Radioactive elements may be concentrated in a deposit and pose the risk of exposure to excessive amounts of gamma radiation. In a natural deposit in Africa, there was so much uranium in one place that a natural nuclear reaction was generated.

Natural occurrences of radioactive isotopes are termed "naturally occurring radioactive minerals" or NORM. If radioactive minerals are removed from these rocks and processed as waste, they can form dangerous concentrations called "technologically enhanced naturally occurring radioactive minerals" or TENORMs. The most common of these elements are uranium, thorium, and radium and they are concentrated by mining and mineral processing, oil and gas production, and drinking water and wastewater treatment.

There are radioactive materials created by human activity. Plutonium is commonly used for defense and its production and use result in tritium. Others are used in medical procedures, sterilization, and industrial applications. Many of these isotopes are produced in research nuclear reactors. These produced isotopes are used in place of natural materials because the uniformity, dosage, and duration of radiation are better controlled. Common examples include cobalt-60, cesium-137, and technetium-99. Exposure is through medical procedures, radiography in metal production, sterilization of some foods, and water or waste.

The most common example of equipment radiation is X-rays. Nuclear power plants produce high quantities of gamma radiation which must be shielded using lead. Some equipment bombards material with high energy particles, generating secondary X-rays. Much of our everyday equipment produces EMF, causing near constant exposure in many professions and recreation.

17.2   Health Impacts of Radiation

Ionizing radiation causes the most health threats. High dose, short exposure results in burns or radiation sickness. Radiation sickness or radiation poisoning is the worst of these and produces various symptoms depending on dose. A 5–10 rems dose can produce changes in blood chemistry. Increased doses of 50 rems produces nausea, 55 rems produces fatigue, and 70 rems produces vomiting. If exposed to 75 rems, hair loss begins in two to three weeks, at 90 rems, diarrhea starts, and at 100 rems hemorrhaging takes place. Radiation exposure of 400 rems is usually fatal in two months, 1000 rems causes destruction of the intestinal lining and internal bleeding begins, which is fatal in one to two weeks of searing pain. At 2000 rems, the central nervous system is damaged and loss of consciousness results in minutes, with death following hours to days later.

Long-term exposure to even low levels of ionizing radiation can damage human health. Most commonly, cancer can occur in a number of organs. Radiation can damage DNA, producing genetic mutations that can be passed on. Damage to a fetus commonly occurs if a pregnant mother is exposed to radiation. Inhalation and ingestion of certain radionuclides over a lifetime can also cause health effects. Radon is radioactive and the most dangerous natural environmental hazard. Radon mostly damages health through alpha decay. It is a colorless, odorless noble gas in most indoor air that is inhaled into the lungs without causing any problems. However, parent radon decays to daughter polonium which can stick to lung tissue. Polonium undergoes alpha decay, emitting a high energy alpha particle that can cause lung cells to mutate which may lead to lung cancer. The US Environmental Protection Agency (EPA) estimates that 25 000 lung cancer deaths per year in the United States result from radon exposure.

Radioactive iodine can concentrate in the thyroid gland through ingestion. This leads to increased occurrence of thyroid cancer. Strontium-90 and radium-226 act like calcium in the bloodstream so they accumulate in the bones and teeth like iodine in the thyroid. Increase of these radioactive isotopes in the bones increases the likelihood of bone cancer.

Solar radiation and especially UVR cause negative health effects with long term of exposure. It causes skin burns and possible eye damage in acute exposure. The burns result in skin damage and sickness called sun poisoning. Long-term exposure to the sun can result in skin cancer to the point where solar radiation is classified as a known human carcinogen. Solar radiation increases skin melanoma, melanoma of the eyes, and non-Hodgkin's lymphoma. UVR is also classified as a known human carcinogen and has been shown to cause DNA damage. UVR is divided into three types UVA, UVB, and UVC. The dangerous types are UVB and UVC, with UVC being the most damaging. UVC, however, is the wavelength that is best absorbed by ozone and most does not reach the surface in the mid latitudes. As a result, UVB accounts for 90% of the DNA damage.

The only sure health effect of EMF is general heating. Some claim that there is an increased incidence of childhood leukemia from EMF from power lines. The National Institute of Environmental Health Sciences and International Association for Research on Cancer classify exposure to ELF fields like those around power lines as possible causes of cancer. However, studies on laboratory animals have found no link.

17.3   Radiation Pollution

There are two types of radiation pollution: direct radiation and dispersion of radioactive materials. There are multiple sources of radiation pollution. Certainly, the most fearsome of the sources is nuclear devices or bombs. During the 1950s and 1960s, there were so many atmospheric detonations of nuclear devices for testing that fallout of radioactive isotopes around the Earth was rampant (Figure 17.2). A total of 504 nuclear devices were detonated at 13 primary testing sites, yielding the explosive power of 440 mt of TNT. The age of groundwater can be designated as pre or post onset of nuclear testing by the resulting spike of tritium in the water. All life on the planet was exposed to excessive radiation during this time, with unknown consequences. Fortunately, through diplomacy, atmospheric tests were mainly discontinued in 1963. Only underground tests were permitted after 1963 (Figure 17.3) and these continued until 1990–1996 depending upon the country.

Nuclear test sites were at remote locations at least 60 miles (100 km) from human settlements. As the cloud rose from the detonation and radioactive materials settled out, distances from the test were classified. Distances of 30–300 miles (50–500 km) from the test are considered local fallout whereas to 300–1800 miles (500–3000 km) is regional fallout and more than 1800 miles (3000 km) is global fallout. Direct doses of as much as 30 Grays can occur as far as 100 miles (160 km) from the test site. Radiation exposure from the fallout is from gamma rays emitted by particles that are inhaled as they settle, on the ground, or from food containing the nuclear fallout. As a result, about 49 000 fallout-related cases of thyroid cancer are likely in the United States, primarily among

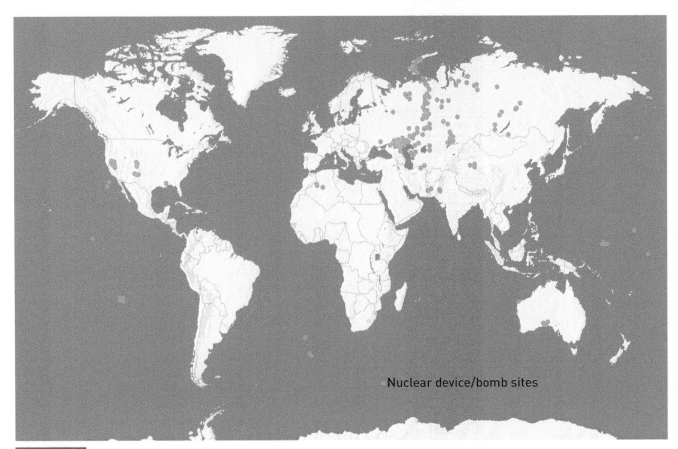

FIGURE 17.2 Map of the world showing the locations of the nuclear device/bomb tests.

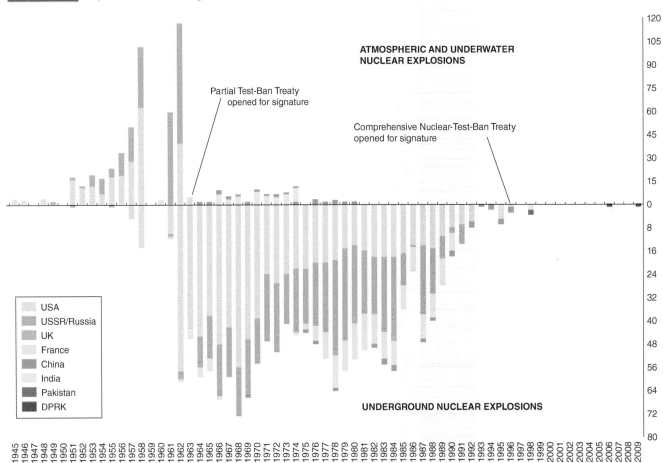

FIGURE 17.3 Timeline showing the number of nuclear tests per year including country conducting the test and whether it was above or below ground.

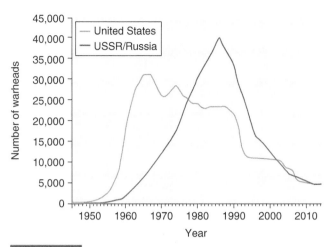

FIGURE 17.4 Graph showing the number of nuclear warheads in the US and USSR/Russia arsenals from 1945 to 2015.

persons who were under 20 between 1951 and 1957. In addition, about 1800 fallout radiation-related leukemia deaths could occur in the United States and an increased rate of bone cancer is also likely. Increases in other cancers are also possible.

It is fortunate that all the nuclear detonations during the Cold War were only for testing. The amount of arming of nuclear weapons primarily by the United States and Soviet Union was enough to eliminate virtually all life on the planet (Figure 17.4). The United States built up arms in the 1960s so quickly that they estimated they could kill every human on the planet 100 times over. As a result, in the late 1960s, they developed Project Plowshare to find productive peacetime uses for nuclear devices and curbed production. The Soviet Union built up their nuclear stockpile in the 1970s in response to United States build-up and actually exceeded their total weaponry by later in the decade.

CASE STUDY 17.1 Project Plowshare

Fracking and Construction Using Nuclear Devices

President Eisenhower gave a famous speech "Atoms for Peace" in December 1953. In response, the Atomic Energy Commission (AEC) established the Plowshare Program in June 1957 to explore the use of nuclear weapons for non-military purposes. There were numerous proposals for experiments and a number were carried out. Between December 1961 and May 1973, 27 nuclear tests were conducted with 35 total detonations. The experiments they conducted were to investigate the use of nuclear weapons (i) for large-scale excavation and quarrying where the explosion would break up and move rock, (ii) enhancing the porosity and permeability of the rock to improve natural gas flow (fracking), and (iii) to produce new heavy isotopes for medical, energy, and therapeutic application that are unobtainable by conventional means. The Soviet Union had a similar program called Nuclear Explosions for the National Economy.

Most of the tests that were conducted at the Nevada Test Site (NTS) were for heavy isotope research but tests were also conducted at other sites for other purposes. The first offsite detonation was Project Gnome in southeastern New Mexico, approximately 25 miles (40 km) southeast of Carlsbad (Figure 17.5). It was designed to create new high-temperature isotopes and to test the feasibility of trapping the energy released by nuclear explosions and then using it to generate electricity. The nuclear device was placed 1184 ft (361 m) underground in a 1115 ft (340 m) long tunnel that was designed to be sealed by the explosion. It was detonated on 10 December 1961 and had a yield of 3.1 kt of TNT. The explosion created a cavity 160–170 ft (49–52 m) wide and 60–90 ft (18–27 m) high with a floor of melted rock and salt.

Project Gnome was publicized to bring attention to the program and, as a result, had more than 400 observers from the press, the public, and foreign nations. They described the ground as having "jumped" 8 ft (2.2 m) into the air. However, the cavern failed to seal as planned and radioactive waste flooded the shaft. Smoke and steam emerged in about two to three minutes after the detonation so some radiation was released. Unfortunately, the radioactive steam vented over the press and observers who were meant to confirm the safety of the project.

A shaft was drilled into the explosion cavity and study teams entered it on 17 May 1962. Even after six months of cooling, the temperature in the spherical cavity was about 140 °F (60 °C). There were stalactites of melted salt, and the walls were covered in salt (Figure 17.5a). The radiation colored the salt blue, green, and violet. The radiation was only five milliroentgen, which was considered safe. The explosion melted 2400 tons (2177 mt) of salt and produced 28 000 tons (25 400 mt) of rubble mixed with the molten salt. Pockets of molten salt still at temperatures of 1450 °F (790 °C) were found within the rubble. High-melting-point radionuclides were found in the solidified melt rock.

The next experiments were to test the feasibility of using nuclear weapons to excavate a channel or trench from surface deposits of soil, rock, and sediment. On 6 July 1962, the code name "Sedan" test involved the detonation of a 104 kt (435 tj) device at the NTS. The explosion excavated more than 12 million tons (11 tg) of soil but raised a cloud of radioactive dust 12 000 ft (3.7 km) into the air (Figure 17.5b). The radioactive dust cloud was blown northeast and then east toward the Mississippi River by prevailing winds, resulting in widespread radioactive fallout. It also resulted in uninhabitable land, relocation of communities, and widespread tritium-contaminated water but these impacts were ignored and downplayed at the time.

Another off-NTS test was called Project Gasbuggy and it was conducted in northern New Mexico by Lawrence Livermore Radiation Laboratory and El Paso Natural Gas Company and funded by the AEC. It was designed to determine if nuclear explosions could be used to "frack" or fracture rock formations to enhance natural gas production. This underground nuclear detonation was carried out on 10 December 1967. The detonation was 21 miles (34 km) southwest of Dulce, and 54 miles (87 km) east of Farmington, New Mexico in a well drilled into sandstone with natural gas deposits (Figure 17.5c). The device had a yield of 29 kt of TNT (120 TJ) and installed at a depth of 4227 ft (1288 m).

A crowd of people watched the detonation from a nearby butte. Once again there was radioactive material released and it was deposited on the surface nearby. Project Gasbuggy succeeded in fracturing the rock in the well for more than 200 ft (61 m) in some cases and it increased natural gas production. The well containing the device produced 295 million cubic feet (8.4 million m³) of natural gas. However, it was radioactive and could not be used for home gas appliances.

Two additional nuclear fracking experiments were subsequently conducted in western Colorado. They were Project Rulison in 1969 and Project Rio Blanco in 1973. Project Rulison was an underground 40-kt nuclear test in a well 8 miles (13 km) southeast of Grand Valley, Colorado. The detonation was on 10 September 1969 and the depth of the device was approximately 8400 ft (2600 m). The detonation itself did not result in the release of radioactive elements into the atmosphere. Subsequent tests of natural gas from the well in 1970 and 1971 released some radioactivity. Project Rulison was successful in producing large amounts of natural gas as well. The safety and production encouraged the next test.

Project Rio Blanco was the next nuclear fracking test and it took place on 17 May 1973 in Colorado, 36 miles (58 km) northwest of Rifle. It was a complex arrangement of three 33-kt nuclear devices that were detonated simultaneously in a single well at depths of 5838, 6230, and 6689 ft (1779, 1899, and 2039 m). The blast liberated some natural gas but was considered an engineering failure which helped to end the fracking experiments.

The fracking tests produced large quantities of natural gas but the radioactivity left it contaminated and unusable for cooking and heating homes. Even though public radiation exposures from use of the gas were less than 1% of background, the public refused to use any gas containing radioactivity. As a result, no natural gas was ever sold.

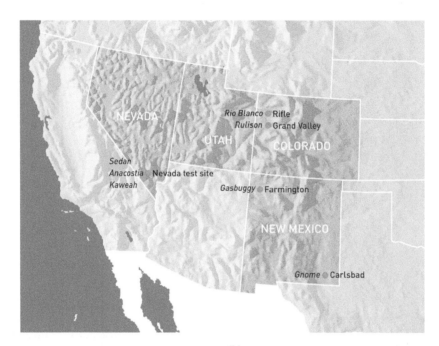

(a)

(b)

FIGURE 17.5 Map of the southwestern United States showing the locations of several of the Project Plowshare nuclear device detonations. (a) Photograph of the underground cavern produced by the Gnome experiment detonation near Carlsbad, New Mexico. *Source:* US Department of Defense / Wikipedia / Public Domain. (b) Photo of the Sedan experiment detonation to use nuclear devices for excavation. *Source:* Federal Government of the United States / Wikipedia Commons / Public Domain. (c) Diagram of the device, engineering, and monitoring of the underground Gasbuggy experiment. *Source:* Diagram courtesy of Los Alamos Labs.

(c)

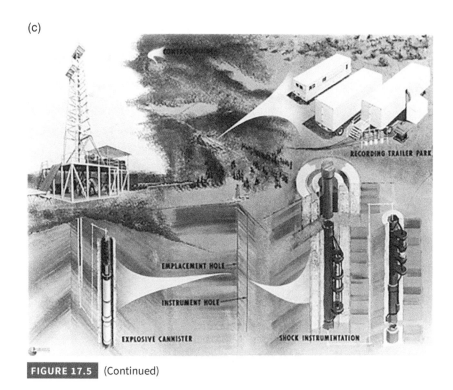

RECORDING TRAILER PARK

EMPLACEMENT HOLE

INSTRUMENT HOLE

EXPLOSIVE CANNISTER

SHOCK INSTRUMENTATION

FIGURE 17.5   (Continued)

The public lost interest in nuclear tests on American soil by the mid-1970s both for environmental impacts and safety. As a result, Project Plowshare was terminated in 1977. The Department of Energy started a remediation project for the sites in the 1970s and completed operations in 1998. A buffer zone was established around the sites by the state of Colorado. By January 2005, radioactivity levels were normal at the surface and in groundwater, though deeper migration of radioactive groundwater was not ruled out.

The other main use for nuclear energy is in generation of electricity. Nuclear powerplants have several big advantages such as large power output and no generation of greenhouse gases. On the other hand, nuclear powerplant accidents can cause catastrophic impacts to public health and the environment unequaled by any industrial accident. For this reason, there is large-scale uneasiness and even fear among residents in the vicinity of the plants. Uncontrolled reactions in powerplants can cause nuclear meltdowns that have been called the "China Syndrome" alluding to a reactor melting through the Earth and coming out the opposite side.

Most nuclear powerplants work similarly using steam turbines that generate energy by converting liquid water to steam like coal or gas powered powerplants. The difference is that nuclear powerplants do not burn any fuels so emit no chemical by-products to the atmosphere. The way nuclear powerplants generate heat is using radioactive fuel, most commonly uranium because it is plentiful and easier to control. Uranium must have a large amount of the isotope uranium-235 to be a good fuel so it is first put through the process of uranium enrichment. Once uranium is enriched, it can be used in power plants for three to five years. Once it is spent, it is still radioactive and must be carefully disposed of. Much of it is recycled into other types of fuel for reuse in different applications.

The enriched uranium fuel is manufactured into ceramic pellets. Each pellet produces the same amount of energy as 150 gal (568 l) of oil. The pellets are installed in 12-ft (3.7 m) metal fuel rods. A bundle of more than 200 of these fuel rods is a fuel assembly. The reactor core provides the power to the plant and contains several hundred fuel assemblies each having thousands of uranium fuel pellets (Figure 17.6). The assemblies are immersed in vats of cooling water.

The heat in the reactor core is provided by nuclear fission of the uranium atoms in the fuel rods. The uranium atoms are bombarded with neutrons that splits them apart to form two smaller atoms plus additional neutrons. Some of the released neutrons then hit other uranium atoms in other fuel rods, causing fission in them as well which releases more neutrons. These neutrons drive more fission reactions producing a chain reaction process. The nuclear fissions in the chain reaction also releases large amounts of heat to the cooling water raising the temperature to about 520 °F (271 °C). This hot water is removed from the reactor, where it generates the steam that drives the turbines for electricity production.

The vigorous nuclear reaction must be controlled to prevent a meltdown and the disaster it generates. This is done using control rods that block the passage of neutrons. The control rods are pipes surrounding the fuel rods made of neutron-absorbing material like silver and boron that can be raised or lowered to speed or slow the reaction as needed. Once the steam drives the

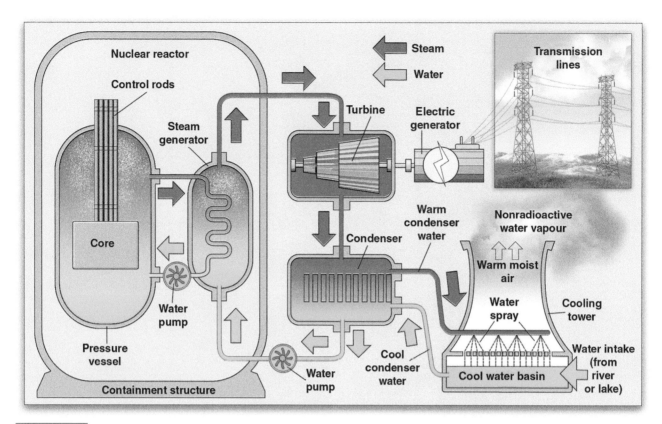

FIGURE 17.6 Diagram of a standard nuclear powerplant to generate electricity. The core contains the nuclear fuel rods which generates the heat to create steam.

turbine, it condenses back to liquid water in a cooling tower, which is a separate structure at the power plant, and water from nearby ponds, rivers, or the ocean is added as needed.

There are about 400 commercial nuclear reactors in more than 30 countries worldwide. The commercial nuclear reactors in the United States are light-water reactors, meaning they use water both as a coolant and to control neutron activity. The two types of light-water reactors are pressurized-water reactors, which use the cooling water to heat separate, non-cooling water for steam, and boiling water reactors which produce steam directly inside the reactor vessel from cooling water. More than 65% of reactors in the United States are the safer pressurized-water reactors.

In addition to not producing greenhouse gas or adding contaminants to the atmosphere the fuel requirements and resource consumption of nuclear power plants are far less than fossil fuel plants. For example, one uranium pellet contains the same energy as 1.1 tons (1 mt) of coal. Nuclear reactors use about 29.8 tons (27 mt) of fuel per year compared with a similar-sized coal power plant which requires more than 2.8 million tons (2.5 million mt) of coal to produce the same amount of electricity.

CASE STUDY 17.2 Chernobyl Nuclear Disaster, Russia

Worst Civilian Nuclear Accident

The Chernobyl nuclear disaster was the most devastating human caused environmental catastrophe in history. It displaced more than 130 000 people and thousands of square miles of productive farmland were contaminated. In addition to the 30 deaths directly caused by the explosion and fire at the Chernobyl nuclear powerplant, the health of some five million people was compromised.

The Chernobyl nuclear power plant was located at Pripyat, Ukraine (Figure 17.7a). It was among the largest and oldest of the 15 public electric generating nuclear plants in the Soviet Union. It was

located 80 miles (128.8 km) north of the city of Kiev, with its population of two million people. Pripyat had a population of 45 000 and was a mere 5 miles (8.1 km) from the nuclear plant.

The plant included four water-cooled, graphite-moderated nuclear reactors and a fifth was under construction. The Soviet Union utilized this type of reactors because they are relatively inexpensive to construct and operate, the plant can be expanded in modular increments to meet demand, and they could use recycled fuel from military reactors. However, these reactors tended to be unstable and susceptible to power surges unless carefully monitored. Also, the reactors required more than 3000 fuel rods that were not built within a reinforced containment structure. Instead, they were contained within a graphite moderator in a honeycomb design that permitted the coolant flow. Because graphite melts at such high temperatures and has such a high heat capacity, a containment structure was not considered necessary. At low and short duration of heat produced by uranium fission, this type of reactor could operate safely even with interruptions in coolant flow. Several Western countries used these reactors for research purposes, but none used them for commercial power production.

Several safety features were installed on the Chernobyl reactors. A sealed metal container of inert gas surrounded each reactor to keep oxygen away from the 1300 °F (704.4 °C) graphite moderator. Extensive shielding was also installed to protect plant operators from radiation emitted during fission. The shielding consisted of a concrete slab

under the reactor, sand, and concrete barriers around the reactor, and a large steel-reinforced concrete slab on top of the reactor.

On 26 April 1986, at about 1:20 a.m., reactor no. 4, which went online in 1983, began to have its power reduced as part of a routine operation test. The test determined how long the turbine and generator could run in an emergency reactor shutdown. The operators turned off the primary and secondary emergency water-cooling system and disconnected the power and automatic reactor shutdown systems. In addition, only six control rods were in place instead of the 30 required in the procedure. These were in direct violation of several plant safety protocols. This was not unexpected as Chernobyl had a history of plant mismanagement, poor training, and labor unrest.

During the test, the reactor dropped below stable levels, resulting in a power surge. As the temperature in the core increased, the fuel

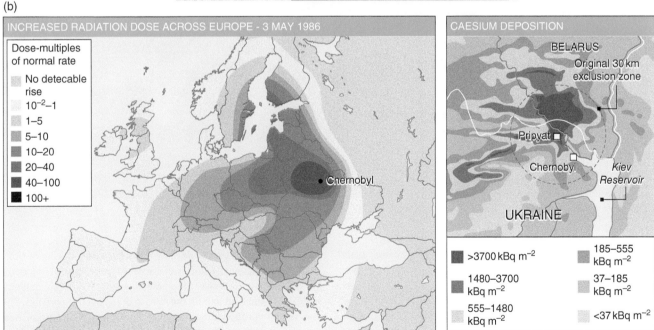

FIGURE 17.7 (a) Map of Ukraine and surrounding countries showing the location of the Chernobyl nuclear reactor. Inset map shows location relative to the world. (b) Map of the radiation released across Europe by the Chernobyl accident (left) and the cesium deposition near the accident site (right). (c) Photo of the destroyed Chernobyl reactor from the accident. *Source:* Joker345 / Wikimedia Commons / CC BY-SA 4.0. (d) Photo of the new sarcophagus around the destroyed Chernobyl reactor to protect it. *Source:* Courtesy of the United Nations.

(c)

(d)

FIGURE 17.7 (Continued)

rods melted and exploded. In response, the operators immediately activated the cooling system. As the cooling water contacted the melting fuel rods, a second explosion blew the 1000-ton (907.2-mt) steel cover off the reactor. The cover was attached to fuel rods and they released radioactive isotopes into the atmosphere. This explosion shot fragments of burning fuel and graphite from the reactor and air rushed into core, causing the remaining graphite to burst into flames. For the next two days, flames shot 1000 ft (304.8 m) into the air.

The fire took several days to extinguish and it took more than 10 days to curb the release of radioactivity. The firefighters were plant workers who faced certain death either quickly from the fire or slowly by radiation sickness. As news of the catastrophe spread, helicopters were brought in to shower the smoldering core with boron, dolomite, sand, clay, and lead to curb the release of radioactive particles. They flew more than 1800 sorties and dumped 5000 tons (4536 mt) of material to control the main radiation leak. Thirty-one people died almost immediately in the explosion and fire and more than 500 people were hospitalized. Pripyat was evacuated on 2 May as the radiation levels rose. Over the next 10 days 130 000 people within 20 miles (32.2 km) of the reactor evacuated their homes and farms. This area was designated as the exclusion zone. Radioactive substances are considered safe for human contact after 10 half-lives of decay. For the cesium-137 that was released, it will be 300 years before the land can be safely used.

The plume of radiation moved north and contaminated over 1 million acres (0.4 million ha) of farmland in Belarus and western Russia (Figure 17.7b). It continued to sweep west across Europe, depositing radioactive fallout which settled around the entire Northern Hemisphere. The amount of radioactivity released in the accident was 200 times more than that released by the Hiroshima and Nagasaki bombs. At least five million people in Belarus, Russia, and Ukraine were exposed to radiation.

After the accident was under control, 200 000 personnel were assembled to assist in cleanup operations. The cleanup force was later increased to 600 000. They were lured to the project by high salaries and special bonuses. The operation lasted from 1986 and 1987 to stabilize the demolished reactor. These personnel received doses of radiation of around 100 millisieverts (mSv). It is estimated that one in 100 people would develop cancer or leukemia from this exposure. At least 20 000 of these people received radiation doses of at least 250 mSv, and some were exposed to more than 500 mSv.

Within seven months of the accident, a reinforced concrete structure or sarcophagus was erected to encase the remains of reactor no. 4 (Figure 17.7c). The structure holds about 180 tons (163.3 mt) of uranium with a total radioactivity 7×10^{17} Bq. It shifted and cracked and the building filled with rain water and snowmelt. Contamination spread internally and increased the risk of radioactive waste being released to groundwater. The structure is so unstable that a windstorm or a small earthquake could cause it to collapse. This would result in a catastrophic release of radioactive dust and debris.

To address this urgent situation, a containment shell, called the New Safe Confinement structure, was constructed and moved into position around the sarcophagus in 2016 (Figure 17.7d). When it was moved into position, it was the largest land-based object ever moved by humans. This New Safe Confinement structure was designed to keep the sarcophagus safely contained for 100 years. This was thought to permit the site to be safely remediated. However, in 2022, Russia invaded Ukraine, threatening the safety of the area and exposing Russian soldiers to excess radiation.

The other main source of pollution caused by the use of nuclear energy is waste. Depending upon the usage, this waste may be very radioactive and dangerous or less so. These wastes are classified in six types: (i) low-level wastes with radioactivity ≤0.01 Ci/kg, (ii) intermediate-level wastes with radioactivity from 0.01–10 Ci/kg, (iii) high-level wastes with radioactivity ≥10 Ci/kg, (iv) uranium tailings, (v) spent fuel rods, and (vi) transuranic (TRU) wastes.

Transuranic wastes are produced as a biproduct of nuclear weapons manufacture and plutonium in reactors. Spent nuclear fuels are pallets of uranium dioxide fuels in the fuel assembly and are high-level wastes. Uranium mine tailings are among the most hazardous radioactive wastes. They are by-products of uranium mining. Uranium ore is treated with chemicals for extraction and leaves a sludge of highly radioactive by-products. These wastes are localized, rare, and contained in other classifications.

High-level radioactive waste is discarded by the military and some nuclear power plants. It is very dangerous and must be carefully handled to protect both humans and the environment. The most common high-level radioactive waste is depleted fuel rods from nuclear power plants. The amount of waste and its radioactivity has decreased greatly in recent years. Further, the nuclear power industry fully contains and manages its waste. The majority of high-level waste from nuclear power plants is not that radioactive. If a family's electrical usage for one year was supplied by nuclear power, only about 5 g of high-level radioactive waste would be produced. The spent reactor fuel can be recycled for further energy production in addition to direct disposal. Many countries use recycled fuel to partially fuel their reactors.

CASE STUDY 17.3 Lake Karachay-Mayak Complex, Russia

Radioactive Water Contamination

Listed as the most polluted lake in the world, Lake Karachay, Russia is really that polluted (Figure 17.8a). This lake is located 9.3 miles (15 km) east of Kyshtym at the Cold War secret complex of Chelyabinsk-65. This complex housed the Mayak Chemical Combine, which operated the USSR's first nuclear reactors and produced the radioactive material for their first atomic bombs. The complex was opened in 1948 and produced nuclear materials for the next four decades. It also discharged more than 123 million curies (Ci) of radioactive waste into storage areas that leaked much of it into the environment, during this period. The facility was not even on maps until 1989 and was named Chelyabinsk-40 before it was admitted to. It may have exposed as many as 400 000 people to dangerous levels of radioactivity.

Construction of the facility began in November 1945 and the first nuclear reactor began operation in June 1948 (Figure 17.8b). The full facility grew to approximately 77.2 square miles (200 km²) during operations. The reactors produced plutonium for nuclear weapons from 1948 until 1 November 1990, after which it reprocessed power reactor fuel for plutonium and uranium recovery. The facility included five graphite-moderated, water-cooled reactors, and two light water-cooled production reactors. The graphite reactors produced a combined 6565 MW thermal and were used for plutonium production. The light-water reactors each had a capacity of about 1000 MW thermal and produced tritium and other isotopes.

When the facility opened in 1948, high level nuclear waste was simply diluted and dumped into the nearby Techa River designated as medium-level waste. By 1951, the massive environmental

(a)

(b)

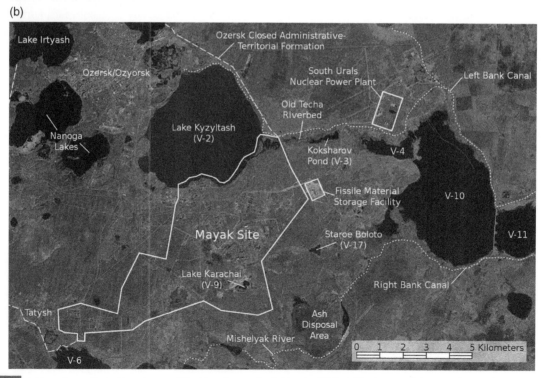

FIGURE 17.8 (a) Location of Lake Karachay and the Mayak complex in Russia. (b) Satellite image of the Mayak complex and surrounding area showing location of several features. *Source:* Courtesy of NASA. (c) Map showing the river system from the Mayak complex to the Arctic Ocean. Inset shows the details of surface water around the Mayak complex.

(c)

FIGURE 17.8   (Continued)

contamination became apparent. By then, 2.8 billion cubic feet (78 million m³) of high-level nuclear waste containing 2.75 MCi (million Ci) of beta activity was discharged into the river, 3.7 miles (6 km) below the source.

The Techa River is 149 miles (240 km) long and flows into the Iset' River, which, in turn, flows into the Tobol River (Figure 17.8c). This river system is about 600 miles (1000 km). The Tobol joins the Irtysh River, which flows into the Arctic Ocean. There is extensive contamination of the Techa riverbed and floodplain, exposing residents to radiation. The 1200 inhabitants of the village of Metlino are 4.3 miles (7 km) downstream from the plant. The riverbank there yielded gamma radiation of up to 5 R/h compared with a natural background of 10–20 microroentgen/h. Ninety-nine percent of the radioactive waste dumped into the Techa deposited within the first 22 miles (35 km).

In 1949, 38 villages with 28 100 people had their main source of drinking water from the Techa River. From 1953 to 1960, 7500 people were relocated and 5000 others were relocated later. The water supply for the remaining residents was switched to deep wells, and the floodplain was fenced. These 28 100 people who received the highest radiation were found to have increased leukemia morbidity and mortality 5–20 years after exposure. It is estimated that 65% of these residents

suffered from chronic radiation sickness. The 1200 residents of Metlino had the highest exposure at about 140 rem and the 7500 relocated residents had 3.6–140 rem exposure.

To address this pollution, four dams were built on the Techa River in 1951, 1956, 1963, and 1964 to impound some of the radioactive waste released. The impoundment had a combined area of 32.4 square miles (84 km²) and a volume of 13.4 billion cubic feet (380 million m³). These impoundments now contain about 193 000 Ci of strontium-90 and cesium-137 radioactivity. At least 124 000 people living along the Techa-Iset'–Tobol River system were exposed to high levels of radiation.

In 1951, the diluted high-level waste was dumped into the 111-acre (45-ha), 6.5–10 ft (2–3 m) deep Lake Karachay. In 1953, the process was changed to precipitating out insoluble fission products and storing them in stainless steel waste tanks but the rest was still dumped in the lake. This practice was still occurring in 1992 when waste was being added at 800 000 Ci per year to maintain the water level of the lake. By 1990, 500 MCi had been released to Lake Karachay and the lakebed had accumulated 120 MCi of long-lived radionuclides. In 1990, the lake had radiation exposure rate of about 600 R per hour. This level provides a lethal dose of radiation within 30 minutes.

To address this danger, between 1978 and 1986, the lake was filled with more than 5000 concrete blocks to prevent the shifting of contaminated sediments. This infilling of rock and soil continued to November 2015 when it had been completely filled in. It was then monitored before adding the final layer of rock and dirt which was completed in December 2016. Monitoring has showed a solid reduction in surface radiation after 10 months.

The history of the lake was not without incident. On 29 September 1957 at 4:20 p.m. local time, the Kyshtym Disaster occurred. One of the steel waste settling tanks exploded with a force equivalent to 70–100 tons (63.5–90.7 mt) of TNT, blowing a 3-ft (1-m)-thick concrete lid 82 ft (25 m) away. It also ejected 70–80 tons (63.5–72.6 mt) of radioactive waste containing more than 20 MCi of radioactivity in comparison with 51.4 MCi released in the Chernobyl accident. About 90% of this radioactivity deposited around the tank. However, about 2.1 MCi, formed a half mile (1 km)-high radioactive cloud that drifted across Chelyabinsk, Sverdlovsk, and Tumensk Oblasts, reaching Kamensk-Uralskiy in 4 hours, and Tyuman in 11 hours. It covered 8880 square miles (23 000 km^2), in a track 186 miles (300 km) long and 18.6–30 miles (30–50 km) wide, contaminated at more than 0.025 Ci per square mi (0.1 Ci/km^2). It covered 217 towns and villages with a total population of 270 000. It was later named the East Ural Radioactive Trace.

Another incident of regional contamination occurred in 1967 when a hot summer followed by a dry winter caused an extensive drought. Lake Karachay shrank and the lakebed was exposed. From 10 April to 15 May, a strong windstorm lifted contaminated dust into a large cloud that extended more than 46.6 mi (75 km). About 600 Ci was spread over an area of 695–1042 square miles (1800–2700 km^2) at greater than 0.025 Ci per square mile (0.1 Ci/km^2). This directly impacted 41 500 people in 63 villages. Some sources claim the impact was much greater, as many as half a million people.

The final major environmental problem is still ongoing. The dense radioactive contaminated water in the lake is driven into the groundwater system around the lake by gravity. This produced an aureole of radioactive groundwater that is spreading away from the lake north and south at about 262 ft (80 m) per year. It had migrated about 1.6–1.9 miles (2.5–3 km) from the lake, covering an area of 3.9 square miles (10 km^2) by 1994. It was estimated that more than 177 million cubic feet (5 million m^3) of radioactive water containing well over 5000 Ci had been forced into the groundwater system. This plume will continue to spread and pollute other water sources indefinitely.

The difference between high-level and intermediate-level radioactive waste is that high-level is heat-producing and intermediate is not, though both have high levels of radioactivity. It typically includes resins, sludge, metal fuel holders, and other items from the operation of nuclear power plants.

About 99% of radioactive waste is low-level. Most low-level waste is from hospitals from medical treatments, radiation equipment, materials exposed to radiation, and handling of radioactive materials. The waste includes syringes, bottles, cloth, and tools. Low-level radioactive waste is also from research laboratories and manufacturing facilities. Industry, universities, and the military all conduct research using radioactive substances. They also include luminous watch dials and reactor coolant residues. Very low-level radioactive waste is a subset of low-level waste that mostly comes from dismantling operations.

It might be assumed that low-level waste is not dangerous to humans or the environment. However, it must be carefully handled and under the right conditions, it can be very dangerous. This was the situation in Orange (Glen Ridge), New Jersey with the "Radium Girls," who were workers who painted luminous watch dials. This situation was brought to infamy in several books and a movie in 2018.

CASE STUDY 17.4 The Radium Girls

Severe Health Effects from Low-Level Radioactive Waste

Radium is a radioactive element in the decay series primarily of uranium 238 but also thorium 232. It also glows in the dark (luminescent) and is highly prized for many industrial and everyday applications. In addition, radium is a good source of gamma radiation for medical, research, and industrial uses. Radium was discovered by the Nobel Prize winner, Marie Curie, in 1898, but it was not isolated as an element until 1911. It was used in a variety of common products such as tonic, toothpaste, ointments, and elixirs in the early 1900s.

The U.S. Radium Corporation made use of this strong luminescence by opening a factory in Orange, New Jersey in 1917 producing luminous paints, watch and clock faces and other radium products for military and medical purposes (Figure 17.9a). Upon opening, they hired approximately 70 women to perform a variety of jobs requiring the handling of radium. The scientists in the factory were very careful handling the radium but assured the women that it was safe. The job paid well and was considered classy with the glow it produced on exposed clothing and hair. Some of the women even painted their nails, teeth, and faces for fun and ate lunch at their worktables, further increasing exposure.

The main job for the total of 300 women in New Jersey factory and many other factories across the United States and Canada that would be later established, was painting the luminous parts of watch faces. Although luminous dials are considered low-level nuclear waste and far safer than uranium or plutonium waste, this was a case where it had disastrous health effects.

The women painted the numbers and arms of watch and clock faces using a fine paint brush. The numbers and letters were so small that it required a very sharp tip on the brush to draw them properly. The women mixed concentrated radium with glue to make the

(a)

FIGURE 17.9 (a) Map of Essex County, New Jersey showing the location of Orange. Inset shows location of Essex County in New Jersey. (b) Photo of an exhumed body of a radium girl in darkness, glowing from the radium in the bones. *Source:* otisarchives2 / Flickr / CC-BY 2.0. (c) Photographs of a radium girl from before working in a radium factory (left) and after exposure to radium produced "radium jaw" that developed into a gross deformity (right). *Source:* INEWS.

luminous paint without any protection. They were then encouraged to use their lips and mouth to bring the brush to a point to get sharper lines on the faces. This became the accepted method of application.

Between 1917 and 1926 while the New Jersey factory was open, luminous dial factories were opened all across the United States and Canada, employing more than 4000 women in all. At peak, production

the New Jersey factory produced upwards of 55 200 dials per year, per dial painter. Most other facilities had similar exposures.

Because the job was prestigious and paid well, the health issues probably became apparent later than they should have. The ingested radium substituted for calcium in the bloodstream of the women and was deposited in calcium-rich body parts such as teeth and bones

(Figure 17.9b). The women began experiencing tooth pain, loose teeth, mouth lesions, ulcers, and tooth extractions that did not heal. This would later be termed "radium jaw." The women also suffered uneven or lack of menstruation, sterility, stillbirths, and doctor-ordered terminations of pregnancy because the fetus did not develop. The women also developed extreme deformities such as collapsing spines, bone fractures, and shortening of limbs (Figure 17.9c).

The first dial painter from the New Jersey factory to die was in 1923 and her jaw had fallen away from her skull before her death. By the next year, 50 dial painters were gravely ill and about a dozen had died. The company attempted to shift blame but quickly succumbed and closed the factory in 1926. By 1929, 23 deaths of dial painters were attributed to radium poisoning. By 1931, there had also been five deaths from bone cancer among 18 deaths of dial painters. This was used as proof radium paint also caused bone cancer. By 1959, deaths of dial painters had reached 45. Even the inventor of radium dial paint died in November 1928 from radium poisoning.

Part of the rapid downfall of this unregulated industry was a highly publicized lawsuit. One dial painter attempted to file a lawsuit in 1925, three years after her symptoms began. Finally, in 1927, she and four other ill dial painters successfully filed a lawsuit against the US Radium Corporation. They became known as the "Radium Girls." By the time they managed to schedule a court appearance in January 1928, two of the plaintiffs were bedridden, and none could raise their arms to take an oath. The company stalled the next hearing to the point that all of the Radium Girls were too ill to attend. They then offered them an out of court settlement of $10 000 each and $600 per year while they were living in addition to paying all of their medical expenses if they did not hold the company liable. The women had no choice but to accept the offer. By the 1930s, all of the Radium Girls were dead.

The publicity of the case energized the first stringent protection of workers from environmental hazards. In 1928, the Surgeon General of the United States and Public Health Service instituted requirements of rigid and continuous inspection of radium factories. Further, they required hourly limits to the distribution of radium, the use of fume hoods when working with radium, and medical monitoring of all exposed employees. Partly also as a result of this case, in 1949, the US Congress passed legislation that made occupational disease compensable, and it extended the time for workers to diagnose illnesses and to file a claim for damage. By 1968, the use of radium in dials ended.

The Radium Girls story, however, was not over. In 1980, the New Jersey Department of Environmental Protection measured radon concentrations of 500 picocuries per liter (pCi/l) in the old factory compared with the US EPA action level of 4 pCi/l. Radon is the daughter product of radium in the uranium decay series. Radon is considered the most dangerous natural environmental hazard and is responsible for as many as 25 000 lung cancer deaths in the United States per year. In this case, the radon was not natural but the result of soil contamination. The soil was contaminated with an average of 200–300 pCi/g of radium with local concentrations of 5000 pCi/g. Apparently, the company had been dumping radium waste on the ground outside of the factory without concern for the environmental impact. As a result, the US Radium property was declared as a Superfund site by the EPA in 1983. Remedial actions lasted from 1997 to 2006 and involved removal of the buildings and stripping of the top 15 ft (4.6 m) of soil. Company officials had apparently also dumped radium-tainted waste in several nearby towns as well. This caused a rash of high indoor radon levels in houses throughout the area. It is not clear if all have been identified.

Solid Waste Disposal

CHAPTER OUTLINE

Polluted Earth: The Science of the Earth's Environment, First Edition. Alexander Gates.
© 2023 John Wiley & Sons, Inc. Published 2023 by John Wiley & Sons, Inc.
Companion website: www.wiley.com/go/gates/pollutedearth

Words you should know:

Hazardous waste landfill – A highly regulated and specified landfill of waste that is designated as hazardous to human health.

Landfill – A repository of waste of any kind, also a dump.

Leachate – Liquid waste produced by filtering water through solid waste or leaking out of solid waste.

Methane – Flammable gas produced by the decay of organic material that can accumulate in landfills and be mined and sold as natural gas.

Municipal waste landfill – A landfill composed primarily of household waste that is handled with less care than hazardous waste.

18.1   Dumps, Landfills, and Waste Disposal

The largest human structure ever built by volume is the Fresh Kills Sanitary Landfill in Staten Island, New York. When active, it was 146 ft (45 m) high, 146 ft (45 m) deep, and covered an area of 296.5 acres (120 ha). Humans generate huge amounts of waste that requires disposal (Figure 18.1). Some waste is relatively benign and just results in an eyesore, but other waste is toxic. Disposing of this waste has taken various forms, some of which are somewhat respectful of the environment whereas most methods are dangerous and some have resulted in environmental disasters. These disposal sites are grouped under the term "landfill" but include several other names.

Landfills have been used by communities for thousands of years. The first waste dumps were documented in Athens, Greece around 500 BCE but could have been used well before then. The simplest of landfills are called dumps or middens in historical times. In these, waste is dumped on the surface and degrades naturally. Municipal dumps are centralized locations for public waste disposal. In rural areas, household waste was dumped in a convenient place, typically behind an outbuilding or rock wall. Depending on the waste, these dumps range from minimal environmental impact to a serious public health threat. The US Congress enacted the Resource Conservation and Recovery Act (RCRA) in 1976, requiring all dumps to be converted to sanitary landfills (Figure 18.2). However, dumping is still practiced in urban and rural areas despite the illegality.

RCRA was not the first legislation regarding landfills. In response to the vast amounts of public and industrial waste generated, in 1953, the US Public Health Service (USPHS) and American Public Works Association (APWA) recommended guidelines for waste collection and disposal for small communities. The guidelines, however, were just recommendations. In 1961, the USPHS issued recommendations for sanitary landfill operations, on a state basis. The Solid Waste Disposal Act was enacted in 1965, allowing the USPHS to incorporate sanitary landfill practices into waste repositories. The US Environmental Protection Agency (EPA) took over solid waste oversight in 1970. RCRA moved oversight of waste disposal from advisory to regulatory. Full regulatory authority over landfills and preparation of landfill materials was granted when the US Congress enacted the 1984 RCRA Hazardous and Solid Waste Amendments.

FIGURE 18.1   Photo of garbage in a municipal landfill. *Source:* Aonprom Photo / Adobe Stock.

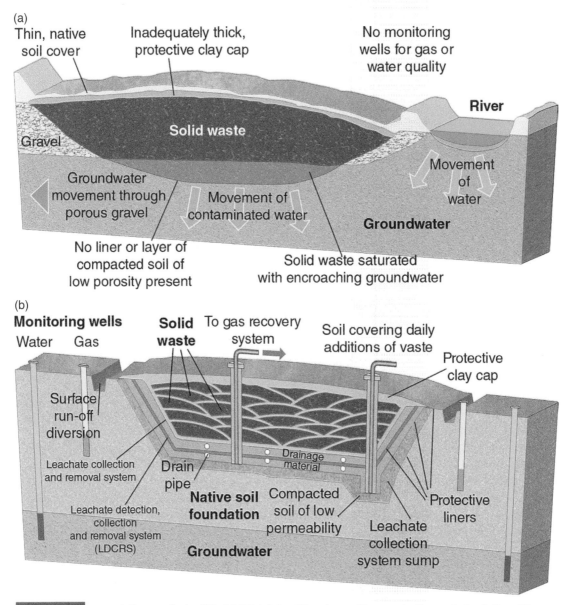

(a)

Thin, native soil cover

Inadequately thick, protective clay cap

No monitoring wells for gas or water quality

River

Gravel

Solid waste

Movement of water

Groundwater movement through porous gravel

Movement of contaminated water

Groundwater

No liner or layer of compacted soil of low porosity present

Solid waste saturated with encroaching groundwater

(b)

Monitoring wells

Water   Gas

Solid waste

To gas recovery system

Soil covering daily additions of vaste

Protective clay cap

Surface run-off diversion

Leachate collection and removal system

Drain pipe

Drainage material

Leachate detection, collection and removal system (LDCRS)

Native soil foundation

Compacted soil of low permeability

Protective liners

Leachate collection system sump

Groundwater

FIGURE 18.2 General diagrams for landfills. (a) Old style landfill or dump with no protection against pollution of the environment. (b) Modern sanitary landfill with a liner, impermeable cap, leachate collection system, and monitoring wells.

Placement of a landfills is almost always a contentious issue. Few residents want landfills near their property because of the odor and reduction of property values. Residents typically battle legislators for years and the result can be landfills in less than optimal locations. Ideally, landfills should be in large enough lots to contain the waste and the monitoring and administrative facilities. Landfills also should be far from populated areas but close to transportation routes. Landfills cannot be located on hilltops or steep slopes because they are unstable or in wetlands or any areas where they could impact surface or groundwater. The waste and leachate of landfills can pollute the water supply and damage the local ecology. There is a conflict of needs as depressions are the best features for containment of the bulk waste but are closer to the water table. A source of soil to cover the waste is also necessary, so bedrock surfaces are undesirable. Some rock types are also not ideal yet are commonly used as landfill sites. Limestone develops karst topography, which includes cave systems. The roofs of the caves can collapse, forming sinkholes. If a sinkhole forms beneath a landfill, the waste winds up in the water supply. Hazardous waste could cause a health crisis for the residents. In some areas, such as Florida, there is no choice but to site landfills on limestone bedrock and crises have happened.

There are other inappropriate locations that have been used as landfills with unfortunate results. Abandoned mines are one such location. Mines contain shafts and other passageways that lead directly to the groundwater system with no filtering. In addition, many mines contain highly reactive minerals. If liquid wastes or dissolved solids react with these minerals, they could remove dangerous compounds and release them to the groundwater system. Coal mines have an even more dangerous aspect in that they can be ignited, causing uncontrollable underground fires that can never be extinguished. This was the case in the former town of Centralia, Pennsylvania as in Case Study 18.1.

CASE STUDY 18.1 Centralia Pennsylvania Coal Mine Fires

Incineration of Solid Waste

There was a lot of garbage from a Memorial Day parade and celebration in Centralia, Pennsylvania (Figure 18.3a). Municipal workers decided to burn the waste in the new town dump located in an abandoned

coal strip mine. The fire was controlled at the surface but beneath, it spread into a mine shaft off of the strip mine and ignited an underground coal seam. There was plenty of air remaining in the shafts of the many old mines in this coal mining area. This allowed the fire to spread unchecked underground. This burning a pile of garbage began

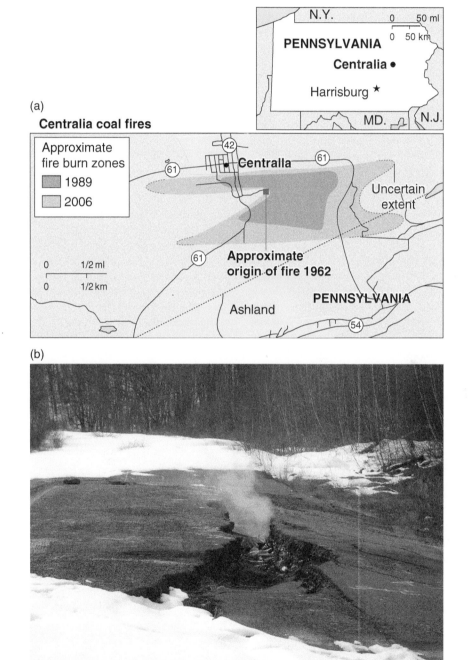

FIGURE 18.3 (a) Map of Centralia, Pennsylvania showing the location of the waste pit where the underground coal fire began and the approximate amount that it had spread by 1989 and 2006. Inset map shows the location of Centralia in the mid-Atlantic states. *Source:* https://www.ritebook.in/2017/09/centralia-pennsylvania-town-of-eternal.html. (b) Abandoned and deformed road in Centralia showing large fissures emitting smoke from the underground fires. *Source:* LaesaMajestas / Wikimedia Commons / CC BY-SA 4.0.

the demise of the town of Centralia and threatened the surrounding towns. The fire burns today and it can never be extinguished.

Centralia was a typical town in the coal country of Pennsylvania where the coal mining industry kept it relatively prosperous with high employment. The population of Centralia was above 2000 residents with 500–600 living just outside of town. The peak population was 2761 right before the fire began.

The sequence of events leading to the fire began in early 1962 when the Centralia town council voted to abandon the current overfilled town landfill for a new location in an abandoned strip mine on the edge of town. The pit had multiple shafts and adits that intersected with old underground mines and where coal seams were mined but abandoned when they became uneconomic. By Pennsylvania law, these openings were to be sealed before the pit could be used as a landfill to prevent leakage to the underground system. The town performed all of the required sealing and was issued a landfill permit by the state. They began using the site as a landfill upon opening.

The trash burning that began the problem took place at the end of May 1962. It was supposedly conducted by volunteer firemen hired by the town but municipal workers were involved as well. The fire, later referred to as "The Incident" by residents, was extinguished when it was done, or so they thought. The fire flared up several days later and the fire department was called to put it out again. While in the pit, the firemen discovered the one unsealed hole beneath the burning garbage. This one hole allowed the fire into a century's worth of interconnected mine shafts and began the problem.

The fire spread through the shafts and tunnels, burning coal dust, debris, and the remaining coal in the rock. There is not enough oxygen in the mine air for the coal to burn completely. This incomplete burning produces a lot of carbon monoxide. The other pollutants that result from partial burning include particulate, PAHs, benzene, and others but they rarely reach the surface. Residents quickly began experiencing symptoms of carbon monoxide poisoning. There were unexplained headaches, nausea, and flu-like symptoms that struck entire families for extended periods of time. The Pennsylvania Department of Environmental Resources began drilling monitoring wells to observe the fire by July 1962.

The fire continued to spread and evidence of it became apparent to residents. By May 1969, the first three families left Centralia because their homes were unsafe. In 1979, a gas station owner noticed that the gasoline at the pumps seemed warm. He lowered a thermometer into the underground gasoline storage tank and determined that the gasoline in it was 172 °F (77.8 °C). In 1981, the situation gained national attention when a sinkhole opened under the feet of a 12-year-old boy. The hole was 4 ft (1.2 m) wide and 150 ft (45 m) deep. The fall would have killed him but he grabbed some roots and hung on until he could be rescued. Many other sinkholes started opening all over town.

There were several efforts to confront the fire but none were effective. The first attempt was in 1969 to stop the spread of the fire. Trenches were dug and clay caps were installed around the area where the fire started, but planning was poor as was execution and the project was never completed.

By 1981, more than $7 million had been spent in several attempts to extinguish the fire. The mines were flushed with a water and fly ash mix, excavated to remove the burning material, trenched and backfilled with soil and drilled to determine the extent of the fire. In 1983, there was enough public attention that the state proposed to dig a trench 500 ft (154 m) deep around the town to contain the fire and then to purchase all of the homes and businesses and evacuate the residents. The total cost was $660 million and the chances that it would work were poor. However, the US Congress allocated $42 million to buy out and evacuate the residents and let the fire burn itself out with the hope that it would not spread.

The purchasing of homes and businesses began immediately, with the final purchase of 26 homes just west of Centralia in 1991. However, the State of Pennsylvania claimed eminent domain the following year and condemned all the remaining homes. Route 61, the main road through town, cracked and melted in a few places and repairs were unsuccessful (Figure 18.3b). As a result, the road was closed in the mid-1990s and traffic was detoured around the area. The U. S. Postal Service canceled the town zip code in 2002.

The fire is now estimated to underlie an area of more than 400 acres (162 ha) and growing. There are warning signs posted around the town as well as metal steam vents. Smoke and gas from the fire escape from these vents, and there are several sinkholes. It is estimated that it will take at least 250 years for the fire to go out.

18.2 | Common Landfill Types

Most urban and suburban areas operate landfills to dispose of municipal solid waste (MSW). In 1996, about 230 million tons (208.7 million mt) of solid waste was generated in the United States, 60% of which went to landfills, 20% was incinerated, and 15% was recycled. There were 75 000 onsite industrial landfills, 5800 municipal landfills, and at least 40 000 abandoned municipal landfills. By 2018, the total generation of MSW was 292.4 million tons (265.3 million mt), 50% of which was disposed of in landfills, 23.6% was recycled, 8.5% was composted, 11.8% was incinerated, and 5.8% went to other food management including animal feed, donations, and other non-human uses. The amount of waste sent to landfills was about the same in 2018 as it was in 1990 and actually decreased until 2010 before increasing again. On a global basis, in 2015, about three billion people lived in urban areas, and produced more than 1.4 billion tons (1.3 billion mt) of solid waste per year. By 2025, the number is predicted to be more than 2.4 billion tons (2.2 billion mt) of waste per year.

In municipal landfills, solid waste is deposited into an excavated pit underlain by a plastic or rubber liner (Figure 18.2b). The waste is compacted using heavy equipment and covered with at least 6 in. (10 cm) of dirt each day. In some landfills, leachate is removed from below and above the landfill using collection pipes. Other landfills require soil and rock underneath them to sufficiently purify the water before it enters the groundwater system. A second plastic sheet and/or thick, impermeable clay cap covers the landfill after it is full to prevent infiltration of rainwater and snow thaw from above. Infiltrating water can dissolve certain waste to produce leachate that might enter the groundwater system if the liner leaks. If the landfill fills with infiltrating water it could overflow and spill into the groundwater system as well. The surface of the completed landfill is built into a mound to drain surface water

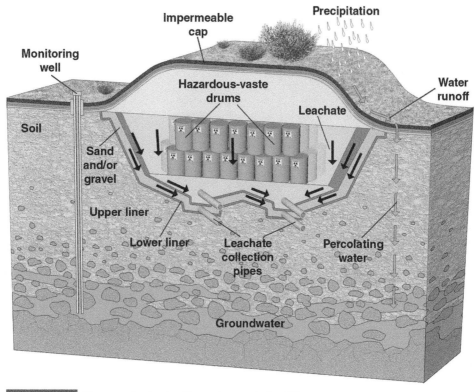

FIGURE 18.4   Diagram of a standard hazardous waste landfill.

away from the landfill. A final 24 in. (40 cm) thick soil cover on the landfill is covered with vegetation to stabilize the slope and absorb minor rainfall. Wells are drilled into the landfill to vent methane gas produced by decay of organic waste. This methane (natural gas) is mined in newer landfills and sold to utility companies. Monitoring wells are drilled around the landfill to determine groundwater quality. Raw leachate reaching groundwater could contaminate the local water supply with pollutants and disease.

Hazardous waste must be much more carefully disposed of than municipal waste (Figure 18.4). Hazardous waste landfills are lined with a thick, chemically resistant, and impermeable geotextile-plastic liner under them. Leachate is removed from the liner and below the landfill using a collection pipe that transports it into a storage and/or treatment facility. Unlike municipal waste, hazardous waste is carefully placed in the landfill and lined on the top and bottom and not compacted. Rock and gravel stabilize the waste to prevent it from being knocked over or crushed. The cap of the hazardous waste landfill is similar to the municipal landfill but includes both an impermeable sheet and a thick clay cap. There are more monitoring wells in hazardous waste landfills and they are frequently sampled.

Hazardous waste landfills are filled with industrial chemicals, pesticides, some radioactive waste, some biological waste, and some military refuse. The materials are highly regulated because they are regarded as extremely dangerous to human health and the environment. They may be regulated on a national security level but must be disposed of separately from commercially produced chemicals. The placement of the materials in the landfill must also follow strict guidelines with proper tolerances.

The bright outlook on MSW is that the amount in sustainable practices like recycled, composted and other food management is steadily increasing while the unsustainable disposal practices are decreasing or remaining about the same. In 2018, the recycled MSW was more than 69 million tons (62.6 million mt). Paper and paperboard accounted for 67% of this, metals comprised 13%, and glass, plastic and wood made up 4–5%. The most-recycled waste in 2018 were corrugated boxes, mixed paper products, newspapers, lead-acid batteries, major appliances, wood packaging, glass containers, tires, and consumer electronics. The composted MSW in 2018 was 25 million tons (22.7 million mt). In 2018, 17.7 million tons (16 million mt) of food was managed as animal feed, anaerobic digestion, bio-based materials/biochemical processing, donation, and wastewater treatment.

18.3   Pollution from Landfills

Landfills are defined as concentrations of pollution so there are no shortage of situations where they have impacted public health (Figure 18.5). If they are improperly located, the solid waste can wind up in inappropriate locations such as stream and rivers, wetlands, karst terranes with caves, and ecologically sensitive areas. These unfortunate situations are widespread on a global basis

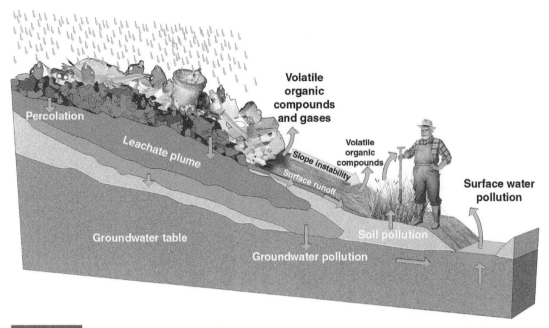

FIGURE 18.5   Diagram showing threats to the environment and public health from a landfill that does not have safeguards.

though many communities have been making extraordinary efforts to remove solid waste from the environment. In reality, even though solid waste is an eyesore on the streets of many urban areas, it is actually less of a threat to the natural environment as those areas are already damaged and marginally fertile.

The first problem with solid waste is that piling it up can be gravitationally unstable and subject to mass wasting. With small piles, the worst that can happen is the spreading of the edges. Large piles of waste without proper support or landfills on hillslopes, however, can result in dangerous waste slides or falls. The classic example of the deadly failure of a waste pile is the Aberfan, Wales disaster of 1966. The area hosts a large coal mining industry that employs most residents. Waste from coal mining was dumped into large piles uphill from the town of Aberfan. One pile was dumped on top of a spring which kept the base continuously wet. Compounding this, a period of heavy rain soaked the waste pile, making it unstable and ultimately causing it to fail. It flowed down the hill and into the town as a slurry. The direction of the flow was the worst possible as it engulfed a junior school and a row of nearby houses. In all, it caused the deaths of 116 children and 28 adults. There are plenty of other examples, but the 2000 Payatas waste slide disaster (Case Study 18.2) was among the worst.

CASE STUDY 18.2   2000 Payatas Waste Slide Disaster

Mass Wasting of Solid Waste

Smokey Mountain was one of the major landfills for Manilla, Philippines and it operated for more than 50 years. It was really just a 164-ft (50-m) high dump of more than 2.2 million tons (2 million mt) of garbage. Much of the garbage was flammable and commonly caught fire, resulting in the smoke shroud and thus the name. Some of the fires were intense enough to cause death of local residents. By the early 1970s, the capacity of the area had been reached and dumping was progressively switched to the Payatas open dumpsite in Lupang Pangako in Payatas, Quezon City (Figure 18.6a) even though Smokey Mountain was not officially closed until 1995.

Prior to the opening of the dumpsite, Lupang Pangako was a ravine surrounded by farming villages and rice paddies. Ironically, Lupang Pangako translates to "Promised Land." A community of destitute people lived in the Smokey Mountain landfill. They eked out an existence working as scavengers in the waste. These people relocated

to the Payatas dumpsite after the Smokey Mountain closed in 1995. This group numbered as high as 30 000 people. Many other people had previously been relocated to Lupang Pangako by the government in 1988 and 1989 after their homes in the Cubao, Tatalon, and Pinyahan areas were condemned and demolished. This group numbered around 15 000 residents and brought the population of the area to about 80 000.

The Payatas is located in northeast metro Manila and gradually developed a reputation as the new Smokey Mountain. At peak usage, approximately 1700 tons (1540 mt) of MSW was dumped at the Payatas landfill per day, which constituted about 15–20% of the total waste generated in metro Manila. Some sources estimate that the dumped waste reached as much as 11 023 tons (10 000 mt) in a day. The dumpsite was designed to cover about 24 acres (9.7 ha) of land. Of the approximate 514 425 tons (466 679 mt) of waste annually dumped, 39% is recyclable and 61% is non-recyclable. About 12 674 tons (11 498 mt) is recovered by the scavengers living in the dumpsite for direct use, recycling, animal food, resale, or composting. Recyclable materials include paper, aluminum cans, plastics, glass, and iron. Thousands of people lived in shanties around the mountain, which had grown to 74 acres (30 ha) and rising as high as seven storeys.

The rapid growth of the Payatas waste mountain made it mechanically unstable and prone to collapse. Although it was deemed full and ordered to cease operation in 1998, it remained active. As a result, the first recorded incident of a garbageslide or collapse occurred on

3 August 1999 during a typhoon. The heavy rain contributed to the collapse. Only pigs were killed but 32 families lost their homes.

The primary cause of the instability is low waste density, which resulted in rapid water infiltration into the pile. The low waste density was the result of its composition, which included a high component of plastic waste and organic material and a lack of paper, glass, and metals, which are dense. This low density waste is not easily compacted, even with using heavy equipment. The reason these wastes are so high

(a)

FIGURE 18.6   (a) Map of the Manilla, Philippines area showing the location of Payatas. Inset shows the location of Manila in the Philippines. *Source:* Left figure inset on right. (b) Photo showing the Payatas landfill with the garbageslide into the surrounding neighborhood. *Source:* Patrick Roque / Wikimedia Commons /CC BY-SA 4.0.

(b)

FIGURE 18.6   (Continued)

is the lack of recycling systems for organic waste and a high rate of recycling of reusable material. The reason for this is the efficient work of the scavengers. The heavy rainfalls coupled with low density waste reduces the stability.

The government finally voted to act on the closure of the Payatas landfill, now 60–108 ft (18–40 m) high, in June 2000. However, before any closure could be addressed, on Sunday, 9 July the mountain of garbage began to slide (Figure 18.6b). The next day, at around 4:30 a.m., on 10 July the upper landfill slope failed and 1.6 million cubic yards (1.2 million m³) of waste slid down and buried cottages, scavengers and part of Lupang Pangako under 33 ft (10 m) of waste and then burst into flames. The slide covered an area of about 322 917 square feet (30 000 m²). The slide destroyed about 100 makeshift shanties of the scavengers which housed about 800 families as well as several houses in the main community. It also buried a lot of people. It is suspected by reputable sources that more than 1000 people were killed but officially the death toll is 234, with up to 300 people missing. Another source estimates 705 deaths. In addition, 655 families were made homeless.

As a result of the impact on the community, at least 500 families with more than 2000 people were evacuated to the Lupang Pangako Elementary School, which became housing for 40–60 families. The

President of the Philippines ordered the immediate closure of the landfill. It was reopened just a few weeks later to prevent an epidemic because there was not enough room in the municipal landfills to dispose of Manila's waste and garbage was piling up on the city streets. In addition, the local inhabitants requested that it reopen because without it they had no other livelihood.

The disaster, however, compelled the government to close all open municipal dumpsites by 2001 and even controlled dumpsites by 2006. In 2004, the landfill was converted to the Payatas Controlled Disposal Facility, an engineered sanitary landfill. It was also converted from a waste to an energy facility where methane gas generated from waste is converted into electricity. It was the first clean energy project using solid waste management in the Philippines and Southeast Asia. The converted facility closed in December 2010 and a new landfill with stricter waste management practices was opened nearby in January 2011. That facility was closed in 2017 when it reached capacity.

The final chapter of the disaster occurred in January 2020 when the Quezon City Regional Trial Court ordered Quezon City to compensate the families of 59 victims who filed for civil damages a sum of 6.49 million pesos. Because the dump was municipal, the city was liable for the damages.

Solid waste on the surface of the earth damages the quality of the soil and leaks into the subsurface, contaminating the groundwater system. Because they can be seen, piles of drums and other solid waste are typically regarded as the most polluted areas by the public and outcry to address them tends to be loud and persistent. For this reason, these areas tend to be remediated quicker and more completely. Buried solid waste can be even deadlier but because it cannot be seen, it commonly requires impact on water quality and public health before it is addressed. For this reason, it takes several reports and growing impact to make the public and elected officials aware of the problem. For this reason, improperly handled and disposed of solid waste tends not to be addressed as quickly as surface waste. Many more buried solid waste repositories, both dangerous and properly contained, remain undiscovered and not remediated.

In unusual circumstances, buried solid waste can emerge at the surface. There are a number of ways this can happen. Erosion can unearth buried waste. If the landfill was not sited properly, it may be in the path of erosive agents such as storm runoff, streams, glaciers, or mass wasting like creep, debris flows, or landslides. This is an especially important issue with the continuing climate change and resulting sea-level rise. Many coastal landfills which were previously secure are now becoming jeopardized. Many of the superfund sites containing the most dangerous of hazardous waste in coastal areas are now designated as potentially being excavated by storms and sea-level rise. These could cause serious threats to public health. Increase in the height of the water table can also cause some types of waste to float toward the surface. This can occur through buoyancy with light waste or partially filled barrels and other containers being forced up by the force of gravity. Typically, this requires the soil encasing the waste to be water laden. Even if the waste is not light, if the soil is wet enough, it can flow through solifluction and be winnowed from around the waste, thereby exposing it. Freezing and thawing of the soil can vertically displace the soil and its contents through frost heave. The water in the soil expands when it freezes pushing the soil surface upward. When the soil thaws, it recedes back down. With repeated freezing and thawing, deeper buried objects and materials can be pushed upwards.

CASE STUDY 18.3 Love Canal, New York

Waste Disposal Disaster

Love Canal was a town southeast of Niagara Falls, New York on a site that was designed and begun excavation as a canal to bypass the falls by a man named Love in 1834 (Figure 18.7a). The problem was that the development company went bankrupt after excavating a trench 3000 ft (914.4 m) long, 60 ft (18.3 m) wide, and 40 ft (12.2 m) deep. This excavation was called Love Canal and encompassed an area of about 16 acres (6.4 ha). It was a swimming hole for neighborhood children until the 1930s, when it was purchased by the Niagara Power and Development Corporation.

The city of Niagara Falls and several communities began using the excavation as their municipal dumpsite. However, the town also allowed dumping of chemical wastes from its active petrochemical industry in addition to the household refuse. In 1942, a manufacturer of industrial chemicals, fertilizers, and plastics began to also use the canal for waste disposal. In 1946, Hooker Chemical purchased the canal to dispose of their waste. Between 1947 and 1952 Hooker Chemical dumped more than 43 million pounds (19 505 million kg) of industrial chemicals into the canal. Hooker then allowed the United States Army and several contractors to dispose of waste from chemical warfare experiments into the canal. By 1952, the canal excavation was filled to capacity. At this point, Hooker installed a high-quality hard-packed, dense, clay/ceramic cover over the area to protect the waste and prevent infiltration by precipitation.

More than 250 different chemicals were buried in the excavation. They were mostly pesticides, and organic chemicals such as chlorinated hydrocarbons, hexachloro-cyclohexane (HCH), benzene, chlorobenzenes, chloroform, trichloroethylene (TCE), methylene chloride, benzene hexachloride, phosphorous, and polychlorinated biphenyls (PCB). There was also 130 pounds (59 kg) of dioxin in the waste. In the early 1950s, this land covering one of the largest hazardous wastes repositories in the world was converted into baseball and football fields.

The geology of the Love Canal site made it reasonable for such disposal. The area is a layer of glacial till consisting of silt, sand, and gravel in a matrix of dense clay. The clay matrix makes the unit relatively impermeable to fluid migration, which is ideal for waste disposal. Liquid chemicals and the leachate from waste in the hollowed-out, partially completed canal, would not infiltrate the groundwater system to any degree. Further, the bedrock underlying the till is also relatively impermeable in case of any minor leakage. Using Love Canal as a waste repository was reasonably responsible under lax environmental controls at the time. The tight fitting, well-designed clay and ceramic cap Hooker installed was also ahead of its time in environmental protection. The only thing that was absolutely required to prevent problems was to maintain the integrity of the cap.

After World War II, the population of Niagara Falls grew quickly as did the need for additional land for new schools and housing. The city approached Hooker Chemical to buy the Love Canal tract that it had used as a chemical waste dump, to build a school. Hooker refused but was threatened to have it condemned through imminent domain. In 1953, they sold the property to Niagara Falls for one dollar which was the minimum amount for the contract to be legally binding. In the purchase agreement, Hooker disclosed the information on dumping, warned about the danger of developing the site, and declared itself absolved from liability involving future use of the property.

Construction of the 99th Street School began as soon as the property was purchased. However, building plans were quickly revised when several pits filled with chemicals were found. The location of the school was switched to directly on top of the landfill. Further, the well-designed cap that Hooker Chemical installed to isolate the waste was removed to build the foundations.

The Love Canal section of Niagara Falls grew quickly. By 1957, there were about 200 houses, a large park, a creek, and the 500-student elementary school. The sewers and water lines were installed that also breached the cap in numerous places. Families purchasing homes knew nothing about the chemical waste landfill or the potential risks.

The breaching of the landfill cap, by the school and by sewer, gas, water, and other utility lines, allowed infiltrating runoff and precipitation water into the landfill and waste. By the 1960s, Love Canal residents were reporting noxious odors and unusual liquids bubbling up in their yards and seeping through their basement walls. The problem was that the impermeable till encasing the waste was also impermeable to infiltrating water. Breaching the cap allowed the landfill to fill with water like a bathtub and all of the waste and toxic chemicals were floating up to the surface (Figure 18.7b).

Niagara city officials attempted to downplay the issue by attributing the odors to nearby chemical and industrial plants, but the problems only accelerated. Dogs and cats began to develop skin lesions and, after playing on the school's ball field, children developed symptoms of skin irritation similar to poison ivy. In 1978, the situation reached the boiling point. Record amounts of rainfall that year filled the waste pit to capacity and it started overflowing. Liquid wastes and contaminated groundwater flowed freely into the Love Canal neighborhoods. On August 1, 1978, the front page, lead article in the *New York Times* was about the leachate that was flowing from the surface of the landfill at Love Canal and entering streams, sump pumps, and low-lying areas throughout the community and onto the grounds of the elementary school.

Rusting and leaking drums of pesticides "floated" through the soil and to the surface of backyards and playgrounds, buoyed by rising water table in the pit (Figure 18.7c). A swimming pool was lifted from its foundation by the pressure of the rising groundwater. It was found to be floating on liquid chemical waste. Noxious puddles of toxic liquid appeared in ditches and other low-lying areas all around the area. Many children were treated for chemical burns after playing outdoors. The smell of industrial solvents was present all over the area both outside and inside of houses.

The city of Niagara Falls and state of New York reacted slowly to the situation. The Love Canal Homeowners Association had been investigating the health concerns of its residents since 1975. They documented high rates of cancer and birth defects in the area. The findings

were denounced by Occidental Petroleum, who had purchased Hooker Chemical, and their own elected officials, who maintained that the waste was safe. However, a reporter from the local newspaper, sampled liquid from a sump in Love Canal and proved that it contained the chemicals that Hooker Chemical buried.

Media and public pressure grew so rapidly that on August 7, 1978, President Carter declared Love Canal an emergency area and approved emergency financial aid. This was the first time emergency aid was approved for an environmental disaster. The US Senate also issued a statement recommending that Federal aid be allocated to address the environmental damage. New York Governor Carey informed Love Canal residents that the state of New York would purchase the 200 homes most impacted by the pollution. By the end of the next month, 98 families were evacuated, and 46 families were in temporary housing. In another

(a)

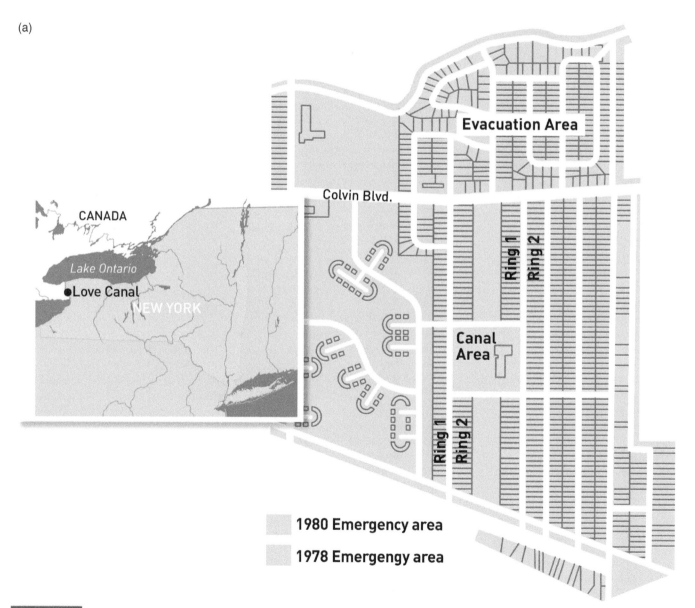

FIGURE 18.7 (a) City map of Love Canal showing the evacuation zones. Inset shows the location of Love Canal in New York and the surrounding area. *Source:* https://thelovecanal.files.wordpress.com/2014/06/ar-120918960-jpgmaxw602maxh602alignvtopq80.jpeg. (b) Photo showing the polluted water emerging from the buried waste enclosure underground in Love Canal. *Source:* https://thelovecanal.files.wordpress.com/2014/06/love-canal-barrels-of-waste1.jpg. (c) Photo of drums of chemical waste excavated from Love Canal. *Source:* https://thelovecanal.files.wordpress.com/2014/06/afgf_love_canal_protest_kids_march.jpg. (d) Crowds of protesters attempting to get action on the pollution in Love Canal. *Source*: EPA (Environmental Protection Agency) / Wikimedia Commons / Public Domain.

(b)

(c)

(d)

FIGURE 18.7 (Continued)

month 221 families from the most contaminated areas had been evacuated and the elementary school was closed. It was later demolished.

Government officials, however, continued to underestimate the resolve of the residents. Residents of Love Canal whose homes had not been purchased were still very concerned about the health of their families and felt excluded from the process (Figure 18.7d). Property values had plummeted greatly decreasing the value of their assets. In May 1980, the U. S. EPA published results of blood tests of Love Canal residents that demonstrated exposure to the toxic chemicals caused chromosome damage. This exposure greatly increased their risk of both developing cancer and experiencing reproductive problems. The residents were so enraged by the findings and the delay to lack of government action, they held EPA officials hostage at gunpoint for two hours before the situation could be defused. This escalation of civil disobedience by otherwise respectable people convinced President

Carter to evacuate all Love Canal families pending appropriation of funds for relocation. In the end, nearly 900 families were relocated and their homes were purchased.

The cleanup was massive and extremely expensive. The waste was not removed because the risks to workers and nearby residents during excavation, packaging, and shipping of the toxic waste were too great. Instead, a geotextile cap covered with clay was installed over the 16-acre (6.4-ha) landfill with an additional 40-acre (16-ha) contaminated area buffer. A barrier to surface drainage and a leachate collection system were installed around the landfill to collect leachate that leaks out. About 3 million gallons (11.4 million l) of contaminated surface and groundwater is recovered and treated annually. Glacial and stream sediments impacted by leachate were excavated and treated before being shipped to a disposal site. The site is frequently checked for hazardous chemicals through a system of groundwater monitoring wells.

The residential area southwest of the landfill was demolished. There is now an 8-ft (2.5-m) high, barbed-wire topped, chain-link fence surrounding that area and the landfill. More than 260 homes west and north of the site that had been evacuated were repaired and sold to new owners and 10 new apartment buildings were constructed. Recreational buildings and facilities are built right up to the fence. There are not even street signs for Love Canal, and it does not appear on maps. Instead, the area has been renamed Black Creek Village.

Finally, certain rock types and structures can cause vertical displacement of waste. In particular, if waste is buried in a limestone terrane, it will be impacted by karst. If a sink hole opens under a solid waste landfill, it can be split apart and exposed on the surface. If the entire landfill falls into a cave beneath it, the waste will be exposed within the cave instead of on the surface but equally dangerous. Even surface settling caused by dissolution and removal of bedrock beneath the landfill can expose waste. Such settling commonly causes cracking of foundations on buildings and homes in these areas.

Both waste on the surface and buried can be infiltrated by water from precipitation and thaw. This forms a fluid that is rich in decomposition products and dissolution of waste and it is termed leachate. The foul-smelling liquid that drips from a garbage truck is like leachate but less dangerous than leachate from a landfill. Fluid that is emitted directly from the waste is also leachate. The toxicity of the leachate depends on the type of waste involved. Hazardous waste landfills produce very dangerous leachate in comparison to that from municipal landfills. Pipes are installed into and beneath modern landfills to remove the leachate. The pumped leachate is treated using a wastewater treatment system and returned to the groundwater or it is removed, depending upon the toxicity.

If leachate enters the groundwater system, it can seriously impact the quality. A remediation project must be initiated to restore the system. To do this requires drilling of recovery wells and construction of a water treatment plant which is expensive. Depending on the type of waste in a modern landfill, patching the underlying liner or removing the waste may be required as well.

The development of leachate from a municipal landfill has three stages. The first stage is aerobic decomposition which consumes all available oxygen in the waste. The chemical reactions are exothermic which increases the temperature in the landfill. The second stage is called the acetogenic stage. Anaerobic bacteria break waste down through fermentation of sugars to acetic acid and other volatile fatty acids and produce carbon dioxide and hydrogen gas. These acids dissolve the waste and increase the inorganic pollutants content in leachate. Sulfate-reducing bacteria later biodegrade the fatty organic acids. The third stage is the methanogenic stage. After the sulfate is consumed, the microorganisms that produced methane and carbon dioxide from volatile fatty acids become dominant. It is in this stage that methane can be mined from the landfill and sold for commercial uses. The decomposition rates eventually decrease as the available organic material runs out and the landfill becomes inactive.

The composition of the leachate varies depending on the type of waste, age of the landfill, climate, precipitation, and other factors. The organic content of leachate is defined in biochemical oxygen demand (BOD) and chemical oxygen demand (COD). These terms measure the amount of oxygen required to oxidize the available organic matter by microbial biodegradation (BOD) and chemical oxidation (COD).

Environmental Industry and Cleanup

Polluted Earth: The Science of the Earth's Environment, First Edition. Alexander Gates.
© 2023 John Wiley & Sons, Inc. Published 2023 by John Wiley & Sons, Inc.
Companion website: www.wiley.com/go/gates/pollutedearth

Words you should know:

Bioremediation – Remediation of pollution using biologic methods, especially microbes and plants.

Brownfields – Federal legislation designed to make profoundly polluted sites clean enough to serve a purpose that does not require full remediation.

Delineation – Mapping the extent, types, and concentrations of pollutants in a medium.

DNAPL – Dense Non-Aqueous Phase Liquids, which are pollutants that are dense enough to sink through water and are not soluble in it.

Environmental site assessment – The steps or phases in evaluating a polluted site.

Ex situ remediation – Remediation of polluted soil or water by removing it from the ground to another place for treatment.

In situ remediation – Remediation of polluted water or soil in the undisturbed ground.

LNAPL – Light Non-Aqueous Phase Liquids, which are pollutants that are light enough to float on water and are not soluble in it.

Permeable reactive barriers – Chemically reactive material that is buried deep enough that groundwater can flow through it. The pollution in the water is neutralized by reacting with the barrier.

Phytoremediation – Remediation of pollution using plants.

Remediation – The repair and neutralization of water and/or soil at a polluted site.

19.1   The Environmental Industry

There is a large industry that assesses or delineates and cleans or remediates polluted sites. These companies are hired by the government to address sites on public land or those so badly polluted that they require oversight, other companies who are dealing with pollution on their property, private property that has pollution issues, or to prepare property for development or sale among other reasons. Environmental scientists, geologists, and civil and environmental engineers are hired by the companies to perform this delineation and remediation. An employee may begin as a technician and progress through the ranks to project manager and finally company executive. Depending on the company and the size of the sites, these professionals will handle one or more sites. Superfund sites and other major projects require multiple professionals for long periods of time. On the other hand, professionals may handle portfolios of multiple projects including underground storage tanks (USTs) at gas stations, various development projects, leaking home heating oil tanks, and real estate transactions.

Depending on the size of the company, the technicians perform the hands-on delineation and remediation of the sites whereas the project managers oversee these site projects and interact with the clients who commission the work. They may oversee several projects and even generate work for the company by meeting with potential clients. In smaller companies, the project manager may also be the one out in the field doing the work. Technicians and project managers may need licensure to certify projects. They can obtain Professional Engineer, Professional Geologist, and some form of Site Remediation Professional licensure depending upon the state. These are obtained by scoring a certain level on written exams and having a certain number of years of practical experience. A licensed environmental professional must endorse the remediation plan and action. Depending upon the jurisdiction, the plan and report of action must be submitted to and approved by a federal, state, or local environmental agency. In some jurisdictions, the licensed professionals can endorse the work but they are then liable for any problems.

The executive branch of an environmental company is like any other company. They manage personnel, manage offices, develop long-term plans, manage finances, and solicit clients and work. These clients are commonly major companies with environmental divisions also managed by professionals with an environmental background. These are generally senior-level positions that take many years of experience to obtain. Scientific staff or company executives may also testify in court as expert witnesses in trials involving negative environmental and public health impacts.

Environmental industry careers directly improve the local and national environment and public health. They tend to be stable, lucrative, and rewarding in that the work directly helps communities to be safer and to thrive. They also facilitate development and property sales. With the rapid increase in public concern for the environment, the field is expanding. The only restriction is that environmental remediation is regulatory. Companies would not invest money in it if they were not forced to. For that reason, the field is also dependent on the political leanings of the powers at the time.

19.2 | Steps in Site Evaluation

There are specific steps in the approach to remediating a polluted site though they vary depending upon the type of pollution, type of area impacted, local regulations, and intended usage among others. The process is an Environmental Site Assessment (ESA) and the steps are measured in phases dictated by the American Society of Testing and Materials (ASTM). The reasons to begin an ESA can include purchase of property, application for a loan from a new lender, buyout or redistribution of a partnership owning property, application for a land use permit, an owner's concern about the pollution history of a property, a regulatory agency concern about pollution on a site, or divestiture of properties among others.

The work begins with a Phase I ESA that determines the value of the property and if there is outward indication of soil, surface, or groundwater contamination both visually and historically. This process typically includes an onsite visit to view present conditions including signs of a chemical spill, die-back of vegetation, etc., signs of potentially hazardous substance or petroleum product usage such as above-ground or USTs, storage of acids, etc. and signs of an environmentally hazardous history. It involves evaluation of risks from neighboring properties to distances ranging from 1/8 to one mile (0.2–1.6 km) depending on jurisdiction. Municipal and county files are examined for land usage and permits. Files at state water boards, county health departments, and other public agencies are searched for water quality and soil contamination issues. Aerial photographs and technical maps are studied for drainage patterns and topography. In some cases, past and present owners, site managers, tenants, and neighbors are interviewed about the property history.

In many cases, the file searches and historical research are outsourced to other companies that specialize in them. Non-Scope Items in a Phase I ESA can include inspection or record searches for lead paint, asbestos, lead pipes, mercury, emerging contaminants, mold and mildew, radon, wetlands, and endangered species if such issues are suspected. In some areas, they may include flooding and landslide, volcanic, and earthquake hazards.

There is also a Limited Phase I ESA which omits one or more of the usual tasks such as the site visit or some of the document or map searches. If the field visit is omitted, the Phase I ESA can also be termed a Transaction Screen.

If the Phase I ESA identifies an environmental problem, a Phase II ESA is legally required in the project or if the client wants to proceed to ensure the safety of the site, a Phase II is initiated. A Phase II is "intrusive" to the site and involves collecting samples of soil, groundwater, or building materials for quantitative analysis of pollutants. A comprehensive Phase II ESA involves extensive sampling to determine the full extent of contamination and all migration pathways. The most common pollutants investigated are petroleum, heavy metals, pesticides, solvents, asbestos, and mold. In contrast, a Limited Phase II ESA is a sampling program to confirm the presence of a specific pollutant. Ultimately, a Phase II is designed to determine if there is a significant pollution concern or health issue at a property.

Not all Phase II investigations are the same. There are numerous possible tests that are assembled to best address the found and possible issues determined in the Phase I investigation. Some of the possible tests that might be performed include sampling of surface soil and water, underground soil sampling using borings, groundwater sampling and analysis from drilled wells, analysis of any materials in drums, sampling of floor drains and dry wells, testing of transformers for polychlorinated biphenyls (PCBs), geophysical surveys to identify buried tanks, pipes, and drums, soil removal to expose buried tanks or waste, pressure tests of USTs for leaks, indoor air and water sampling and analysis among other possibilities. The Phase II only determines if pollutants are present. It does not delineate the areal and vertical extent of them. Therefore, the Phase II may not provide enough information for a remedial action plan or cost estimate. That requires determination of the complete extent of the pollution. As a result, additional testing is usually necessary. The results of the Phase II investigation are included in a report on the site which also makes recommendations to address the risks to health and safety as well as regulatory compliance for the area.

The Phase II ESA is the highest level recognized by the ASTM, so things can get less clear after that. A Phase III ESA can also be conducted, if needed, though the term is informal. If no pollutants are identified in a Phase II investigation, a Phase III ESA is not conducted. If they are found, using this information, a Phase III ESA will be designed to determine the lateral and vertical extent of pollutants on the site and how fast they are spreading. These are maps showing the extent and levels of pollutants in a plume extending from the point source (Figure 19.1). It is a highly detailed investigation that is necessary before any remedial action can be initiated. Like a Phase II, Phase IIIs vary greatly depending on the site and pollution. They may include detailed groundwater testing, subsurface soil assessment, sediment testing, wetland delineation, and detailed hydrogeologic modeling, all designed to assess the pollutant pathways in the soil and groundwater. This can involve installation of multiple testing and monitoring groundwater wells, soil-vapor probes, and extensive soil and rock boring. The Phase III analysis facilitates the development of a Remediation Action Plan (RAP) which needs to be approved by local, state, and federal environmental agencies in most cases. It also contains a full scope of options and time and costs of the cleanup.

Some areas even use Phase IV ESAs but they are also not part of ASTM. A RAP may also be developed in this phase. The RAP is based on the findings of the other three phases and are used to develop a cost-effective cleanup of the site. Remedial methods included in Phase IV may include excavation and disposal of soil and rock, groundwater treatment, mitigation measures, chemical

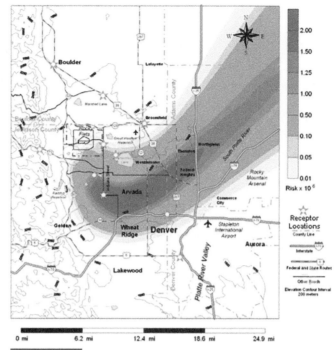

FIGURE 19.1 Map of a pollutant plume in the groundwater in an urban area with pollutant levels shown by color.

oxidation, and the biotreatment of soils. Remediation continues until concentrations of contaminants meet acceptable levels. This remediation usually includes the property being entered into Leaking Underground Storage Tank (LUST) programs and Site Remediation Programs (SRP).

19.3 Remedial Methods

The choice of remedial methods to be used depends on many factors including the pollutant, whether it floats or sinks in water and the type material polluted, its geometry and relationship to other materials, and where it is located relative to communities. Another major constraint is how clean the site needs to get. The goal of environmental remediation is to remove all pollutants and restore the land to its natural state. However, the cost of remediation and monitoring may be so high that companies commonly just pay the taxes, fence off the site, and let it sit. Many of these dangerous sites are in urban areas where they slow economic growth, decrease property values, and threaten public health. In 1995, the US Environmental Protection Agency (EPA) introduced the Brownfields program to allow companies to partly remediate these severely polluted properties.

If a site meets the EPA criteria of severe pollution, it can be partly remediated for certain purposes. It cannot be a playground or a park, but it can be a parking lot or a warehouse. To meet Brownfields, the most toxic pollutants must be removed and the rest must be reduced to certain levels. The site must pass certain criteria before development is permitted. Brownfields increases the tax base, reduces development pressures, and improves and protects the environment. It is an innovative method for urban renewal in older industrial cities.

19.3.1 Underground Storage Tanks (USTs)

Once the level of remediation is determined, then the remedial actions are designed depending on the site characteristics. Many sites have USTs, which are metal or fiberglass tanks used for storing liquids like gasoline, home heating oil, and other liquid chemicals including benzene, xylene, ethylbenzene, and other VOCs. Commercial and home heating oil USTs and gas station tanks were previously made of steel. The UST rusts through the bottom and produces leaks (Figure 19.2). The fuels and liquids infiltrate and percolate through the soil and accumulate on the water table. The spilled liquid accumulates under the tank or moves away in a plume (Figures 19.1 and 19.2). The plume can flow under houses or other buildings, where fumes can enter the indoor air through cracks. The pollutants can wind up in surface waters if the plume enters a stream, spring, or lake and spread farther.

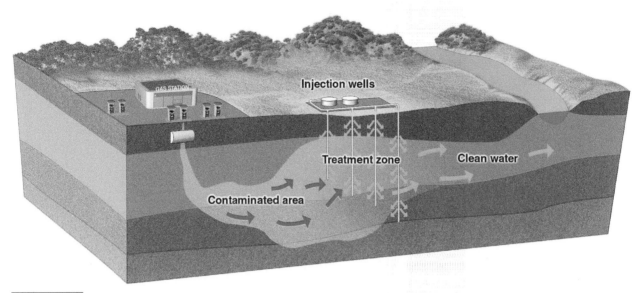

FIGURE 19.2 Block diagram showing a pollutant plume leaking from an underground storage tank at a gas station and wells remediating the pollution.

The first step in remediating a plume from a leaky tank is delineating its extent. Soil around the tank is sampled using an auger or drill rig and analyzed for pollutant at a commercial lab. If the leak has spread to the water table, groundwater samples are also collected from a drilled well and analyzed. Other wells are drilled strategically away from the tank to determine the groundwater flow direction, the extent of the plume and the distribution of pollutant within the plume. These data are used to produce a map of the pollution plume that is used to guide the remediation of the site (Figure 19.1).

Remediation depends on a number of factors like the pollutant type, the type of development around the site, surface water bodies, and the severity of pollution and extent of affected area. The UST is excavated, removed from the ground, and disposed of. If the leak is small, only removal of small amounts of soil may be required. For larger leaks that reach the water table, wells are installed and the pollutant is recovered using pumps. Impacted soil is excavated and removed from the site or, if the pollution is minor and not dangerous, it may be washed or aerated onsite and returned to the excavation as fill. Even if a residential UST has not leaked, if a new tank is installed inside the house or if the heating system is changed to gas or electric, the UST should be removed or it can produce a number of dangers like collapse.

In gas stations, the fuel that is pumped into vehicles is stored in USTs. If they leak, plumes can be much larger and more dangerous than residential tanks (Figure 19.2). Pipes to the tank or to the pump can also leak. Gasoline UST leaks can be large and dangerous because gasoline flows quickly and contains harmful additives. As a result, remediation can be more difficult and require monitoring after the cleanup. Pollution from leaking gas station USTs is commonly the most dangerous environmental problem in communities with no major industry. It can lead to involved and expensive lawsuits. Leaking industrial USTs can pose an even bigger threat and cause greater damage. If the pollutants in the plume are denser than water, dense non-aqueous phase liquids (DNAPLs), they sink through the groundwater which make more complex situations for remediation. If large USTs leak, they can be severe enough to qualify as Superfund sites.

In 1988, the USEPA and many state environmental agencies established laws that required the upgrades or replacement of industrial USTs. The upgrades include double walls on the UST and piping and alarm systems that detect leaks. These USTs now rarely leak, though the pipes that fill or drain them may leak. Military bases also use USTs but they may contain weapons or weapon components as well as fuel and solvents.

19.3.2 Groundwater Remediation

Groundwater can be remediated through extraction and treatment at the surface (ex situ) or underground treatment (in situ). The selection of which to use is dependent on the contaminant, the site conditions, the remediation objectives, and the cost. In ex situ remediation, the contaminated groundwater is pumped from the aquifer at a steady rate to the treatment system where the pollutants can be removed. This is essentially cleaning polluted water in a treatment plant and then releasing it back to the environment. In some cases, a slurry wall may need to be installed to prevent or slow the flow of groundwater so it can be treated (Figure 19.3). In this case, a trench is dug through the water table in front of the plume and filled with impermeable clay, cement, or even steel. Further, the waste generated in the treatment system must be safely disposed of, which may cause other problems. Depending on the situation, ex situ can be the best option but it can be expensive and generate other pollution, and on some sites, it may not work well.

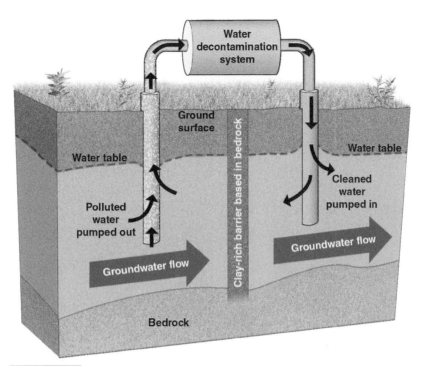

FIGURE 19.3 Diagram showing polluted groundwater being extracted, treated, and discharged across a clay-rich slurry wall.

In some cases, flow in the aquifer needs to be enhanced for complete treatment in a timely manner. This can be accomplished using hydrofracturing or fracking, in which water and sand are injected under very high pressure into the bedrock. This fractures the rock and the sand props the fractures open so the groundwater and gas flow better. This works well in sedimentary rock, but in crystalline rock fracturing requires explosive. This is only done where absolutely necessary. Soil can even be fractured to enhance fluid and gas flow to a collection system. This process is called pneumatic fracturing, which is accomplished using air pumped into the soil at high pressure.

In contrast to ex situ remediation, in situ groundwater remediation involves introducing treatment methods directly into the aquifer to neutralize the pollutant. These methods are preferred by regulatory agencies because they do not generate wastes. They are also less impactful to the land and they tend to be less expensive and generally safer. If the pollutant is distributed over a large area or if the source of the plume is under a building or other structure, in situ is the only solution. However, they can require more detailed site information and if subsurface conditions are not well characterized, they may not remediate the pollution as anticipated. This could result in multiple treatment methods being required, extended treatment periods, and incomplete remediation.

Steam or hot water extraction is among the most effective in situ remedial techniques but it can be expensive. Steam is injected into the pollutant plume where it vaporizes VOCs and semiVOCs or SVOCs. A vacuum tube collects the vaporized pollutants to an above-ground treatment system where it is recovered and treated. In some systems, hot water is used instead of steam. The hot water washes the pollutants into extraction wells where they are pumped to the surface. There, the water is treated and either discharged or reinjected.

In air sparging, air is injected into the polluted groundwater where it volatilizes contaminants and especially VOCs (Figure 19.4). The vapors then bubble out of the aquifer and into the vadose zone and are vacuumed out using soil-vapor extraction. This is one of the most common remediation methods for groundwater and soil contaminated with VOCs. Air sparging systems must generate sufficient flow rates to get air into the aquifer but not high enough to cause uncontrolled release of vapor into the atmosphere or a basement. Deeper pollution and less permeable soils make air sparging less effective. Air sparging only works on VOCs that volatilize easily.

Dual-phase extraction (DPE) or multiphase extraction or vacuum-enhanced extraction is considered an in situ method because it does not involve soil excavation though it is really ex situ technology. DPE pumps floating liquid contaminant, impacted groundwater, and contaminant vapors at once from underground to a surface treatment system. They can be single or multiple pump systems depending on the amount of contaminant removed.

In-well air stripping or in-well vapor extraction is an in situ vapor stripping that involves three steps. First, a specially designed circulating well is installed into the polluted groundwater. Second, air or gas is bubbled through the groundwater, which strips out VOCs in vapor form. The stripped gases are collected by a vacuum tube and into a treatment system. Third, the groundwater circulates and flushes out residual VOCs from the soil or aquifer. In-well air stripping is best for VOC pollutants that do not adhere well to soil. Vertical variations in soil and sediment like layering or beds can be problematic for the treatment.

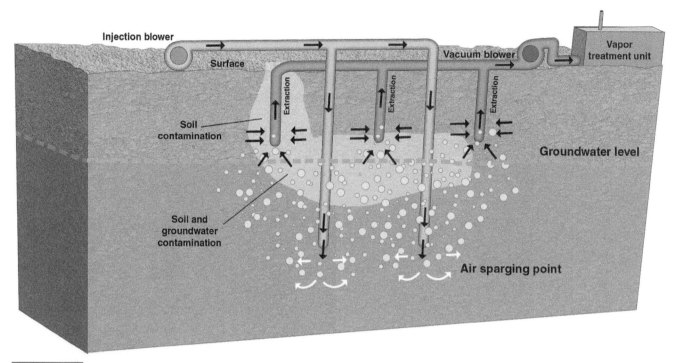

FIGURE 19.4 Diagram showing groundwater pollution being remediated using air sparging with air being pumped in from the left and vaporized pollutants being removed to the right.

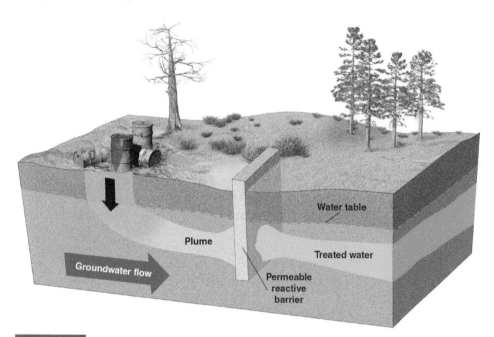

FIGURE 19.5 Block diagram showing leaking drums of waste producing a pollutant plume that is being remediated using a permeable reactive barrier.

A more invasive remedial method is a permeable reactive treatment wall or permeable reactive barrier (Figure 19.5). These are installed like slurry walls with extensive trenching through the water table and into the aquifer. The difference is that these walls are completely permeable, allowing the groundwater to pass through them as it flows. However, they are chemically reactive and neutralize the pollutant as it passes through. The wall is made of granulated metal or other remedial compounds like activated charcoal or zeolite and it is installed perpendicular to the flow path of the migrating contaminated groundwater. As the groundwater passes through the permeable wall, the pollutants react with it and are absorbed or neutralized. There are no operational or maintenance costs and no above-ground treatment systems, so the post-installation costs are minimal and they do not impact the use of the land above them. In some cases, trenching is not even needed. Instead, the reactive materials are filled into closely spaced wells called barrier well injection. These systems can be installed quickly with only minor surface disruption and the reactive materials can be replaced as often as needed.

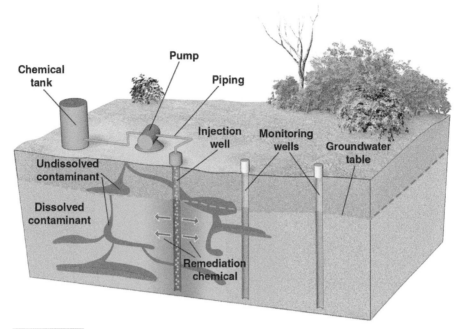

FIGURE 19.6 Diagram showing in situ chemical treatment of groundwater pollution.

The reactive medium slowly degrades or saturates with removed groundwater pollutant and must be replaced. The wall can also lose its permeability as pollutants precipitate or bacteria clog the pore spaces. The required treatment wall can be too expensive if the plume is too broad or lasts too long. Conditions like depth to bedrock, the thickness of the plume, and crossing subsurface utility lines among others may also make installing a reactive treatment wall impractical. Specialized trenching equipment is required for walls greater than 80 ft (24.4 m) high.

In situ chemical oxidation (ISCO) is often used for remediation of pollutants (Figure 19.6). Chemicals are injected into the soil and polluted groundwater. These chemicals reduce the pollutants to carbon dioxide and water. ISCO is most used at sites where there is residual pollution or if there are time constraints on the remediation job. ISCO requires groundwater flow through the polluted area to deliver the remedial chemicals. It is most used at sites with stratified, low permeability soils or where the pollutants are difficult to treat. The most used chemicals are Fenton's reagent (hydrogen peroxide), potassium permanganate, and, less commonly, ozone.

If groundwater is flowing slowly, both injection and extraction wells may be required. ISCO is best used on VOCs, polycyclic aromatic hydrocarbons (PAHs), petroleum products, and explosive compounds.

CASE STUDY 19.1 Otis Air Force Base and Joint Base Cape Cod

Multiple Plumes and Remediation Techniques

In the mid-1970s, a US Geological Survey researcher was investigating the environmental impact of a sewage treatment plant at the Otis Air Base in Cape Cod, Massachusetts (Figure 19.7). He was staying in a hotel in Falmouth. He turned on a faucet and the water foamed as it came out. The reason? Falmouth had a water supply well in the middle of a pollutant plume. Groundwater testing later revealed that there were at least eight pollutant plumes under the military base, many extending into civilian communities and causing health crises. People living in and around the military reservation had a 15.4% higher mortality rate from heart disease, a 14.1% higher mortality rate from respiratory disease, and a 7–13% higher mortality rate from cancer than those living in other parts of the cape. They were more likely to suffer from liver, bladder, kidney, and breast cancer. Further, babies born to women exposed

to perchlorethylene (PCE) in the area had an increased risk of birth defects. As a result, parts of the base were designated as EPA Superfund sites and will take many years to remediate using multiple techniques.

Cape Cod, Massachusetts is a coastal area composed of sand and glacial deposits that host a fragile groundwater system. The Joint Base Cape Cod (JBCC), previously the Massachusetts Military Reservation (MMR), is a 22 000-acre (8900-ha) base used for military training activities since 1911. The JBCC was established in 1935. It is located over the sole source aquifer for Cape Cod, which provides drinking water for 200 000 residents and 500 000 seasonal residents.

Part of the aquifer is contaminated by fuel and chemical spills, training, disposal, and other activities at JBCC. Two environmental remediation programs at JBCC address groundwater pollutant plumes. One program is a Superfund site on the Otis Air National Guard Base in the southern JBCC. The second program addresses

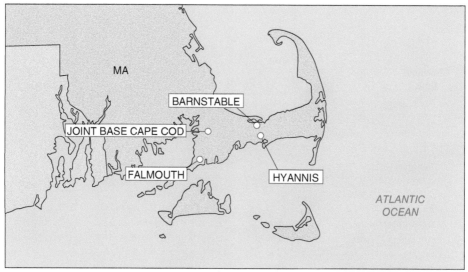

FIGURE 19.7   Map of the Joint Base Cape Cod (JBCC) showing the locations of pollutant plumes. *Source:* Courtesy of AFCEE (2010). Inset: Map of eastern Massachusetts and Cape Cod with the location of JBCC.

pollution from the northern Camp Edwards/Impact Area and managed by the US Army.

The Otis Air National Guard Base was designated as a Superfund site in 1989. The pollution resulted from chemical and fuel spills, landfills, fire training, and drainage structures. Pollutants were first detected in 1984 and, as a result, numerous remediation projects of many types have been undertaken. Nine groundwater pollution plumes are undergoing extraction and treatment remediation. Treatment involves treating the polluted groundwater using chemical methods of granular activated carbon, ion exchange resin, or both to remove pollutants and returning treated water to the aquifer. The systems treat 8.5 million gallons (32.2 million l) per day.

In February 1997, the EPA ordered the National Guard to investigate the effects of military training on groundwater quality. In May 1997, for the first time in US history, military training was suspended on a military base in response to environmental and public health issues. The EPA stopped military training at Camp Edwards, including the use of explosives, propellants, flares, and lead bullets. They found groundwater and soil pollution from military training using munitions, from unexploded ordnance, and from disposal of munitions and hazardous waste. There are seven groundwater pollution plumes also using extraction and water treatment plant remediation of 4.1 million gallons (15.5 million l) per day. These plumes contain fuel, trichloroethylene (TCE), PCE, the explosive hexahydro-1,3,5-trinitro-1,3,5-triazine (RDX), thallium, toluene, perchlorate, lead, and benzene among others so a variety of methods are being used. All treated water is being returned to the aquifer.

In January 2000, the EPA ordered the National Guard to remove unexploded ordnance and to clean up polluted groundwater and soils. This order on a US military base was the first of its kind. In January 2001, the EPA further ordered the use of a base detonation chamber to destroy more than 2500 rounds of ammunition dug out of pits on the firing ranges.

Many soil and groundwater pollution remediation projects have been implemented since the 1990s. About 25 areas were remediated by 2000–2002. More than 100 000 tons (90 718 mt) of soil were treated onsite using chemical methods. Municipal water supply wells were fitted with treatment systems and some 6.7 million gallons (25.4 million l) of polluted groundwater per day is being treated both in situ and ex situ using chemical water treatment plants.

The pollution of Camp Edwards/Impact Area was first investigated in 1996 and more than 15 000 acres (6070 ha) have been investigated to date. The investigation includes 14 remediation units, 1200 monitoring wells in 700 locations and more than 100 000 groundwater and soil samples. As a result, more than 120 000 tons (108 862 mt) of polluted soil has been excavated and chemically treated ex situ. About 300 acres (121 ha) was cleared of unexploded ordnance, resulting in 600 tons (544 mt) of munitions-related scrap metals having been recycled. More than 27 750 tons (25 174 mt) of soil was removed and disposed of. Seven groundwater plumes primarily polluted with RDX and perchlorate are treated by 16 treatment systems. About 4.1 million gallons (15.5 million l) of polluted groundwater are treated each day by ex situ water treatment. By July 2021, seven pollutant plumes were being treated by separate ex situ groundwater treatment systems at 7.4 million gallons (28 million l) per day in total. All treated water is returned to the aquifer.

Recently, the emerging firefighting pollutant Per- and polyfluorinated substances (PFAS) was found in two of the groundwater plumes in addition to several sites within the flight line and a small, off-base plume. The Air Force is remediating PFAS at the sites and is providing bottled water to homes with PFAS concentrations which exceed the limits. They also installed a treatment system to a public water supply well. In addition, the emerging pollutant 1,4-Dioxane occurs in four pollutant plumes.

Soil vitrification is a less commonly used physical–chemical remedial technique. It is most commonly used to stabilize radioactive waste but it can also be very effectively used to remediate other dangerous pollutants like arsenic. It kills all biologic pollutants and dissociates all organic pollutants as well. Basically, vitrification involves melting the polluted soil into an impervious glass. It can be applied both in situ and ex situ. In ex situ, polluted soil is excavated and heated in a furnace using plasma, direct electricity, combustion, induction, or microwave though electric energy is the most common technique. Once in the furnace, electric power is introduced until the soil reaches 2012–2552 °F (1100–1400 °C) depending on the soil and pollutant. The molten soil and waste is quenched or allowed to cool to a solid and stored in a disposal facility. In situ vitrification involves inserting graphite electrodes directly into the polluted soil to be treated. Electricity is applied through the electrodes until the temperature reaches 2912–3632 °F (1600–2000 °C). This charge melts the soil, vaporizing any water, organics, and light elements, and it cools into an impervious underground mass. The problem is that the cost is high and it is extremely disruptive in situ.

A relatively new and largely experimental remedial technology is nanoremediation. It uses engineered micro- and submicroscale materials to remediate pollution. The basic idea is that the nanomaterials are so small that they have a high surface-to-volume ratio, allowing them to better adsorb and react with pollutants. They are especially effective at remediating persistent compounds such as pesticides, chlorinated solvents, halogenated chemicals, heavy metals, and even emerging pollutants such as pharmaceuticals and personal care products. There is a large variety of highly versatile nanomaterials, making them able to effectively treat polluted water, soil, or air. Engineered nanoremedial materials are in categories of metal, carbon, polymer, and silica-based and they can either be injected directly into a polluted medium or used as a filter. Metal based materials include iron oxide, zinc oxide, and titanium dioxide and are mainly used for water purification. Carbon-based materials include activated carbons, carbon nanotubes (CNTs), including single-walled nanotubes (SWCNTs) and multi-walled nanotubes (MWCNTs), and graphene and graphene oxide. They are nanoporous and have physicochemical properties that are effective for water treatment to remove pollutants like heavy metals, fluorides, textile dyes, and pharmaceutical products. Polymer-based nanoparticles are often used in nanomembranes which filter nanoparticle pollutants. The polymer, chitosan, is commonly used for nanofiber membranes. Graphite oxide and silica nanomaterials with surface modification to enhance reactivity are commonly used to reduce air pollution from exhaust gases.

19.4 | Bioremediation of Pollution

In some cases, the best method to degrade pollution in soil or water is to utilize plants, fungi, and/or microbes or their enzymes, which constitutes bioremediation. The environmental industry uses bioremediation as an advanced technology. It is controlled decay and/or chemical transformation to neutralize or remove pollutants. A subdivision of bioremediation is phytoremediation if plants are used to remediate the problem. There are other types of bioremediation including bioaugmentation, biostimulation, bioventing, and biotransformation.

If the naturally occurring microbes in the soil degrade the pollution, it is natural attenuation. In some cases, the subsurface conditions are altered to increase microbial activity. Air pumped into soil can increase productivity of aerobic microorganisms (Figure 19.8). Certain nutrients like methane, phosphorus, and nitrates injected into soil can also help microorganisms flourish. The stimulation of naturally occurring microorganisms through the addition of nutrients is called enhanced bioremediation. Additional bacteria may be introduced into the soil to enhance remediation and some bacteria may be genetically engineered to be more productive. If there are multiple pollutants, a cocktail of multiple microbial species can be injected into the plume or soil with or without stimulation. In some instances, soil is removed from the site and bioremediation is performed ex situ.

There are a number of bioremedial methods that are commonly used in in-situ groundwater remediation. Bioventing is an in situ method of bioremediation in which growth of pollutant degrading bacteria is encouraged by injection of air and/or nutrients into the plume (Figure 19.8). Bioventing is usually added to a DPE system as light non-aqueous phase liquids (LNAPLs) and groundwater are removed. The combination of these two methods is called bioslurping and it is used to remediate polluted soil and groundwater at once. This is most commonly used to remediate soil and groundwater polluted with automobile and aviation fuels. Less permeable soil makes it more difficult to draw out LNAPL and push in air.

Another bioremedial method is monitored natural attenuation (MNA). Many remedial processes operate in the natural environment without intervention. Under certain conditions, these may be sufficient to remediate pollution while protecting public health and the environment. This approach is usually used in conjunction with other remedial measures or as the final step in a larger remedial action. It cannot be used on persistent and toxic pollutants and it requires knowledge of the types and extent of pollutants, the rate and direction of groundwater flow relative to the populated areas, and the ability of microbial activity to remove the pollution within a reasonable period. The only stipulation is that the remedial progress must be regularly monitored.

Ex situ remediation of soil and even some water can also be accomplished using bioremediation. Sometimes heavy metals and radioactive element pollution can be addressed using plants or phytoremediation. Some plants can partition certain types of elements to allow some to be taken up while blocking others. There are some plants that absorb heavy metals and radioactive elements making them excellent for phytoremediation. These plants are planted on polluted soil and remove many of the pollutants as they grow. This leaves the problem of what to do with the tainted plants. If they are burned, the pollutant goes into the smoke particulate and can be inhaled or fall out or washed out by rain in another area. Tobacco plants phytoremediate heavy metals and radioactive elements, which is one of the reasons that tobacco smoke is dangerous.

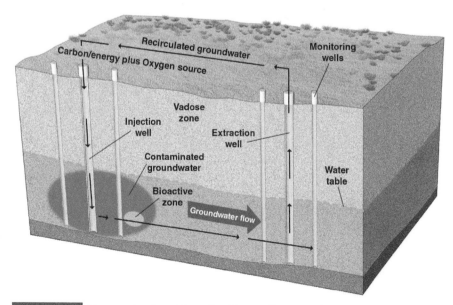

FIGURE 19.8 Diagram showing a bioventing bioremediation configuration of a pollutant plume.

CASE STUDY 19.2 Ensign Bickford, Connecticut

Phytoremediation of Polluted Soil

The Ensign Bickford Company facility is 15 miles (24 km) northwest of Hartford, Connecticut (Figure 19.9). It has been developing, producing, and testing explosives for the US military at this 350-acre (142-ha) site since 1836. The site contains a 200-acre (81-ha) heavily wooded western area that contains a storage facility for raw materials and finished products, and process buildings. The 150-acre (61-ha) eastern area houses the manufacturing and waste management activities. Explosives, pyrotechnics, and primers were developed and tested on a three-acre (1.2-ha) site in the eastern area designated the Open Burn/Open Detonation (OB/OD) site. Open-air testing was done in the past but was replaced by computer modeling and laboratory testing.

This testing of explosives left extensive pollution on the site. The water table here is shallow (2–4 ft, or 0.6–1.2 m, below surface) and most of the OB/OD area sits in the 100-year flood zone in the floodplain of the Farmington River. The OB/OD area soil contains 635 mg/kg to more than 4000 mg/kg of lead. Lead concentrations in soil in residential areas are limited to 100 mg/kg or less. Ensign Bickford initiated a phytoextraction remediation project on a 2.4-acre (1-ha) site in the OB/OD area to address the soil pollution with the approval of the EPA and Connecticut Department of Environmental Protection (CTDEP).

Phytoextraction is also called phytoabsorption or phytoaccumulation, and it is the process where soluble pollutants are absorbed through plant roots and stored in the upper part of the plant. The pollutants are not chemically altered, just relocated from the soil to the leaves and stems. This works well on heavy metal and radioactive element pollutants. Once grown, the plant is harvested and disposed of by burning or the metals and radioactive elements are chemically extracted from the plant.

Heavy metals are commonly phytotoxic but plants known as hyperaccumulators can tolerate between 100 and 1000 times more heavy metals than common plants. Hyperaccumulator plants concentrate the vast majority of heavy metals in their roots rather than their leaves and stems. The phytoextraction takes place in the upper root zone, which is primarily in the top 2 ft (0.7 m) of soil. This is relatively shallow which limits the use of phytoextraction as a remedial method in thicker polluted soils. Deep tilling has been used to bring deeper polluted soil closer to the surface and within the range of hyperaccumulating plants with some success.

Another limitation of phytoextraction is the chemical attachment of heavy metals onto clay particles in the soil. Clay particles carry a negative charge whereas most heavy metals are positively charged. As a result, the metals chemically attach to the clay particles. To make the metals absorbable to the plants, a chelator is added to the soil

and it bonds with the metal ions. The preservative ethylene diamine tetra-acetic acid (EDTA) is mixed with the soil and it removes the metal from the clay, allowing it to be absorbed by plants. This application is carefully monitored to make sure that the metals in solution do not wind up in the groundwater instead of the plant root system.

The OB/OD site was subdivided into five areas based on pollution level. The environmental professionals tilled phosphorus, nitrogen, and potassium-based fertilizers and lime into the lead-bearing soil. An overhead water spraying system was installed that delivered additional fertilizer. The hyperaccumulator, Indian mustard, was then planted in the polluted soil and harvested, followed by planting of another hyperaccumulator, sunflower, which was also harvested before a final planting and harvesting of more Indian mustard took place. All of this was done within a six-month growing season. As a result, lead levels in the soil decreased in four of the five areas of the site by nearly 25%, and no soil contained lead above 4000 mg/kg. The sunflower and Indian mustard plants extracted lead at about 1000 mg/kg. The harvested plants were incinerated. The phytoextraction project along with other remedial efforts kept polluted groundwater on the Ensign Bickford property and it is becoming cleaner.

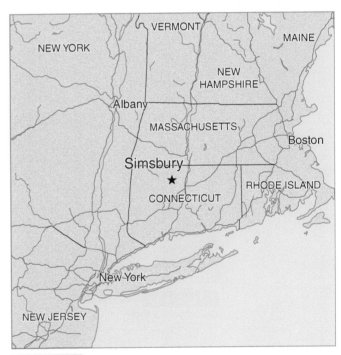

FIGURE 19.9 Map of southern New England showing the location of Simsbury, Connecticut.

Another form of phytoremediation is rhizofiltration or phytofiltration and it can be biotic or abiotic. It best treats broad and shallow plumes with low levels of metals in the parts per billion (ppb) range. As very shallow polluted groundwater flows through the roots of plants, it can interact with them. Some types of plants can take up or fix dissolved pollutants in their roots. Abiotic rhizofiltration occurs by precipitation of dissolved metals from groundwater onto the surface of the roots. Biotic rhizofiltration occurs by nutrient uptake, drawing dissolved pollutants from the groundwater through the roots and into the plant.

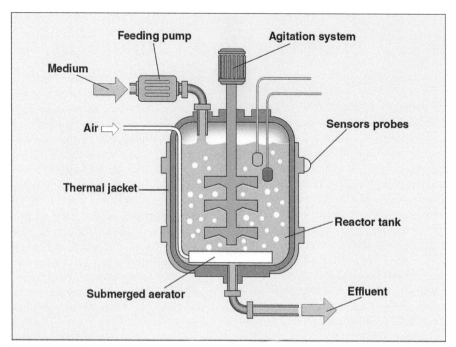

FIGURE 19.10 Diagram of a standard bioreactor with input of polluted water on the upper left and output (effluent) of treated water to the lower right.

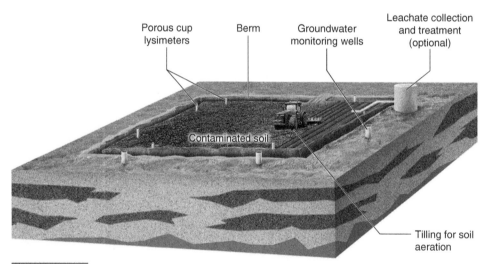

FIGURE 19.11 Diagram showing land farming bioremediation method. *Source:* Courtesy of US Environmental Protection Agency.

In rhizofiltration, pollutants saturate the plant and its roots, leaves, stems, and branches are harvested and disposed of in a waste treatment facility. It is typically used where groundwater is polluted with lead, chromium, cadmium, or other metals or radionuclides like uranium, cesium, and strontium. Plants used for rhizofiltration have demonstrated the ability to remove these metals from groundwater and have fast-growing root systems. The plants are germinated in greenhouses, with their roots in water to increase their root mass. The plants are then transplanted at a polluted site.

Phytovolatization withdraws pollutants from groundwater and exhales them to the atmosphere. The plants take up water with the pollutant through their roots, transmit it through their vascular system to their leaves, and there it is released. This process can even chemically alter the pollutant to a less toxic state which is more desirable. Once released to the natural environment, many pollutants are degraded through photodegradation from the sun or other natural processes.

There are also several common ex situ bioremedial techniques depending upon the pollution and polluted medium that are used in addition to ex situ chemical and physical techniques. For polluted groundwater pumped from an aquifer, water treatment systems are used to remove the pollutants before releasing to the environment. Depending on the situation, technology similar to a wastewater treatment plant may be used which are a filtration/bioremedial technology (Figure 19.10). That technology is activated sludge which operates using microbes suspended in the wastewater. There are many types including rotating biological contactors, packed bed reactor, biological fluidized bed reactors, continuous stirred tank bioreactors, upflow anaerobic sludge blanket reactors, and photobioreactors and membrane bioreactors (MBRs). The more advanced systems are microfiltration MBRs. These systems combine a suspended growth bioreactor with solids removed through filtration, making them both biological and physical. They are efficient at removal of nitrogen, phosphorus, bacteria, biochemical oxygen demand, chemical oxygen demand, and total suspended solids.

There are two MBR types, one with internal/submerged membranes and the other with external/side stream membranes in a separate unit. The submerged MBRs use bubble aeration to enhance mixing and are considered superior.

A bioremediation treatment process for polluted soil that can be in situ or ex situ is land farming, land treatment, or land application (Figure 19.11). In situ land farming is performed in the upper soil horizons and ex situ is performed in biotreatment cells. In ex situ land farming, excavated polluted soils, sediments, and/or sludges are mixed into other soil and spread into a thin layer over a broad area. This mixture is turned over or tilled regularly to promote aeration. This aeration removes most of the lighter petroleum pollutants, like gasoline, by evaporation. This is a chemical or physical process. Heavier, less volatile petroleum like heating and lubricating oils do not evaporate during aeration and are removed through microbial activity. The mixtures of polluted soils are treated with fertilizer, lime, and water that enhance the microbial activity. This enhanced activity degrades the heavy and remaining light petroleum products.

Is It Too Late?

CHAPTER OUTLINE

Polluted Earth: The Science of the Earth's Environment, First Edition. Alexander Gates.
© 2023 John Wiley & Sons, Inc. Published 2023 by John Wiley & Sons, Inc.
Companion website: www.wiley.com/go/gates/pollutedearth

Words you should know:

1970 Clean Air Act – The first rigorous environmental legislation establishing the most dangerous "criteria" air pollutants.

1972 Clean Water Act – The first rigorous environmental legislation to protect surface and groundwater.

1980 Comprehensive Environmental Response, Compensation, and Liability Act (CERCLA) – The US federal act establishing the National Priorities List to identify and remediate profoundly polluted superfund sites.

Conservationist – A person who takes action to maintain natural areas and protect species for future generations.

Claire Patterson – The scientist and pioneer credited with eliminating lead from the environment and human impact.

Emerging pollutants – Recently identified chemicals primarily in water that are potentially dangerous to the environment and public health.

Environmentalist – A person who advocates for the protection of the environment from misuse through human activity.

Great Pacific Garbage Patch – An enormous mass of floating garbage in the Pacific Ocean pushed together by currents.

Montreal Protocol – A worldwide agreement to reduce and eliminate the use of chlorofluorocarbons to protect and restore the stratospheric ozone layer.

20.1 | Believers and Non-Believers

Now that you are at the end of this book, you should have a solid general understanding of many of the environmental challenges that face humankind over the next few decades. These are not all of the pollution issues. There are microplastic beads that are in most water and ingested by most living things. Scientists do not know the long-term effects of this pollutant. There are pharmaceuticals and personal care products (P&PC) that occur in 97% of public water supplies because they are not removed by water treatment plants. These too may have long-term effects but no one is sure what they will be. There are "emerging pollutants" that are newly discovered on a regular basis and evaluated as to their health impacts.

In addition, there is a Great Pacific Garbage Patch or plastic island, which is a mass of floating garbage that is primarily plastic and covers an estimated 0.6 million square miles (1.6 million km²) of ocean (Figure 20.1). Rotating ocean currents or gyres sweep the trash into this patch, which floats between Hawaii and California. The area is about three times the size of France and it is rapidly expanding. Covering so much ocean, it shades the sunlight so phytoplankton become less productive, producing less O_2 and less food for the other ocean life since they are the base of the food chain. There is a second patch in the western Pacific as well. There are even more problems.

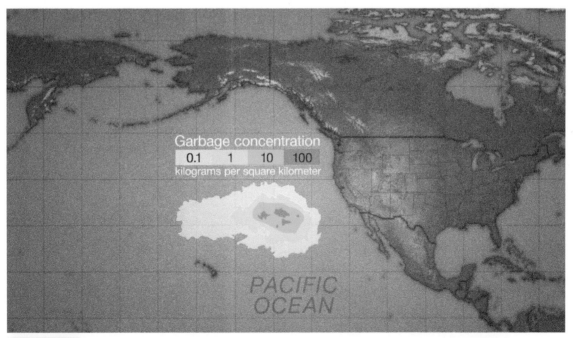

FIGURE 20.1 Map of the eastern Pacific Ocean showing the location and density of the Great Pacific Garbage Patch. *Source:* https://www.cbsnews.com/news/the-great-pacific-garbage-patch-isnt-what-you-think-60-minutes-2019-08-18 / Accessed on December 24, 2022.

People can take one of several views of these apparently overwhelming issues. They can claim that this is all fake news and ignore them. They can claim that the Earth and its resources were put here to serve humans and we do not need to worry about these insignificant issues. This is getting more difficult to maintain as even the Pope has expressed concern and encouraged people of the Catholic faith to protect the environment. Environmental concerns are generally not taken up by religions.

People who are concerned about the environment fall into several groups. There are environmental activists who march in protests and attempt to influence people and politicians to take action. Perhaps environmentalists take action, donate to causes, and are also involved in influencing people and the government as advocates. Conservationists take action to do their part to help preserve the planet for future generations but are less involved with influencing anyone. The vast majority of people, however, are concerned about the environment but do not know what to do. Some people and even scientists believe that it is too late to take action. They believe that we are past the "tipping point", or essentially the point of no return, and that the human race is doomed because of the damage we have done. This could be true, but there are a number of examples of improvement that should give us hope, if we take action.

20.2 | A Not-So-New Hope?

People that grew up in an industrialized country from post-World War II to the mid-1970s were so bombarded with pollution, it is astonishing that they survived as long as they have. The air, water, and soil were badly polluted around most urban areas and even in some suburban and rural areas. Rampant atmospheric nuclear bomb tests led to radioactive fallout worldwide. Industry treated the natural environment with complete disregard and with only a little more concern for human health. It seemed as if the human race was headed for self-destruction but people worked together and pressured companies and the government to address the horrible conditions.

The persistent organic pesticide DDT is an example of an addressed pollutant problem. Virtually all survivors of World War II across Europe were dusted with DDT to delouse them (Figure 20.2). Trucks sprayed DDT on all streets in most urban and suburban areas in all industrialized countries for mosquito control and children ran behind the trucks through the pesticide mist. It was in most foods to the point where new mothers were advised not to nurse their newborn babies because they would expose the babies to unsafe levels of pesticides including DDT. Large bird-kills occurred in Massachusetts and Long Island, New York among others that were reported in the press. DDT thinned the eggs of many birds to the point that they could no longer reproduce and they almost went extinct, including peregrine falcons, brown pelicans, and California condors. The once plentiful bald eagle was down to about 400 reproducing pairs in 1967 from the 300 000 to 500 000 birds that once inhabited the United States. Overuse of DDT on cranberry bogs in 1959 was even so extreme that traditional cranberry sauce could not be eaten by Americans for Thanksgiving.

However, Rachel Carson worked to inform people about the dangers of persistent pesticides and especially DDT. Despite concerted efforts of chemical companies to discredit her, she attracted the interest of then President Kennedy and presented her concerns to the US Congress. Her book *Silent Spring*, which alerted the public to the dangers of pesticides in 1963, was a sensation. There was a public uproar against the pesticides she named and the newly formed US Environmental Protection Agency got them banned in 1972, just nine years later. One result of the banning of DDT in conjunction with protection efforts, bald eagles recovered spectacularly. By 2019, there were more than 315 000 bald eagles in the United States. Many other once near-extinct birds are also now thriving.

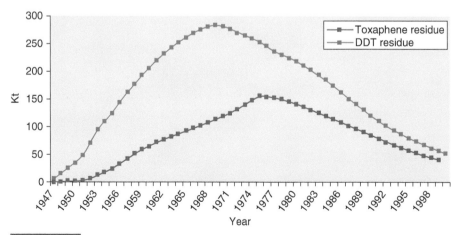

FIGURE 20.2   Graph showing the sharp increase of the persistent pesticides DDT and toxaphene worldwide after World War II and the sharp decrease after regulations were instituted.

CASE STUDY 20.1 Clair (Pat) Patterson

Champion of Lead Banishment

Another success story about removing a dangerous pollutant from the environment is about the dangerous substance, lead. This heavy metal impacts public health in many ways but especially brain function. Yet in the 1960s and 1970s, it was everywhere. It was in gasoline, paint, solder in pipes and electrical devices, and pesticides and even in food. It was in the air, water, and soil. The California Institute of Technology geologist Clair (Pat) Patterson, who had no training in environmental lead, became an unlikely champion to get it banned (Figure 20.3). He showed that lead was being deposited on the Earth's surface at 80 times the normal rate. He showed that Americans were regularly exposed to 100 times the lead of natural levels and within two times the level of lead poisoning. He found that lead was elevated in many canned goods, especially tuna. He then tested lead in human blood and bones and found that most Americans were carrying lead in their bodies at unhealthy levels. Comparing his data with that from mummies, he showed that modern human bones had 700–1200 times the amount of lead of ancient bones.

Patterson was a respected scientist but was attacked personally and professionally as a troublemaker by respected industry professionals. Although he was a quiet, unassuming gentleman, in 1966 he wrote to Governor Pat Brown of California to warn him about lead in the air and was rejected initially but prevailed in the end. He sent a letter to Senator Edwin Muskie, who was the chairman of the Senate

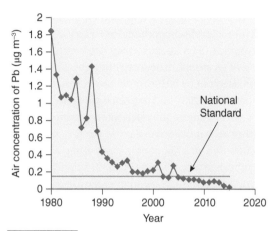

FIGURE 20.3   Graph showing the sharp decrease of lead in the air after action was taken to eliminate it.

Subcommittee on Air and Water Pollution and was invited to testify. He tenaciously used his scientific data from many sources to prove his assertions. By the time Patterson was through in the late 1970s, he had almost single-handedly eliminated all commercial uses of lead. Today, the only real exposure to lead in most industrialized countries is from old paint and old pipes that need replacing.

Perhaps the most impressive global effort to overcome a serious environmental threat is the hole in the stratospheric ozone layer (Figure 20.4). Ozone in the stratosphere protects the Earth's surface from dangerous ultraviolet radiation. This can cause greatly increased incidence of skin cancer. However, this layer was put in great peril by the human wonder chemical of Freon and other chlorofluorocarbons (CFCs). These chemicals revolutionized cooling and refrigeration.

CFCs are light, very stable, and long-lived, which makes them able to reach the stratosphere, 15–25 miles (24–40 km) above the surface without breaking down. There, radiation breaks down the CFCs, producing free chlorine. This participates in the ozone-consuming reactions over and over again. The cycle of ozone destruction continues for as long as two years, consuming an average of 100 000 ozone molecules by a single chlorine atom. This potentially catastrophic effect on the ozone layer was described by Mario Molina and F. Sherwood Rowland in 1974 at the University of California. These two would share the 1995 Nobel Prize in Chemistry with Paul Crutzen

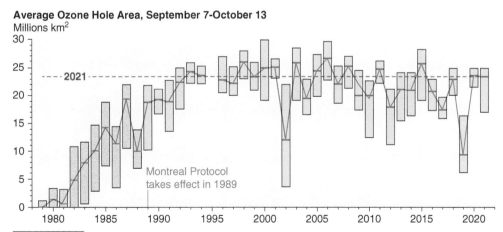

FIGURE 20.4   The annual average size of the Antarctic hole in the ozone layer for the period 7 September to 13 October from 1979 to 2022 including the date of the Montreal Protocol.

for this work. Their prediction was proven in 1985, when a British team found an "ozone hole," over the Antarctic. A smaller hole was found over the North Pole, as well.

The world panicked but officials responded quickly. The United Nations arranged an international treaty called the Montreal Protocol in 1987. The treaty required countries to stop using CFCs and banned all uses of ozone-depleting gases by 1996. HCFCs (hydrochlorofluorocarbons) or HFCs (hydrofluorocarbons) were quickly developed to replace CFCs because they are much less destructive to ozone. However, HCFCs and HFCs are still ozone-depleting gases and the Montreal Protocol requires that they are phased out by 2030. This action is already improving the situation. The size of the hole peaked in 2007 and has since generally receded, though it is not a consistent change. It will take about 100 years before the ozone layer will reach pre-CFC conditions.

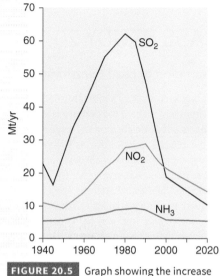

FIGURE 20.5   Graph showing the increase and decrease of acid rain compounds NO_x and SO_2 in Europe from 1940 to 2020.

Another very impressive success story is acid rain. In the 1960s through 1980s, acid precipitation was rampant in most urban areas in developed countries and even invaded suburban areas (Figure 20.5). Residents of many cities suffered from eye and respiratory track irritation through exposure. Many people wore face masks to protect their lungs. Some cities with antiquities limited or eliminated automobile traffic around antiquities because they were rapidly deteriorating from exposure to acid precipitation. Infrastructure and even newer cement on buildings began to decay. Vegetation was stressed from exposure and many water bodies were badly impacted. Many lakes wound up dead with no life at all.

Acid rain in Europe, Japan, and North America was reduced because laws were passed to limit SO_2 and NO_x emissions. In the US, the Clean Air Act of 1970 was the first effort to reduce these criteria air pollutants. In 1991, the Canada–United States Air Quality Agreement was enacted to further reduce acid precipitation and there were similar measures in Europe and Japan. The efforts to eliminate acid precipitation have been very successful. Annual SO_2 emissions have been reduced by over 93% and annual NO_x emissions have been reduced by over 87% since restrictions began.

These are not the only examples. Asbestos was considered a wonder mineral. It is flexible, fireproof, and can be woven into cloth among many other applications. Unfortunately, when inhaled, it can cause mesothelioma (lung cancer) and silicosis (lung disease). In response, it was banned and exposure is now only through contact with older applications. Coal-fired power plants and industry released mercury to the atmosphere, which rained out of the sky for hundreds of miles downwind and some stayed aloft around the globe (Figure 20.6). In the United States, Midwestern powerplants dumped excessive mercury across the whole eastern half of the country. However, the switch to fracked natural gas to generate electricity and better smokestack emission controls on industry have eliminated the problem. Natural gas is much cleaner than coal.

Perhaps the most telling illustration of how quickly pollution can be reversed was during the COVID-19 pandemic (Figure 20.7). The UN reported that many human-related pollutant emissions fell sharply during the pandemic. More than 65% of cities worldwide experienced better air quality and some 84% of countries reported improvements. The worldwide particulate PM2.5 concentrations decreased by 30–40% and SO_2 was 25–60% lower than previous years. Carbon monoxide decreased worldwide relative

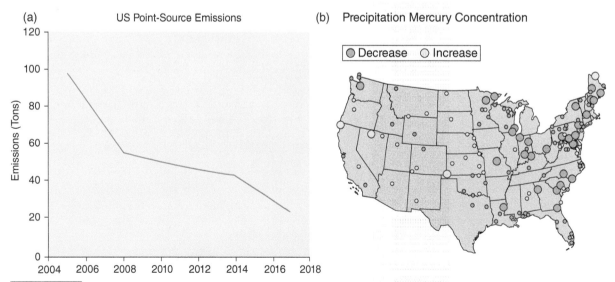

FIGURE 20.6   (a) Graph showing the reduction of mercury emissions from 2005 to 2017. (b) Map of the United States showing the locations and amount of mercury precipitation increase and decrease from air pollution fall out over the same period.

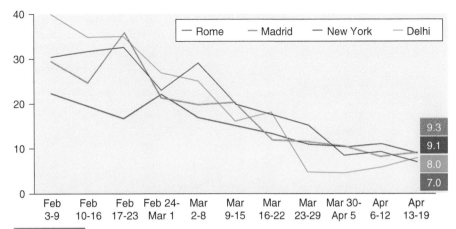

FIGURE 20.7 Graph showing the rapid decrease of NO_2 in the atmosphere in the cities Rome, Madrid, Delhi, and New York over the three-month period from February to April 2020 as a result of the covid lockdown.

to previous years and in South America the decrease was as much as 40%. The countries of China, Spain, France, Italy, and the United States also documented 20–30% reductions in NO_2 in ambient air. Even total CO_2 emissions decreased by 5.4% in 2020 compared to previous years, though the atmospheric concentrations were unaffected. That will take many years. This radical decrease in just one year shows what is possible with concerted effort.

20.3 | Let the Government Do It

Many people think that the government should fix the pollution problems. However, the federal government is "of the people, by the people and for the people" and it is for everyone. There are numerous lobbyists and benefactors who put in a lot of time, effort, and money to get politicians to endorse and support actions that benefit them. In order for laws to be passed to protect the environment, there needs to be concerted public pressure in one of several possible forms. The chances that the government will unilaterally fix environmental problems are not good.

Concerted public pressure is how the environmental movement began and progressed. Rachel Carson began the struggle in the United States and her message resounded with the American public. The media built on this concern by covering the environmental problems and disasters. As a result of this passion, the first Earth Day on 22 April 1970 was attended by 20 million Americans or 10% of the population. The environmentalist wave was so strong that President Nixon, who was no friend of the environment, was forced to sign the US Environmental Protection Agency into law by pure public pressure. Further, the first stringent and sweeping Clean Air Act was signed into law in 1970 (Figure 20.8) and the Clean Water Act was signed into law in 1972. Many of the persistent organic pesticides identified by Carson were also banned in 1972. In 1973, the Endangered Species Act was also signed into law by Nixon. All of these great advancements in environmental protection were accomplished against the wishes of the President.

The great advancements in environmental protection in the 1970s actually deflated public fervor. The attendance at Earth Day declined, as did public pressure. However, in the late 1970s a series of well-publicized environmental disasters brought environmental protection back to public attention and outrage. Probably the last straw in these dozen or so disasters was Love Canal, New York in 1978 and 1979. The poisoning of a whole town by exposure to legacy (not active) pollution showed the country that there could be serious environmental public health time bombs hidden anywhere. It was well covered by the media, fueling outrage, and the federal government did not assess the situation well. They sent federal officials to Love Canal in a very non-urgent manner and they were promptly taken hostage at gunpoint. The situation was resolved peacefully but it caught the full attention of the government and action was taken.

The result was the passage of the next major environmental protection bill, the Comprehensive Environmental Response, Compensation, and Liability Act (CERCLA), enacted by the US Congress on 11 December 1980. This bill established the Superfund program to clean up the potential environmental and public health time bombs. This law levied taxes on chemical and petroleum industries and provided authority to the government to address releases or threatened releases of hazardous chemicals that could endanger the environment or public health. Over the first five years of the program, $1.6 billion in taxes was collected to clean up abandoned or uncontrolled hazardous waste sites. It also established guidelines for handling closed and abandoned hazardous waste sites. It developed procedures to assign liability to persons responsible for releases of hazardous waste. It also authorized short-term removal and long-term remedial actions to reduce the dangers of releases and threats of release of hazardous substances that are not immediately life-threatening. These sites are included on the National Priorities List.

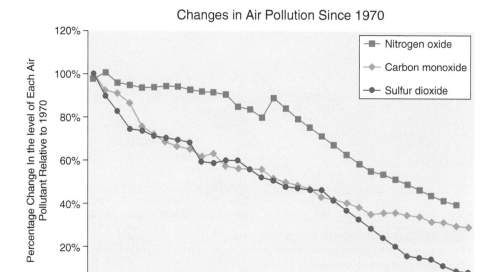

Changes in Air Pollution Since 1970

FIGURE 20.8 Graph showing the decrease of the criteria air pollutants, SO_2, NO_x, and carbon monoxide, in the United States after the enactment of the 1970 Clean Air Act until 2017.
Source: Courtesy of US Environmental Protection Agency / Public Domain.

Sometimes countries do make unilateral efforts to improve the environment. For example, in 2010, 16 of the top 20 most polluted cities in the world were in China. Beijing had 80% acid rain days in 2005 and generally very unhealthy air in the cities. To address this problem, in September 2013, China released their Air Pollution Action Plan which aimed to make marked improvements in air quality between 2013 and 2017. It was designed to reduce PM2.5 particulate by 33% in Beijing and 15% in the Pearl River Delta. The plan was instituted and was relatively successful at least locally. By 2022, China only had two of the top 20 most polluted cities in the world and they were number 19 and 20. China is a large country, with the world's greatest population.

Some individuals also make heroic efforts to preserve and repair the environment. For example, Harrison Ford, the actor famous for the Star Wars and Indiana Jones series, is also an environmental and conservation activist. Ford joined Conservation International, serving as a board member and later the Vice Chair of the Board of Directors. He has donated large amounts of his time and money to their conservation efforts and especially their Global Conservation Fund. He was instrumental in helping to secure the protection of over 40 000 000 acres (16 200 000 ha) on three continents in just 18 months. In all, this group has helped protect more than 2.3 million square miles (6 million km²) of land and sea across more than 70 countries. He has made speeches to the United Nations among many other groups on the dire need for conservation worldwide. He even donated 389 acres (157 ha) of his property for a conservation easement to the Jackson Hole Land Trust in Jackson Hole, Wyoming.

Another champion of the environment is the famous actor, Leonardo DiCaprio. He has worked extensively to fund and bring attention to protection of biodiversity, ocean and forest conservation, and climate change. He established the Leonardo DiCaprio Foundation in 1998 which supports more than 35 innovative conservation projects around the world. The fund protects fragile ecosystems and endangered species. He has donated more than $15 million to environmental organizations. He and his partners pledged $43 million toward a major conservation effort to restore the Galápagos Islands to natural conditions. He is among the most active celebrities in the climate change movement. He produced two web documentaries and the DiCaprio Foundation donated $100 million to fight climate change in December 2018.

There are several organizations working to clean up the planet as well. One example is 4Ocean, that promises to extract one pound of plastic from the ocean for every product sold. Others include World Wildlife Fund and the Sierra Club among others. The government may ultimately need to be involved in any significant environmental effort but it will take public involvement and will to begin the effort.

A

Aldrin   pesticide. Used as a soil insecticide for beetles, root worms, and termites. Banned in 1970, the EPA approved it for termite control in 1972 but even this use was banned in 1987. It is considered more toxic than DDT. It was found in 99% of breast milk samples before the ban. Exposure causes dizziness, headache, irritability, and convulsions, leading to possible brain damage, coma, and death with increasing dosage. The EPA limits aldrin in drinking water to one micrograms/L. OSHA limits exposure to 250 micrograms/m³ of workplace air.

Antimony   heavy metal. It is mixed with other metals to produce alloys and mixed with lead to improve batteries or in place of lead in solder and bullets. It is used in plastics, porcelain, medicine, and oil refining. Ingestion/inhalation produces vomiting and diarrhea, respiratory difficulties, pulmonary edema, and death with increasing dosage. It also causes blood disorders and liver, kidney, and reproductive problems. EPA set the drinking water limit at 6 ppb. OSHA limits antimony in workplace air to 0.5 mg/m³.

Arsenic   inorganic pollutant. It is used in wood preservative, alloys, pesticides, glass, paints, dyes, medicine, and semiconductors. Ingestion causes nausea, vomiting, abdominal pain, and hemorrhage, renal failure, low blood pressure, circulatory collapse and headache, muscular weakness, convulsions, coma, and death with increasing dosage. Repeated exposure causes cancer of the nasal passages, colon, bladder, prostate, and kidneys. It occurs in tobacco, possibly helping to make smoking so deadly. EPA limits arsenic in drinking water at 10 ppb. OSHA limits it to 10 micrograms/l of workplace air.

Asbestos   mineral. Fibrous mineral used for brake linings for cars, pipe insulation, floor and ceiling tiles, plasters and stuccos, fire proofing, fire doors, shingles, fire blankets, clutch plates, gaskets, sheetrock tape, insulation, siding, countertops and gaskets. Health effects from inhaling asbestos includes asbestosis, which is lung scarring, and a lung cancer called mesothelioma. OSHA limits asbestos to 100 000 fibers/m³ of industrial air. EPA limits drinking water to 7 million fibers/l.

B

Barium inorganic   pollutant. It is the main ingredient in oil drilling mud and also filler, extender, and weighting of paints, plastics, and rubber, and also used in car brake and clutch pads, automotive paint, photographic paper, X-ray contrast and shielding, fireworks and munitions, pigments, glass, dyes, leather, and chemical manufacture. Exposure causes vomiting, diarrhea, slow heart rate, high blood pressure, cardiac issues, muscle weakness and tremors, numbness and paralysis, and death from cardiac and respiratory failure with increasing dosage. The EPA drinking water limit is 2 ppm. OSHA limits it to 0.5 mg/l of workplace air.

Benzene   VOC, industrial solvent. It also occurs in tobacco smoke. Inhalation of benzene fumes causes dizziness, headaches, tremors, rapid heartbeat, unconsciousness, and death with increasing dosage. Contact causes rashes, lesions, eye irritation, and damage to the cornea. It causes acute myeloid or erythroblastic leukemias and chronic myeloid and lymphoid leukemias and cancer of the lungs, skin, mammary glands, ovaries, oral cavity, and the Zymbal gland. The EPA limits benzene in drinking water to 5 ppb. OSHA limits benzene in workplace air at 1 ppm

Beryllium   heavy metal with industrial, military, alloy and high tech uses. Inhalation causes berylliosis including rhinitis, pharyngitis, tracheobronchitis, and severe pulmonary symptoms. It damages the lungs, skin, bones, liver, kidneys, spleen, lymph nodes, myocardium, muscles, and salivary glands. It causes lung cancer and possible breast, bone, and uterine cancer. The EPA limits beryllium to 4 ppb in drinking water. OSHA limits beryllium to two micrograms/l of workplace air.

Bromine   inorganic element used for agriculture, dyes, insecticides, pharmaceuticals, and chemical production. Exposure irritates eyes and skin causing burns, the nose, throat and lungs causing coughing, shortness of breath, pulmonary edema, and death with increasing dosage. Chronic exposure causes kidney and brain damage. It likely causes kidney cancer. The World Health Organization recommends a limit of 6 mg/l in drinking water. OSHA limit limits bromine to 0.1 ppm in workplace air.

C

Cadmium   heavy metal in paint, alloys, high-tech devices. Inhalation damages the lungs and can cause death. Ingestion causes vomiting, diarrhea, intestinal damage, and possible death. Long-term exposure causes high blood pressure and liver, brain, and immune system damage. It likely causes cancer. The EPA limits cadmium in drinking water to 5 ppb. The FDA limits cadmium in dye to 15 ppm. OSHA limits worker exposure at five micrograms/m³ of air.

Carbofuran   insecticide; the third most toxic field crop pesticide and especially deadly to birds. It has restricted usage and is banned in some areas. Exposure can cause headache, salivation, nausea, vomiting, diarrhea, abdominal cramps, labored breathing, blurred vision, increased blood pressure, convulsions, coma, and death with increasing dosage. One quarter teaspoon can be fatal. The EPA limits carbofuran in drinking water to 40 ppb. OSHA limits exposure to 0.1 mg/m³ of air.

Carbon monoxide   gas from incomplete burning and used in the chemical industry. Exposure produces headache, muscle weakness, nausea, vomiting, chest pain, confusion, hallucinations, unconsciousness, coma, and death with increasing dosage. In pregnant women, it can produce stillbirths, spontaneous abortion, premature birth, and early postnatal death. OSHA limits workspace air to 50 ppm

Carbon tetrachloride   VOC, industrial solvent and production chemical. Inhalation and ingestion produces headache, vertigo, loss of coordination, and respiratory failure, coma, and possible death at high doses. It also can cause nausea, abdominal pain, and diarrhea. Extended exposure causes liver, kidney, and lung damage. It also causes cancer. The EPA limits carbon tetrachloride to 5 ppb in drinking water. OSHA limits it to 10 ppm in work areas.

Polluted Earth: The Science of the Earth's Environment, First Edition. Alexander Gates.
© 2023 John Wiley & Sons, Inc. Published 2023 by John Wiley & Sons, Inc.
Companion website: www.wiley.com/go/gates/pollutedearth

Cesium radioactive isotope used in oil drilling fluid, optical glass, and as a catalyst promoter. Cesium can damage human cells by radiation. It can cause acute radiation syndrome with nausea, vomiting, diarrhea, bleeding, coma, and death at very high exposures. It can cause cancer. The EPA limits exposure to 4 millirem per year. OSHA limits cesium hydroxide to 2 mg/m³ in workplace air.

Chlordane insecticide of exceptional persistence. It was largely banned in 1983. Exposure causes nausea, vomiting, dizziness, vision problems, convulsions, hemorrhagic gastritis, bronchiopneumonia, respiratory failure, and death in high doses. Long-term exposure causes liver, kidney, heart, central nervous system, endocrine, and immune system damage. It also causes liver and skin cancer, leukemia, and non-Hodgkin's lymphoma. The EPA limits chlordane to 2 ppb in drinking water. OSHA limits it to 0.5 mg/m³ in workplace air.

Chlorobenzene industrial chemical and solvent. Exposure produces skin, eye, and respiratory irritation, headaches, nausea, and vomiting depending upon dosage as well as organ damage from long-term exposure. The EPA limits chlorobenzene in drinking water to ≤0.1 ppm. OSHA limits it to ≤75 ppm in workplace air.

Chloroform VOC, chlorinated solvent used mostly to manufacture refrigerants. Exposure causes nausea, vomiting, and lethargy at low dosage and narcosis, cardiac arrhythmia, and death at high dosage. Long-term exposure causes liver, kidney, and reproductive system damage. It also causes colon, rectal, and bladder cancer. The EPA limits chloroform in drinking water to 1 ppb. OSHA limits chloroform in workplace air to 50 ppm

Chromium heavy metal used in alloys, metal coatings, nutritional supplements, and catalysts. There are several types of chromium; hexavalent is the deadliest. Inhalation causes nose irritation, bleeding, ulcers and holes in the nose, shortness of breath, wheezing, and coughing in allergic people. Contact with hexavalent chromium causes chemical burns. Ingestion causes vomiting, ulcers, convulsions, hemorrhaging, kidney and liver damage, and death. It also causes lung cancer, birth defects and reproductive issues. EPA limits chromium to 0.1 ppm in drinking water. OSHA limits hexavalent chromium in air at 52 micrograms/m³.

coal tar creosote coal tars contain large numbers of compounds including PAHs, phenols, naphthalene, amines, furans, and creosols among others. Creosote is used as wood preservative, road and roofing tar and in smelting among others. Exposure produces vomiting, labored breathing, vertigo, headache, hypothermia, cyanosis, convulsions, and death with increasing dosage. Chronic exposure produces skin and eye lesions, headache, vertigo, vomiting, light sensitivity, corneal damage, and strong allergic reactions. It also causes of the skin, lung, bladder, kidney, digestive tract, oral cavity, larynx, and esophageal cancers as well as leukemia and brain tumors. OSHA limits coal tar pitch volatiles to 0.2 mg/m³ of workplace air.

Cobalt metal and radioactive isotope used in alloys, high tech devices, medical applications. Inhalation produces wheezing, asthma, pneumonia, edema, and lung hemorrhage. Long-term exposure causes asthma, pneumonia, fibrosis, immune system and thyroid damage, and congestion of the liver, kidneys, and conjunctiva. It also causes lung cancer. Radioactive cobalt causes nausea, vomiting, diarrhea, bleeding, coma, and death with increasing dosage. It causes many types of cancer. OSHA limits stable cobalt to 0.1 mg/m³ in workplace air.

Cryptosporidium protozoan in raw sewage and surface water. Infection causes diarrhea abdominal cramps, vomiting, lethargy, and general malaise. In some cases, it has caused renal failure and liver disease.

Cyanide organic compound used in metals, synthetics, medical, rocket fuel, and mining. Cyanide is among the deadliest toxins. Exposure produces shortness of breath, headache, vomiting, convulsions, loss of consciousness, coma, and death from cardiac arrest in a few minutes. It causes severe brain and heart damage in survivors. The EPA limits cyanide in drinking water to ≤0.2 ppm. OSHA limits hydrogen cyanide and most cyanide salts to 10 ppm for workplaces.

D

DBCP (1,2-dibromo-3-chloropropane) agricultural pesticide. Mostly banned by the EPA in 1977 and completely banned in 1985. Exposure causes male infertility. Other effects include headache, nausea, vomiting, eye, nose, and throat irritation, tremors, double vision, seizures, unconsciousness, coma, and death with increasing dosage. Long-term exposure causes liver, kidney, spleen, eye, and bone marrow damage and many types of cancer.

DCE (Dichloroethene) industrial organic chemical used in plastic wrap. Depending on the type, exposure to DCE causes unconsciousness, liver and lung damage, heart failure and death. Long-term exposure causes kidney, liver, and immune system damage and possible birth defects and reproductive problems. Causes kidney, lung, lymph gland, and breast cancer. The EPA limits 1,1-DCE to 7 ppb, cis-1,2-DCE to 70 ppb, and trans-1,2-DCE to 100 ppb in drinking water. OSHA limits 1,1-DCE to 1 ppm and 1,2-DCE to 50 ppm in workplace air.

DDE breakdown product of DDT. Concentrates in human breast milk.

DDT (dichloro-diphenyl-trichloroethane) persistent pesticide. The main target of Rachel Carson's campaign. It was banned in 1972. Exposure causes tremors, seizures and liver damage with high dosage and may cause breast cancer. It is extremely damaging in the environment. OSHA limits DDT to 1 mg/m³ of workplace air.

Dieldrin pesticide. Used as a soil and seed treatment, as a wood treatment, and as mothproofing for wool and directly on insects. Banned in 1970, the EPA approved it for termite control in 1972 but this use was banned in 1987. It is considered more toxic than DDT. It was found in 99% of breast milk samples in the 1960s. Exposure causes headache, and convulsions, leading to brain damage, coma, and death with increasing dosage. EPA limits dieldrin in drinking water to two micrograms/L. OSHA limits exposure to 250 micrograms/m³ of workplace air.

Dioxin (2,3,7,8-tetrachlorodibenzo-p-dioxin) industrial organic intermediary, byproduct, additive to defoliant. Exposure results in chloracne, a severe skin disease, pancreas and liver damage, and inability to metabolize hemoglobin, lipids, proteins, and sugar. Dioxin causes learning disabilities, delayed development, lowered IQ scores and immunological damage. It also causes several types of cancer. EPA limits dioxin in drinking water to 1 ng/l.

E

Escherichia coli group of bacteria that occur in food and water. Symptoms of exposure range from abdominal cramping and diarrhea to blindness, paralysis, kidney damage, and death depending on the type and afflicted individual. EPA limits total coliform (including *E. coli*) to <5% with a goal of 0 ppm

Endosulfan insecticide. It was banned in many countries in 2012. Exposure leads to tremors, vomiting, diarrhea, agitation, possible blindness, convulsions, unconsciousness, and death. Chronic exposure leads to liver, kidneys, blood chemistry, parathyroid gland, and immune system damage.

Endrin pesticide. The EPA restricted use in 1979 and banned it completely in 1991. Exposure irritates skin and eyes and damages vision. It causes vomiting, diarrhea, headaches, muscle tremors, seizures, unconsciousness, coma, and death with increasing dosage. It is highly damaging in the environment. EPA limits endrin in drinking water to ≤2 ppb. OSHA limits endrin to 0.1 mg/m³ of workplace air.

ethylbenzene VOC, component of fuel, used to make styrene (plastics). Exposure causes throat and eye irritation, labored breathing, dizziness, headache, pulmonary effects, kidney and liver damage and death with increasing dosage. Chronic exposure causes blood, central nervous system, kidney, and liver damage. It may cause cancer. EPA limits ethylbenzene in drinking water to ≤0.7 ppm. OSHA limits ethylbenzene to 1 ppm in workplace air.

ethylene oxide (EtO)   organic gas used for sterilization and as an intermediary for antifreeze, textiles, plastics, detergents, and adhesives. Inhalation yields headache, nausea, coughing, shortness of breath, wheezing, and vomiting. Chronic exposure causes white blood cell and breast cancers, non-Hodgkin's lymphoma, myeloma, and lymphocytic leukemia. OSHA limits EtO in workplace air to 1 ppm

F

fluorine   nonmetal, inorganic element gas used for nuclear energy, to synthesize chemicals like CFCs and teflon, in metal smelting, etching, medical, water treatment, and other applications. Fluorine is highly corrosive and deadly. Exposure produces irritation of the eyes and respiratory system, liver and kidney damage, eyes and nose damage and death with increasing dosage. The EPAs limits fluoride in public water supplies to 4.0 mg/l. OSHA limits fluorine to workplace air to 0.1 ppm

formaldehyde   VOC, besides embalming, it is primarily used as an industrial chemical to produce resins. Exposure causes irritation of the eyes, nose, and throat. Vomiting, coma, and possible death result from drinking large amounts of formaldehyde. It likely causes nose cancer and possibly brain cancer. OSHA limits formaldehyde to 0.75 ppm in workspace air.

Furans   organic industrial chemical intermediary for resin production. There are many furans but chlorodibenzofurans or CDFs produce especially adverse health effects. Exposure produces irritation and chemical burns of the skin and eyes. Inhalation irritates the nose, throat and lungs and causes pulmonary edema, headache, unconsciousness, coma, and death with increasing dosage. Chronic exposure damages the thymus, reproductive organs, liver, and kidney. CDF also causes liver and adrenal gland cancer as well as leukemia.

G

Giardiaprotozoan   parasites. Infection causes severe diarrhea leading to dehydration.

H

HCB (hexachlorobenzene)   fungicide and industrial organic chemical. It was banned for agricultural uses in 1965 because it occurred in human breast milk. Exposure causes skin lesions, muscle weakness, convulsions and nerve and liver damage but high doses can be fatal. Infants exposed through nursing had severe liver damage. It may also cause cancer. EPA limits HCB in drinking water to 1 ppb.

HCH (hexachlorocyclohexane)   organic chemical best known for the dangerous pesticide lindane. It concentrates excessively in human breast milk. Exposure causes headaches, sex hormone disruption, blood disorders, convulsions, pulmonary edema, respiratory failure, and death with increasing dosage. Chronic exposure produces kidney, liver, pancreas, testes, ovaries, and mucous membrane damage. It also causes lung cancer and leukemia. EPA limits HCH in drinking water to 0.2 ppb. OSHA limits workplace air to 0.5 mg/m³ of lindane.

Heptachlor   persistent insecticide. It was first limited in 1978 and later banned in 1988. Rachel Carson identified heptachlor as a dangerous pesticide to the environment. Exposure causes irritability, liver damage, convulsions, coma, and death by respiratory failure with increasing dosage. It likely produces liver cancer. The EPA limits heptachlor to 0.4 ppb and heptachlor epoxide to 0.2 ppb in drinking water. OSHA limits heptachlor in workplace air ≤0.5 mg/m³.

Hydrogen sulfide   acidic gas used in textiles but mostly pollutant. Exposure causes eye, nose and throat irritation, dizziness, headache, irritability, insomnia, upset stomach, apnea, convulsions, coma, and death with increasing dosage. Chronic exposure damages lungs, liver, kidneys, and the heart. OSHA limits exposure to hydrogen sulfide at 20 ppm through the day or 50 ppm in one exposure.

I

Iodine (radioactive)   radioactive inorganic solid used in medical imaging and treatment. Exposure increases thyroid cancer and thyroid disease. Chronic exposure also increases leukemia, stomach cancer, and salivary gland cancer. The EPA limits radioactive iodine to 108 pCi/l in water and 100 pCi/m³ in air. The FDA limits it in food to 4600 pCi/kg. OSHA limits it to 50 mSv (5 rem) for the whole body annually.

L

lead   heavy metal. One of the most dangerous pollutants. Previously used in plumbing, gasoline, paint, and many other applications but most uses were banned by 1978. It is now primarily used in batteries. Lead poisoning causes nausea, vomiting, painful stomach cramps, diarrhea, dehydration, discoloration of skin, convulsions, limb paralysis, coma, and death with increasing dosage. Chronic exposure causes neurological, renal, hematological, endocrine and cardiovascular damage and hypertension, reproductive and developmental problems, and cancer. The EPA limits lead in drinking water to 15 ppb with a goal of 0 and lead in air in public spaces to ≤1.5 mcg/m³. OSHA limits lead in workplace air to 50 mcg/m³.

M

MEK (methyl ethyl ketone)   organic industrial chemical used mainly as a solvent. Exposure causes nausea, headache, dizziness, sleepiness, and confusion. High concentrations cause loss of consciousness, pulmonary edema, respiratory failure, cardiac arrest, and death. MEKP is a related compound that is far more dangerous. OSHA limits MEK in workplace air to 200 ppm.

Melamine (1,3,5-triazine-2,4,6-triamine)   industrial chemical used to produce resins in plastics but also as a non-protein nitrogen source for cattle. This use was banned in the US in 1978. Exposure produces eye, skin, and mucous membrane irritation. Chronic exposure causes dermatitis, inflammation of the kidneys and bladder, and kidney and bladder and possible reproductive effects. China used melamine in pet food, killing and damaging many pets in the US in 2007.

Mercury   heavy metal used for many items like light bulbs, batteries, and electronic equipment. Previously, it was used in thermometers, paint, and mining. There is significant biomagnification of methylmercury to humans, which causes poisoning and death in even moderate amounts. It also causes neurological damage as well as damage to the cardiovascular system and the immune system. Metallic mercury causes nausea, vomiting, diarrhea, increases in blood pressure and heart rate, skin rashes, and eye irritation. Chronic exposure to mercury and methyl mercury vapors damages the kidney and brain. The EPA and FDA limit mercury to 2 ppb in drinking water. OSHA limits organic mercury to 0.1 mg/m³ and metallic mercury vapor to 0.05 mg/m³ of workplace air.

methylene chloride   industrial solvent used in hairspray, cleaners, paint removers, and plastics. High doses result in unconsciousness, narcosis, and death. Chronic exposure results in headaches, nausea, paresthesia and fainting, and liver damage. It also causes cancer. The EPA limits methylene chloride in drinking water to 5 ppb. OSHA limits methylene chloride in workplace air to 25 ppm

methyl parathion   insecticide that is restricted and banned in some cases because of its damage to the environment. Exposure causes vomiting, diarrhea, abdominal cramps, headache, eye pain, blurred vision, and general confusion. Higher doses produce slurred speech, fatigue, and eventual paralysis of extremities and respiratory muscles. Death is caused by cardiac arrest or respiratory failure. It has been banned in Indonesia, Sri Lanka, and Tanzania, and it is restricted in China, Colombia, Korea, and Japan.

MIC (methyl isocyanate)   organic chemical intermediary in the production of pesticide. Exposure produces coughing, chest pain, dyspnea, asthma, irritation of the eyes, nose and throat, skin damage, pulmonary edema, blindness, hemorrhages, bronchial pneumonia, and death with increasing dosage. Chronic exposure to MIC results in eye and lung damage. OSHA limits exposure to MIC to 0.02 ppm in workplace air.

N

Naphthalene   a polycyclic aromatic hydrocarbon (**PAH**) used in mothballs, insecticides, medicines and industrial chemicals. Exposure produces vertigo, nausea, vomiting, diarrhea, and blood in the urine. Very high doses can cause death. Chronic exposure can lead to nose, and kidney, liver, thymus, and spleen damage. It can also cause colorectal cancer and cancer of the larynx. The EPA limits naphthalene in drinking water to 0.1 ppm. OSHA limits it to 10 ppm in workplace air.

Nickel   heavy metal used in alloys and other metallurgy and batteries. Exposure causes headache, nausea, respiratory problems, kidney damage, and death with increasing dosage. Chronic exposure causes chronic bronchitis, kidney, liver and blood disorders, immune system and reproductive damage and nasal and lung cancer. The EPA limits nickel in drinking water to 0.1 mg/l. OSHA limits nickel carbonyl to 0.007 mg/m^3 and metallic nickel to 1 mg/m^3 in workplace air.

NOx (nitrogen oxides)   primarily gaseous air pollutant but also used in medical procedures and industrial chemical processes as a liquid. NOx was named one of the six principal air pollutants by the EPA in 1970. By 1998, five of the six were substantially reduced, but NOx had increased by 10%. Exposure irritates the eyes, nose, throat, and lungs and causes shortness of breath, nausea, and vomiting. A delayed response to exposure is that fluid can build up in the lungs from pulmonary edema and lead to loss of consciousness. High levels of NOx can cause swelling of the throat and asphyxiation. Build-up of fluid in the lungs can be fatal. Chronic exposure can cause asthma and permanently damage the lungs and heart. The EPA limits average annual NOx in air to 0.053 ppm. OSHA limits nitric oxide to 25 ppm in workplace air.

O

Ozone (O$_3$)   air pollutant at the earth's surface also used for sanitizing. Exposure produces coughing, throat irritation, wheezing, and nausea. Chronic high exposure causes reduced lung function, chronic lung diseases, asthmatic conditions, inflamed and damaged lung tissue, and damage to the immune system, increasing the likelihood of bronchitis and pneumonia. It also causes damage to the lung tissue, decreased pulmonary function, and premature death. The EPA limits average ozone in air to 80 ppb. The World Health Organization limits ozone to an average exposure of 60 ppb.

P

PAHs (polycyclic aromatic hydrocarbons)   a group of chemicals that are primarily air pollutants but also are used in medicines, dyes, plastics, and pesticides. Exposure to PAHs results in red blood cell and immune system damage. Chronic exposure produces reproductive and developmental problems as well as cancer of the lungs and skin and possibly other organs. The only PAH that EPA limits is benzo(a)pyrene and it is 0.0002 mg/l. OSHA limits PAH to 0.2 mg/m^3 of workplace air.

Particulate   particles of various materials and chemicals. Particulate is one of the six "criteria" air pollutants. PM10 are particles less than 10 μm, and PM 2.5 are less than 2.5 μm. Coarse particulate can lodge in the lung tissues and cause shortness of breath. Fine particulate causes coughing, wheezing and painful and labored breathing. Chronic exposure increases lung diseases such as emphysema, bronchiectasis, pulmonary fibrosis, and lung cancer. Coarse particulate (PM10) is limited to 50 mcg/m^3 of air by the EPA. Fine particulate (PM 2.5) is limited to 15 mcg/m^3 of air.

PCB (polychlorinated biphenyls)   organic industrial chemical used in electric devices like transformers but also plasticizers, home appliances, and many plastic products. PCBs were banned in 1976 in the United States because of their toxicity. Exposure causes skin conditions, anemia, ocular effects, elevation of blood pressure and endocrine and reproductive damage. Chronic exposure causes delayed child development, pneumonia, Epstein–Barr virus, non-Hodgkin's lymphoma, liver cancer and malignant melanoma. The EPA limits PCBs to less than 0.5 ppb in drinking water. The FDA limits PCBs to 0.2 ppm in infant formula, 0.3 ppm in eggs, 1.5 ppm in dairy products, 2 ppm in fish, and 3 ppm in meat. OSHA limits PCBs to less than 1 mg/m^3 in workplace air.

PCE (perchloroethylene)   VOC and DNAPL used mainly for dry cleaning, solvent and chemical production. Exposure produces dizziness, headache, and unconsciousness, eye and upper respiratory irritation, kidney and liver damage and death at high dosage. Chronic exposure produces headaches, nausea, liver cirrhosis, hepatitis, nephritis, cardiac arrhythmia, kidney damage, spontaneous abortions and decreased fertility. It likely causes lung, liver, skin, colon, cervix, larynx, bladder, esophagus, and lymph gland cancer.

PCP (pentachlorophenol)   Semivolatile organic compound used as a biocide. PCP was designated a restricted use pesticide by the EPA and banned for use by the public in 1984. Exposure causes confusion, sweating, hyperthermia with high fever, eyes, nose, and throat irritation, breathing difficulty, spasms and death by respiratory failure with increasing dosage. Chronic exposure results in liver, kidney, blood and nervous system damage, skin rash, weight loss, vision damage, bronchitis, circulatory problems, and heart failure. It also causes Hodgkin's disease, leukemia, and lip, mouth, and pharynx cancer. The EPA limits PCP in drinking water to 1 ppb. OSHA limits PCP in workplace air to 0.5 mg/m^3.

PFAS (Per- and Polyfluoroalkyl Substances)   industrial "forever chemicals" used in firefighting foam, electroplating, electronics, textiles, and paper. Exposure causes damage to the immune, reproductive and endocrine systems, to developmental progress in children and it causes prostate, kidney, and testicular cancers. The EPA issued a health advisory for PFAS in water at 70 ppt.

Phosphorous (white)   inorganic element used primarily for chemical fertilizer, weapons, and chemical processes. Exposure causes cough, mouth damage, jaw and teeth deterioration with intense pain, throat, lungs and eye irritation. Ingestion of white phosphorus produces vomiting, severe diarrhea with loss of blood, extreme fatigue, and stomach, intestines, liver, heart and kidney damage, coma, and death with increasing dosage. White phosphorus burning of skin can result in heart, liver, and kidney damage. OSHA limits phosphorus in workplace air to 0.1 mg/m^3.

Phthalates organic compounds used as plasticizers in plastics. Chronic exposure can cause liver, kidney, lung, endocrine and blood system, thyroid and testes damage. It likely causes cancer. The most dangerous phthalate is DEHP. The EPA limits DEHP in drinking water to 6 ppb. They limit DBP, another phthalate, in drinking water to 34 ppm. OSHA limits DEHP in workplace air to 5 mg/m^3.

Plutonium radioactive element made from uranium and used in weapons, satellites, and powerplants. Inhalation causes scarring of the lungs, lung disease and cancer. It enters the blood stream in the lungs and circulates through the body, it concentrates in the bones, liver, kidneys and spleen, damaging them with radiation and increasing incidence of cancer. The EPA limits exposure to plutonium to 10 mrem/year. OSHA limits whole body exposure to 1.25 rems per quarter.

R

Radium radioactive metal formerly used in luminescent dials but recently used in medical applications, radiography, and other chemical processes. Exposure causes bone damage and anemia among other symptoms. Chronic exposure produces bone, liver, breast, lymphatic, thyroid, pancreatic, uterine, colon, and genital cancer, and leukemia. The EPA limits radium in drinking water to 5 pCi/l. They limit radium 226 in mine tailings to 5 pCi/g in the top 5.9 in. (15 cm).

Radon radioactive gas, primarily naturally occurring. Chronic exposure causes lung cancer. It is estimated that 25 000 people die per year from this exposure. The EPA limits indoor air to 4 pCi/l of radon.

Red tide _Karenia_ brevis dinoflagellates (among others), a type of harmful algal bloom (**HAB**). The algae emit toxins that cause skin irritation, burning eyes, coughing, and sneezing. If impacted shellfish is consumed, it can produce Paralytic Shellfish Poisoning (**PSP**) including burning, numbness, respiratory paralysis, and possible death. It can also cause Amnesic Shellfish Poisoning (**ASP**) including nausea, vomiting, diarrhea and can result in permanent central nervous system damage. Diarrheal shellfish poisoning (**DSP**) causes nausea and vomiting, and diarrhea. Neurotoxic Shellfish Poisoning (**NSP**) causes vomiting, nausea, and neurological damage.

S

Salmonella bacterial infection from contaminated food or water. Infection causes diarrhea, stomach cramps, fever, nausea, vomiting, chills, and headache. It can be life-threatening if the infection spreads.

sulfur dioxide air pollutant compound. It is one of the six principal air pollutants identified by the EPA in 1970. Minor use in food preparation and industrial chemical processes. Exposure irritates the nose, throat, eyes, and lungs and causes coughing, shortness of breath, headache, nausea, fever, vomiting, serious skin and eye burns, convulsions, pulmonary edema, and death with increasing dosage. It is especially dangerous to people with asthma, heart and lung disease, and children and the elderly. EPA set a one year average limit of 0.03 ppm in air and a 24-hour period limit of 0.14 ppm. OSHA limits sulfur dioxide to 5 ppm in workplace air.

T

TCA (1,1,1-trichloroethane) organic compound used as a solvent and a chemical intermediate in industrial processes. Exposure causes loss of consciousness, low blood pressure, liver, nervous system, and circulatory system damage and even death. The EPA limits TCA in drinking water to 0.2 ppm. OSHA limits TCA in workplace air to 350 ppm.

TCE (trichloroethylene) VOC, chlorinated hydrocarbon-solvent, degreaser and chemical component. Exposure produces headache, dizziness, confusion, and unconsciousness. Chronic exposure produces irritability, personality disorders, poor coordination, loss of short-term memory, shortness of breath, pulmonary edema, and possible death with increasing dosage. It also likely causes cancer of several organs. The EPA limits TCE in community water supplies to 5 ppb. OSHA limits TCE to 100 ppm in workplace air.

Thallium inorganic metal used in electronics, pharmaceuticals, glass manufacturing, infrared detectors, and poisons. It was banned in rodent control poison in 1972. Exposure leads to hair loss and nervous system, lung, heart, liver, and kidney damage and death in high doses. OSHA limits thallium in the workplace to 0.1 mg/m^3. The EPA limits thallium to 13 ppb in surface waters.

Trihalomethanes a large group of organic chemicals used primarily as solvents, to produce teflon and as refrigerants. Health impacts vary by individual compound. In general, exposure causes central nervous system, liver and renal damage, cardiac depression, and arrhythmia and death with increasing dosage. They also cause many types of cancer depending on the compound. The EPA limits total trihalomethanes (**TTHM**), to 80 ppb in drinking water.

tritium radioactive form of hydrogen used in luminescent lights in watches, gun sights, numerous instruments and as a radioactive tracer in medical and scientific applications. It is also used as a nuclear fusion fuel. Normally, radiation from exposure to tritium is below background radiation. However, industrial exposure at high concentrations causes radiation sickness and even death by pancytopenia. It can also increase the incidence of cancer. EPA limits radioactivity from man-made radionuclides in drinking water at 4 millirems per year.

toluene VOC used as solvent, flexible foam, coatings. Exposure produces loss of coordination, nausea, eye irritation, impaired reaction time, narcosis, and death by respiratory failure with increased dosage. Chronic exposure produces headaches, confusion and memory loss, weakness, nausea, loss of muscle control, decreased mental ability, and possible damage to the kidneys. Toluene diisocyanate likely produces cancer in several organs. The EPA limits toluene to 1 ppm in drinking water. OSHA limits toluene in workplace air to less than 100 ppm. OSHA limits toluene diisocyanate to 0.02 ppm in workplace air.

Toxaphene pesticide. Pesticide of choice to replace banned DDT in 1970 but so deadly that it was also banned in 1982. Exposure causes nausea, vomiting, restlessness, muscle spasms, convulsions, brain damage, and death by respiratory failure with dosage. Chronic exposure causes headaches, nausea, vision problems, and liver, kidney, adrenal gland, blood system, and immune system, damage. It likely causes cancer. The EPA limits toxaphene in drinking water to 3 ppb. OSHA limits it to 0.5 mg/m^3 in workplace air.

U

Uranium radioactive heavy metal used for weapons, fuel, research. Uranium has both chemical and radiation health impacts. Inhalation irritates the lungs causing coughing and shortness of breath. It damages the kidneys, liver, and blood cells and bones. Chronic exposure causes lung damage and cancer of multiple organs. The EPA limits uranium to 30 μg/l. OSHA limits airborne insoluble uranium in workplace air to 0.2 mg/m^3.

V

vinyl chloride organic industrial compound used primarily to make PVC for all applications currently. Dizziness, unconsciousness, skin blistering and liver, lungs, heart, and blood system damage and death with increasing dosage. Chronic exposure produces major liver, nerve, and immune system damage. It also produces liver, brain, lungs, lymphatic system cancer, as well as in the organs and tissues that produce blood. The EPA limits vinyl chloride to 2 ppb in drinking water. OSHA limits it to 1 ppm in workplace air.

X

Xylene   VOC used as a solvent and to produce many other industrial chemicals. Exposure results in eye, nose, and throat irritation, headache, difficulty in breathing, respiratory failure, pulmonary congestion, and death with increasing dosage. Chronic exposure produces loss of balance, memory loss, headaches, chest pain, fever, lung damage, electrocardiograph abnormalities, pulmonary edema, and liver and kidney damage. The EPA limits xylene to 10 ppm in drinking water. OSHA limits xylene to 100 ppm in work area air.

Z

Zinc   heavy metal used in alloys, galvanizing, vitamins, and many industrial applications. Inhalation of zinc fumes from welding and smelting causes metal fume fever with headache, fever, chills, muscle aches, nasal irritation, cough, reduced lung capacity, nausea, vomiting and fatigue. Inhalation of zinc chloride causes nose and throat irritation, cough, pulmonary inflammation and fibrosis, headache, nausea and vomiting. Ingestion of large amounts of zinc causes nausea, vomiting, diarrhea, and gastric bleeding. Zinc chloride causes pharyngitis, esophagitis, and pancreas damage. Zinc phosphide causes vomiting, hypotension, cardiac arrhythmia, circulatory collapse, pulmonary edema, convulsions, kidney damage, coma, and death within two weeks. Chronic exposure to zinc causes anemia and pancreas, kidney and liver damage. The EPA limits zinc to 5 mg/l in drinking water. OSHA limits zinc chloride to 1 mg/m^3 of workplace air.

Appendix B: Online Videos on Case Studies

Chapter 1

1.1 Rachel Carson (1907–1964)

Short
https://ny.pbslearningmedia.org/resource/envh10.sci.life.eco.silentspring/rachel-carsons-silent-spring/

Long
https://www.pbs.org/video/rachel-carson-voice-of-nature-viqzrt/

1.2 9/11 World Trade Center Disaster New York

Short
https://www.asbestos.com/world-trade-center/
https://www.youtube.com/watch?v=uCk-ZFTEuuE

1.3 Environmental Impact of Cryptocurrency

Short
https://www.youtube.com/watch?v=rujSxh_TdP8
https://www.youtube.com/watch?v=yMsDUEnOw8c

Long
https://www.youtube.com/watch?v=1bxx1jFSlMI

Chapter 2

2.1 Bhopal Chemical Disaster

Short
https://www.youtube.com/watch?v=zgGHBcQQNuM

Long
https://topdocumentaryfilms.com/one-night-in-bhopal/

2.2 Citarum River, Indonesia

Short
https://www.youtube.com/watch?v=G0o7POT7-tQ

Long
https://www.youtube.com/watch?v=GEHOlmcJAEk

2.3 Flint, Michigan Water Crisis

Short
https://www.youtube.com/watch?v=GYiVHh4U4pE

Long
https://www.youtube.com/watch?v=6oVEBCtJgeA

2.4 Cancer Alley, Louisiana

Short
https://www.youtube.com/watch?v=ZB8CbDG7gpk
https://www.youtube.com/watch?v=xFxY454NTYE

Chapter 3

3.1 Exxon Valdez Oil Spill, Prince William Sound, Alaska

Short
https://www.youtube.com/watch?v=3YeeIHJZeSE

Long
https://www.youtube.com/watch?v=dtF-4JvSh8o

3.2 Hole in the Stratospheric Ozone Layer

Short
https://www.youtube.com/watch?v=MgUobxtdm4A

Long
https://www.youtube.com/watch?v=Ll_TR7C4xr4

Chapter 4

4.2 2013 Super Typhoon Haiyan

Short
https://www.youtube.com/watch?v=5MSO-vzv2V0

Long
https://www.youtube.com/watch?v=-BnahLG_DmQ

4.3 1982–1983 El Niño

Short
https://www.youtube.com/watch?v=d6s0T0m3F8s
https://www.youtube.com/watch?v=WPA-KpldDVc

Long
https://naturedocumentaries.org/8733/chasing-el-nino-pbs-1998/

Polluted Earth: The Science of the Earth's Environment, First Edition. Alexander Gates.
© 2023 John Wiley & Sons, Inc. Published 2023 by John Wiley & Sons, Inc.
Companion website: www.wiley.com/go/gates/pollutedearth

4.4   1871 Peshtigo Wildfire

Short

https://www.youtube.com/watch?v=GmsnVwLMWkg

https://www.youtube.com/watch?v=1PWHYj0f--8

4.5   Permian Extinction Event

https://www.youtube.com/watch?v=hDbz2dpebhQ

https://www.youtube.com/watch?v=S8dk19naJsU

Chapter 5

5.1   Bangladesh Arsenic in Soil and Groundwater

https://www.youtube.com/watch?v=1WLDulkwsXY

5.2   2021 Mongolia-China Dust Storm

https://www.youtube.com/watch?v=fy-mxGzQxto&list=RDCMUCg lo7cYi5uhn__RzBBfBbLw

5.3   Radon in Boyertown, Pennsylvania

https://www.youtube.com/watch?v=Y2Zojeqrl1c

https://www.pbs.org/video/radon-4bxop1/

https://www.youtube.com/watch?v=z-ZomIX4wTk

Chapter 6

6.1   Fracking in the United States

Long

https://www.youtube.com/watch?v=hERUwnZjI18&t=247s

https://www.youtube.com/watch?v=Xvz_m5uPV4s

6.2   Vermiculite Mountain Libby, Montana

https://youtu.be/SC02WJSUu2I

6.3   Waste Isolation Pilot Plant, New Mexico

Short

https://www.youtube.com/watch?v=3bo36aKc8EY

Chapter 7

7.1   1783 Laki Fissure Eruptions

https://www.youtube.com/watch?v=Ia3NKygegGY

https://www.youtube.com/watch?v=v2TzbS3qGNk

7.2   2010 Port-au-Prince

Short

https://www.npr.org/sections/health-shots/2012/04/13/150302830/water-in-the-time-of-cholera-haitis-most-urgent-health-problem

Long

https://www.youtube.com/watch?v=02MgwqJk95Y

7.3   2008 Sichuan Earthquake, China

Short

https://www.youtube.com/watch?v=KvBYpYwg9Yw

Long

https://www.youtube.com/watch?v=nzGkeLoTitM

7.4   2011 Tohoku Earthquake

https://www.youtube.com/watch?v=mUBxtTEOiPI

Long

https://www.youtube.com/watch?v=qRKScRgsUaE

7.5   2005 Hurricane Katrina

All Talking

https://www.c-span.org/video/?189138-5/environmental-impact-hurricanes

Racism

https://www.youtube.com/watch?v=d90oYzF8SCc

Long

https://www.youtube.com/watch?v=JEAedjLXw7Q

Chapter 8

8.1   Edwards Aquifer, Texas

Short

https://www.youtube.com/watch?v=awza69Qs3bg&t=47s

Long

https://www.youtube.com/watch?v=Ti4zlvf9GTw

8.3   Woburn Wells G and H, Massachusetts

Slideshow

https://slideplayer.com/slide/1656443/

A Civil Action

Trailer

https://www.youtube.com/watch?v=8zsKyywHPkU

Full Movie

https://www.youtube.com/watch?v=ecBKI_Zi1HU

8.4   Pacific Gas and Electric, Hinkley, California

Follow Up

https://www.youtube.com/watch?v=OhDCTBqlkVM

Erin Brockovich

Trailer

https://www.youtube.com/watch?v=ELzu636Xf6Y&t=15s

https://www.youtube.com/watch?v=OpISHolWtKs

Chapter 9

9.1   Cuyahoga River, Ohio
https://www.youtube.com/watch?v=XvrAd0bM_7I

9.2   Hudson River, New York
https://www.youtube.com/watch?v=hI27Vg3oNyo

9.3   2015 and 2019 Minas Gerais Disasters, Brazil
https://www.youtube.com/watch?v=he2fkMpJCYg
https://www.youtube.com/watch?v=XjiRCWi_zi4

9.4   Lake Uru Uru, Bolivian, Andes
https://youtu.be/XcdHfoIUvGA

9.5   Hackensack Meadowlands, New Jersey
https://www.youtube.com/watch?v=dn8oVCLhi8E
https://vimeo.com/221523063

Chapter 10

10.1   Bunker Hill Complex, Idaho

Short
https://www.youtube.com/watch?v=q6zYK8-Hvqk
https://www.youtube.com/watch?v=lwgVmw93g1Q

10.2   Times Beach Superfund Site, Missouri
https://www.youtube.com/watch?v=G6kshs2ZQcQ

10.4   Usinsk Oil Spill, Russia
https://www.youtube.com/watch?v=YlxKb7KQw34
https://www.facebook.com/watch/?v=323076295525834

Chapter 11

11.1   New Zealand Flightless Birds

Short
https://www.youtube.com/watch?v=Az9InkxkQQ8

Long
https://www.youtube.com/watch?v=jJE3mFNXYdw

11.2   Monarch Butterfly Population Collapse

Short
https://www.youtube.com/watch?v=ozvbazloZtI

Long
https://www.youtube.com/watch?v=maM2gl30cJc

11.3   American Bald Eagle

Short
https://www.youtube.com/watch?v=5oSvIDTWfaE

Long
https://www.dailymotion.com/video/x5ricdf

Chapter 12

12.1   Torrey Canyon Scilly Islands, UK
https://www.youtube.com/watch?v=zmGvuBAHc-Q
https://www.youtube.com/watch?v=IV-EhBesVjg
https://www.youtube.com/watch?v=qPbufQhJLsY

12.2   Deepwater Horizon Oil Spill, Gulf of Mexico

Long, 3 Aspects
https://www.youtube.com/watch?v=vbl7QeqfE-Q
https://www.youtube.com/watch?v=GT4oENdowFE
https://www.pbs.org/video/a-decade-after-deepwater-nof5ng/

12.3   Beaufort Dyke, Irish Sea, Scotland
https://www.youtube.com/watch?v=PKliFnGF2aE&t=151s

12.4   "Ecomafia" in Italy and Somalia

Short
https://www.youtube.com/watch?v=44nyUZAL8TM
https://www.youtube.com/watch?v=plSgjy2jTmw

Long
https://www.youtube.com/watch?v=SabmlxeUBwk

Chapter 13

13.1   1980 Chemical Control Corporation
https://youtu.be/NPgwifGzJRU

13.3   The Great London "Killer" Fog

Short
https://www.bbc.co.uk/archive/great-smog-of-london/zhjx7nb

Long
https://www.youtube.com/watch?v=Vkx-2mT1-q4

13.4   Donora Killer Fog
https://www.youtube.com/watch?v=zOGvQsWW1As
https://www.youtube.com/watch?v=IKDpNTNm1Cc&t=94s

13.5   Sudbury Canada Mines and Smelters
https://www.youtube.com/watch?v=8wB-r1NrpGs

Chapter 14

14.1   Aurul Gold Mining Disaster
https://www.youtube.com/watch?v=eKiLCPolS6U

14.2   Aberfan Coal Waste Disaster

Short
https://www.youtube.com/watch?v=1lzJLww3DvM&t=142s

Long
https://www.youtube.com/watch?v=pSWI5aYjVOY

14.3 2005 BP Refinery Explosion
https://www.dailymotion.com/video/x3c4hu6
https://www.youtube.com/watch?v=goSEyGNfiPM&t=638s
https://www.youtube.com/watch?v=0uq_hHvr814

Chapter 15

15.1 1991 Gulf War Oil Spills
https://www.youtube.com/watch?v=gyOMF4DXF_A
https://www.youtube.com/watch?v=dOMliL4i9iM

From Jarhead
https://www.youtube.com/watch?v=olHaqANv6Uw

15.2 Marine Base Camp Lejeune, North Carolina
https://www.cbsnews.com/news/toxic-water-marine-base-service-members-families-justice/
https://www.nbcnews.com/healthmain/contamination-nc-marine-base-lasted-60-years-1c8880227

15.3 Nevada Nuclear Test Site

Environmental
https://www.youtube.com/watch?v=wJG-S0rMcms&t=129s

History
https://www.youtube.com/watch?v=Z7X7QNDi1mQ

Long
https://www.youtube.com/watch?v=lRPp0RNVhCY&t=3s

Chapter 16

16.1 1930s Central United States Dust Bowl
https://www.youtube.com/watch?v=x2CiDaUYr90
https://www.pbs.org/video/intro/

16.2 Aral Sea Disaster, Kazakhstan, and Uzbekistan

Short
https://www.youtube.com/watch?v=FzvEW1FHc60

Long
https://www.youtube.com/watch?v=dp_mlKJiwxg

16.3 Ogallala Aquifer, Central United States

Short
https://www.youtube.com/watch?v=XXFsS94HF08

Long
https://www.youtube.com/watch?v=7CxA8PeDhlc

16.4 Gulf of Mexico Dead Zone
https://www.youtube.com/watch?v=5LwbeK-QXNs

Solutions
https://www.youtube.com/watch?v=YQd8FUELp7w
https://www.youtube.com/watch?v=mo_rG7KCc4A

16.5 Hurricane Floyd and Hog Waste in Eastern North Carolina
https://www.wral.com/20-years-after-hurricane-floyd-little-has-changed-and-flooding-is-still-a-threat/18629403/

+Environmental Justice
https://www.youtube.com/watch?v=WsUNylsiDH8&t=26s

Chapter 17

17.1 Project Plowshare
https://www.youtube.com/watch?v=yXQhfZanPtE
https://www.youtube.com/watch?v=u13ZAXUYBrc

Long
https://www.youtube.com/watch?v=kpjFU_kBaBE

17.2 Chernobyl Nuclear Disaster, Russia

Short
https://www.youtube.com/watch?v=eB1vfga9Y_c
https://www.youtube.com/watch?v=oUVv3OGjKYY

HBO
https://www.youtube.com/watch?v=s9APLXM9Ei8

17.3 Lake Karachay-Mayak Complex, Russia
https://www.youtube.com/watch?v=A28phMSit4c

Accident
https://www.youtube.com/watch?v=oWqk5CUTvWk

Long
https://www.youtube.com/watch?v=V0uymbOYOgQ

17.4 The Radium Girls
https://www.youtube.com/watch?v=1799d6A3PUg

Movie Trailer
https://www.youtube.com/watch?v=7RxnlriKLRk

Chapter 18

18.1 Centralia Pennsylvania Coal Mine Fires
https://www.youtube.com/watch?v=kcbW7GkvT1o

Movie Trailer
https://vimeo.com/ondemand/centralia/236685619?autoplay=1

Long
https://www.youtube.com/watch?v=IFZqWRtYvpM

18.2   2000 Payatas Waste Slide Disaster
https://www.youtube.com/watch?v=n_NBmlChce0
https://www.dailymotion.com/video/xe39dy

18.3   Love Canal, New York
https://www.youtube.com/watch?v=Kjobz14i8kM

History
https://www.youtube.com/watch?v=TUTF57Chos4

Chapter 20

20.1   Claire (Pat) Patterson
https://www.youtube.com/watch?v=eBLuefl26AQ

Index

Note: Page numbers in *italics* indicate figures, those in **bold** indicate tables.

Aberfan coal waste disaster, 262, *263,* 264
acid mine drainage, 7, 97, 163, 177, 255, 261
acid precipitation, 13, 231, 249–251
    reduction in, 364, *364*
    volcanic eruptions, 104, 106
action, environmental, 13–15
aerosol, from volcanic eruptions, 103–104, 106–107
Agent Orange, 170, 184, 273, 277, *278–279*
agricultural chemicals, 303–307
agricultural pollution, 13, 293–309
    case study: Hurricane Floyd and hog waste in North Carolina,
        308, 308–309
    oceans, 224, 306–307, *306–307*
agricultural threats to environment, 295–303
    case study: Aral Sea disaster, 299–301, *300*
    case study: Ogallala Aquifer, *302,* 302–303
    deforestation, 295–296
    overuse of water, 299–303
    soil preparation, 296–299
air masses, 231, 234–237, *236–237*
air pollutants, 237–242
    criteria, 78, 231, 239–241, 361, 364, *366*
    primary, 231, 240, 244
    secondary, 231, 240, 244
air pollution
    Cancer Alley, Louisiana, 30–33
    natural, 78–84, *79–82,* 242
    9/11 World Trade Center disaster, 10–12, *11–12*
    non-point source, 13
    nuclear testing and fallout, 288–289, *289–290*
    point source, 7–8
air sparging, 349, *350*
aldrin, 170, 205, 303–304, 367
alluvium, 175, 182–183, 189
alpha decay, *82,* 82–83
animal waste, 7, 307–309, *308*
antimicrobial properties of rocks, minerals, and soils, 97–98
antimony, 140, 163, 367
aquatic environments, 6–8
aquiclude, 126, *127,* 128–129, 132
aquifer recharge zones, 125–126, *127, 131,* 301, 303
aquifers
    definition, 125
    Edwards, *131,* 131–132
    fractured rock, 125, *127,* 127–129, 138
    karst, 125, 128, *128,* 132, 138
    Ogallala, 126, 301–303, *302*
    perched, 125, 128, *129,* 188
    unconfined, 126, 128, 132
Aral Sea disaster, 299–301, *300*
architecture of the Earth
    liquid Earth spheres and systems, 39–44
    solid Earth spheres and systems, 37–39

arsenic, 140, 367
    in soil and groundwater, 74–78, *75–77*
    in weapons, 276
artesian system, *127,* 130
asbestos, 10, 39, 87, 367
    banning, 364
    earthquakes, 107
    in minerals, 87
    particulate, 78
    in vermiculite, 93–95, *95*
ash, 239
    coal, 247, 262
    fly, 180, 287, 334
    volcanic, 78, 103–104, 106, 120, 177, 242
asthenosphere, 37–38, *38*
atmosphere, 37, 39, *40,* 44–47
    layers, *44,* 44–45
    ozone layer, 44, 45–47, *46*
atomic weapon, 273, 291
atoms, 87, *87*
Aurul gold mining disaster, 259, *260,* 261
avalanche, debris, 119–120, *120*

bald eagle, 205–207, *206*
Bangladesh, arsenic in soil and groundwater of, 74–75, *75–77*
barium, 91, 93, 137, 140, 222, 267, 367
barrier islands, 5, 211, 212, *212,* 217
bathymetry, 211, *211,* 213
beaches, 211–212, 215, 220–224, *223,* 226
Beaufort Dyke, Irish Sea Scotland, *225,* 225–226
bees, 13, 305
believers and non-believers, 361–362
benzene, 7, 104, 107, 132, 265–267, 273, 277, 287, 367
beryllium, 226, 367
Bhopal chemical disaster, 21–24, *22–23*
bioaccumulation, 193, 204–205
biodiversity, 193–194, 196, 199–201, 207, 226, 267, 366
biological oxygen demand (BOD), 26
biomagnification, 193, 204–207, *205*
bioreactor, *356,* 357
bioremediation, 345, *354,* 354–357, *356*
biosphere, 40, *40,* 44
bioventing, 354, *354*
bitcoin, 14–15
blast furnace, 178, 264, *264*
blockchain, 14
blowout, 222, 255, 267–268
bonding, 88, *88*
BP refinery explosion (2005), 268–270, *269–270*
braided streams, 147, 149–150, *150,* 156, 160, 183–184, 302
bromine, 183–184, 367
brownfields, 345, 347
BTEX chemicals, 7, 265, 267

Polluted Earth: The Science of the Earth's Environment, First Edition. Alexander Gates.
© 2023 John Wiley & Sons, Inc. Published 2023 by John Wiley & Sons, Inc.
Companion website: www.wiley.com/go/gates/pollutedearth